The cdma2000® System
for Mobile Communications

Prentice Hall Communications Engineering and Emerging Technologies Series

Theodore S. Rappaport, *Series Editor*

DOSTERT   *Powerline Communications*

DURGIN   *Space–Time Wireless Channels*

GANZ, GANZ, & WONGTHAVARAWAT   *Multimedia Wireless Networks: Technologies, Standards, and QoS*

GARG   *Wireless Network Evolution: 2G to 3G*

GARG   *IS-95 CDMA and cdma2000: Cellular/PCS Systems Implementation*

GARG & WILKES   *Principles and Applications of GSM*

KIM   *Handbook of CDMA System Design, Engineering, and Optimization*

LIBERTI & RAPPAPORT   *Smart Antennas for Wireless Communications: IS-95 and Third Generation CDMA Applications*

MURTHY & MAJOJ   *Ad Hoc Wireless Networks: Architectures and Protocols*

PAHLAVAN & KRISHNAMURTHY   *Principles of Wireless Networks: A Unified Approach*

RAPPAPORT   *Wireless Communications: Principles and Practice, Second Edition*

RAZAVI   *RF Microelectronics*

REED   *Software Radio: A Modern Approach to Radio Engineering*

STARR, CIOFFI, & SILVERMAN   *Understanding Digital Subscriber Line Technology*

STARR, SORBARA, CIOFFI, & SILVERMAN   *DSL Advances*

TRANTER, SHANMUGAN, RAPPAPORT, & KOSBAR   *Principles of Communication Systems Simulation with Wireless Applications*

VANGHI, DAMNJANOVIC, & VOJCIC   *The cdma2000 System for Mobile Communications*

WANG & POOR   *Wireless Communication Systems: Advanced Techniques for Signal Reception*

The cdma2000® System
for Mobile Communications

Vieri Vanghi

Aleksandar Damnjanovic

Branimir Vojcic

PRENTICE
HALL
PTR

Prentice Hall PTR
Upper Saddle River, New Jersey 07458
www.phptr.com

Library of Congress Cataloging-in-Publication Data

A CIP catalog record for this book can be obtained from the Library of Congress.

Editorial/production supervision: *Mary Sudul*
Cover design director: *Jerry Votta*
Cover design: *Talar Boorujy*
Manufacturing manager: *Maura Zaldivar*
Acquisitions editor: *Bernard Goodwin*
Editorial assistant: *Michelle Vincenti*
Marketing manager: *Dan DePasquale*

© 2004 Pearson Education, Inc.
Publishing as Prentice Hall PTR
Upper Saddle River, New Jersey 07458

Prentice Hall books are widely used by corporations and government agencies for training, marketing, and resale.
The publisher offers discounts on this book when ordered in bulk quantities. For more information, contact Corporate Sales Department, Phone: 800-382-3419; FAX: 201-236-7141;
E-mail: corpsales@prenhall.com
Or write: Prentice Hall PTR, Corporate Sales Dept., One Lake Street, Upper Saddle River, NJ 07458.

Other product or company names mentioned herein are the trademarks or registered trademarks of their respective owners.

Printed in the United States of America
2nd Printing

ISBN 0-13-141601-4

Pearson Education LTD.
Pearson Education Australia PTY, Limited
Pearson Education Singapore, Pte. Ltd.
Pearson Education North Asia Ltd.
Pearson Education Canada, Ltd.
Pearson Educación de Mexico, S.A. de C.V.
Pearson Education — Japan
Pearson Education Malaysia, Pte. Ltd.

CONTENTS

PREFACE

Spread spectrum communications techniques have been used in military applications since the 2nd World War but have found a widespread commercial use only in the last ten years or so. To cope with the accelerating demand for mobile communications in the early 1990s, the introduction of spread spectrum techniques in cellular communications afforded a bandwidth efficient digital technology that could accommodate, within a given wireless spectrum allocation, a larger population of mobile users than other analog or digital technologies.

Spread spectrum systems exploit the noise-like characteristics of the spread signal waveform to allow multiple simultaneous transmissions using a common bandwidth. This is accomplished by means of spreading codes that are unique to each user and have mutually low correlation so that the multiple access signals can be separated at the receiver by means of despreading. Due to the use of spreading codes to achieve multiple access capability, this technology was named code-division multiple access (CDMA). Several favorable properties of spread spectrum signals can be exploited in the context of CDMA. Firstly, the wide-band characteristic of the spread signal enables to resolve and constructively combine the multipath components at the receiver, thus mitigating channel fading. Also, the wide-band nature of spread spectrum signals allows employing powerful forward error correction codes without the bandwidth expansion penalty that is incurred in narrow-band technologies. In the context of cellular CDMA, spread spectrum allows for universal frequency reuse, which increases overall network capacity and eliminates the task of frequency planning. Finally, spread spectrum allows for soft handoff, a technique which improves performance at the cell boundary, and increases cell range and capacity.

The first cellular CDMA system was pioneered by QUALCOMM Inc., whose efforts led to the adoption of the IS-95 CDMA standard by the Telecommunication Industry Association (TIA) in 1993. The IS-95 standard and its associated core network protocols are collectively known as cdmaOne™. Since then, the ever increasing demand for bandwidth efficiency, higher data rates and new services has motivated the constant evolution of the CDMA standard. IS-95 was primarily designed for voice services and to support low speed data applications. The data capabilities have since then improved, achieving higher data rates with increased bandwidth efficiency. At the same time the support of voice services has also improved with the adoption of more efficient vocoders. The milestones in the CDMA standards evolution are illustrated in CDMA air-interface standards evolution. (only the radio interface standards are included in this figure). An important milestone was achieved in 1999, when the IS-2000 CDMA standard (also referred to as CDMA 1X), developed under the auspices of standard development organizations of several countries, was approved by the International Telecommunication Union (ITU) within the IMT-2000 initiative, as one of the standards for the 3rd generation mobile communications. Standards evolution exploited the flexibility afforded by CDMA in multiplexing multiple channels, which has allowed the adoption of *revolutionary* concepts without disrupting backward compatibility. Among them, the concept of fast forward link data rate adaptation with fast scheduling and hybrid ARQ was first introduced in IS-856, also referred to as High Rate Packet Data (HRPD), a CDMA system optimized for data only transmission that achieves very high forward link data rates and bandwidth efficiency. Similar concepts have been recently adopted in IS-2000-C, which allows for both circuit switched voice and high speed forward link data applications. IS-2000-C is sometimes referred to as 1X Evolution for Data and Voice (1X EVDV). The IS-856 and the IS-2000 standards, together with the associated core network and service protocols, are collectively known as the cdma2000 standard. At the time of the completion of this manuscript, IS-2000 revision D and IS-856 revision A are being standardized. The main feature of the new revisions is efficient support for high speed reverse link packet data through hybrid ARQ operation.

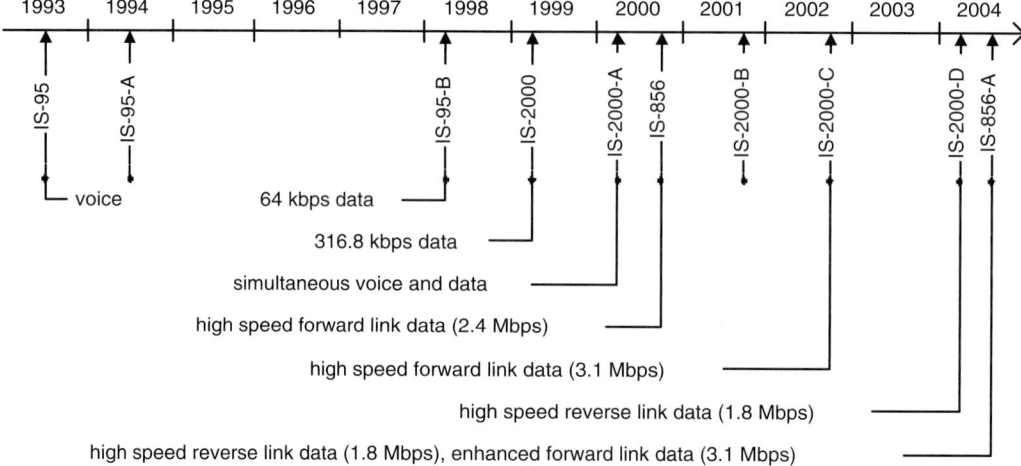

Figure 1   CDMA air-interface standards evolution.

The cdma2000, specifications comprise thousands of pages and pose a daunting challenge even to the experienced practitioners. More importantly, with the ever increasing number of radio channel configurations, functionalities and applications supported by cdma2000 specifications, the motivating CDMA concepts tend to become obscure. In light of the above, the aim of this book is twofold. Firstly, it is to present how the principles of spread spectrum communication in general and CDMA in particular are applied to the cdma2000 standards. Secondly, it is too navigate the reader through the maze of specifications and distill their fundamentals into a manageable, but still comprehensive description of cdma2000 1X.

This book approaches cdma2000 mainly from a radio access network perspective, and focuses on the mobile station and base station interoperability procedures as specified in the IS-2000-C revision of the standard. In addition, this book also describes network architecture and services, and how these services are realized, end-to-end, throughout the various network interfaces. Chapter 1 introduces the main concepts of spread spectrum techniques applied to CDMA cellular systems. Chapter 2 and 3 set the stage from a network perspective, describing the overall cdma2000 network architecture and the services it supports. Chapter 4 describes the functions performed by the CDMA modem and summarizes the IS-2000-C physical layer protocol. Chapter 5 and 6 describe the media access and signaling layer IS-2000-C protocols, respectively. The following chapters, from Chapter 7 to Chapter 9, describe soft handoff, power control, and packet data transmission techniques both from a protocol and an implementation perspective, giving practical guidelines and examples on how to implement these functionalities. Chapter 10 provides an analytical framework to estimate CDMA cell capacity and cell range, together with numerical examples that are useful to the practitioner. Finally, Chapter 11 describes how services are realized, end-to-end, in the cdma2000 network.

Given its scope and depth, we believe this book to be not only an invaluable aid to those that approach CDMA systems for the first time, but also an useful reference to the practitioners, system designers, and network operators. The book content and its structure also lend itself to be used for specialized courses and as a secondary academic text for courses in mobile communication systems and CDMA.

This leaves us with the pleasant task of acknowledging the contributions of the many individuals who reviewed this book. We would like to thank Alpaslan Savas, and our colleagues at QUALCOMM Inc., Baaziz Achour, Sanjeev Athalye, Tao Chen, Walid Hamdy, Duncan Ho, Jack Holtzman, John Ketchum, Nikolai Leung, Jack Nasielski, Joe Odenwalder, Ragulan Sinnarajah, and Edward Tiedemann. Thanks to Don Rayner for his contributions to the cover design. To all of the above we express our sincere thanks. We would also like to acknowledge the entire QUALCOMM team of engineers whose relentless efforts over past fifteen years have made CDMA the technology of choice for the present and future mobile systems. Last but not least, we are indebted to our families and soulmates for their support and patience during the many evenings and weekends we spent writing this book.

Introduction to CDMA

S pread spectrum transmission techniques, in most general terms, employ larger bandwidth than absolutely necessary to transmit a given amount of information. This is in contrast to the traditional approach in communications where a system designer attempts to maximize the amount of transmitted information per unit bandwidth, which is important for most communication channels and in particular when the bandwidth is scarce. The spectral redundancy introduced by means of spread spectrum transmission is exploited in a variety of communications systems. For example, spread spectrum is used to combat jamming or to facilitate low probability of intercept in military communications, and is used for accurate ranging and positioning applications. Of interest to us, however, is the capability of spread spectrum to allow multiple simultaneous transmissions using a common bandwidth. In such systems, separation of users is achieved in the code domain and they are commonly referred to as code-division multiple access (CDMA) systems. In this chapter we review the fundamentals of spread spectrum techniques applied to cellular CDMA systems, and the reasons why such systems exhibit superior performance and bandwidth efficiency.

1.1 Direct Sequence Spread Spectrum

In broadest terms, we can categorize spread spectrum techniques as frequency hopping, time hopping, direct sequence, and hybrid techniques. Here we focus on direct sequence techniques as those are the ones used in cellular CDMA systems. It is easy to understand the principle of direct sequence spread spectrum (DSSS) by considering the simplified system illustrated in Figure 1.1, a baseband equivalent of binary phase shift keying (BPSK). The data signal $d(t)$, composed of a sequence of rectangular pulses of duration T (see A in Figure 1.1), is spread (B in Figure 1.1) by multiplying it by pseudorandom sequence of short rectangular pulses (usually called chips) of duration T_c and amplitudes +1 or -1, $a(t)$, thus forming a noiselike broadband signal. In the receiver the data signal is recovered by despreading (C in Figure 1.1), which consists of multiplying the received signal with synchronized replica of $a(t)$. After that point, a matched filter for data signal can be used to optimally detect the data signal. Since the bandwidth of the signal is proportional to the rate at which it changes, it should be apparent why these operations are called spreading and dispreading. Namely, when the signal $d(t)$ of bandwidth W is multiplied by $a(t)$ of N times larger bandwidth, where $N = T/T_c$ is the spreading factor, the composite signal $d(t)a(t)$ has also N times larger bandwidth than $d(t)$. After synchronous multiplication of received signal $d(t)a(t)$ with $a(t)$ we perform the despreading operation and recover the data signal $d(t)$, because $a(t)^2 = 1$. Power spectral densities of original data signal and spread signal for pulse shapes in Figure 1.1 are shown in Figure 1.2, assuming 8 chips per information bit. For rectangular pulses, the power spectral density has the form $[\sin(x)/x]^2$. As it can be seen, the first lobe of power spectral density of signal $d(t)a(t)$ at point B is wider than the first lobes of the signal before spreading (point A) and after dispreading (point C). In this particular example, the first lobe of the spread spectrum signal is eight times wider than the first lobe of the information signal. Although somewhat simplistic, the presented model succinctly captures the essence of DSSS and corresponding spreading and despreading operations.

Spread spectrum signals in general, and DSSS in particular, are suitable for multiple access applications. This multiple access capability is normally achieved by assigning different spreading waveforms to different communication pairs and is the subject of the next section.

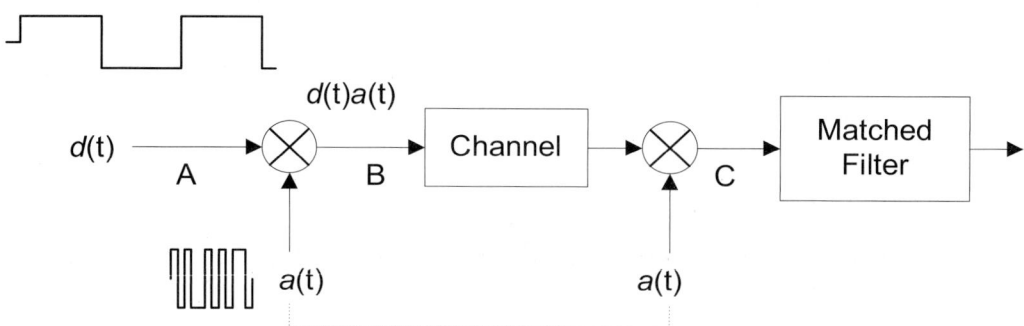

Figure 1.1   Simplified block diagram of a baseband DSSS system.

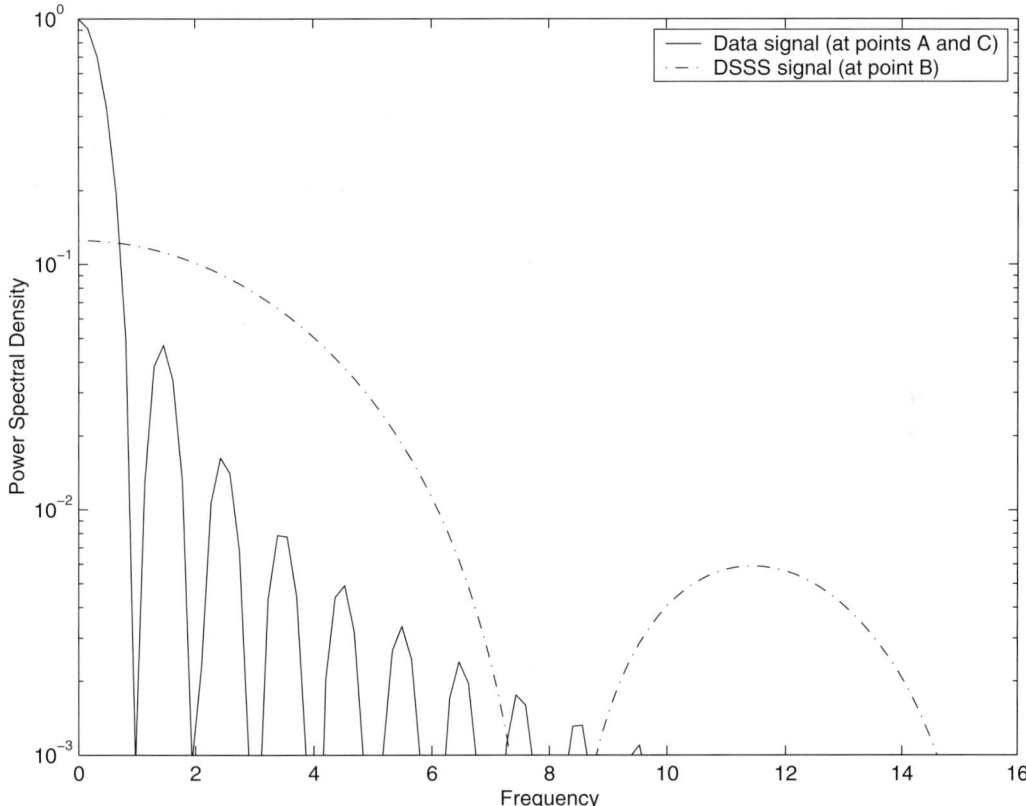

Figure 1.2   Power spectral densities of data and corresponding DSSS signals from Figure 1.1.

1.2 Code Division Multiple Access

CDMA is a multiuser communications method that employs spread spectrum signals. Here we concentrate on CDMA with DSSS. A fundamental issue in CDMA is how this technique affords multiple simultaneous transmissions using a common bandwidth. The underlying principle is that of distributing relatively low-dimensional data signals in a high-dimensional environment. This is accomplished by means of spreading codes or signature waveforms. The signature waveforms are unique to each user and have mutually low correlation so that the multiple access signals can be separated at the receiver by means of despreading. It is important to notice that all users use the entire available bandwidth simultaneously, thus contributing to the background noise. This additional interference that limits capacity is called multiple access, or multiuser interference. To minimize multiuser interference, each user's transmit power is controlled to the minimum required in order to maintain a requried signal to noise ratio at the receiver for the desired quality of service.

In the section that follows, we discuss the characteristics and generation of the spreading codes used in CDMA systems. To assess the impact of multiuser interference on the performance, we then introduce an analytical model for CDMA and derive the signal-to-noise ratio performance. Finally, we discuss the role of power and rate control and 'RAKE' receiver in CDMA systems.

1.2.1 Pseudonoise Spreading Sequences for CDMA

As previously discussed, DSSS signals suitable for CDMA applications are noiselike waveforms with mutually low correlation. That leads toward the use of random spreading sequences, that is, sequences of independent uniformly distributed random variables over some alphabet (usually binary). However, since for spread spectrum transmission to work both transmitter and receiver need synchronized replicas of the same sequence, these random sequences would have to be transmitted to communicators and stored in communication devices. This suggests that, for practical reasons, spreading sequences have to be generated algorithmically, with a relatively small number of stored parameters. To be used in CDMA applications, these deterministic sequences are designed to mimic randomness with the following key properties:

1. Relative frequencies of '0' and '1' are each 1/2;
2. In any period, half of the runs of consecutive 0s or 1s are of length 1, one quarter are of length 2, one eighth are of length 3, and so on, as one would expect in a coin-flipping experiment;
3. If the sequence is shifted by any number (nonzero) of elements and wrapped around, the resulting sequence will have an equal number of agreements and disagreements with the original sequence.

A deterministic sequence that closely resembles a random sequence with the properties above is called a pseudonoise (PN) sequence. Most known families of PN sequences are maximal length linear shift register sequences, or m-sequences, Gold, Kasami and bent sequences [1][2]. A simple way to generate PN sequences is by means of shift register generators with modulo-2 feedback connections, as shown in Figure 1.3.

All elements are binary (0 or 1) and additions are modulo-2. When $a_i = 0$ there is no feedback from the i-th stage and when $a_i = 1$, the feedback is present. In each clock interval the register content is shifted for one position to the right and based on the previous register state and feedback coefficients, the new value for b_1 is calculated. The all 0s state is not allowed, otherwise the register will remain in that state forever.

The PN sequences used in cdma2000 systems are m-sequences. M-sequences have period $2^r - 1$ (all binary r-tuples except all 0s) and are specified by primitive polynomials of length r. They closely achieve the key randomness properties discussed above. Specifically, m-sequences have the following properties:

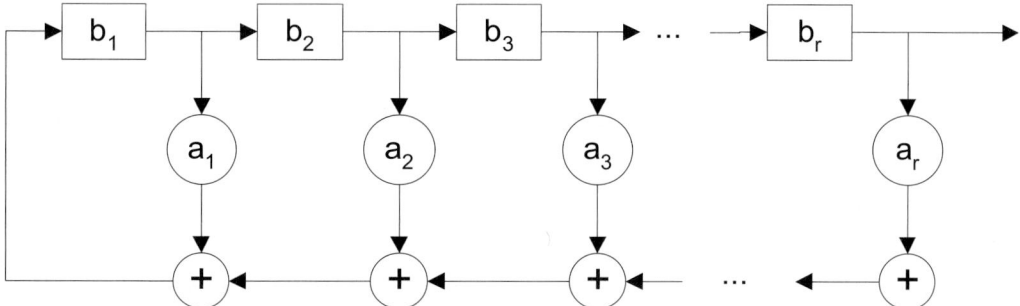

Figure 1.3 Shift register sequence generator.

1. There is an approximate balance of zeros and ones in that there are 2^{r-1} ones and $2^{r-1}-1$ zeros;
2. In any period, half of the runs of consecutive 0s or 1s are of length 1, one quarter are of length 2, one eight are of length 3, and so on;
3. If we define the antipodal sequence $b_k' = 1 - 2b_k$ and $L = 2^r - 1$, than the autocorrelation function satisfies:

$$R_{b'}(l) = \sum_{k=1}^{L} b_k' b_{k+l}' = \begin{cases} 1, & l = 0, L, 2L, \\ -1/L, & \text{otherwise} \end{cases} \tag{1.1}$$

As an example, consider a 4-stage shift register generating an m-sequence of period 15, shown in Figure 1.4. If we initially load the register with the sequence 0001, it will cycle through the states shown in the figure. The last bit in each state corresponds to the output bit in that cycle, so that one full period sequence is 100011110101100. After 15 cycles, the states and output bits repeat and so on [3][4].

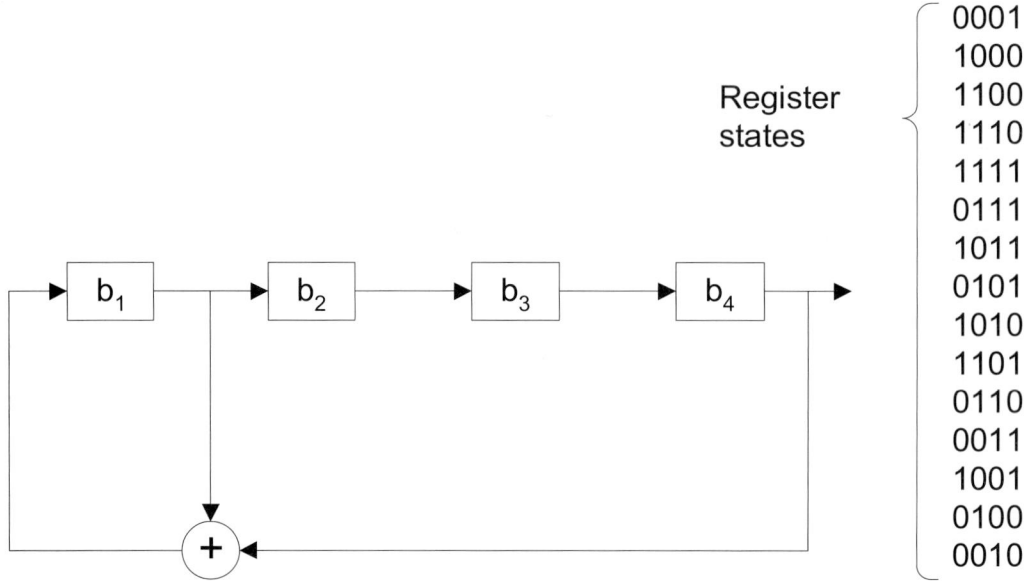

Figure 1.4 Example of m-sequence generator of length 4, period 15.

1.2.2 DS CDMA Performance—A Passband Model

In the 'standard' problem of digital transmission, the set of M signaling waveforms $\{s_i(t), \ 0 \le t \le T, \ 1 \le i \le M\}$, known to both transmitter and receiver, is used to transmit $(\log_2 M)/T$ bits/sec. If, for example, $s_i(t)$ is sent, the received signal is $r(t) = s_i(t) + n_w(t), \quad 0 \le t \le T$, where T denotes the symbol duration and $n_w(t)$ is additive, white Gaussian noise (AWGN) with two-sided power spectral density $N_0/2$ W/Hz. It is well known that the signal set can be completely specified by a linear combination of $D \le M$ orthogonal basis functions. The dimensionality, D, of the signal waveforms, is approximately equal to $2B_DT$, where B_D is the total (approximate) spectral occupancy of the employed signal set. If the total *available* bandwidth is B_N, corresponding to an N–dimensional signal space, the maximum number of simultaneously active users, each one using D dimensionality, with orthogonal multiplexing is $K = N/D$. With quasi-orthogonal multiplexing, however, it is possible to accommodate even more than N/D users in the same bandwidth, but with some mutual multiple access interference. In addition to sharing the bandwidth, the quasi-orthogonal users share interference as well.

A passband model of the DSSS CDMA system is presented in [5]. We will base the performance analysis on this model. Consider K simultaneous users that employ BPSK transmissions. The users' transmissions are assumed to be asynchronous in both timing and phase. Figure 1.5 depicts the transmitter model for the k-th user.

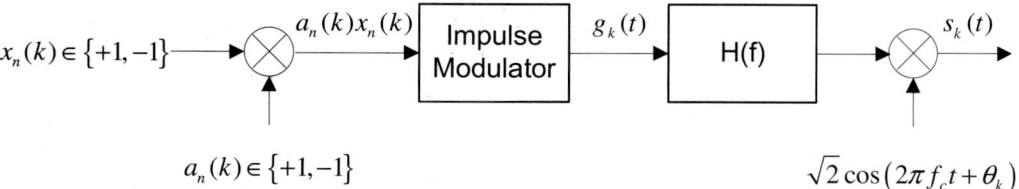

Figure 1.5   The block diagram of a passband DS CDMA transmitter.

$x_n(k)$ represents the antipodal modulation bit of the k-th user. The subscript n denotes the chip index. The chip rate is $1/T_c$ and the energy per chip is $E_c(k)$. $H(f)$ is the transfer function of the chip shaping filter, corresponding to the impulse response $h(t)$. f_c and θ_k correspond to the carrier frequency and phase, respectively. The transmitted signal of the k-th user can be written as:

$$s_k(t) = \sqrt{2E_c(k)}\sum_n x_n(k)a_n(k)h(t - nT_c)\cos(\omega_c t + \theta_k).\qquad(1.2)$$

The received signal in the multiple access system can be written as:

$$r(t) = s_k(t) + \sum_{i \neq k} s_i(t) + n(t),\qquad(1.3)$$

where the second term corresponds to other users interference and the third term represents receiver thermal noise, modeled as AWGN of two-sided power spectral density $N_0/2$. The receiver model is shown in Figure 1.6.

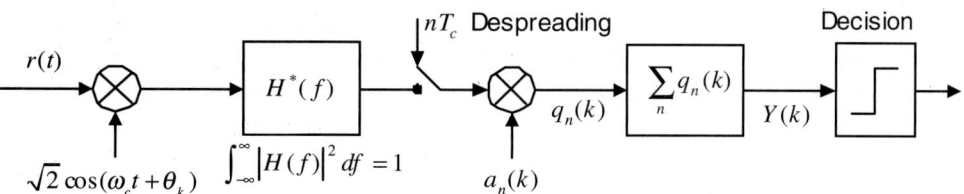

Figure 1.6   The block diagram of a passband DSSS CDMA receiver.

After multiplying the chip matched filter output in the *n*-th interval with the corresponding chip value, $a_n(k)$, we get:

$$q_n(k) = \sqrt{E_c(k)}x_n(k) + v_I(nT_c) + v_{MA}(nT_c) + v_N(nT_c),$$ (1.4)

where the first term is the desired signal contribution and the second, third and fourth terms represent zero-mean interchip interference, multiple access interference and noise, respectively. After despreading, the chip samples are accumulated over the bit period forming the decision variable:

$$Y(k) = \sum_{n=1}^{1/(RT_c)} q_n(k)$$ (1.5)

where R is the bit rate. Following the derivation in [5], one can obtain the first and second order moments of $Y(k)$. For the mean we obtain:

$$E\{Y(k)1xu(k)\} = \pm 1/(RT_c)\sqrt{E_c(k)}$$ (1.6)

and for the variance we obtain:

$$\mathrm{Var}\{Y(k)1xu(k)\} = 1/(RT_c)(V_I + V_{MA} + V_N)$$ (1.7)

where V_I, V_{MA} and V_N represent the second moments of the interchip interference, multiple access interference and noise, respectively, given by:

$$V_I = E_c(k)\sum_{\substack{m=-\infty \\ m \neq 0}}^{\infty} \left[\int_{-\infty}^{\infty} |H(f)|^2 \cos(2\pi fmT_c)df \right]^2,$$ (1.8)

$$V_{MA} = \frac{1}{2T_c}\int_{-\infty}^{\infty} |H(f)|^4 df \sum_{\substack{i=1 \\ i \neq k}}^{K} E_c(i)$$ (1.9)

and

$$V_N = \frac{N_0}{2}.$$ (1.10)

At the last stage in the receiver, a bit decision is made using the output of the chip accumulator according to $\hat{x}_n(k) = \text{sgn}\left[Y(k)\right]$. Under these circumstances, a measure of performance which is monotonically related to the signal energy to noise plus interference density ratio (SNIR) can be expressed as:

$$\text{SNIR}_k = \frac{\left[E\{Y(k)|x_n(k)\}\right]^2}{2\text{Var}\{Y(k)|x_n(k)\}} = \frac{E_c(k)/RT_c}{2(V_I + V_{MA} + V_N)} \tag{1.11}$$

So far we have not made a specific assumption about the chip shape. If we assume ideal bandlimited, Nyquist chip shape defined by the transfer function:

$$H(f) = \frac{1}{\sqrt{W}} rect\left(\frac{f}{W}\right), \tag{1.12}$$

where $W = 1/T_c$, Eq (1.11) reduces to:

$$\text{SNIR}_k = \frac{E_b(k)}{N_0 + \frac{1}{N}\sum_{\substack{i=1 \\ i \neq k}}^{K} \frac{E_c(i)}{E_c(k)}} \tag{1.13}$$

where $N = 1/RT_c$ and $E_b(k) = NE_c(k)$ is the energy per bit for the k-th user. In the denominator, the first term is due to the thermal noise and the second term is due to the multiple access interference. It can be seen from the second term that multiple access interference is suppressed by the factor N (spreading factor), which explains why the spreading factor is often called the processing gain. If we let $E_c/N_0 \to \infty$ (multiple access interference dominates) and assume $E_c(i) = E_c(l)$, for all i, l, then:

$$\lim_{E_c/N_0 \to \infty} SNIR_k = \frac{N}{K-1} \tag{1.14}$$

This last result reveals the fundamental difference between orthogonal and quasi-orthogonal multiplexing; even for vanishingly small thermal noise, the SNIR is finite. Hence, the number of simultaneous transmissions, for K >>1, is $K \leq N/\text{SNIR}$, where SNIR corresponds to the desired transmission quality.

1.2.2.1   Quadrature Spreading

In the previous section we considered BPSK modulation and BPSK spreading. The expression for variance of multiple access interference in Eq. (1.9) hides the fact that interference from a single interferer is dependent on the phase difference between the desired user signal and the interfering users' signals. Specifically, the second moment of the interference present in the k-th user's signal, conditioned on the phase offsets with the i-th user's signals, $\Xi = \{\theta_i\}, i = 1,...,K, i \neq k$, is given by:

$$V_{MA}(\Xi) = \sum_{\substack{i=1 \\ i \neq k}}^{K} V_{MA}(i|\theta_i) = \frac{1}{T_c} \int_{-\infty}^{\infty} |H(f)|^4 \, df \sum_{i=1, i \neq k}^{K} E_c(i) \cos^2 \theta_i \qquad (1.15)$$

Only after averaging over the random, uniformly distributed phase offsets, $0 < \theta_i \leq 2\pi$, one obtains Eq. (1.9). This approach is valid in the case of a large number of interferers. However, for a small number of interferers a particular phase realization may significantly affect performance. For example, when an interferer is in-phase with the desired signal it causes maximum interference, because $\cos^2 \theta_i = 1$. When the interferer is in quadrature relative to the desired signal, it causes no interference to it. On average, the interference factor due to phase averaging is equal to 1/2.

As shown in [1][5], by using Quadrature phase shift keying (QPSK) spreading, *i.e.*, different spreading sequences in the in-phase and quadrature branches of the modulated signal, the impact of the random phase offsets is eliminated without averaging and the i-th user contribution to the total interference from the in-phase and quadrature branches is:

$$V_{MA}(i|\varphi_i) = \frac{E_c(i)}{2T_c} \int_{-\infty}^{\infty} |H(f)|^4 \, df \qquad (1.16)$$

With BPSK modulation and QPSK spreading, the k-th user transmit signal is given by:

$$s_k(t) = \sqrt{E_c(k)} \sum_n x_n(k) h(t - nT_c) \left[a_n^{(I)} \cos(\omega_c t + \vartheta_k) + a_n^{(Q)} \sin(\omega_c t + \vartheta_k) \right] \qquad (1.17)$$

while the transmitter and the receiver for the k-th user are illustrated in Figure 1.7 and Figure 1.8, respectively.

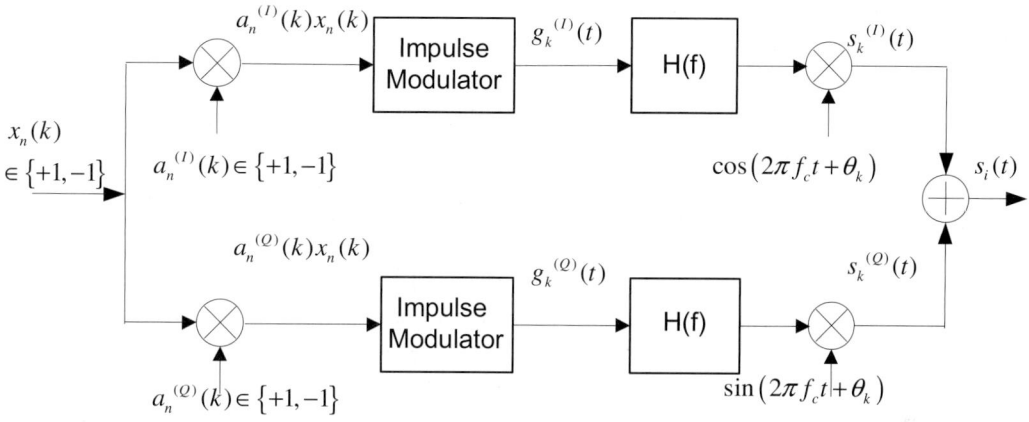

Figure 1.7   Block diagram of CDMA transmitter with QPSK spreading.

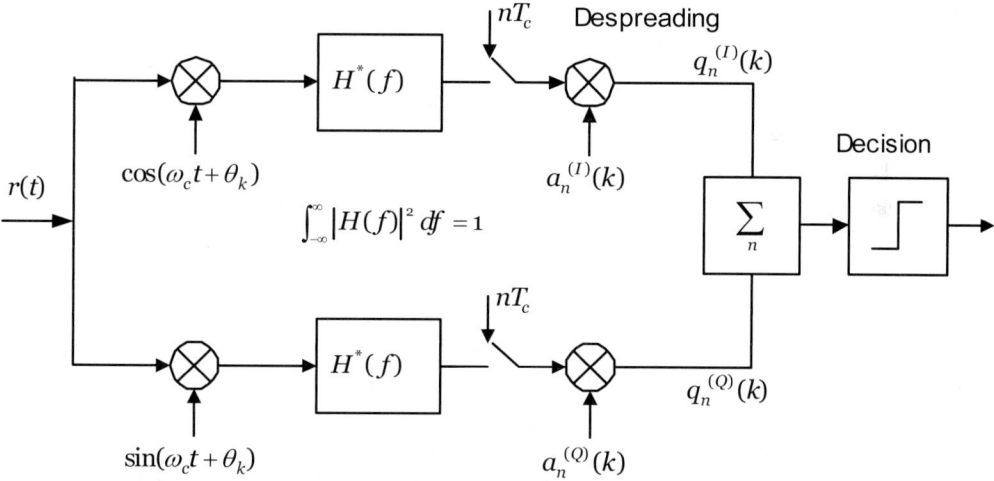

Figure 1.8   Block diagram of CDMA receiver with QPSK despreading.

1.2.3 Power Control

In CDMA systems, employing a correlation-type detector it is important to have undesired signals at the same or lower power level as the desired signal to achieve good performance. Based on the communication system design, modulation type, forward error correction (FEC) coding, diversity, and similar, a CDMA system can tolerate a certain amount of interference for a given transmission quality. This amount of interference can correspond to a few strong interferers or many small equal power interferers. To maximize the capacity it is desired to have all received signals at the minimum power level required to achieve the target SNIR. In cellular communications this is normally achieved by using power control. In the mobile station to base station direction (reverse link), a combination of open and closed-loop power control is used. In the base station to mobile station direction (forward link), closed-loop power control is used. Open-loop is less critical than in the reverse direction.

At the mobile station, open-loop power control employs measurements of the average received signal power to estimate the forward link path loss and shadowing. As path loss and shadowing are largely frequency independent within a given frequency band, the reverse link path loss is approximately equal to that of the forward link. The forward and reverse link channel reciprocity is exploited by open-loop power control that continuously adjusts the transmit power by an amount inversely proportional to changes in the received signal power, thus compensating for changes of propagation loss and shadowing.

However, in systems with frequency division duplexing (FDD), using different frequencies for transmit and receive directions, the multipath fading exhibits frequency selectivity in that different fading is observed on the two different frequencies. That is, small changes in multipath delays at radio frequencies cause large different variations in phase, which ultimately results in different fading on the two different frequencies [6]. To combat such a non-reciprocal multipath fading a closed-loop mechanism must be employed.[1]

Figure 1.9 depicts the general principle of closed loop power control for the forward link in a CDMA system. The mobile station measures the received SNIR and compares it with the desired SNIR. Depending upon whether the measured SNIR is above or below the desired SNIR, respectively, the mobile station instructs the base station to decrease or increase the transmit power by a predetermined amount (usually 1 dB). The power control command is composed of one bit that is sent on a reverse link control sub-channel. The frequency of transmission of the power control commands varies depending on the configuration of the traffic channel in use. The maximum power control rate, which is used in most scenarios, is equal to 800 bps. Such rate represents a trade-off between forward link power control performance and reverse link power control channel overhead, as the two have an opposite effect on the forward and reverse link capacity, respectively. Due to the finite granularity of power adjustment, limited rate of the power control sub-channel and channel dynamics, power control may not be able to track fading variations in moderate and fast fading channels. In Figure 1.9, the dotted line illustrates how received signal power may still show variations around the desired signal level. As discussed in the following sections, FEC coding and interleaving compensate for the power control inefficiency because they provide maximum coding gain against fading when fade variations are fast. For slow fading, the closed-loop power control mechanism can completely neutralize the impact

---

1.   There are other practical considerations that require the use of closed power control. Those will be discussed in detail in Chapter 8.

of fading. Usually, the worst performance is achieved at medium speeds where power control is not fully effective and FEC coding and interleaving are not able to provide full coding diversity for moderately fast fading.

A similar closed-loop power control mechanism is employed to regulate the mobile station's transmit power on the reverse link, except now the base station and mobile station roles are reversed. The signal levels are measured at the base station, which instructs the mobile stations to adjust their power up or down so that all received signal power levels at the base station are approximately the same.

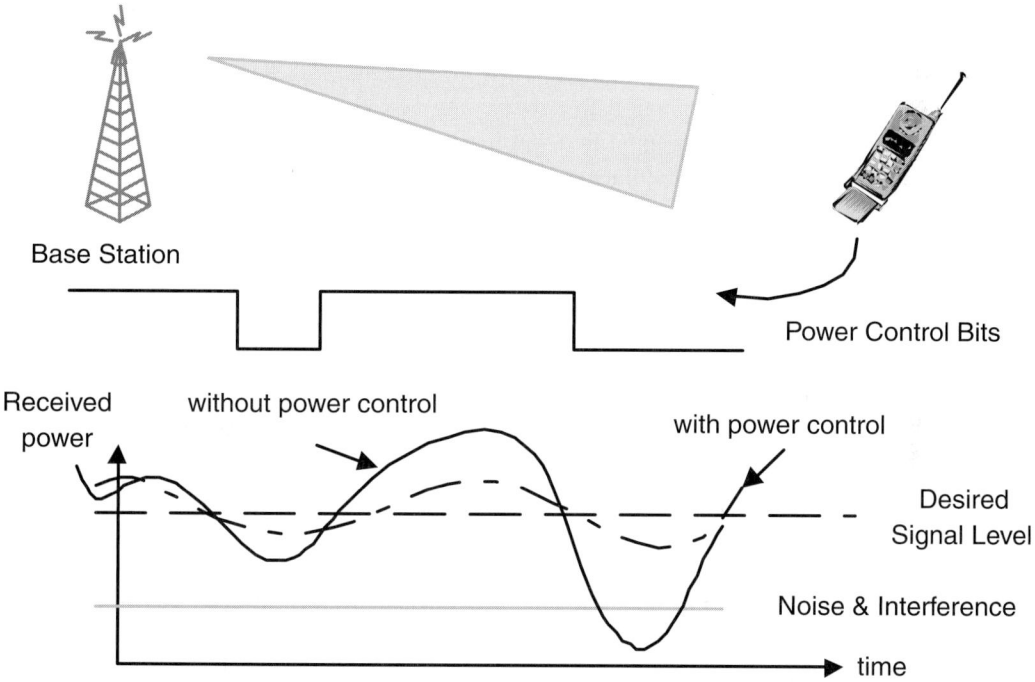

Figure 1.9 Principle of operation of closed-loop power control.

1.2.4 Rate Control and Multiuser Diversity—Is Fading Really Bad?

In delay tolerant applications such as packet data applications, the system can adaptively assign transmission data rates to maximize the throughput for given transmit power resources and channel conditions, that is, achievable received SNIR. This form of channel adaptation, aided by channel state information fed back to the transmitter from the intended receiver, is referred to as *rate control*. Supporting transmission of delay tolerant data with rate control is thus fundamentally different from real time data with stringent delay and quality of service requirements, because the latter necessitates power control to guarantee that a constant bit rate and FER are achieved.

But is rate control desirable compared with power control? Yes, as rate control allows achieving a higher average throughput than power control for the same average transmit power. We shall demonstrate this using a simplified, yet insightful, example. From Eq. (1.13), we can express the received bit energy to total (noise and multiple access interference) interference power spectral density as a function of the transmit power, P_{TX}, and the channel gain, α, as

$$\frac{E_b}{N_t} = \frac{W}{R_b} \frac{\alpha P_{TX}}{P_I} \tag{1.18}$$

where P_I is the received interference power in the band W. If we consider rate control with a fixed amount of transmit power, $P_{TX,0}$, the average data rate, or throughput, is given by:

$$E\{R_b\} = \frac{W}{(E_b/N_t)_{req}} \frac{P_{TX,0}}{P_N} E\{\alpha\} \tag{1.19}$$

where $(E_b/N_t)_{req}$ represents the received bit energy to interference power spectral density required to attain the desired data rate. Now consider transmission over the same channel using power control so that the bit rate is constant, $R_{b,0}$. Under this scenario, the average transmit power is given by:

$$E\{P_{TX}\} = \frac{(E_b/N_t)_{req}}{W} P_N R_{b,0} E\left\{\frac{1}{\alpha}\right\} \tag{1.20}$$

We want to compare relative performance of power control and rate control by comparing the average transmit power required by the two systems to achieve the same average data rate. Then, by letting $R_{b,0} = E\{R_b\}$, we obtain from Eq. (1.19)-(1.20):

$$\frac{E\{P_{TX}\}}{P_{TX,0}} = E\{\alpha\} E\left\{\frac{1}{\alpha}\right\} \geq 1 \tag{1.21}$$

The inequality in Eq. (1.21) holds for any non-negative random variable, in which case the harmonic mean is always larger than the inverse of the mean. The equality holds only in case α is constant, that is, when the channel is not fading. Then, Eq. (1.21) says that rate control is more power efficient than power control.[2] We must point out that the model discussed above is highly simplified, because it is assumed that the $\left(E_b/N_t\right)_{req}$ is independent of the data rate, and that the power and rate control are ideal (*i.e.,* power control range is infinite and rate adaptation has infinitesimal resolution). In practice, the advantage of rate control over power control depends on the implementation details of power and rate control, and also on the degree of variability experienced by the channel fades.

Beside rate control, delay tolerant applications allow for yet another throughput improvement opportunity. The system designer can exploit temporal signal variations due to fading of different users to opportunistically schedule their transmissions when they experience good channel conditions. The concept of scheduling users opportunistically, based on their corresponding channel qualities with the base station, is called multiuser diversity. Once the user experiencing the best channel quality is selected, the base station will estimate the maximum spectral efficiency that can be achieved over that link. As an illustration, if the available modulation schemes are QPSK, 8-PSK, and 16-QAM, which correspond to 2, 3 and 4 bps/Hz, respectively, the system will select the one with the largest alphabet size that can still operate below specified FER under such channel conditions.

This approach is fundamentally opposite of what communications engineers have traditionally done. Conventional methods to deal with fading are to design countermeasures, such as FEC coding or diversity, or to incorporate a fading margin (to use extra power) into the link budget, or to use both, which is the prevailing approach. This is still a valid approach for circuit switched, delay-sensitive services. However, in the case of delay tolerant, packet-switched services, one can argue that fading is 'good' not 'bad.' What do we mean? To illustrate this point, consider the scenario depicted in Figure 1.10 that shows the evolution over time of SNIRs in a fading channel for three different users. We can see that when one user has a bad channel, another user may have a good channel. If we always give the use of the channel to the user with the best SNIR at that time, we could achieve much better performance, corresponding to the envelope of the three SNIR curves. This can be considerably better than what is achievable for one particular user under average conditions. That's why in this case we say that fading is 'good,' as long as we can take advantage of it. At the top of Figure 1.10, we show the time slot allocation corresponding to the user with the best SNIR. User 2 (dashed line) is assigned the channel initially, then user 1 (solid line), followed by user 3 (dashed-dotted line), and so on. It should be noted that the described algorithm for channel assignments may result in unequal use of the channel for different users, as well as variable delays from user to user. It is possible to add additional criteria, such as the fair use of channel or delay constraint. In principle, one can trade delay for throughput. If more delay variation is acceptable, it is possible to wait longer for

---

2.   This concept will be revisited in Chapter 8 where we estimate the excess transmit power due to power control.

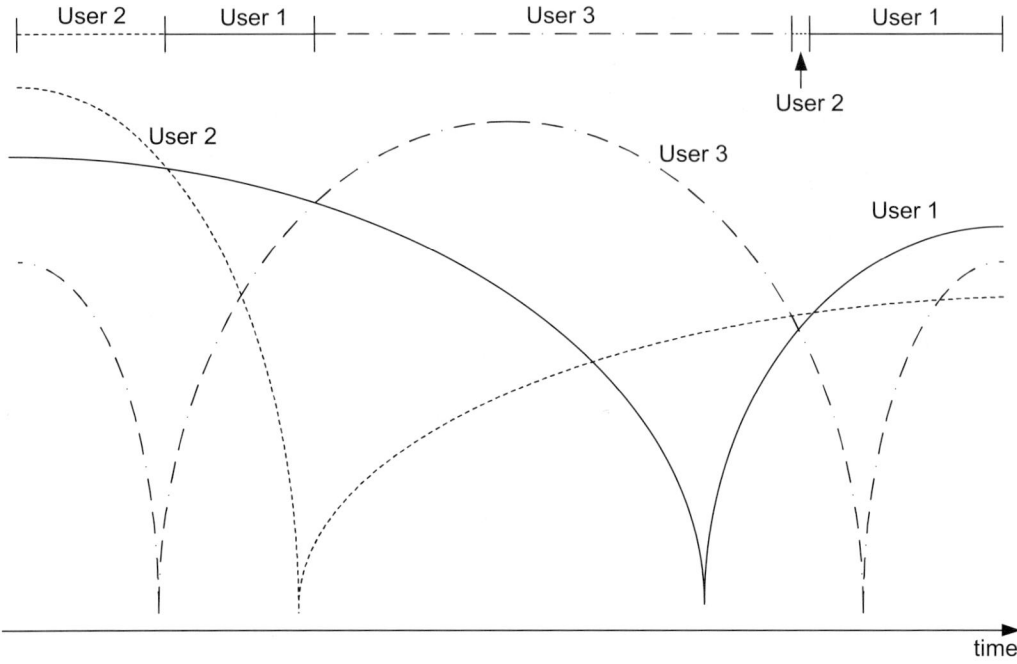

Figure 1.10 Scheduling of users over time based on received SNIR.

a particular user to get a good channel. Although the variations of the proposed algorithm will ultimately affect the average throughput, the basic philosophy will remain. We discuss in more detail the operational aspects of multiuser diversity and adaptive rate assignment in the context of the Forward Packet Data Channel (F-PDCH) in Chapters 9 and 10.

1.2.5 Multipath and RAKE Receiver

Multipath propagation in mobile channels is a major source of signal impairments such as amplitude fading, rapid carrier phase variations and intersymbol interference. When all multipath components arrive with approximately the same time delay at the receiver, the constructive or destructive signal superposition results in frequency-flat signal fading. However, if mutual delays of multipath components are not negligible relative to the bit duration, then the multipath phenomenon may give rise to frequency-selective channel behavior.

Thanks to the redundant bandwidth (*i.e.,* spreading), a DSSS receiver synchronized to a particular multipath component signal will suppress the energy of all multipath components that are delayed by more than one chip duration, proportionally to the spreading gain. This is similar to the suppression of multiple access interference, discussed in the previous section. To illustrate this further, consider the k-th user received signal comprising two multipath components with amplitudes α_1 and α_2 [3][4]:

$$r(t) = \alpha_1 s_k(t - \tau_1) + \alpha_2 s_k(t - \tau_2) + n(t) \qquad (1.22)$$

where $s_k(t)$ is as in Eq. (1.17) and $n(t)$ accounts for both thermal noise and other users' interference. If the receiver is synchronized to the multipath component with delay τ_1, it will suppress the signal delayed by τ_2, because the autocorrelation of spreading sequences for delays $\tau_2 - \tau_1 > T_c$ is in general quite low. Alternatively, one can say that the spread data signal at delay τ_1 appears as a signal with a different spreading sequence to the receiver synchronized to the delay τ_2. However, instead of synchronizing onto and capturing only one multipath component, one can employ several receiver branches, each synchronized to a distinct multipath component, at least one chip duration, i.e., T_c seconds apart from other multipath components, so that each branch, referred to as 'RAKE finger,' captures the signal energy in the corresponding multipath component and suppresses multipath energy at other delays. Finally, the receiver can combine the received signal energy from all branches (RAKE fingers), with the proper synchronization in time. This explains where the term RAKE comes from. Namely, this multi-branch receiver collects multipath signals analogous to a garden rake collecting leaves. Not only does the RAKE receiver achieve the energy gain relative to a single branch receiver, but it also achieves a multipath diversity gain as each distinct multipath component experiences different fading, in general. The diversity gain refers to a reduction in SNIR variance. It is important for delay sensitive applications that cannot take advantage of channel sensitive scheduling, because the FER is a convex function of SNIR and smaller SNIR variance implies lower required SNIR.

To further illustrate the operation of RAKE fingers, consider a multipath profile example in Figure 1.11. Multiple clusters of multipath signal components often correspond to significant reflecting structures, man made or natural. Within each cluster there are many multipath components with approximately the same time delays, typically giving rise to a frequency non-selective fading within each multipath cluster. It is beyond the scope of this book to go into details of multipath fading phenomenology as there are many books that address this subject thoroughly [6][7].

Let us now assume that we have a receiver with three RAKE fingers, whereby each finger is synchronized to one of the multipath clusters in Figure 1.11. As discussed earlier, each RAKE finger will capture the signal energy in its corresponding multipath cluster and suppress multipath energy from other multipath clusters. Here we assume that the multipath spread within each cluster is small relative to Tc, so that multipath components within a cluster are not resolvable.

This description is succinctly captured in Figure 1.12. The output of each RAKE finger contains the captured desired multipath signal plus multipath interference from the other two clusters that are suppressed proportionally to the processing gain (spreading factor). Finally, the outputs of RAKE fingers are combined to produce a multipath diversity gain. In Figure 1.12 we show maximal-ratio combining (MRC), which is optimum in the sense of maximizing the SNIR at the combiner output [1][5]. Specifically, if the signal amplitude for the i-th RAKE finger is α_i, $i = 1$, 2, and 3, then each RAKE receiver output is multiplied by its corresponding signal amplitude before adding all signals together. This agrees well with the intuition that we give more weight to a more reliable signal (with better SNIR) and less weight to a less reliable signal. The

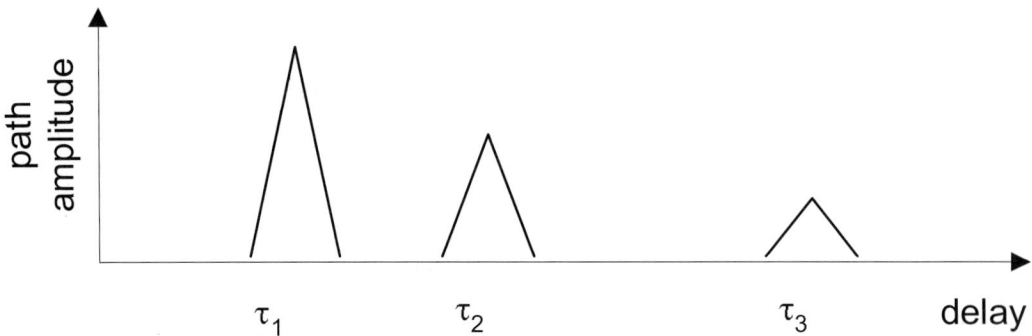

Figure 1.11 An example of multipath profile representative of wireless mobile channels.

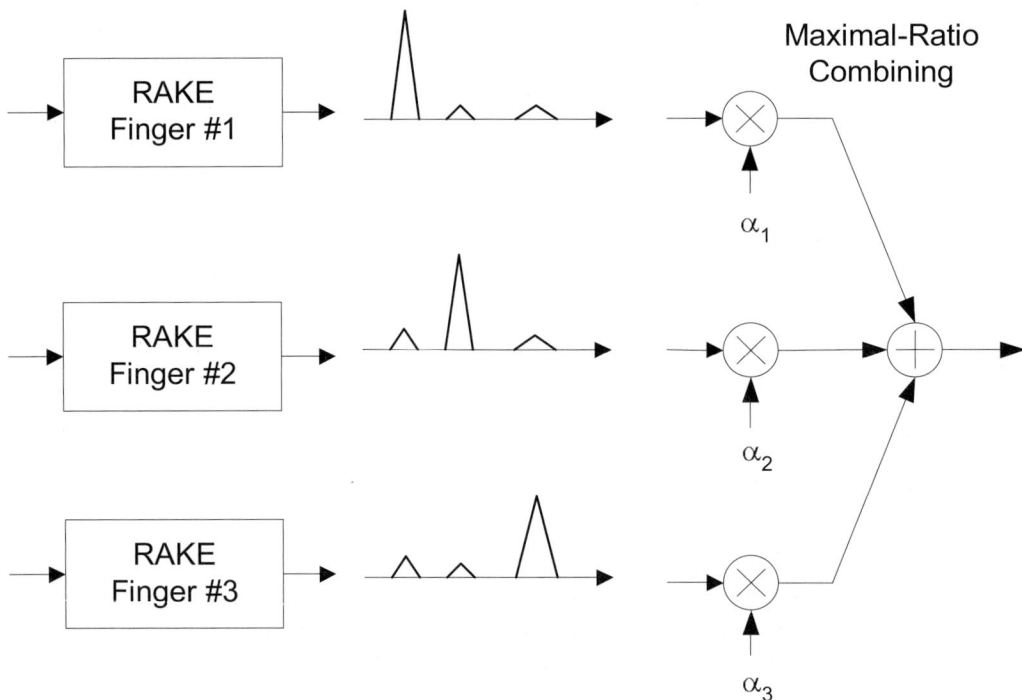

Figure 1.12 Illustration of RAKE combining.

combining is coherent, *i.e.,* after phase estimation in each RAKE finger, the phase offset is removed prior to RAKE combing. Alternately, the phase removal can be implemented jointly with RAKE combining.

The described RAKE with MRC requires an estimation of signal amplitudes to perform the weighted combining operation. The signal amplitude is efficiently estimated by using the known pilot signal, which is usually done in conjunction with phase estimation for coherent demodulation. Specifically, at time instant i, the output of the pilot signal despreader, y_i can be written as:

$$y_i = A_i e^{j\theta_i} + n_i \tag{1.23}$$

where A_i and θ_i represents the signal amplitude and phase, respectively, and n_i accounts for the aggregate of thermal noise and multiple access interference. Assuming that the channel is constant over N sampling intervals and that *noise* is Gaussian, it is a simple exercise to obtain the maximum likelihood estimator of A and θ, as shown in [8], given by

$$\begin{aligned}
\hat{A}_{ML} &= abs\left(\frac{1}{N}\sum_{i=1}^{N} y_i\right), \\
\hat{\theta}_{ML} &= \arg\left(\frac{1}{N}\sum_{i=1}^{N} y_i\right).
\end{aligned} \tag{1.24}$$

The variance of the estimates is inversely proportional to the pilot's signal-to-noise ratio and number of samples N. Thus, by increasing N in scenarios with slow channel variations, *i.e.,* by narrowing the bandwidth of estimation loop, we can improve channel estimation. Conversely, in channels with faster variations we are forced to use correspondingly smaller N (smaller window size, or wider loop bandwidth) to track channel variations, thus obtaining a noisier estimate.

1.3 Forward Error Correction and Interleaving

The FEC coding has become a standard component of most existing digital communications systems [6]. It plays an important role in improving reliability of transmission in the presence of different channel impairments. FEC is especially important in fading and interference channels, where its gain may be more pronounced than in additive white Gaussian noise. Likewise, it is a crucial ingredient of CDMA systems and contributes considerably to the spectral efficiency of these systems.

A simplified block diagram of a coded spread spectrum system is shown in Figure 1.13. New blocks that facilitate efficient error correction are the FEC encoder and interleaver in the transmitter and deinterleaver and FEC decoder in the receiver. The FEC encoder adds coded redundancy to the information data stream, based on which the FEC decoder can perform reliable decoding and neutralize the impact of channel errors. The interleaver/deinterleaver pair helps to achieve coding diversity by providing approximately independent channel outputs, which are discussed in more detail subsequently.

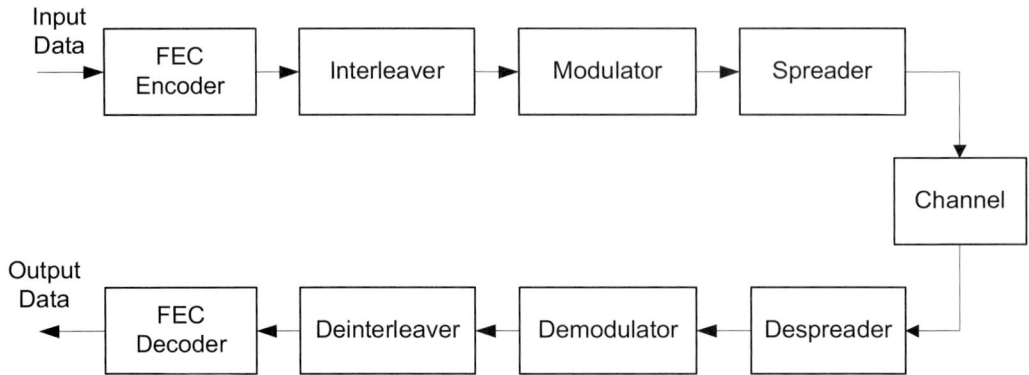

Figure 1.13 Block diagram of coded spread spectrum system.

In traditional digital communications, the gain due to FEC does not come without a price. When coding redundancy is introduced, we either need to increase the bandwidth to keep the same information rate, or reduce the information rate if the bandwidth is fixed.[3] In DSSS and CDMA however, FEC does not introduce bandwidth expansion even if we maintain the fixed information rate, because the signal bandwidth is determined by the chip rate. The FEC enables the system to operate at significantly lower SNIR than in the case of uncoded transmission for the same residual error rate. The difference in SNIR refers to the FEC coding gain.

However, in CDMA systems, the spreading gain per coded symbol is reduced, which diminishes the ability to suppress interference. For example, consider an uncoded system with the processing gain equal to N (N chips per information symbol). Given the same information rate, if we introduce FEC of rate $r=1/2$, the coded symbol duration will be a half of the information symbol duration and, thus, the spreading factor per coded symbol will be $N/2$. The reduced spreading gain diminishes the ability of a spread spectrum system to suppress interference, but the coding gain more than compensates for the loss. Moreover, the lower the code rate, the larger is the coding gain of the employed code and despite the reduction in the spreading gain, the detection performance of the correlation receiver improves [9][10].

Because of the FEC coding gain, coded CDMA systems can operate at lower SNIR and achieve the same bit error rate performance as in the case of uncoded CDMA. In other words, we can keep increasing the number of users (*i.e.,* the amount of multiple access interference), until we reach this lower SNIR enabled by FEC. The FEC coding gain results in an increase in multiple access capacity (or spectral efficiency of the system), which is quite important in mobile communications systems because of limited available frequency spectrum. Thus, it is

3.   Of course, this is not the case for the trellis-coded modulation [6] where a coding gain is achieved by virtue of joint coding and modulation design without bandwidth expansion at the expense of increased modulation order.

important to use as powerful FEC codes as is practically appropriate to maximize capacity. Convolutional and turbo coding techniques are adopted for the cdma2000 air interface. In practice, the convolutional codes outperform turbo codes for short payload sizes, while turbo codes offer stronger coding gains for packets larger than about one to two hundred bits.

Any communication and signal processing technique that would enable a CDMA system to operate at a lower SNIR, for a given transmission quality, will have a direct impact on a capacity improvement.

1.3.1 Convolutional Coding

The FEC encoder for convolutional codes can be implemented using linear shift registers with modulo-2 adders, similarly to the case of PN code generators. Consider a simple convolutional code [11] defined with generators in octal notation given by $g_0 = (7)_{oct}$ and $g_1 = (5)_{oct}$. These generators define a convolutional code of rate $r=1/2$ and constraint length $K=3$ and the corresponding encoder is shown in Figure 1.14. Each input bit produces two output bits in each clock interval, hence the rate $r=1/2$. Moreover, each output bit is a function of current and previous information bits, in this case two previous bits.

The convolutional encoder in Figure 1.14 is a finite-state machine. In this case, the encoder is characterized by four states: 00, 01, 10 and 11. In general, the encoder has 2^{K-1} states. The trellis diagram of the convolution code, given in Figure 1.14, is illustrated in Figure 1.15. The first bit of the branch label denotes the input information bit and the last two bits denote the output coded bits for the corresponding state transition. The new state is determined by the previous state and the incoming information bit. However, the output coded bits are functions of the encoder state, information bit and modulo-2 connections to the output.

To illustrate the operation of this convolutional encoder, consider the following information sequence: 1, 1, 0, 0, 0. The corresponding coded sequence is: 1, 1, 0, 1, 0, 1, 1, 1, 0, 0. It should be noticed that the last two 0s in the information sequence flush the encoder (i.e., bring it

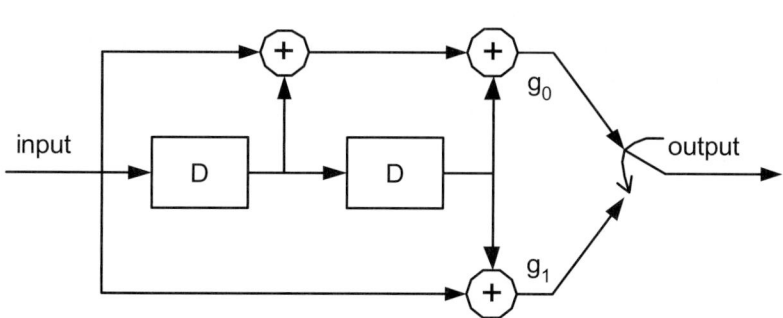

Figure 1.14 Convolutional encoder with $r=1/2$, $K=3$; $g_0 = (7)_{oct}$ and $g_1 = (5)_{oct}$.

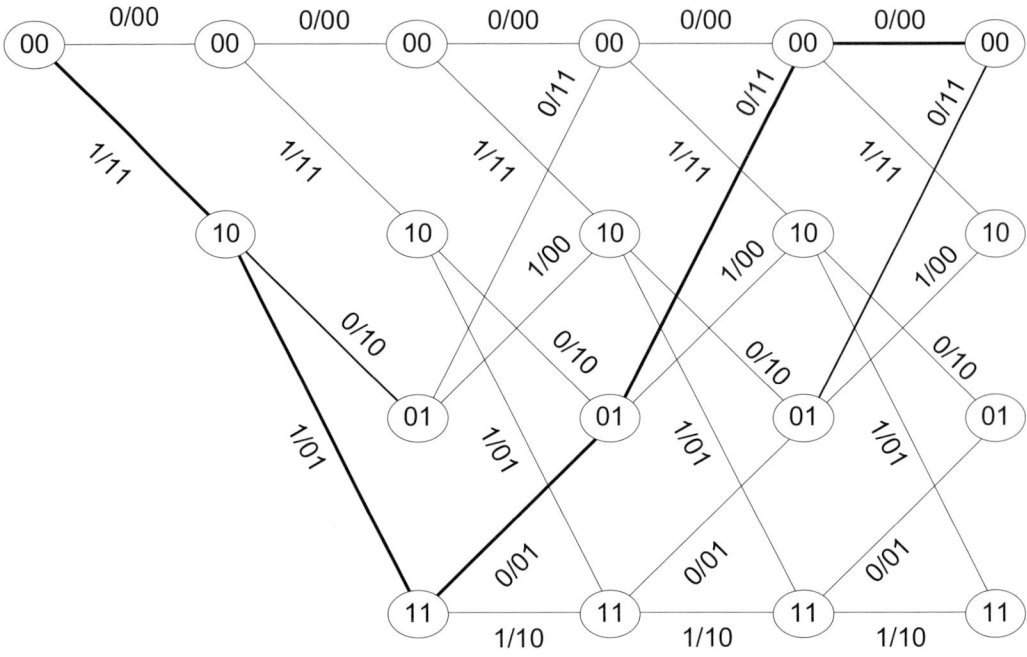

Figure 1.15 Trellis representation for convolutional code $g_0 = (7)_{oct}$ and $g_1 = (5)_{oct}$.

again to the state 00). The path corresponding to the information sequence 1, 1, 0, 0, 0 of our example above is marked in the trellis by the bold line.

1.3.1.1   *Decoding of Convolutional Codes*

The trellis diagram is extremely useful in decoding convolutional codes and was instrumental in developing the Viterbi algorithm [11] for decoding of convolutional codes. The Viterbi algorithm exploits the Bellman's optimality principle to reduce the number of paths that need to be kept during decoding to one per state. That is, if two or more paths merge in a given state, we need only to keep the path with best metric (survivor) to facilitate maximum likelihood decoding, and remaining paths can be discarded. The path metric, in general, depends on specific modulation scheme chosen. To illustrate the path metric calculation, consider binary antipodal modulation with demodulator outputs, $y_{k,m}$ given by:

$$y_{k,m} = a_{k,m} x_{k,m} + n_{k,m} \qquad (1.25)$$

where index k corresponds to the k-th branch or the k-th information bit, index m corresponds to the m-th coded bit of that branch, $a_{k,m}$ represents the coded symbol amplitude, coded symbol

$x_{k,m}$ is either +1 or −1 and noise, $n_{k,m}$ is zero-mean Gaussian with variance $N_0/2$. The maximum likelihood path metric, for the *i*-th path, $PM^{(i)}$ is given by [6]:

$$PM^{(i)} = \sum_k \mu_k^{(i)} \tag{1.26}$$

Assuming *M* coded bits per branch:

$$\mu_k^{(i)} = \log P(y_{k,1}, \quad , y_{k,M} \mid x_{k,1}^{(i)}, \quad , x_{k,M}^{(i)}) \tag{1.27}$$

where $P(\cdot \mid \cdot)$ denotes the conditional probability. For assumed binary antipodal signaling Eq. (1.27) reduces to the correlation metric:

$$\mu_k^{(i)} = \sum_{m=1}^M y_{k,m} a_{k,m} x_{k,m}^{(i)}. \tag{1.28}$$

It is important to notice that in order to form the optimum metric, one needs to know the signal amplitudes. When signal amplitude is constant over the coded sequence, the optimum metric becomes:

$$\mu_k^{(i)} = \sum_{m=1}^M y_{k,m} x_{k,m}^{(i)}. \tag{1.29}$$

The metric given in Eq. (1.29) can be used when amplitudes are not known or when their measurements are noisy, but with some degradation in performance relative to the optimum decoding case.

1.3.2 Interleaving and Deinterleaving

As mentioned earlier, interleaving and deinterleaving are used in conjunction with coding and decoding to maximize coding diversity. Their role is to change the order of coded bits on the channel as compared to their order in the coded sequence. That is, the interleaver/deinterleaver pair destroys the channel memory to enable random error correcting codes to operate efficiently (*i.e.*, to disperse burst errors and enable the decoder to correct them successfully). This is especially important in the case of correlated fading or other channel impairments that could cause burst errors.

The simplest form of interleaving is block interleaving. A block interleaver consists of *N* columns and *M* rows. The coded symbols are written into the interleaver row-wise and read out to the channel column-wise. The deinterleaver in the receiver performs reverse operation to reestablish the original order of coded bits prior to decoding. It should be noted the two symbols that are adjacent on the channel are separated N positions apart in the coded sequence, so that the decoder would effectively see approximately N times faster fading over the coded sequence than it really is on the channel. In addition to the technique described above, interleavers used in cdma2000 utilize intra-raw permutations to maximize the performance of the employed FEC code. The interleavers used in cdma2000 are described in Chapter 4. The price for block interleavers is additional memory and delay introduced in both the transmitter and receiver.

1.3.3 Turbo Coding

Since their invention in the early 1990s, turbo codes [12] have attracted considerable interest because of their potential to approach the Shannon's capacity limit within a decibel or so, particularly when coupled with large interleavers. To illustrate the encoding operation for turbo codes, consider an 8-state (K=4) parallel concatenated convolutional code (PCCC) defined with feed-forward generator polynomial $g_0=(13)_{oct}$ and feedback generator polynomial $g_1=(15)_{oct}$. The transfer function of constituent codes is given by:

$$G(D) = \left[1, \frac{g_1(D)}{g_0(D)}\right] = \left[1, \frac{1+D+D^3}{1+D^2+D^3}\right] \cdot \tag{1.30}$$

The block diagram of the encoder is shown in Figure 1.16. From the first component code, both systematic bits and parity bits are transmitted, whereas from the second component code, after interleaver, only parity bits are transmitted while the interleaved systematic bits are completely punctured. In this example, the overall code rate is 1/3. Rate 1/5 turbo code can be obtained with rate 1/3 constituent encoders and so on. The dashed lines indicate connections for termination of encoded sequence (i.e., trellis termination). It should be noted that for a code with memory m (in this case m=3), $4m$ termination bits are transmitted. That is, termination bits are also sent for the punctured systematic bits of the second component decoder.

Usually some kind of a pseudorandom interleaver is used. The interleaver choice can significantly impact the residual error rate. It is widely claimed that a fraction of a decibel gain in the region of practical interest as well as some improvement in error floor can be achieved by using an optimized pseudorandom interleaver [13].

Higher code rates can be achieved by puncturing parity bits. Puncturing of systematic bits results in more loss than puncturing of parity bits and is not normally recommended. The code rates $r<1/3$ can be achieved by employing the constituent encoders with $r<1/2$.

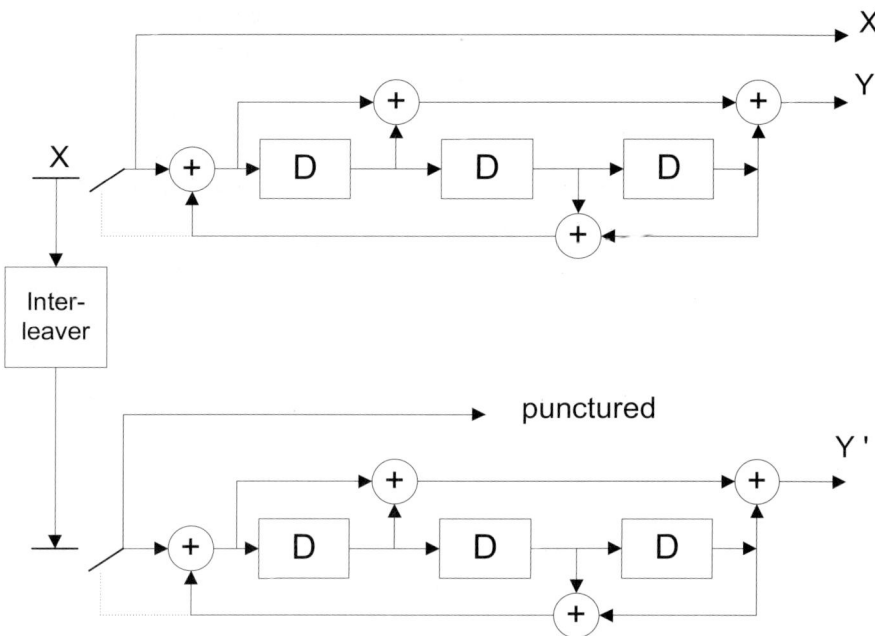

Figure 1.16 Block diagram of the basic r=1/3, 8-state PCCC encoder.

1.3.3.1   Decoding of Turbo Codes

The optimum decoding of turbo codes is extremely complex. However, close to optimum performance can be achieved with soft-input soft-output (SISO) iterative decoding. The name turbo comes from this concept of iterative decoding, where in every iteration the extrinsic information obtained in each iteration improves previous estimates. The core of SISO decoding is symbol-by-symbol maximum a posteriori probability (MAP) algorithm. Unlike Viterbi algorithm, which estimates the maximum likely bit sequence, the output of the MAP algorithm is the a posteriori probability for each of the bits. The 'soft' output of the algorithm enables SISO iterative decoding.

Consider again the channel output as in Eq. (1.25), except that now we drop the subscripts for simplicity:

$$y = ax + n, \tag{1.31}$$

where a is the coded bit fading amplitude, x is the coded bit +1 or −1 and n is zero-mean Gaussian random variable of power $N_0 / 2$. Here we assume that $E(a^2) = E_s$ so that the average signal-to-noise ratio per coded bit is equal to $2E_s / N_0$. Note that we can decompose the symbol amplitude, a into the fading part α and nonfading part $\sqrt{E_s}$ so that $a = \alpha\sqrt{E_s}$, $E[\alpha^2] = 1$. The input to the SISO decoder is based on the channel log-likelihood ratio $L(x \mid y)$. The channel log-likelihood ratio is a function of a posteriori probability $P(x \mid y)$, and can be expressed as:

$$
\begin{aligned}
L(x \mid y) &= \ln\frac{P(x=+1 \mid y)}{P(x=-1 \mid y)} = \ln\left[\frac{P(y \mid x=+1)}{P(y \mid x=-1)}\frac{P(x=+1)}{P(x=-1)}\right] \\
&= \ln\frac{\exp\left[-\dfrac{1}{N_0}(y-a)^2\right]}{\exp\left[-\dfrac{1}{N_0}(y+a)^2\right]} + \ln\frac{P(x=+1)}{P(x=-1)} \\
&= L_c y + L_a(x)
\end{aligned}
\tag{1.32}
$$

where $L_c = 4\alpha E_s / N_0$ represents the channel reliability and $L_a(x)$ is a priori log-likelihood ratio for x. Normally, equally likely bits are assumed, so that $L_a(x) = 0$. For non-fading channels, *i.e.,* for channels with constant signal amplitude, $\alpha=1$.

Let's now examine the inputs and outputs of the main building block of iterative SISO decoding, shown in Figure 1.17. In addition to soft channel outputs, $L_c y$, the input to the SISO decoder are the information bits a priori log-likelihoods $L_a(u)$, where u denotes the information bit. In the first iteration, a priori log-likelihoods are equal to 0 for equally likely bits. However, after the first iteration, we use the extrinsic information, $L_e(u)$ from the other component code as a priori information. The extrinsic information of information bits can obtained as:

$$
L_e(u) = L(u) - L_c y - L_a(u)
\tag{1.33}
$$

where $L(u)$ denotes the reliability of the information bit u after SISO decoding. Therefore, the extrinsic information is the difference between a posteriori values for the information bits obtained by the SISO decoder in a given iteration, and channel log-likelihood and a priori log-likelihood ratios. An interested reader is referred to [14] and references therein for details on MAP decoding.

The principal block diagram of turbo decoder for rate 1/3 turbo code is shown in Figure 1.18. Although the output of either decoder can be used to make decisions, one full iteration consists of decoding by SISO #1 and SISO #2, so that we use a posteriori log-likelihood output of SISO #2, after a desired number of iterations. The interleaver and deinterleaver correspond to the turbo code interleaver and deinterleaver (see Figure 1.16), not to the channel interleaver. In the first iteration, the uppermost input to SISO #1 is zero, because we assume that all information bits are equally likely.

In general, several iterations of turbo decoding are performed, up to a point of diminishing coding gain returns. A posteriori log-likelihood ratios at the output of decoder SISO #2 can be used as estimates of decoded information sequence (after quantization to two levels, +1 and −1).

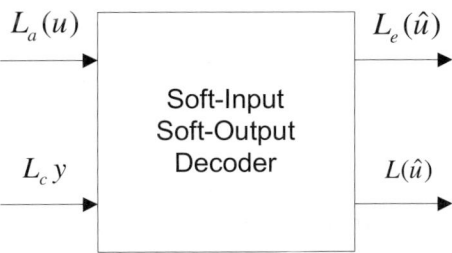

Figure 1.17 Block diagram of SISO decoder for iterative decoding.

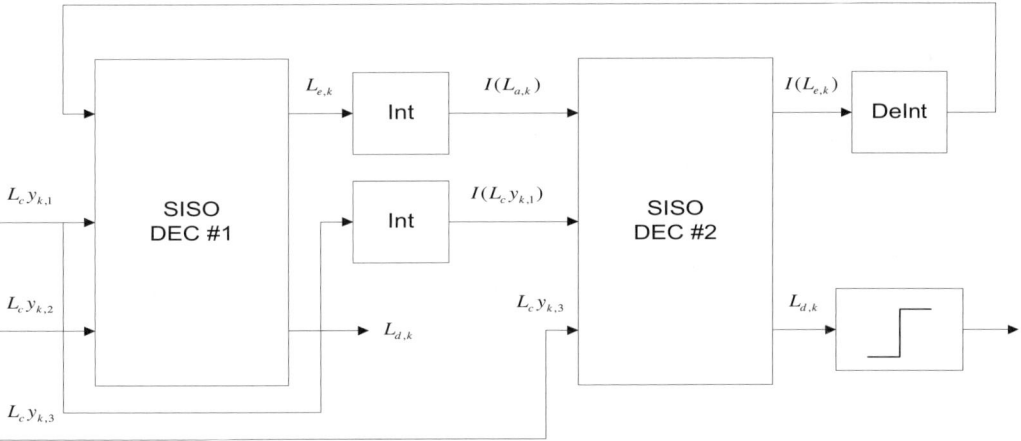

Figure 1.18 Block diagram of turbo decoder for PCCC decoding.

1.3.4 Hybrid ARQ and Code Combining

Hybrid Automatic Repeat Request (HARQ), in general, is an approach of error control where both FEC and error detection coding are employed for improved performance [15]. Incremental redundancy refers to a technique where additional coded bits are transmitted if previous packet transmissions of the same information packet were unsuccessful. Importantly, the previous transmissions are preserved in soft form and combined with retransmission resulting in the energy gain, due to additional energy received in incremental redundancy coded bits, and the coding gain because of essentially decreased coding rate with each retransmission. Assuming equal energy per retransmission, the k-th retransmission energy gain, relative to the aggregate of previous k transmissions, is $k/(k-1)$. Obviously this gain is most pronounced in the second transmission. If the code rate was r in the first transmission, the effective code rate after k transmissions is r/k, each retransmission contributing to a stronger overall FEC capability. In addition, if channel conditions change from transmission to transmission, this technique also provides a form of time diversity. Moreover, this is a form of adaptive time diversity and coding, only when it is needed (*i.e.,* when decoding of all previous transmissions of the corresponding information packet failed). It is an extremely efficient way to implement a retransmission protocol, because no transmission is wasted as in standard ARQ protocols.

A somewhat simpler form of packet combining is called diversity combining. In this case exactly the same replicas of the original transmission are used for retransmissions. This results in no incremental coding gain, but the energy and diversity gains are the same as in the case of incremental redundancy.

1.4 CDMA in Cellular Communications

So far we have focused on the physical layer characteristics of DSSS for CDMA applications and emphasized their unique robustness to interference and multipath. Yet, most of the gains of CDMA are realized when this multiple access technology is employed in a cellular communication network. As we shall see in this section, the basic operating principle of CDMA, that is, all users in the network share a common temporal and frequency allocation, allows for significant network performance improvement.

1.4.1 Universal Frequency Reuse

One of the most important advantages of CDMA relative to other multiple access techniques, such as frequency division multiple access (FDMA) and time division multiple access (TDMA), occurs in cellular communications due to CDMA's inherent better frequency reuse [5]. In CDMA, it is possible to have universal frequency reuse, that is, a frequency reuse of 1. One of the factors that enable universal frequency reuse is that the co-channel interference, i.e., the interference caused by nearby cells using the same carrier frequency, is to a large extent neutralized by virtue of the processing gain of the spread spectrum signals. This is in contrast to FDMA and TDMA where co-channel cells must be at sufficient distance from the desired cell to assure a relatively low level of interference, thus preventing reuse of the same frequency at least in the immediately adjacent cells.

Figure 1.19 depicts an interference scenario characteristic of the forward link in a CDMA cellular system. There are two sources of interference suffered by any given user.

The first source of interference, referred to as *in-cell interference*, comprises the signals transmitted by the serving base station to other users in the same cell. As we discuss in Chapter 4, the cdma2000 forward link uses orthogonal codes to multiplex signals transmitted to multiple users; hence, the in-cell interference is suppressed by despreading if the channel has a single path. In case of multipath, however, the loss of orthogonality between delayed replica of the transmitted signal causes a residual in-cell interference to be present after despreading. Still, orthogonal spreading provides some advantage. For example, with two equal strength paths that are resolvable by the RAKE receiver, the relative in-cell interference is 3 dB lower than what would be received if random spreading sequences, rather than orthogonal, were used. Note that orthogonal spreading is possible on the forward link since that is a one-to-many link in which all users signals are transmitted synchronously at chip level. Chip level synchronization among users' transmission cannot be easily achieved on the reverse direction, which is a many-to-one link with uncoordinated transmitters.

The second source of interference, referred to as *other-cell interference,* comprises the signals received from all surrounding base stations, in the 1st tier of adjacent cell, 2nd tier, and so on. Of course, due to the dependency of propagation path loss with distance the closest base station cause most interference and normally the interference from base stations in the 2nd tier and beyond can be neglected.

The relative amount of in-cell and other-cell interference depends on the mobile station location. The in-cell interference is independent of location. The other-cell interference is instead location dependent. It is larger, on average, at the cell edge, where the signal from the serving base station has the same average strength as signals from the two closest adjacent base stations, than it is in a location close to the serving base station. For a given target SNIR, the base station must allocate more transmit power to users located at the cell edge than to those located in the proximity of the base station. As the base transmit power is limited, the forward link capacity will depend on the user's distribution throughout the cell area and channel characteristics. In Chapter 10 we introduce a statistical model of the interference that allows estimating the forward link cell capacity.

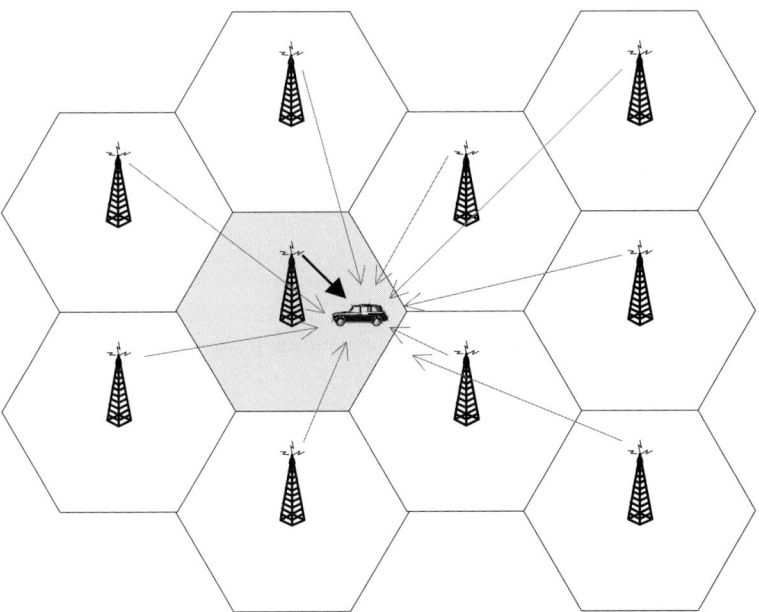

Figure 1.19 Forward CDMA link interference geometry.

Figure 1.20 shows the interference scenario for the reverse link in a CDMA cellular system. The desired base station, in the shaded cell, receives signals from all mobile stations in its coverage area and also from all mobile stations in all surrounding cells that are served by other base stations. It is of interest to estimate the relative strength of the in-cell and other cell interference. All users in the cell of interest are power controlled by the same base station so that their transmitted signals are received by the base station with equal power, on average, regardless of the users' locations. Therefore, the relative in-cell users interference is equal to K-1, where K is the number of users in the cell. Other cells users are instead power controlled by other base stations. The amount of interference received by the base station of interest will then depend on the distance of the other cell's user to its serving base station relative to its distance to the base station of interest. Consider another cell's user located at distance r_0 from its serving base station, and distance r_1 from the base station of interest. Assuming that the path loss increases with distance as r^m, $m > 0$, the relative interference due to such single other cell user is, on average, $(r_0/r_1)^m$. The exponent m, called the propagation loss exponent, typically varies between 3 and 5 depending mainly on terrain morphology. If $r_0 < r_1$, the relative other cell's user interference can be quite low. Consider for example a user served by the base station of a cell adjacent to the cell of interest, and located half-way between the cell edge and the serving base station. Assuming the cells' radii to be identical, the ratio of distances is one third. Then, with $m = 4$, the relative interference is ~ -20 dB. To compute the total other cell interference we have to account for all users in the cells of the 2nd tier, 3rd tier, and so on. As we show in Chapter 10, in a typical scenario the total interference from all mobile stations outside the desired cell is equal to a factor $f = $ 50-60% of the interference generated by the mobile stations within the desired cell. Therefore, the relative total (in-cell plus other-cell) interference is equivalent to that caused by $(K-1)*(1+f)$ in-cell users. The factor f depends on cell layout and channel propagation, and can be estimated using the models discussed in Chapter 10. Here it suffices to notice that, in the presence of other cells interference, the maximum reverse link cell capacity becomes ~ $N/[(K-1)*(1+f)]$, where N is the processing gain. Clearly, reducing f can provide considerable gain in capacity. That is the subject of the next section.

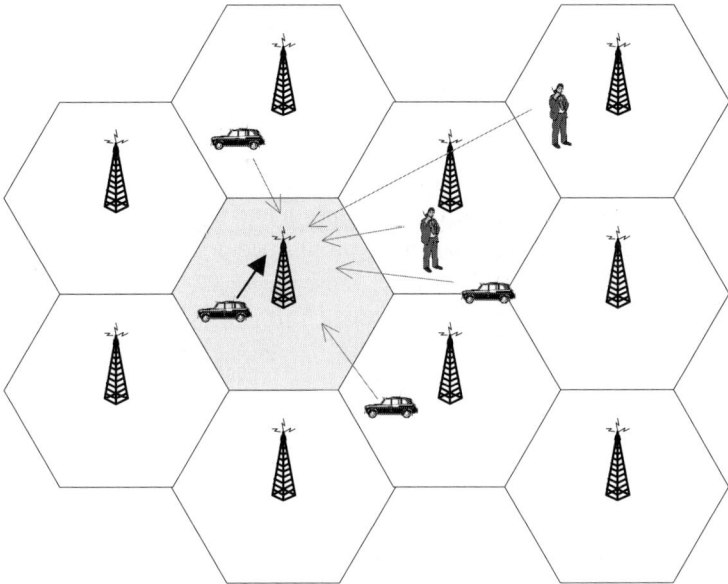

Figure 1.20 Reverse DS CDMA link interference geometry.

1.4.2 Cell Sectorization

Cell sectorization is a mechanism of using multiple directional antennas at base stations, with corresponding receivers, to illuminate only a fraction of a cell. In Figure 1.21 we show an example of cell sectorization with three sectors per cell. In this case, each sectoral antenna uses a 120° beamwidth directional antenna pattern in the horizontal plane. Assuming that the antenna radiation pattern is such that the gain is virtually zero outside the beam, the interference outside of the 120° sector radiation angle is essentially suppressed, which reduces the relative other cell interference by a factor of 3 and improves CDMA capacity by the same factor. Sometimes four or six sectors per cell are deployed, corresponding to 90° and 60° radiation patterns, respectively. In these cases the capacity improvement is proportional to the factors 4 and 6, respectively.

In practice, the antenna radiation pattern exhibits a smooth roll off outside its main beam, thus the gains achieved through sectorization are reduced from the ideal ones, especially when more than three sectors per cell are used. Furthermore, the advantages of sectorization are achieved to the fullest extent when the distribution of users throughout the network area is uniform. If interfering users are instead concentrated on a small area of an adjacent cell or hot spot, the interference seen by one antenna whose beam spans the hot spot would not decrease despite sectorization, and no capacity benefit is achieved. Three-sector cells are widely deployed in CDMA cellular systems, except for indoor or pico-cell systems where distributed antenna systems may be better suited.

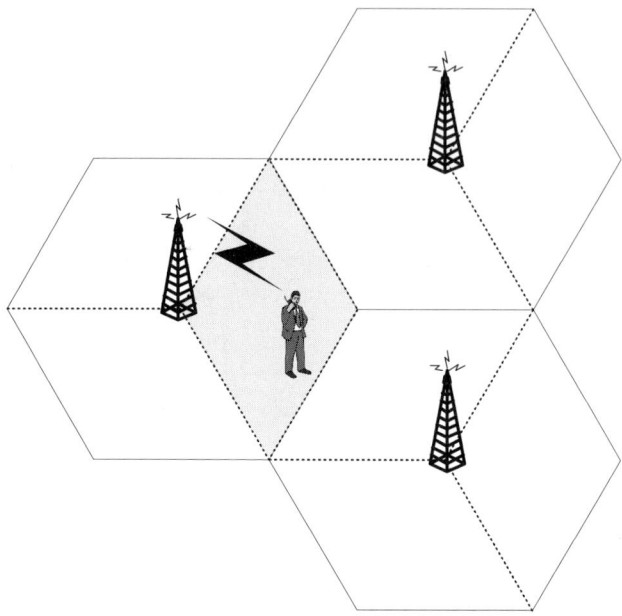

Figure 1.21 Example of sectorization with three sectors per cell.

1.4.3 Soft Handoff

Soft handoff allows the mobile station to commence communication with a new base station without interrupting communication with the current serving one. This 'make-before-drop' technique is made possible by the universal frequency reuse of CDMA. It differs drastically from hard handoff techniques employed by earlier cellular technologies, in which a connection with the target base station can be established only after releasing the connection with the serving one. Soft handoff can be seen as a form of macro-diversity, which results in multiple benefits. A diversity gain is provided to mobile stations when they need it the most (i.e., near the edge of coverage). This reduces variations in received SNIR and, thus, improves the communications reliability. In addition, the binary decision to hand over to the adjacent base station (i.e., the one with a stronger signal) is avoided in the region of weak signals by exploiting the diversity effect. This in turn reduces the frequency of handoffs in the border region compared to hard handoff techniques. Sometimes repeated handoffs between two base stations are called a 'ping pong' effect and soft handoff effectively eliminates it.

When involved in soft handoff, each one of the two or more base station communicating with the same mobile station transmit synchronously a signal that is modulated by the same data sequence but spread by codes that are unique to each base station. Then, effectively the mobile station receives multiple copies of the same signal, possibly delayed due to the differential propagation delay. The mobile station can thus use the RAKE receiver for combining the signals from the base stations in soft handoff just as it does with multipath. That explains why soft handoff can be interpreted as a form of transmit diversity. In this case, the price paid for soft handoff is increased interference in the network, because the same information signal is transmitted two or more times, generating additional interference in the network. Typically, the diversity gain provided by soft handoff outweighs the loss due to the additional interference injected into the network.

There are important differences between soft handoff on forward and reverse links. On the reverse link, the signal transmitted by the mobile station can be received by all base stations involved in the soft handoff. Each base station independently demodulates and decodes a block of bits. The blocks of bits are sent to a centralized network entity that selects the block with the best quality metric among those received by the base stations in soft handoff. This selection process, called selection combining, is a form of diversity that improves performance. The diversity gain of selection combining is smaller than that achieved on the forward link with MRC. However, the diversity gain on the reverse link is achieved without requiring additional interference being injected into the network, as in the forward link case.

Finally, we distinguish between soft and softer handoffs. The soft handoff corresponds to the case when adjacent base stations are involved in handoff. On the other hand, the softer handoff refers to the case when two sectors of the same base station support handoff. The main difference, and advantage, of softer handoff is that the signals received from the sector antennas can be fed into the same demodulating element and combined with MRC, before the resulting block of bits is sent to the network controller for selection combining with those, if any, received from other base stations in soft handoff.

1.5 Antenna Diversity in CDMA

Diversity techniques have been known for a long time to be effective fading countermeasures. The basic idea is to provide multiple independent, or nearly independent, channels between the transmitter and receiver. When one channel is in a deep fade, the other/s may be experiencing good channel conditions, and as such provide uninterrupted communication. While some forms of diversity techniques, such as time and frequency diversity, require additional channel resources, the space or antenna diversity does not.[4] In the case of antenna diversity, we can have receive or transmit diversity or a combination of both.

1.5.1 Receive Diversity

Receive diversity corresponds to a situation when the receiver is equipped with multiple receive antennas. In general, the benefits of the receive diversity are twofold: for a given amount

---

4.   It does require, however, additional hardware (antennas and receivers).

of transmit energy, when using L antennas, the combined received SNIR is up to L times that achieved with a single receive antenna (energy gain); furthermore, the L-order diversity decreases the received SNIR variance relatively to its mean, which further decreases the mean combined SNIR required for a given quality of service (diversity gain). Such gains are akin to those achieved with the RAKE receiver in the presence of multipath. Specifically, assuming that the signal amplitude fades independently at each of the L antennas, and that the noise plus interference components received at each antenna are uncorrelated, the combined received SNIR after MRC is L times that achieved with a single receive antenna, while the standard deviation grows only proportionally to the square root of L. In an interference limited systems such as CDMA, an L-fold increase in SNIR produces an L-fold increase in system capacity. In a fading channel the capacity improvement is even larger because of the diversity gain. The diversity gain caused by decreased SNIR variance effectively decreases the fade depth and fade rate, thus reducing the probability of outage that occurs when power control is not able to cope with a deep fade, a phenomenon commonly occurring in Rayleigh fading channel. Even under the assumption of perfect power control, a smaller SNIR variance implies a lower required average SNIR (and larger system capacity) because the FER is a convex function of the SNIR.

Because of the substantial gains it offers, receive diversity is important for delay sensitive applications that cannot take advantage of channel sensitive scheduling, such as voice. Indeed, since the early deployment of CDMA cellular systems the base stations have been typically equipped with two receive antennas, each with vertical polarization. Such form of receive diversity is called dual-spatial diversity because it relies on the physical separation of the two antennas to ensure that the received signals are uncorrelated. Because of zoning restrictions and costs, more than two receive antennas have not found practical use. Instead of increasing the antenna count, a practical approach to further increase the diversity order is to use two cross-polarized antennas, thus obtaining a quad-spatial/polarization diversity system [17]. Both dual and quad diversity antenna systems are illustrated in Figure 1.22.

On the mobile station side, on the other hand, use of receive antenna diversity is a more recent trend, fueled by technology advancements that allow cost-effective integration of a second antenna in a hand-held device. Note that engineering texts will insist the diversity antennas be placed much farther apart than a hand-held phone would allow. That is true to extract the maximum gain possible, but field tests have proved that the diversity gain available with mobile station receive antennas less than a centimeter apart can greatly increase forward link capacity [18].

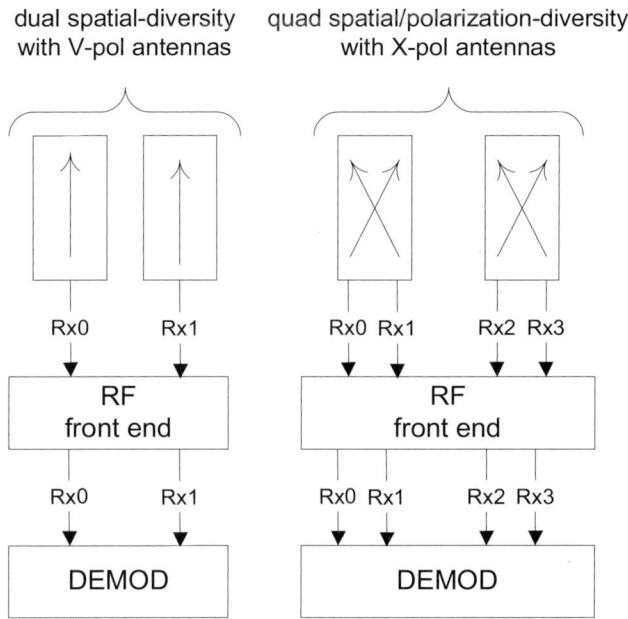

Figure 1.22 Dual and Quad receive antenna diversity at the base station.

1.5.2 Transmit Diversity

Unlike receive diversity that may apply both to the forward and reverse links, transmit diversity applies to the forward link only. In cdma2000 two forms of transmit diversity are supported: space-time spreading (STS), and orthogonal transmit diversity (OTD). Both STS and OTD use two separate transmit antennas at the base station.

Transmit diversity using STS is a variation of the scheme that was originally proposed by Alamouti in [19], called space-time coding (see Figure 1.23). Space-time coding is essentially a complex orthogonalizing transformation over two symbol intervals, each of duration T. In this case all symbols are transmitted by both antennas, according to the algorithm shown in Table 1.1. Specifically, in the 1st symbol interval, antenna 0 transmits symbol s_0, while antenna 1 transmits symbol s_1. In the 2nd symbol interval, antenna 0 transmits $-s_1^*$, while antenna 1 transmits s_0^*, where * represents the complex conjugate operation.

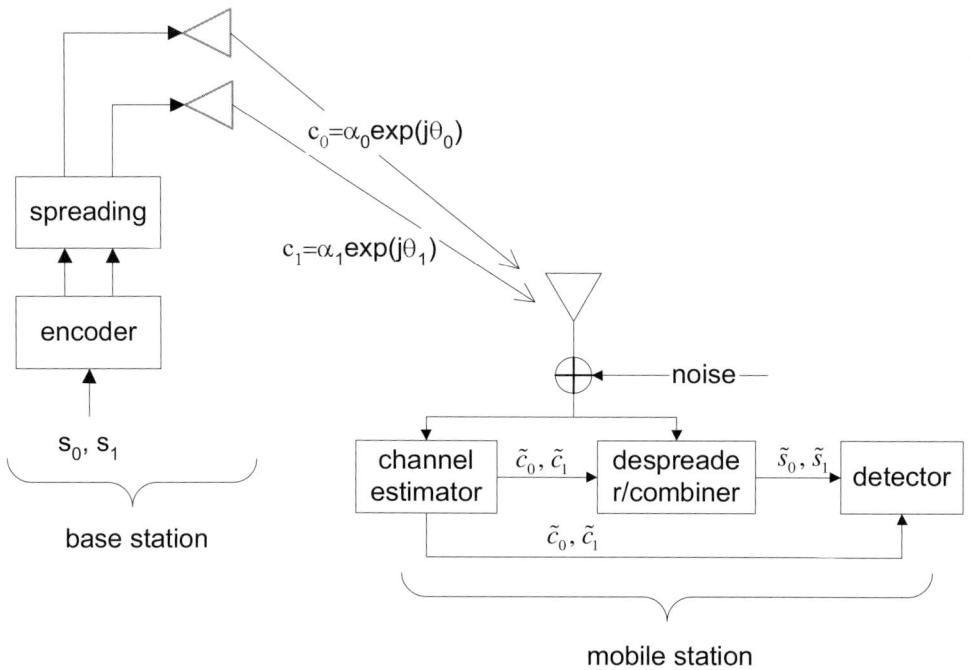

Figure 1.23 Transmit diversity using space-time coding or STS (spreading/despreading applies to STS only).

Table 1.1   Encoding and transmission sequence with space-time coding.

	Antenna 0	Antenna 1
time $0 < t \leq T$	s_0	s_1
time $T < t \leq 2T$	$-s_1^*$	s_0^*

Due to path loss and fading, the signals transmitted by antenna 0 and 1 are received at the mobile station corrupted by the complex multiplicative distortion $c_i = \alpha_i \cdot e^{j\theta_i}$, $i = 0$ or 1, and by additive noise, n, with variance $N_0/2$. Assuming the channel to be stationary over the period $2T$, the signals received in the two consecutive time slots at the mobile station are given by:

$$r_0 = c_0 s_0 + c_1 s_1 + n_0,$$
$$r_1 = -c_0 s_1^* + c_1 s_0^* + n_1. \qquad (1.34)$$

At the mobile station, the received signals in the two consecutive time slots are processed according to the linear transformation in Eq. (1.35), with c_i, $i = 0$ and 1, representing the channel estimates:

$$s_0 = c_0^* r_0 + c_1 r_1^* ,$$
$$s_1 = c_1^* r_0 - c_0 r_1^* .$$

(1.35)

Assuming ideal channel estimation, plugging Eq. (1.34) into (1.35) we obtain :

$$s_0 = \left(\alpha_0^2 + \alpha_1^2\right) s_0 + c_0^* n_0 + c_1 n_1^*$$
$$s_1 = \left(\alpha_0^2 + \alpha_1^2\right) s_1 + c_0 n_1^* + c_1^* n_0$$

(1.36)

Note that n_0 and n_1 are uncorrelated as they correspond to two different sampling intervals. Thus, the received instantaneous symbols' SNIR is found to be:

$$\text{SNIR}_0 = \text{SNIR}_1 = \frac{\left(\alpha_0^2 + \alpha_1^2\right)|s|^2}{N_0}$$

(1.37)

STS is conceptually equivalent but uses orthogonal spreading to achieve intra-antenna orthogonality, hence the term space-time spreading. The STS transmission sequence is illustrated in Table 1.2, where w is a spreading code of duration T.

Table 1.2   Encoding and transmission sequence with STS

	Antenna 0	Antenna 1
time $0 < t \le T$	$s_0 w - s_1^* w$	$s_0^* w + s_1 w$
time $T < t \le 2T$	$s_0 w + s_1^* w$	$-s_0^* w + s_1 w$

Effectively, if we account for symbol repetition and consider the sequence $W_0 = [w, w]$ and $W_1 = [w, -w]$, the STS scheme is such that antenna 0 transmits the signal $s_0 W_0 - s_1^* W_1$ and antenna 1 transmits the signal $s_0^* W_1 + s_1 W_0$ over the repeated symbol period $2T$. Thus, the received signal is:

$$r = c_0 \left(s_0 W_0 - s_1^* W_1\right) + c_1 \left(s_0^* W_1 + s_1 W_0\right) + n$$
$$= W_0 \left(c_0 s_0 + c_1 s_1\right) + W_1 \left(c_1 s_0^* - c_0 s_1^*\right) + n$$

(1.38)

At the mobile station, the received signal is despread by the sequence W_0 and W_1 and processed according to the following transformation:

$$s_0 = c_0^* r W_0 + c_1 r^* W_1 ,$$
$$s_1 = c_1^* r W_0 - c_0 r^* W_1 . \tag{1.39}$$

If we plug Eq. (1.38) into Eq. (1.39) and account for the fact that W_0 and W_1 are orthogonal functions of duration $2T$, the resulting decision variables are:

$$s_0 = 2\left(\alpha_0^2 + \alpha_1^2\right)s_0 + c_0^* n_0 + c_1 n_1^*$$
$$s_1 = 2\left(\alpha_0^2 + \alpha_1^2\right)s_1 + c_0 n_1^* + c_1^* n_0 \tag{1.40}$$

Note that n_0 and n_1 are white noise projections on orthogonal functions of duration $2T$, and are, therefore, uncorrelated random variables of variance $N_0/2$. Thus, the received instantaneous symbols' SNIR is found to be:

$$\text{SNIR}_0 = \text{SNIR}_1 = \frac{2\left(\alpha_0^2 + \alpha_1^2\right)|s|^2}{N_0} \tag{1.41}$$

With OTD, the transmission sequence on the two diversity branches is summarized in Table 1.3.

Table 1.3   Encoding and transmission sequence with OTD

	Antenna 0	Antenna 1
time $0 < t \leq T$	$s_0 w$	$-s_1 w$
time $T < t \leq 2T$	$s_0 w$	$s_1 w$

Assuming the channel to be stationary over the period $2T$, the OTD signals received in the two consecutive time slots at the mobile station are given by:

$$r_0 = c_0 s_0 w - c_1 s_1 w + n_0 ,$$
$$r_1 = c_0 s_0 w + c_1 s_1 w + n_1 . \tag{1.42}$$

At the mobile station, the received signals in the two consecutive time slots are despread and processed to obtain the decision variables as in:

$$s_0 = c_0^* \left(r_0 w + r_1 w \right),$$
$$s_1 = c_1^* \left(-r_0 w + r_1 w \right). \tag{1.43}$$

From Eq. (1.42)-(1.43), it follows that the decision variables are of the form:

$$s_0 = 2\alpha_0^2 s_0 + c_0^* \left(n_0 + n_1 \right)$$
$$s_1 = 2\alpha_1^2 s_1 + c_1^* \left(n_0 + n_1 \right) \tag{1.44}$$

where n_0 and n_1 are uncorrelated as they correspond to two different sampling intervals. The received symbols SNIR with OTD is then equal to:

$$\mathrm{SNIR}_0 = \frac{2\alpha_0^2 |s|^2}{N_0}, \; \mathrm{SNIR}_1 = \frac{2\alpha_1^2 |s|^2}{N_0} \tag{1.45}$$

We can now compare the SNIR achieved with the different transmit diversity schemes and draw important conclusions on their performance relative to a system that does not employ transmit diversity. First, one must note that a fair comparison has to be made on the basis of equal total transmit power. Under such constraint, the per-antenna symbol energy $|s|^2$ that appears in the SNIR formulas above is not identical. Figure 1.24 illustrates the amount of power allocated to each symbol per each antenna for each diversity technique in order to keep the same amount of transmitted energy per symbol over the $2T$ signaling period. Without transmit diversity, each symbol is transmitted at full power, P, but only for a period of T. With OTD, the symbol duration is $2T$ so that the transmit power is reduced to $P/2$. The same amount of power is needed for space-time coding. However, in STS, since we transmit both symbols on both antennas for a period of $2T$, the amount of power required is $P/4$. Then, if $|s|^2_{\text{non-TD}}$ is the per-antenna symbol energy without transmit diversity, we have that:

$$|s|^2_{\text{non-TD}} = 2|s|^2_{\text{OTD}} = 2|s|^2_{\text{STC}} = 4|s|^2_{\text{STS}} \tag{1.46}$$

Therefore the *average* SNIR is identical for all diversity schemes and it is the same as that achieved without transmit diversity. That is, unlike receive diversity, transmit diversity does not provide any energy gain. However, space-time coding, STS and OTD achieve diversity gain. Examining again the instantaneous SNIR formulas reveals the fundamental difference between STS and OTD. With STS, even when one antenna experiences a deep fade, both symbols may be

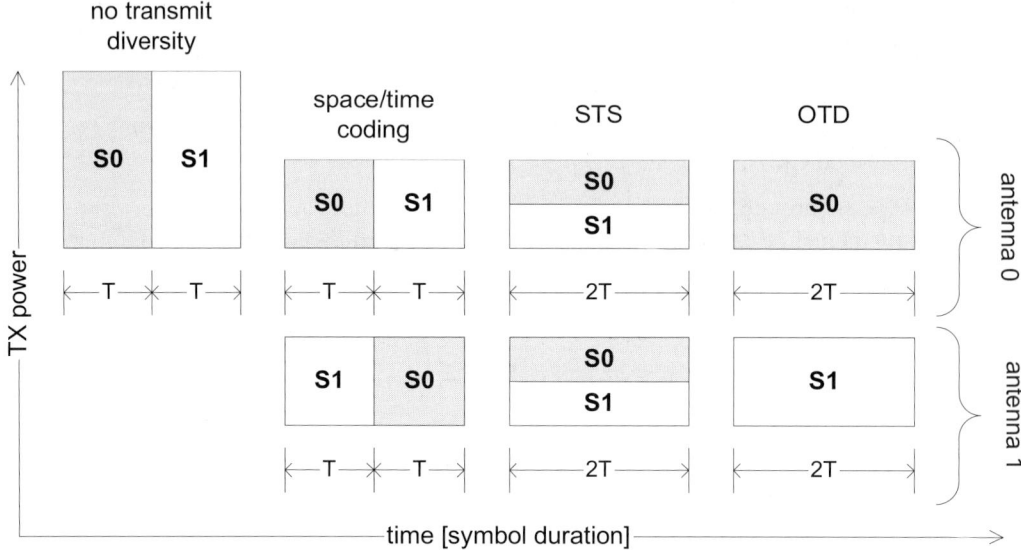

Figure 1.24 Power allocation per symbol and per antenna with and w/o transmit diversity.

recovered if the channel seen from the other antenna is good. With OTD, at least one symbol is erased when one antenna experiences a deep fade. The relative performance of STS and OTD will then depend on the FEC encoder that is employed. When the code rate is high, STS will significantly outperform OTD because small coding gain cannot compensate for the diversity loss. When the code rate is lower, STS and OTD performances are similar, but in general STS outperforms OTD.

References

[1] Simon, Marvin, J. Omura, R. Scholtz and B. Levitt, *Spread Spectrum Communications Handbook*, McGraw-Hill, New York, NY, 1995.

[2] Fan P. and M. Darnell, *Sequence Design for Communications Applications*, Research Studies Press Ltd., 1996.

[3] Pickholtz, Raymond, D. Schilling and L. Milstein, "Theory of spread spectrum communications—a tutorial," *IEEE Trans. on Communications.*, COM-30(5), May 1982.

[4] Pickholtz, Raymond, D. Schilling and L. Milstein, Revisions to "Theory of spread spectrum communications—a tutorial," *IEEE Transactions on Communications*, COM-32(2), Feb. 1984.

[5] Viterbi, Andrew J., CDMA—Principles of Spread Spectrum Communications, Addison Wesley, Reading, MA, 1995.

[6] Proakis, John, *Digital Communications*, McGraw-Hill, New York, NY, 1995.

[7] Rappaport, Theodore S., *Wireless Communications—Principles and Practice*, Prentice Hall PTR, Upper Saddle River, NJ, 2002.

[8] Kay, S.M., *Fundamentals of Statistical Signal Processing: Estimation Theory*, Prentice Hall, Englewood Cliffs, NJ, 1993.

[9] Hui, Joseph, "Throughput analysis for code division multiple accessing of the spread spectrum channel," *IEEE Journal on Selected Areas in Communications*, SAC-2, pp. 482-486, July 1984.

[10] Viterbi, Andrew J., "Very low rate convolutional codes for maximum theoretical performance of spread-spectrum multiple-access channels," *IEEE Journal on Selected Areas in Communications*, SAC-8, No. 4, May 1990.

[11] Viterbi, Andrew J. and J.K. Omura, *Principles of Digital Communication and Coding*, McGraw-Hill, 1979.

[12] Berrou C. and A. Glavieux, "Near optimum error-correcting coding and decoding: Turbo codes," *IEEE Transactions on Communications*, Vol. 44, October 1996.

[13] Vucetic Branka and J. Yuan, *Turbo Codes—Principles and Applications*, Kluwer Academic Publishers, 2000.

[14] Hagenauer, Joakim, E. Offer and L. Papke, "Iterative decoding of binary block and convolutional codes," *IEEE Transactions on Information Theory*, Vol. 42, No. 2, March 1996.

[15] Wicker, S. B., *Error Control Systems for Digital Communication and Storage*, Prentice Hall, Englewood Cliffs, NJ, 1995.

[16] Gilhousen, Klein S., I.M. Jacobs, R. Padovani, A.J. Viterbi, L.A. Weaver and C.E. Wheatley, "On the capacity of a cellular CDMA system," *IEEE Transactions on Vehicular Technology*, Vol. 40, No. 2, pp. 303- 312, May 1991.

[17] Aydin, Levent, E. Esteves, and R. Padovani, "Reverse Link Capacity and Coverage Improvements for CDMA Cellular Systems Using Polarization and Spatial Diversity," in the proceedings of *International Conference on Communications*, New York, July 2002.

[18] Agashe, Parag, and Roy Davis, "Mobile station receive diversity in cdma2000: simulations and field test results," October 2002, white paper available at www.cdg.org.

[19] Alamuti, S. "A simple transmit diversity technique for wireless communications," *IEEE Journal on Selected Areas in Communications*, Vol. 16, No. 8, October. 1998.

CHAPTER 2

Architecture

The cdma2000 wireless network can be viewed as a collection of logical entities and the associated interfaces. A logical entity is an implementation-independent representation of a function or set of functions, and it may not correspond to a physical device. A *reference point* is the point of connection between adjacent logical entities. A reference point is also called an interface if the logical devices it connects correspond to separate physical devices. The reference point is specified in terms of a set of procedures and associated signaling that define the operational responsibility of the connected network entities. The cdma2000 interfaces are fully standardized, or *open*, which allows the operators to deploy a network consisting of components from different manufacturers. A high-level representation of the cdma2000 network reference model[1] is depicted in Figure 2.1. It comprises three major parts: the core network (CN), the radio access network (RAN) and the mobile station (MS). The core network is further decomposed in two parts, one interfacing to external networks such as the Public Switched Telephone Network (PSTN) and the other interfacing to the Internet Protocol (IP) [1] network, or Internet. The interfaces toward the PSTN and the IP network are called the analog interface (A_i) and the packet interface (P_i), respectively. The part of the core network interfacing the PSTN supports messages and protocols defined in IS-41[2] [2] standard. Throughout the chapter, this part of the core network is also referred to as the IS-41 network. The part of the core network interfacing the IP network supports the IS-835 wireless IP network standard [3], and it is also referred to as the packet core network (PCN). The core network interfaces toward the RAN through a group of A inter-

---

1.   The cdma2000 reference model in Figure 2.1 does not entirely apply to High Rate Packet Data (HRPD) [10], which is also part of cdma2000 family but is not treated in this book.

2.   The IS-41 standard is also a standard of the American National Standard Institute referred to as ANSI-41.

faces, specified in IS-2001 [4] standard. The RAN is connected to the mobile station over a U_m interface, specified in IS-2000 [5][6][7][8][9] standard.

The organization of the material of this chapter follows the network decomposition depicted in Figure 2.1. The network entities are described from a functional viewpoint, together with their interfaces and associated protocols. When applicable, a description of the most common physical implementations is also given.

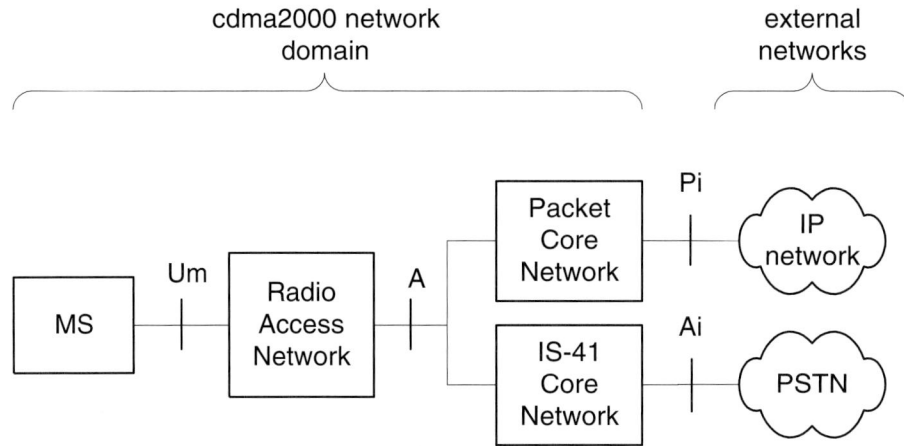

Figure 2.1   High-level reference model for cdma2000 wireless network.

2.1 Mobile Station

The mobile station terminates the radio path on the user side of the network and enables subscribers to access network services over the U_m interface. Figure 2.2 depicts a simplified reference model. For simplicity, the figure does not show the User Identity Module (UIM), which contains user subscription information. The mobile station without UIM is commonly called mobile equipment (ME). After neglecting this detail, according to the model, the mobile station consists of the mobile terminal 2 (MT2), which provides connectivity to the network, and the terminating equipment 2 (TE2), which is the data-processing device.[3] The MT2 and TE2 are connected over the R_m interface. MT2 refers to a phone and TE2 to an external device such as laptop computer. For voice applications, MT2 is sufficient and TE2 is unnecessary, because in addition to network connectivity, MT2 commonly provides circuit-switched voice service. How-

---

3.   Mobile terminal configurations other than MT2 are also possible. For example, MT0 refers to a mobile terminal without any external interface.

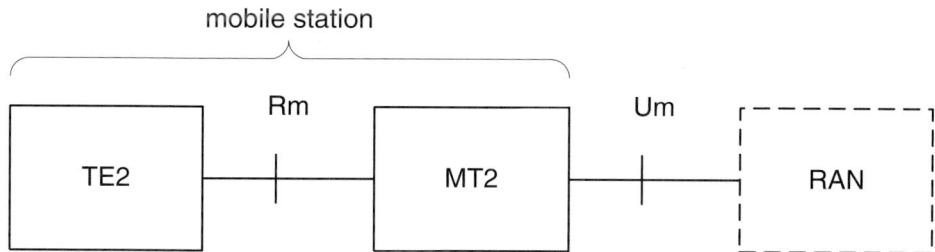

Figure 2.2 Data-capable mobile station reference model (dashed boxes represent external entities).

ever, as it is explained in the sections that follow, in order to get packet data service, an external device may be needed. Depending on the R_m interface model, MT2 alone may or may not support packet data service.

2.1.1 Reference Points

The interfaces internal and external to the mobile station are summarized in Table 2.1. An overview of these reference points is given in the following sections. More detailed description of the U_m interface is provided in later chapters.

Table 2.1   MS Internal and External Interfaces

Interface	Entities using the interface	Applicable standard
R_m	TE2-MT2	IS-707
U_m	RAN-MS	IS-2000/IS-707

2.1.1.1 U_m Reference Point

The mobile station is connected to the RAN over the air interface, or U_m reference point. The IS-2000 protocol architecture is shown in Figure 2.3. In this section, we briefly describe the IS-2000 protocol layers; they are covered in detail in later chapters.

The upper layer of the U_m interface consists of user services, such as packet data service, voice over IP, and circuit-switched voice and data, and of the IS-2000 Layer 3 signaling protocol. The IS-2000 Layer 3 supports a variety of functions that can be broadly categorized as follows: broadcast of system information; mobility management; resource management; and establishment and maintenance of connections between the mobile station and the RAN. The IS-2000 Layer 3 is discussed in Chapter 6, "IS-2000 Layer 3 Protocol."

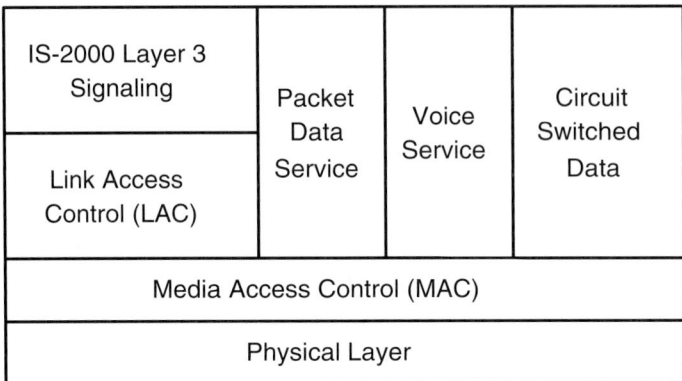

Figure 2.3 U_m protocol layer structure.

The Link Access Control (LAC) layer provides transport services over logical channels for Layer 3 signaling. Logical channels are so called because they determine *what* data is transported rather than *how*, the latter being only a property of the physical channel onto which the logical channel is mapped. Hence, logical channels hide the upper layers from the characteristics of the physical channels. Logical channels can be either dedicated or common. Dedicated channels are mapped to physical channels assigned to a single user. Dedicated channels are used to carry user-specific traffic such as voice. Common channels are mapped to physical channels shared by multiple users. Common channels are used to carry traffic destined to multiple users, such as overhead system information, or traffic destined to a single user that has not yet been assigned to a dedicated channel, such as page messages. The LAC is further divided in sublayers performing distinct functionality such as authentication, integrity, addressing, assured mode signaling or automatic repeat request (ARQ), segmentation/reassembly, and utility functions.

The Media Access Control (MAC) layer maps the logical channels to physical channels and coordinates the use of physical resources. The MAC is also responsible for enforcement of the negotiated Quality of Service (QoS) level by mediating conflicting requests from competing services. In addition, MAC handles error-correction mechanisms for erroneously decoded data. Error correction is provided through the Radio Link Protocol (RLP) and hybrid ARQ (H-ARQ) protocols. RLP is not explicitly handled by the MAC layer. RLP is separately described in the IS-707 [11] standard but is conceptually part of the MAC. LAC and MAC are discussed in Chapter 5, "IS-2000 Layer 2 (Media and Signaling Link Access Control Layers)."

The physical layer provides transmitter and receiver functionality on physical channels. The physical layer protocol specifies channels' modulation characteristics and their numerology. In addition, the physical layer protocol specifies radio channels' related functions, such as link maintenance, power control, handoffs, rate control, and many others. The physical layer channels are discussed in Chapter 4, "IS-2000 Physical Layer," while physical layer functions like power control and handoffs are described in later chapters.

2.1.1.2 R_m Reference Point

The reference point between MT2 and TE2 is called the R_m interface and it has two protocol options. The first option, the Relay layer R_m protocol option, is illustrated in Figure 2.4. The figure shows that the TE2 is responsible for packet data service, such as support of the Point-to-Point Protocol (PPP), IP, and upper layers. The MT2 entity interfaces to the RAN through the U_m. In Figure 2.4, the U_m Layer 2 protocol corresponds to the IS-2000 LAC/MAC and IS-707 RLP, while Layer 1 refers to the IS-2000 physical layer.

The second option is the Network layer R_m interface protocol option, shown in Figure 2.5. There are independent link layer connections between TE2 and MT2 and between MT2 and the core network. The link between MT2 and TE2 is implemented either through PPP or Serial Link IP (SLIP) [12]. TE2 acts as if it is locally connected to a router.

Therefore, in case of the Relay layer option, MT2 is simply a modem. MT2 just relays packet data to and from the RAN. MT2 alone cannot provide any packet data service because it does not support IP. However, the implementation of the Relay model is simple, and circuit-switched voice is supported. The Network layer option allows the MT2 to have its own IP address, and therefore MT2 is capable of providing packet data service. The implementation of the Network model is more complicated than the implementation of the Relay model. Still, it is a more attractive option than the Relay model because it allows packet data applications to run on MT2. An example of a Relay model is a phone that supports only circuit-switched voice but contains an external interface that connects to a laptop. An example of a Network model is a phone with an external interface and an integrated Web browser.

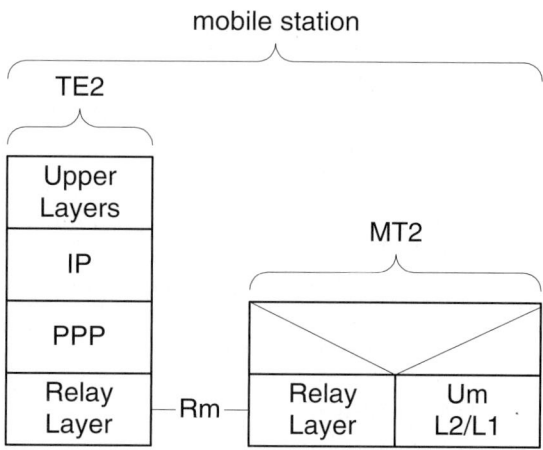

Figure 2.4   Relay layer R_m interface protocol option.

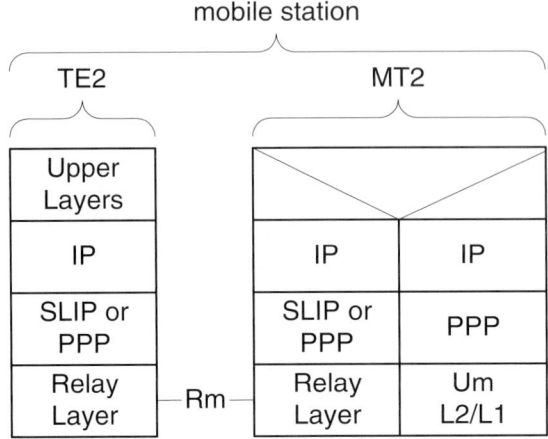

Figure 2.5   Network layer R_m interface protocol option.

2.2 Radio Access Network

The cdma2000 RAN provides radio bearers between the core network and the mobile station for the transport of user data and nonaccess stratum signaling, thus enabling mobile stations to access the services offered by the PSTN and Internet. The main RAN functions include establishment, maintenance, and termination of radio channels; radio resource management; and mobility management. The RAN consists of the base station (BS) and packet control function (PCF). The base station is further decomposed in one control and one or multiple radio-terminating equipment portions named base station controller (BSC) and base transceiver station (BTS), respectively. In the IS-2000 standard, the term *base station* also refers to a cell. Throughout this book, when clear from the context, we use the term "base station" to indicate either a cell or the aggregate of BTS and BSC.

The RAN architecture[4] and interfaces are shown in Figure 2.6. The RAN is connected to the core network through a group of A interfaces. The A interfaces are categorized as those carrying packet data user traffic and associated signaling, terminated in a core network entity called packet data serving node (PDSN), and those carrying circuit-switched (CS) voice and data traffic and associated signaling, terminated in a core network entity called mobile switching center (MSC). PDSN and MSC are described in detail in Sections 2.4.1.1 and 2.3.1.1. The reference point between the BSC and BTS is called the A_{bis} interface. The reference point that connects two BSs is called the A_{ter} or A3/A7 interface.

---

4.   A RAN entity, not shown in Figure 2.6, is the radio network manager responsible for operation administration and management (OA&M) functions. Its associated reference points are not standardized.

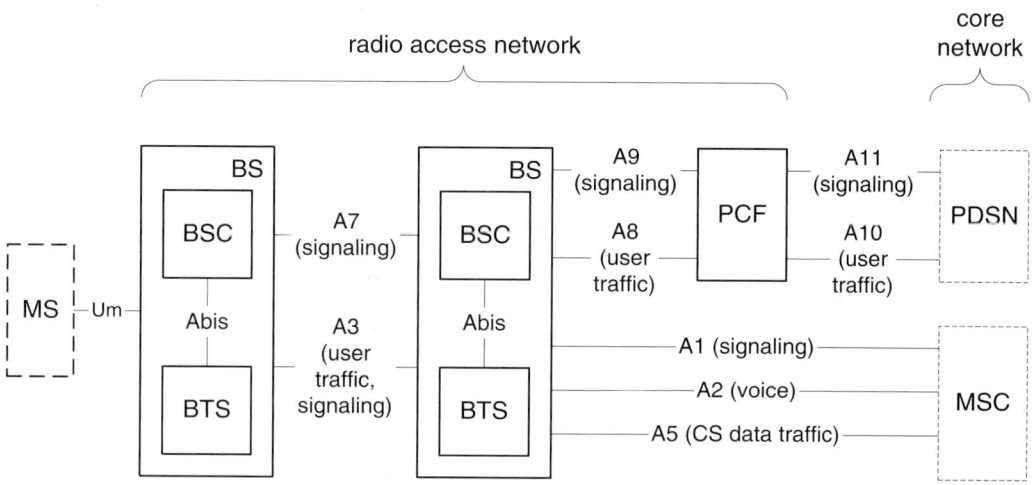

Figure 2.6 RAN reference model (dashed boxes represent external entities).

2.2.1 RAN Logical Entities

In this section we provide a high-level overview of the logical entities that comprise the RAN. More detailed descriptions of their functions are given in later chapters.

2.2.1.1 *Base Transceiver Station*

The BTS is responsible for terminating the radio links with the mobile station. The BTS also typically terminates the IS-2000 LAC/MAC protocols for common channels, although in some implementations such protocols are terminated at the BSC. In case of dedicated channels, the BTS exchanges physical layer frames with the BSC over the A_{bis} interface. One exception is in case of those physical channels used for very high data rate transmission (the F-PDCH, see Chapter 4). For such channels, BTS tasks also include local radio resource management, scheduling, link adaptation, and H-ARQ Type II operation (see Chapter 4). The BTS is typically equated to the physical site of the wireless network where the antennas are located. A BTS may comprise multiple cells, each associated with a set of collocated antennas but distinguished coverage areas. Note that hereafter we often refer to cells and sectors interchangeably. A 3-cell BTS configuration is most common, but configurations with up to 6-cell BTS are also found in the field. When implemented in a multicell configuration, the BTS is responsible for combining the reverse-link signal and multicasting the forward-link signal for those mobile stations that are in softer handoff with two or more of its cells (see Chapter 7, "Handoffs").

2.2.1.2   Base Station Controller

The BSC is responsible for several functions that can be broadly categorized as call-processing and supplementary services, radio resource management, mobility management, and transmission facilities management.

The BSC represents the termination point for the U_m Layer 3 signaling protocol. The BSC also terminates the Layer 2 (LAC/MAC) protocol for dedicated logical channels, except those mapped to physical channels employing H-ARQ that, as previously noted, are terminated at the BTS. In some implementations, the BSC may also terminate the LAC and part of the MAC protocols for the common logical channels. The LAC/MAC protocol termination point on dedicated channels is implemented in an entity called the selection/distribution unit (SDU). An SDU instance is created with each active mobile station when the associated radio bearer is mapped onto a dedicated logical channel. The SDU is so called because, in case of soft handoff (see Chapter 7), it is responsible for selecting the "best" incoming air interface reverse-link data frame from the receivers involved in the soft handoff. It also distributes copy of the same air interface forward-link data frame to each transmitter involved in the soft handoff. The SDU performs LAC/MAC functions on the dedicated channel, performs power control of the physical channels to which the logical channel is mapped, and handles traffic to and from the vocoder. The vocoder (see also Chapter 3, "Applications and Services") is the physical device used to compress the Pulse Code Modulated (PCM) 64-kbps bit stream carrying voice traffic received from the MSC. A common BSC implementation includes a vocoder-bank subsystem that is pool-shared by all the SDUs.

The BSC is typically implemented as a highly scalable, modular device capable of supporting from a few tens up to hundreds of BTSs. The BSC is connected to multiple BTSs via the A_{bis} interface. The physical A_{bis} interface consists of a set of trunks collectively called *backhaul*. The BSC is connected to a single MSC via the A1/A2/A5 interface. The physical interface with the MSC consists of a set of trunks collectively called *fronthaul*.

2.2.1.3   Packet Control Function

The PCF is a logical entity necessary for the support of packet-switched data calls. It connects the RAN with the PDSN. The PCF maintains the status of the radio resources associated with a packet data call. As we describe in Chapter 3 and in Chapter 9, "Packet Data Operation," packet data calls are characterized by bursty traffic. Active periods during which data packets are exchanged between the mobile station and the core network are followed by inactive periods during which packet transfer is suspended. If the inactive period is prolonged, the BSC may command the mobile station to enter the dormant state and then release the radio resources associated with such mobile stations. To make this process transparent to the core network, the PCF maintains the status of the packet data calls. When data packets destined to a dormant mobile station are received from the PDSN, the PCF buffers them until radio resources are set up. This typically happens when the packet call transitions out of dormancy into active state. The PCF also collects the air link–related accounting information and forwards it to the PDSN. A PCF can be a standalone device serving multiple BSCs, or it can be implemented within one BSC.

2.2.2 RAN Reference Points

Table 2.2 summarizes the RAN internal and external interfaces and applicable standards. In the sections that follow we provide a high-level overview of the associated reference points. The description of the A10/A11 interface is differed to Section 2.4.2.1. This interface is built upon mobile IP [13], and we discuss this reference point after the concept of mobile IP is introduced.

Table 2.2   RAN Internal and External Interfaces

Interface	Entities using the interface	Applicable standard
U_m	BSC/BTS-MS	IS-2000
A_{bis}	BSC-BTS	IS-828
A3/A7	BSC-BSC/BTS	IS-2001
A8/A9	BSC-PCF	IS-2001
A10/A11	PCF-PDSN	IS-2001
A1/A2/A5	BSC-MSC	IS-2001

2.2.2.1 A_{bis} Reference Point

Data exchange between the BSC and BTS takes place over the A_{bis} reference point. The A_{bis} interface protocol is specified in the IS-828 [14] standard. It can be divided into the user plane that carries traffic data and the application control plane that carries signaling messages. The protocol architecture for the A_{bis} was designed after the A3/A7, described in the next section, and reuses the same protocol stack.

Although standardized, the A_{bis} interface is commonly implemented as a proprietary interface. The main obstacle to the deployment of an open A_{bis} interface, which would allow interoperability of a BTS and a BSC of different manufacturers, is the lack of an open OA&M interface within the BS.

2.2.2.2   A3/A7 Reference Points

The A3/A7 or A_{ter} interface allows a mobile station to be in soft handoff with multiple BSs. When two or more BSs contribute radio links to the call in soft handoff, the SDU is located in one and only one of them. The BSC where the SDU is located is called the source BSC. The A7 interface carries signaling traffic between the source and target BSC used to control allocation and release of radio resources on the target BSC. The A7 procedures also include facility, or trunks, management. The A7 interface physically terminates at the BSC, although logically it is an interface between two BSs. The A3 interface carries both signaling and user traffic. A3 signaling is used to establish and remove A3 traffic connections and for call-specific operational procedures. On the user plane, the A3 carries data frames between the MAC entity at the source BSC—that is, the SDU—and the physical layer entity at the BTS—that is, the modem. Note that both ends of the A3 interface may be physically terminated at the BSC. The target BSC is then responsible to route A3 traffic to the BTS via its own A_{bis} interface. The protocol architecture for the signaling part of the A3/A7 interface is shown in Figure 2.7.

The Asynchronous Transfer Mode (ATM) [15] provides the A3/A7 interface with connection-oriented packet-switched transport of both signaling and user traffic. ATM was selected because it efficiently handles both low-latency constant-stream traffic and bursty traffic. ATM can be decomposed into two layers: ATM Adaptation Layer (AAL) and ATM layer. Relevant to the A3/A7 interface are AAL type 2 (AAL2) and AAL type 5 (AAL5). AAL2 is intended for the bandwidth-efficient transfer of low-rate traffic with stringent delay requirements. It multiplexes the streams from multiple calls to provide low delay and high utilization of the ATM data unit, or 'cell,' payload. The Service-Specific Segmentation and Reassembly (SSSAR) sublayer of AAL2 is used for segmentation and reassembly of AAL2 data units. AAL5 allows for efficient transport of bursty data traffic. AAL5 accepts a variable-length data unit from the layer above and produces a frame that is a multiple of 48 bytes. The frame is then divided into multiple 48-byte payloads and passed to the ATM layer. The ATM layer accepts the 48-byte data unit from the AAL and adds a header containing virtual circuit information that provides sequential delivery of ATM cells across connections established between the BSs. The ATM header also indicates the cell type, which can be used to differentiate QoS for distinct connections. The A3 traffic, for example, requires low latency and may be given priority over A7 traffic. Transmission Control Protocol (TCP) [16] is used on top of IP and ATM to provide reliable transport service to the A3 signaling subchannel and the A7 interface.

The A3/A7 interface physical layer is a T1, E1, T3, or OC3 digital interface with bit rates ranging from 1.544 Mbps to 155.52 Mbps.

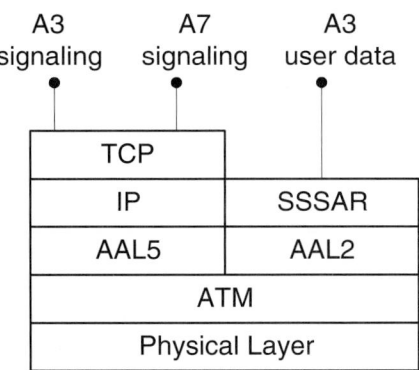

Figure 2.7 A3/A7 interface protocol architecture.

2.2.2.3   A8/A9 Reference Points

The A8/A9 or $A_{quinter}$ reference point connects the BSC with PCF. The relationship between the PCF and BSC is one-to-many. The A9 interface carries signaling traffic, while A8 serves user data traffic. The protocol architecture for these two interfaces is shown in Figure 2.8.

The A9 link is used to set up and tear down A8 connections. The A9 signaling also enables handoff between two BSCs that do not necessarily belong to the same PCF. If its protocol stack includes User Data Protocol (UDP) [17] instead of TCP, the A9 signaling application uses time-based retransmission to recover from errors. Alternately, TCP could be used to provide reliable service.

The A8 interface encapsulates users' data traffic and transmits it over the IP network between the BSC and PCF. The encapsulation method, also called *tunneling*, is based on the Generic Routing Encapsulation (GRE) [18]. The GRE header contains a key that uniquely identifies each user's traffic. In the forward direction, the PCF tunnels and the BSC detunnels user's packets. Similarly, in the reverse direction, the BSC encapsulates and the PCF performs detunneling. The standard does not specify the physical and the link layers for the A8/A9 reference point.

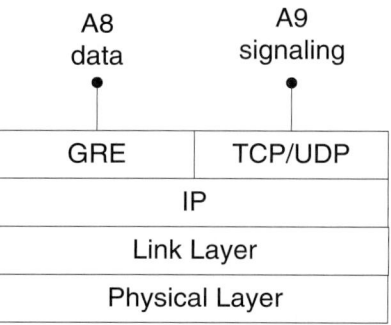

Figure 2.8   Protocol architecture of A8/A9 reference point.

2.2.2.4   A1/A2/A5 Reference Points

The reference point between BSC and MSC is implemented with three separate interfaces: A1, A2, and A5. The protocol stacks for the A1/A2/A5 interface are shown in Figure 2.9.

The A1 interface is used for signaling. Its application service layer, the Base Station Application Part (BSAP), is divided into two parts: Base Station Management Application Part (BSMAP) and Direct Transfer Application Part (DTAP). The BSMAP messages are those that are processed by the BSC and that require actions on its part. Examples include call-processing messages that require base station resources to be established or released, or messages used to manage the fronthaul trunks. Conversely, DTAP messages are not processed by the BSC; rather, they are piped through to and from the MSC or mobile station. In the forward direction, A1 DTAP messages are those that are received from the MSC and are destined to a mobile station, which the BSC maps to the corresponding IS-2000 Layer 3 messages. In the reverse direction, messages received from the mobile station and destined to some application at the MSC are mapped to the corresponding A1 DTAP messages. Examples of DTAP messages include those used by the MSC to perform call supervision (e.g., enabling/disabling the alerting or ring-back tones) or to enable supplementary services (e.g., call waiting or three-way calling). The A1 interface uses Signaling System 7 (SS7) [19] for reliable transport of messages. The use of SS7 for signaling transport between two network elements may seem at odds with the fact that SS7 is a packet-switched technology and as such is not intended for point-to-point transport. SS7 was selected mainly because it was a readily available technology when the cdma2000 A interface was designed. SS7 protocols play an important role in the IS-41 core network architecture and interfaces. A discussion of its main characteristics is deferred to Section 2.3.2.2. However, for better understanding of Figure 2.9, let us now point out that Mobile Transfer Part (MTP) Level 2 is the SS7 link layer protocol, while MTP Level 3 and Signaling Connection Control Part (SCCP) are SS7 network layer protocols. The A2 interface is used to transfer PCM voice over DS0 channels, while A5 interface transports circuit-switched data. The A5 uses the Intersystem Link Protocol (ISLP) as an adaptation layer between the variable-rate circuit-switched data and the constant bit rate of a DS0 channel. The physical layer is a T1 or E1 digital interface. The physical layer is also referred to as MTP 1.

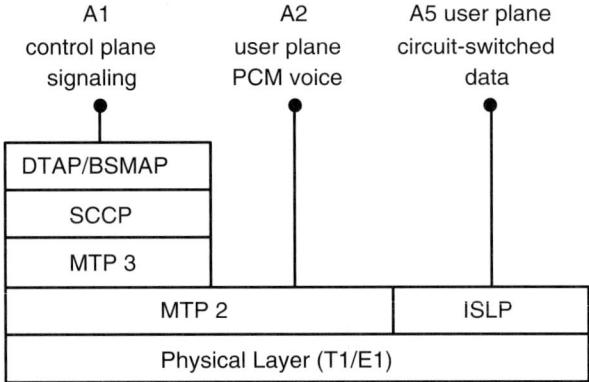

Figure 2.9   A1/A2/A5 protocol architecture.

2.3 IS-41 Core Network

The reference model showing most relevant logical entities and reference points of the IS-41 network is presented in Figure 2.10. A good reference on IS-41 can be found in [20]. The entities that belong to the core network are the MSC, home location register (HLR), visitor location register (VLR), authentication center (AC), message center (MC), and short message entity (SME). Also shown in the figure are the base station and mobile station. One MSC is connected to one or more base stations, forming a system. MSCs are interconnected with each other, forming a network. The IS-41 network implements specialized mobility management and call-processing functions that enable users to access the PSTN and Integrated Service Digital Network (ISDN) for services such as voice and circuit-switched data.

Mobility management represents a broad group of functions that allow the mobile station to obtain services while in motion. The functions can broadly be divided into automatic roaming and authentication. Automatic roaming refers to the ability of the mobile station to obtain service outside the home area without requiring special actions on the part of the user. Functions categorized as automatic roaming include the mobile station service qualification, location updating, and state management. The service qualification consists of validating the information provided by the mobile station and management of service profile information. For example, after examining the

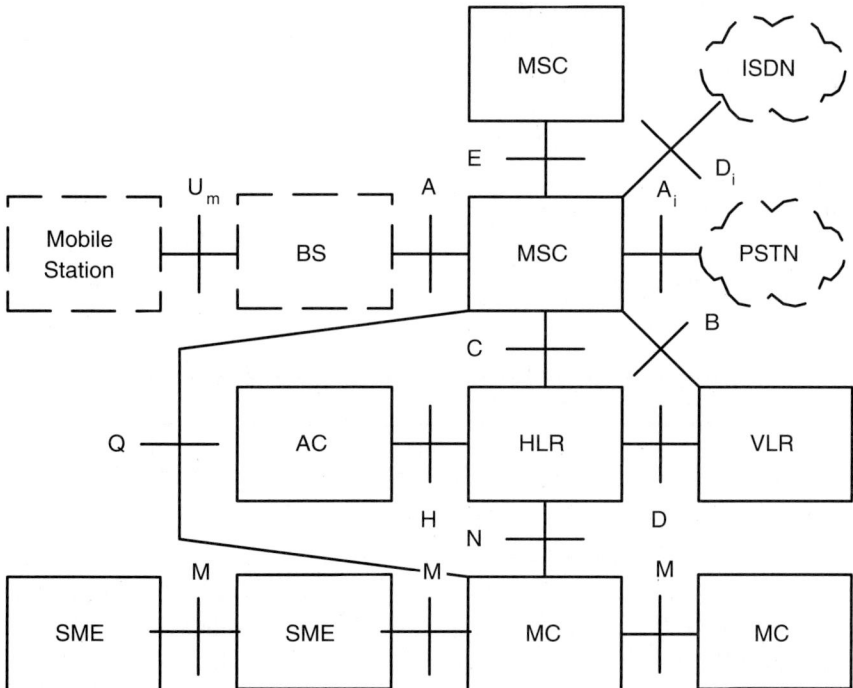

Figure 2.10 IS-41 network reference model.

mobile station identification, access may be granted or not, depending on the service profile information. Location updating management consists of functions that maintain the mobile station location. The location information consists of the network and MSC identification numbers. The state management represents functions that maintain the status of the mobile station. The status can be either active or inactive. When in the active state, the mobile station can receive calls. It cannot receive calls when it is in the inactive state. The inactive state typically means that the mobile station is outside the coverage area. Authentication is the procedure by which the mobile station and the IS-41 network exchange information to verify the identity of the mobile station with a strong degree of confidence. After successful authentication, provided its subscription allows it, the mobile station is authorized to access the system. Mobility management features of the IS-41 network are necessary even for packet data calls, switched through the PDSN, because the IS-41 protocol authorizes access to the cdma2000 network for all calls. Intersystem handoff is also a mobility management function. It is the ability of the mobile station to maintain calls when moving between two cells that belong to two different MSCs.

Call processing represents a group of functions that establish, maintain, and release calls to and from the mobile station. The two main functions are call setup and call release. The call setup can further be decomposed into mobile station-originated and mobile station-terminated calls. Certain functions are invoked for mobile station-originated calls and not invoked for mobile station–terminated calls and vice versa. Many of the call-processing functions involve mobility management. For example, mobile station-originated call processing includes authentication and service qualification. Mobile station-terminated calls, however, require mobile station location and state information. Other functions include obtaining routing information, paging, requesting the radio traffic channel setup, and applying call features such as call waiting, call forwarding, and three-way calling. Many of the call-processing functions are not needed for packet data calls. However, the IS-41 network is still responsible for paging the mobile station and requesting the establishment of the radio traffic channels.

2.3.1 IS-41 Core Network Logical Entities

In this section a high-level overview of the logical entities that comprise the IS-41 core network is given. We list the functions that each of these entities perform and describe the interaction among them.

2.3.1.1 Mobile Switching Center

The MSC performs interworking between the RAN and PSTN. The MSC enables establishment and maintenance of circuit-switched connections between a mobile station and the PSTN. The circuit-switched traffic consists primarily of voice traffic, although data services are possible (see Chapter 3). Calls originated from the PSTN and destined to a mobile station are routed to the home MSC based on the dialed digits. In this respect, the MSC is seen by the PSTN as a local office. Because the mobile station the call is destined to may not be located in its home system coverage area, the MSC routes such calls to the MSC currently serving the mobile station using the inter-MSC trunks. The MSC supports traditional call control signaling such as ISDN User Part (ISUP) to manage trunks toward the PSTN. The MSC is also responsible for the call-processing functions toward the mobile station. The MSC controls intersystem handoff that enables continuous service when the MS moves between cells belonging to different systems. The MSC typically controls multiple BSs. Its capacity is expressed in terms of Erlangs of traffic. The MSC is typically designed as a scalable system, and its capacity varies according to the implementation and configuration, but capacity of tens of thousands of Erlangs is common.

2.3.1.2   Home Location Register

The HLR represents the primary database for subscriber data. Subscribers are associated with the HLR on the basis of the geographical area they typically reside—that is, based on the home network. The HLR stores information such as location, status, and identification information. Location information retrieved from the HLR, for example, is used to route PSTN-originated calls to roaming subscribers. Status information retrieved from the HLR allows the network to invoke specialized call treatments, such as when announcements are played to the originator of a call destined to an inactive mobile station. Identification information is used to associate a mobile station identification number to the dialed digits (see Chapter 6 for details on mobile station identification). The HLR also contains information about the features that the mobile station is subscribed and/or authorized to, and the ones that are currently activated. The former are controlled by the network operator based on the subscription agreement, while the latter can be activated or deactivated on demand by the user. One HLR may serve one or more MSCs. The subscriber capacity of an HLR varies according to the implementation, but capacity of millions of subscribers or more is common. The HLR is implemented as a highly reliable and available system because a single HLR failure may affect a large number of subscribers. To achieve high availability, the HLR is implemented as a redundant system in which the pair of replicated components operate either in load-sharing mode or in active/standby mode.

2.3.1.3   Visitor Location Register

The VLR is a local database that maintains records of the mobile stations currently served by the MSCs, or one of the MSCs, with which the VLR is associated. These records are temporary in the sense that once the mobile station moves in the service area of a new system, the previous serving VLR is informed and the record associated with the mobile station is removed from its database. The VLR stores information such as user location, status, and service information retrieved from the HLR. The VLR may also be delegated to perform authentication procedures on behalf of the authentication center. To that purpose, the VLR database maintains secret data shared with the authentication center, such as authentication keys. Although according to the reference model a VLR may serve multiple MSCs, it is most commonly collocated with the MSC, forming a single physical entity. Like the HLR, the VLR is also implemented as a highly reliable and available system.

2.3.1.4   Authentication Center

The authentication center is the entity that processes and manages the authentication information such as encryption and authentication keys associated with the mobile station. The authentication center prevents unauthorized users from accessing the network by verifying and validating the MS identity. It may serve one or more HLR, but it may also be located within and indistinguishable from the HLR. The authentication center is managed by the network operator to store or change authentication keys using secure protocols. These secure protocols, left unspecified by IS-41, are of paramount importance because the entire authentication process is dependent on the authentication keys remaining secret.

2.3.1.5   Message Center

The message center is the entity that stores and forwards short messages in support of Short Message Service (SMS). The messages are forwarded to and from the mobile station if the mobile user is available. If the mobile user is unavailable, the message is stored at the message center and the delivery attempt is postponed. Undelivered messages stored at the message center may expire and be discarded depending on message delivery options specified by the message originator or the message center implementation. Upon successful message delivery, the message center also triggers transmission of delivery receipt messages to the message originator if so requested by the originator. The message center performs signaling procedures to support mobile station location and status query. In case of messages destined to mobile stations located in other systems, the message center maps the destination address to the associated message center and routes the message to it.

2.3.1.6   Short Message Entity

The SME is a logical entity that can originate or terminate short messages, or do both. The SME can be either internal or external to the IS-41 network. SME internal to the IS-41 network can be implemented within the HLR or MSC, or as an independent physical device. Commercial mobile stations that are SMS-capable are considered external SMEs.

2.3.2 IS-41 Reference Points

Table 2.3 summarizes the IS-41 internal and external interfaces and applicable standards. In the sections that follow we provide a high level overview of the associated reference points.

Table 2.3   IS-41 Core Network Internal and External Interfaces

Interface	Entities using the interface	Applicable standard
A1/A2/A5	BSC-MSC	IS-2001
B	MSC-VLR	IS-41
C	MSC-HLR	IS-41
D	HLR-VLR	IS-41
E	MSC-MSC	IS-41
H	HLR-AC	IS-41
M	MC-MC, SME-MC, SME-SME	IS-41
N	MC-HLR	IS-41
Q	MC-MSC	IS-41
A_i	MSC-PSTN	IS-93
D_i	MSC-ISDN	IS-93

2.3.2.1   IS-41 Core Network Internal Reference Points

The protocol architecture of the reference points within the IS-41 network is shown in Figure 2.11. It consists of the IS-41 application services layer and the IS-41 data-transfer services layer. The application services layer is further decomposed in the Mobile Application Part (MAP), which is defined by the IS-41 standard, and the Transaction Capabilities Application Part (TCAP), which is actually a part of the SS7 protocol that enables reliable transfer and remote database access. The IS-41 MAP defines *intersystem* operations in support of subscriber mobility management. Intersystem operations consist of signaling messages and associated functional procedures. The intersystem operations defined in IS-41 are vastly transaction oriented—that is, they involve exchange of information based on a query or command and the response to such query or command. SS7 provides standard transaction-based protocol mechanisms and was therefore selected for inclusion in the IS-41 protocol stack.

The IS-41 application relies on SS7, a reliable packet-switched protocol, also for reliable data-transfer service. The IS-41 data-transfer services layer consists of the following SS7 protocols: MTP Level 1, 2, and 3, and SCCP.

MTP Level 1 is a physical-level protocol. It specifies the use of one DS0 channel on a channelized digital T1 link. In broadband links, it may use a full DS1 channel. MTP Level 2 is a link-level protocol. It provides for error detection, error correction, flow control, and sequential

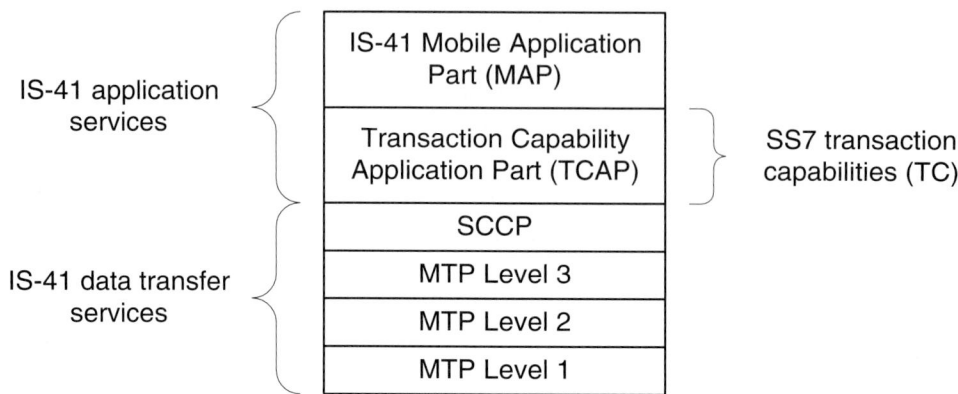

Figure 2.11 IS-41 protocol architecture.

frame delivery. Retransmission of messages that are either lost or received in error is allowed. MTP Level 3 is a network protocol that provides two categories of functions: message handling and network management. The message handling functions are responsible for determining if a message is addressed to the receiving node or needs to be forwarded. If a message needs to be forwarded, a routing function determines the outgoing link based on the routing table maintained by the administrator. A routing table must indicate the primary route to the destination as well as the alternative route or routes. The network management is used to reroute traffic to alternative links or divert traffic from the specific nodes, which happens when a link fails or a node becomes unavailable. Note that MTP Level 3 can only route a message to an adjacent node. MTP Level 3 implements flow control.

SCCP provides end-to-end addressing so that a message can be routed through the network nodes. SCCP provides MTP with enough information so that MTP can route a message from one node to the other. Although SCCP can support four different class of service, in IS-41 networks only one class is used: basic connectionless service, or Class 0. Class 0 service does not guarantee sequential delivery.

The nodes in the SS7 network are called signaling points. There are three signaling point types: Service Switching Point (SSP), Signaling Transfer Point (STP) and Service Control Point (SCP). The SSP creates and sends messages and queries. It can switch as well as originate and terminate calls. The STP acts like a router, while the SCP is an interface to a database. The IS-41 network nodes coincide with SS7 nodes. For example, the MSC is an SSP, while the HLR is typically viewed as an SCP.

2.3.2.2 A_i and D_i Reference Points

The reference point between the MSC and PSTN is A_i, while the interface between MSC and ISDN is D_i. Note that A_i and D_i interfaces are actually the same. They are specified by the IS-93 [21] standard. The distinction in names is made due to the anticipated need for differentiation between the two in the future. Presently, however, the PSTN supports both analog and digital signaling protocols and there is no need to distinguish between A_i and D_i interfaces. The A_i and D_i interfaces use ISUP for signaling. PCM voice is transmitted as user data over the A_i interface on a DS0 link. Unrestricted digital information is user data transmitted over the D_i interface.

2.3.3 IS-41 Interworking Example: Voice Call Origination

In this section we present an exemplary IS-41 interworking procedure. The call flow example depicted in Figure 2.12 refers to a simplified, circuit-switched voice call setup procedure, which encompasses some of the nodes and functionalities discussed so far.

Referring to Figure 2.12, a mobile station originates a circuit-switched voice call destined to a land party by sending IS-2000 Layer 3 signaling message on the reverse access channel. The message contains the mobile station identity and an indication that circuit-switched voice call is requested. The BTS acknowledges the reception and forwards the message to the BSC over the A_{bis} link. On receipt of the message, the BSC interprets and reformats its content and sends a request for service to the MSC over A1 interface.

If the network decides to verify the mobile station's identity, the VLR (assumed to be collocated with the MSC) forwards the mobile station identity and associated authentication infor-

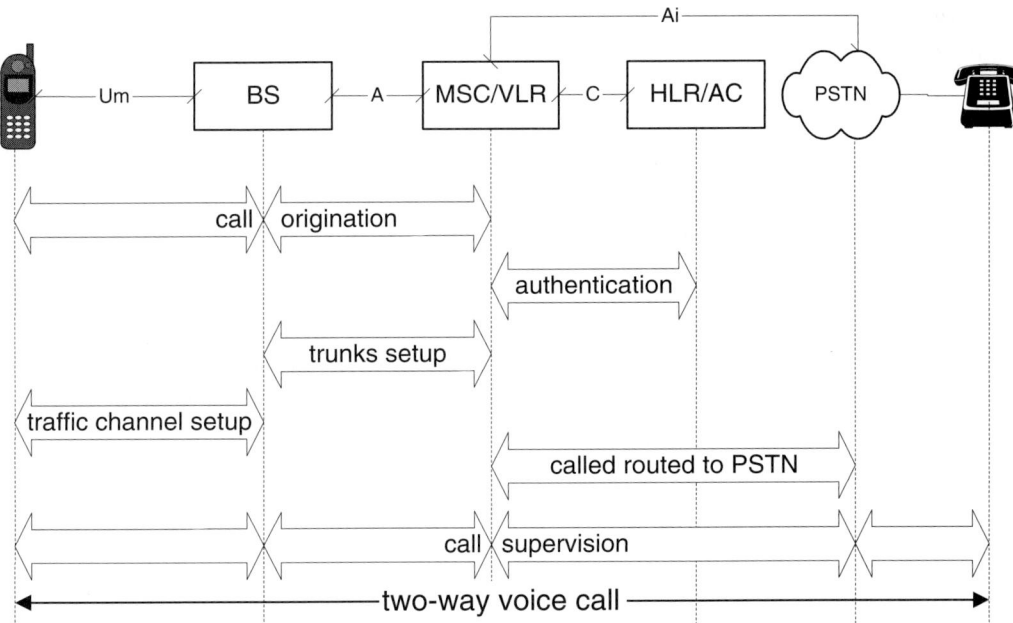

Figure 2.12 Voice call origination example.

mation to the HLR/AC. The authentication center successfully authenticates the mobile station and replies to the VLR request.

The MSC sets up the voice call. The call setup includes allocation of the A2 link between the MSC and BSC, and allocation of a PSTN trunk over the A_i interface.

The BSC proceeds with allocating a circuit on the A_{bis} interface, radio resources at the BTS, and an SDU for the call. Then, the radio channel between the BTS and mobile station is established by means of IS-2000 Layer 3 signaling messages.

In parallel with the establishment of the MSC-BSC trunks and the radio link, the MSC routes the call to the PSTN, which alerts the party to which the call is destined. The MSC performs call supervision on the mobile station side (e.g., ringback tone), while the PSTN performs call supervision on the land party side (e.g., alerting). Once the land party answers the call, the PSTN sends a connect message to the MSC that in turn disconnects the ringback tone at the mobile station. Two-way conversation can now begin.

2.4 Packet Core Network

The logical entities that comprised the PCN are PDSN and authentication, authorization, and accounting (AAA) servers. AAA servers can be classified into three categorizes: home AAA (HAAA), visited AAA (VAAA) and broker AAA (BAAA). The PCN is attached to RAN through an A10/A11 interface. The RAN point of attachment is PCF.[5] The main purpose of the PCN is to provide the mobile station access to the IP network. In order to access an IP network, a mobile station must obtain an IP address. The PCN network can provide two access methods: simple IP [1] and mobile IP [13]. The distinction between the two is the procedure by which an IP address is obtained and by which packets are routed to and from the mobile station. Regardless of how the IP address is obtained, the mobile station could retain an IP address it has been assigned after releasing the packet data call. This feature is called an *always-on connection*. With the always-on feature enabled, the mobile station does not have to go through a relatively lengthy procedure of obtaining an IP address whenever a new packet data call is originated. The access service provided by the PCN can be summarized in the following:

Simple IP access. With this method, the access service provider dynamically assigns an IP network address to the mobile station. The IP address is topologically correct with a network-dependent geographical area and remains unchanged as long as the user remains in it.[6] If the mobile station moves out of this area, it must obtain a new IP address that is topologically correct with the new area. The PCN reference model for simple IP access is shown in Figure 2.13. Note that P-P refers to an interface between two PDSNs, and P_i refers to packet interface based on IP and other

5.   PCF is explained in Section 2.2.1.3.
6.   Always-on connection is assumed.

protocols from the Internet Engineering Task Force (IETF). These reference points are discussed in more detail in Section 2.4.2.

Mobile IP access. With this method, a mobile station can maintain an IP network address as it moves from one network to another. The mobile station's home network, not the access service provider network, assigns the IP address to the mobile station. The PCN reference model for mobile IP access is illustrated in Figure 2.14. When mobile IP–based access is supported, the PCN also performs the functions of the foreign agent (FA) and the home agent (HA). The foreign agent function is performed by the PDSN. More details on the functions performed by the PCN logical entities are given in Section 2.4.1.

A brief overview of mobile IP and other IETF protocols relevant for understanding of the PCN is provided in Section 2.6. In the sections that follow, we describe the functions provided by the logical entities and depict the reference points.

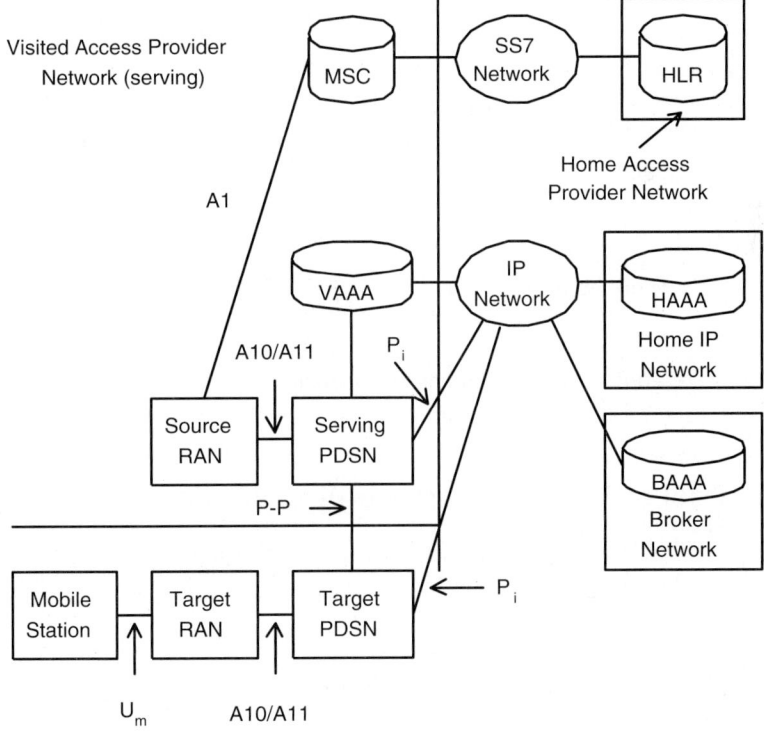

Figure 2.13 PCN reference model for simple IP access.

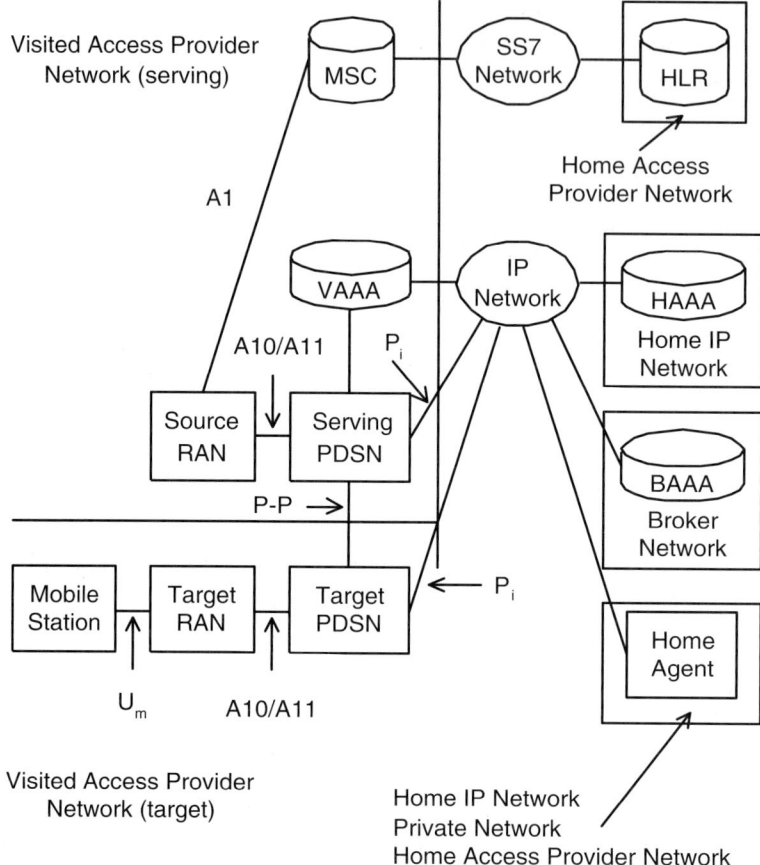

Figure 2.14 PCN reference model for mobile IP access.

2.4.1 Packet Core Network Logical Entities

The PCN network provides simple and mobile IP access service. In this section we describe the functions performed by the PCN logical entities for both access methods. We also explain the interaction among them.

2.4.1.1 *Packet Data Serving Node*

The PDSN establishes, maintains, and terminates the PPP [22][23]. Recall that at the mobile station side, PPP is terminated at TE2 or MT2 for relay and network models, respectively. The PDSN performs all the functions provided by the service provider's network access server (NAS). It is associated with the local AAA server to which it passes the authentication information, received from the mobile station that enables the mobile station's authorization. Based on the response from the local AAA server, the PDSN grants or denies packet data service to the mobile station. It also processes and passes the accounting information received from the PCF to the local AAA server. The PDSN has a publicly visible IP address, and it acts as a router by forwarding packets to and from the attached IP network. The PDSN marks and processes packets according to the QoS profile. For example, based on IP addresses and port numbers of the source and destination, the PDSN filters the incoming traffic to the appropriate call or service instance. After the traffic is appropriately filtered, the RAN can provide adequate QoS over the U_m interface. For traffic originating at the mobile station, the PDSN could mark the outgoing IP packets with the appropriate DiffServ [24] codes so that adequate treatment is given to the packets in the external network. In the case of simple IP service, the PDSN may provide an IP address to the mobile station or support address assignment from the local AAA server. For mobile IP access, the PDSN supports foreign agent functions, which are explained below.

2.4.1.2   *Packet Data Serving Node as Foreign Agent*

In order to support mobile IP operation, the PDSN provides foreign agent functions, such as supporting mobile IP registration procedures and detunneling and forwarding packets to the mobile node (MN)[7] or mobile station. The foreign agent optionally establishes, maintains, and terminates secure communication with the home agent using the Internet Key Exchange (IKE) [25] protocol. After security association (SA) between the PDSN and home agent has been established, secure communication is provided by IP Security (IPSec) [26][27]. The foreign agent could also support authentication of mobile IP registration messages using HA-FA authentication extension. The PDSN allows dynamic mobile station's home address and home agent address assignments. For example, during mobile IP registration, a mobile station may request a particular IP address to be its home address, or it may let the home agent assign the IP address dynamically. Similarly, a mobile station may indicate the address of its home agent. If the IP address is not specified, it is dynamically assigned by the PDSN after the PDSN obtains the address from the HAAA of the mobile station. If reverse tunneling is negotiated with the mobile station, the PDSN routes packets received from the mobile station to the home agent.

2.4.1.3   *Home Agent*

The home agent has a publicly routable IP address, and it is associated with the HAAA server. The home agent intercepts packets routed to the mobile station and tunnels the packets to the PDSN or foreign agent. When reverse tunneling is supported by the foreign agent, the home agent receives reverse-tunneled packets from the foreign agent, detunnels the packets, and forwards them to the destination. The home agent also optionally establishes, maintains, and terminates secure communication with the foreign agent using IKE. After security association

---

7.   In mobile IP terminology, roaming mobile station is called the mobile node.

between the foreign agent and home agent has been established, secure communication is provided by IPSec. The home agent could also support authentication of mobile IP registration messages using HA-FA authentication extension. Upon a request, the home agent assigns a dynamic home address to the mobile station.

2.4.1.4   Authentication, Authorization, and Accounting

As we have already mentioned, the AAA servers can be divided into three groups: VAAA, HAAA, and BAAA. The HAAA authenticates and authorizes the user based on the requests from the PDSN or VAAA, and it provides the user's profile to the PDSN. For mobile IP access, HAAA could also provide key information to the home agent and VAAA (VAAA passes the key to the foreign agent). The key information could be used for IKE as a preshared secret or for HA-FA authentication extension. Another optional function of the HAAA is dynamic home agent assignment to the mobile station. Remember that during the mobile IP registration procedure, the mobile station may request dynamic home agent assignment from the foreign agent. The foreign agent passes this request to the HAAA, and the HAAA responds with the home agent's IP address. For both simple and mobile IP access, VAAA passes authentication requests from the PDSN to the HAAA and forwards authorization responses from the HAAA to the PDSN. VAAA stores accounting information and provides the user's profile to the PDSN as received from HAAA. BAAA forwards requests and responses between the VAAA and the HAAA if they do not have bilateral associations.

2.4.2 *Packet Core Network Reference Points*

The internal and external PCN reference points are all based on IETF protocols. Three reference points are defined: the R-P reference point, which is the interface between the RAN and PDSN; the P-P reference point, which is the interface between two neighboring PDSNs; and the P_i reference point, which is the interface between the PDSN and external IP network.

2.4.2.1   R-P Reference Point

The R-P interface is also called the A10/A11 or A_{quater} interface. It connects the PCF, which is considered part of RAN, to the PDSN, which is considered part of PCN. The A10/A11 interface is specified in the IS-2001 standard, but it is explained at this point because it is built on mobile IP. The PDSN acts as a home agent, and the PCF behaves as a foreign agent and mobile node. The protocol architecture for the A10/A11 interface is shown in Figure 2.15. The A11 interface is for signaling, while the A10 is for user data traffic.

The A11 link serves to set up, tear down, and refresh A10 connections. It also passes accounting information to the PDSN. A11 signaling uses time-based retransmissions to recover from errors.

The tunneling is based on the GRE protocol. The GRE encapsulates the data traffic and sends it over the IP network between the PDSN and PCF. The GRE header contains a key that

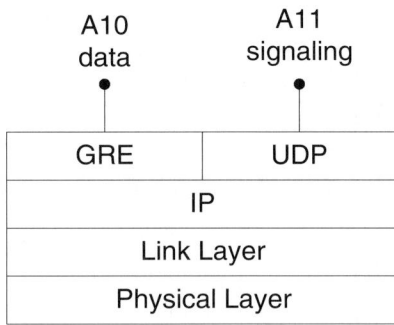

Figure 2.15 Protocol architecture for R-P or A10/A11 interface.

uniquely identifies each user's traffic. In the forward direction, the PDSN encapsulates or tunnels and the PCF detunnels each user's packets. Similarly, in the reverse direction, the PCF encapsulates and the PDSN performs detunneling. The outer header contains the IP address of the PDSN and PCF. The inner header in the forward direction contains the IP addresses of the source—for example, a host—and the destination, the mobile station. The standard does not specify the physical and the link layers for this reference point.

2.4.2.2 P-P Reference Point

The P-P interface provides fast handoff between two PDSNs. The fast handoff enables mobile stations to move from the geographical area associated with one PDSN to the area of another PDSN without PPP renegotiation. Since PPP renegotiation is not necessary, the mobile station does not experience interruption of service. The source PDSN remains the serving PDSN. The forward traffic received at the source PDSN is tunneled through the P-P interface to the target PDSN, which forwards the traffic to the mobile station over a corresponding R-P connection. Fast handoff is triggered after the source RAN initiates the handoff procedure via MSC to the target RAN that is not reachable by the serving PDSN. During the handoff procedure, the user data is bicasted until the service or call instance bearer on the target side is successfully completed.

After all active service instances transition to the dormant state, the P-P connection is released and a new PPP session is established at the target PDSN. As it is for the R-P interface, the protocol stack at the P-P interface is mobile IP–based. It consists of signaling and user traffic. The signaling is used to set up, tear down, and refresh connections that carry data traffic. The serving PDSN acts as a home agent, and the target PDSN behaves as a foreign agent and mobile node. The protocol architecture for the P-P interface for signaling and user data is shown in Figure 2.16 and Figure 2.17, respectively. Since the P-P interface may be over a public Internet, IPSec is mandatory. The standard does not specify the physical and the link layers for this reference point.

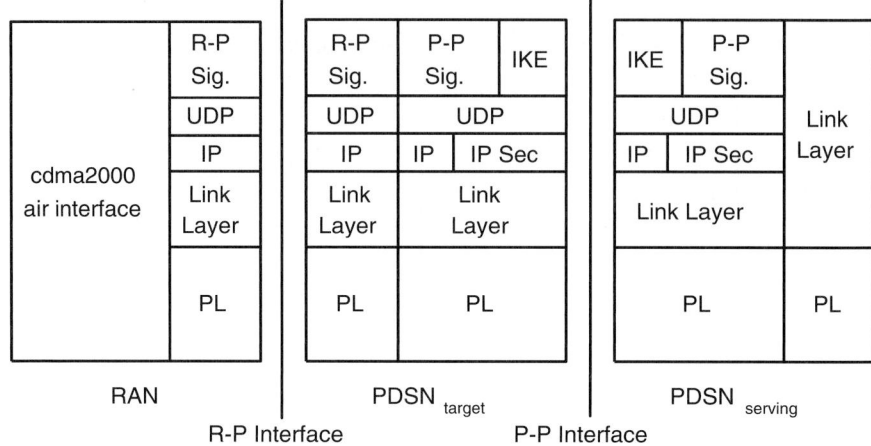

Figure 2.16 Protocol architecture for signaling traffic over a P-P interface.

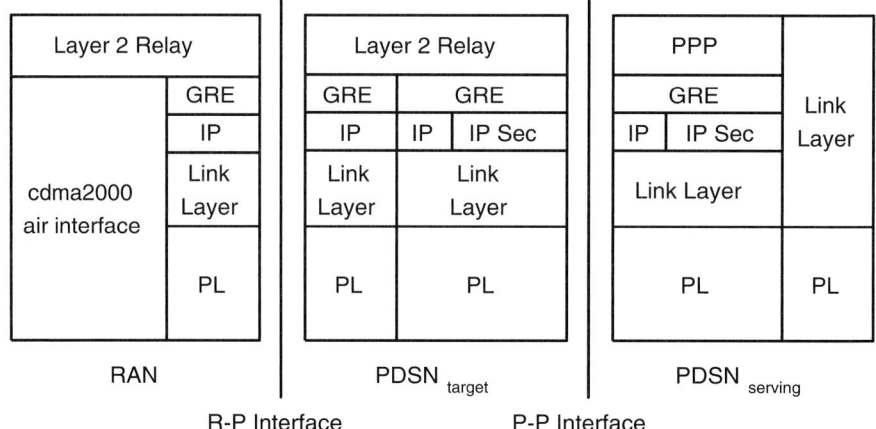

Figure 2.17 Protocol architecture for user's data traffic over a P-P interface. The PPP terminates at the mobile station, which is not shown in the figure.

2.4.2.3 P_i Reference Point

The P_i reference point connects the PCN with an external IP network. It also connects PCN entities. The protocol architecture depends on which network entities the interface connects and whether simple IP or mobile IP is implemented. The protocol stack for the P_i reference point between the PDSN and AAA servers is shown in Figure 2.18. In this case, Remote Access Dial In User Service (RADIUS) [28][29] is the AAA protocol. For simple IP access, the main

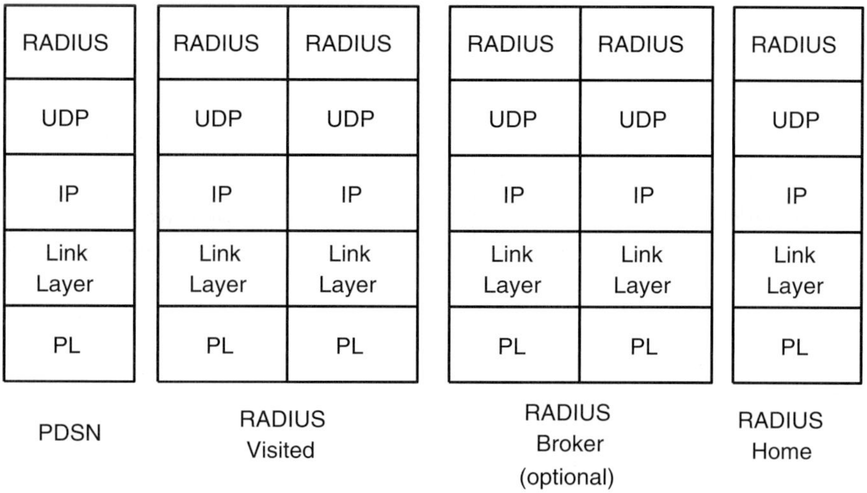

Figure 2.18 Protocol architecture for Pi interface control signaling between PDSN and AAA server.

function of the RADIUS infrastructure is passing accounting records and information necessary to authenticate the mobile stations that attempt to access the network and actual access authorizations or rejections. In case of mobile IP access, in addition to accounting records and authentication and authorization information, the RADIUS infrastructure provides other functions, such as exchanging the preshared secret keys for IKE between home agent and foreign agent, distributing dynamically assigned home agent IP addresses, and indicating reverse tunneling. The reference model in Figure 2.18 assumes that the VAAA server communicates with the HAAA server via the optional proxy server or the BAAA.

In the case of simple IP access, the protocol stack for PDSN-to-end host connection is illustrated in Figure 2.19. The PPP establishes the link between the mobile station and PDSN. From the PPP and IP perspective, RAN is just a pipe. Even though a mobile station can support multiple services, there is only a single PPP instance established with the PDSN. In the case of always-on service, the PPP remains active as long as the mobile station is powered on.

The protocol architecture for mobile IP user data and mobile IP control and IKE is shown in Figure 2.20 and Figure 2.21, respectively. In addition to the mobile station-to-PDSN link, which is in terms of the protocol stack equivalent to the one shown in Figure 2.19, Figure 2.20 illustrates the tunneling on the link between the home agent and foreign agent or PDSN. As the figure shows, in order to provide secure communication, IPSec could be employed after security association has been established. The tunneling between the home agent and the foreign agent is mandatory in the forward direction. It could also be implemented in the reverse direction if reverse-link tunneling is negotiated. Figure 2.21 shows that the mobile IP signaling uses UDP as a transport layer protocol. As we already discussed, IKE is used to establish a security association between the PDSN and home agent.

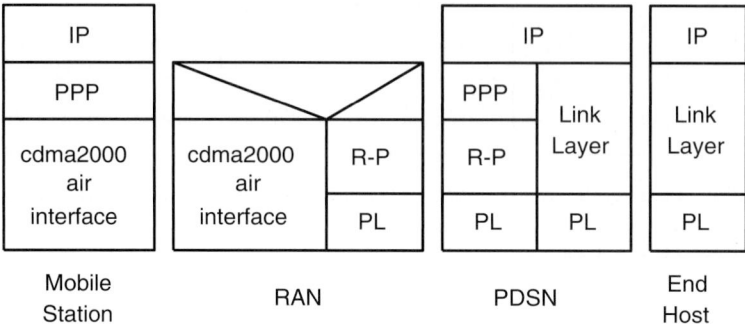

Figure 2.19 Protocol architecture for Pi interface for simple IP user data between PDSN and end host in IP network.

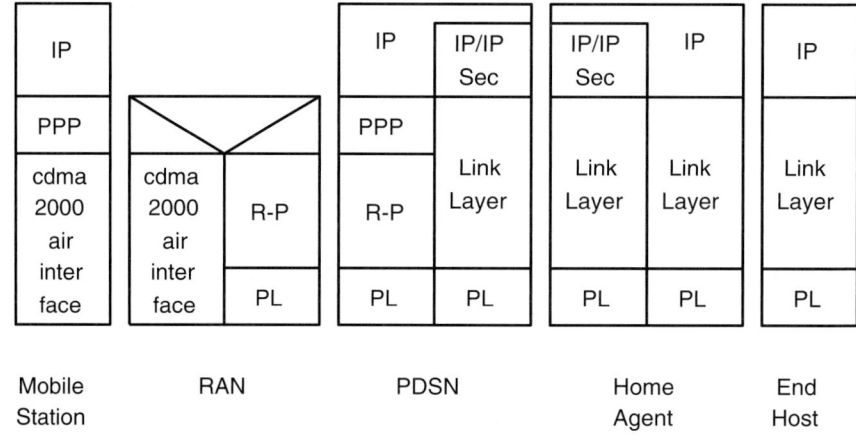

Figure 2.20 Protocol architecture for Pi interface for mobile IP user data between PDSN and end host in IP network.

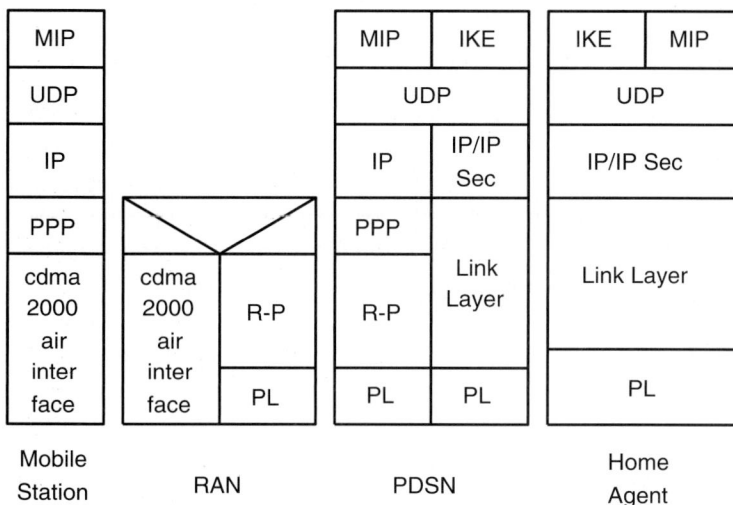

Figure 2.21 Protocol architecture for Pi interface for mobile IP (MIP) control signaling and IKE between PDSN and home agent.

2.5 PCN Interworking Examples

In this section, we present simplified packet-switched call setup, call-flow diagrams for two different scenarios. The first example shows how a mobile station connects to the PCN using simple IP access. The second example we shows how a mobile station performs mobile IP registration.

2.5.1 Packet Data with Simple IP

Let us assume that the traffic channels are not set up; that is, there is no active service at the mobile station. A simplified call-flow diagram for simple IP access is shown in Figure 2.22. The packet data service is requested by the mobile station with the origination message indicating such service request. As is done for circuit-switched calls, the mobile station is authenticated and authorized for access by the MSC, using the user's record stored at the HLR. After the MSC authorizes the mobile station to access the wireless network, it sends a message to the BSC to request allocation of radio resources.

In parallel with radio resource allocation, BSC typically sets up an A8 connection to the PCF, which sets up an A10 link with the PDSN. After all radio and PCN resources are allocated and RLP protocol is initialized, the mobile station begins PPP negotiation with the PDSN.

During the Link Control Protocol (LCP) phase of PPP, the link is configured. For example, in this phase PPP compression and the authentication mechanism that will be performed when

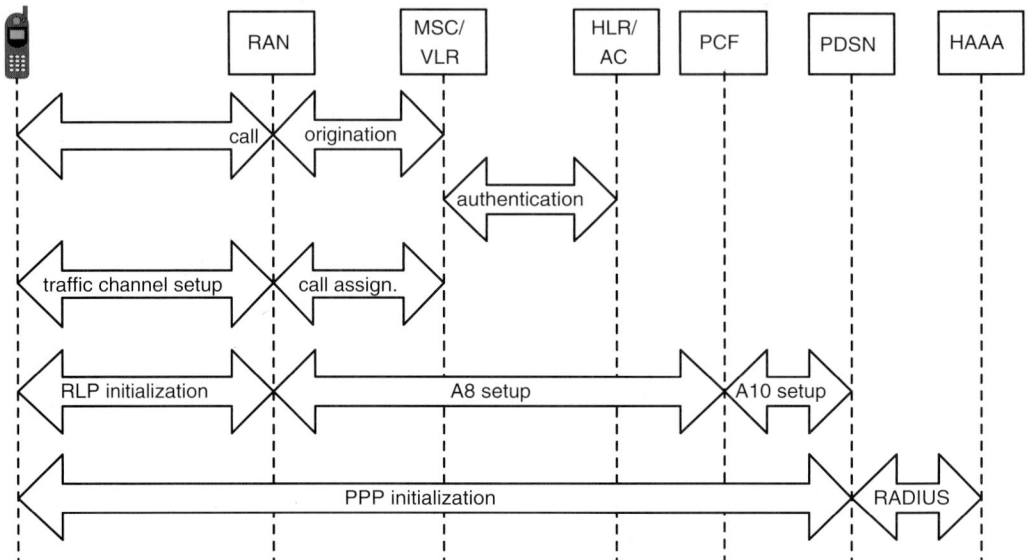

Figure 2.22 Simple IP packet data call setup.

the LCP phase ends are negotiated. After the LCP phase is completed, the PPP authentication and authorization phase begins. The Challenge Handshake Authentication Protocol (CHAP) [30] is the preferred authentication method. The PDSN acts as a network access server (NAS). It sends a cryptographic challenge to the mobile station. The mobile station responds with the response to the CHAP challenge and the Network Access Identifier (NAI) [31], which is in the form of *user-name@service-provider.com*. On reception of the response, the PDSN, acting as a RADIUS client, communicates the data to HAAA, which authenticates and authorizes the mobile station for service.

After the authentication and authorization phases are completed, the PPP Network Control Protocol (NCP) phase begins. NCP is a generic name, and for simple IP access, the NCP refers to Internet Protocol Control Protocol (IPCP) [32]. It includes the negotiation of the mobile station's IP address, TCP/IP header compression technique, and the Domain Name Server (DNS) IP address. IPCP is the last phase of the PPP before the actual data exchanged can start.

2.5.2 Packet Data with Mobile IP

The procedure for mobile IP access follows the same steps as for simple IP service until the PPP initialization phase. A simplified diagram for the remainder of the call flow is shown in Figure 2.23.

The PPP initialization is different. For example, the PPP authentication does not utilize CHAP. The authentication of the mobile station is performed during the mobile IP registration procedure. Also, during the IPCP phase, the IP address is not assigned to the mobile station.

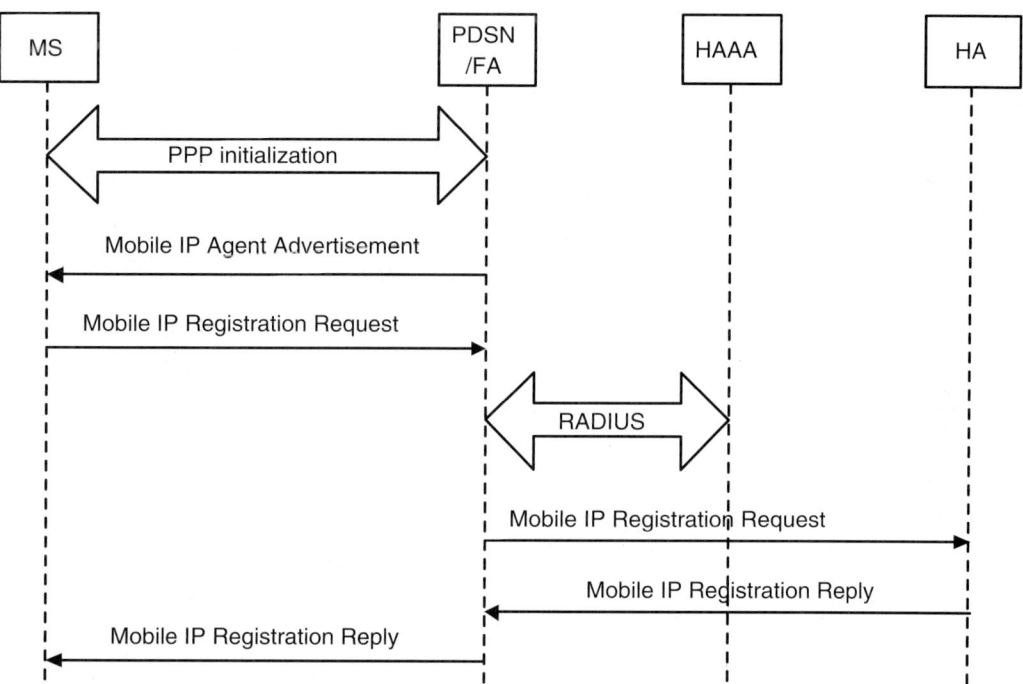

Figure 2.23 Mobile IP packet data registration.

Rather, the mobile station uses its permanent home address.[8] However, as in the case of simple IP access, the PDSN negotiates TCP/IP header compression with the mobile station. It may also negotiate the DNS IP address.

In the case of mobile IP, the PDSN acts as a foreign agent, and when a PPP link is established, the PDSN sends the mobile IP agent advertisement message to the mobile station. This message includes an MN-FA challenge extension [33]. The purpose of the extension is to allow the FA to challenge and HAAA server to authenticate the mobile station. The mobile station responds with the mobile IP Registration Message that includes the MN-FA challenge extension, the MN-AAA authentication extension [33], and the NAI extension [34]. These extensions enable the mobile station's authentication. The PDSN relays the authentication data to the HAAA server, which authenticates and authorizes the mobile station. On successful authorization, the PDSN forwards the mobile IP registration request message to the home agent. The home agent processes the message and updates the binding list in order to associate the mobile station's permanent IP address with the new care-of address. After all the processing is completed, the home agent sends the mobile IP registration reply message to the foreign agent. The

8.   A mobile station could also request dynamic IP address allocation from the home agent.

foreign agent updates the list of the visiting mobile stations and passes the message to the corresponding mobile station. From this point on, the mobile station can send and receive IP packets using mobile IP.

2.6 IETF Protocols

A basic proficiency in IETF protocols is needed for understanding the interaction between the PCN logical entities and associated interfaces. This section gives a brief overview on the subject. The relevant protocols are described only in terms of the general concepts behind them and the functions they provide. For more detailed description, refer to the corresponding requests for comments (RFC).

2.6.1 Point-to-Point Protocol

PPP is a link layer protocol that encapsulates higher layer data, such as IP packets, over point-to-point links. PPP is described in [22]. It is commonly employed with High-level Data Link Control (HDLC) framing [23]. The protocol typically connects a personal computer (PC) to the NAS of an Internet service provider (ISP) over a telephone line and a modem. In PCN networks, the PPP connects the mobile station to the PDSN, which performs the NAS function.

The PPP has three phases. The first one is LCP phase. During LCP, the link is configured. It is recommended that the authentication be performed with CHAP [30], because CHAP provides cryptographically strong authentication. An alternative to CHAP is the Password Authentication Protocol (PAP). PAP authentication merely consists of supplying an unencrypted user name and a password. For CHAP to work, it is necessary that the two parties have somehow established a secret key. The NAS initially sends a challenge to the user in the form of a random number. The user responds with NAI (for example, *user-name@service-provider.com*) and a cryptographic checksum of the challenge based on the secret key. The challenge and the response are passed to the AAA server, which computes the cryptographic checksum itself. If the checksum matches, the user is authenticated. The CHAP phase is followed by NCP. In this phase, the NAS configures the user's network layer, such as IP. The header compression algorithms for layers above PPP are negotiated during the NCP phase. The actual data-transfer phase begins after NCP has been completed.

2.6.2 Internet Protocol and Internet Protocol Security

IP version 4 is described in [1]. It is a network layer protocol used to interconnect various types of networks into a single, globally routable network. The main task of IP is to provide the routing function, where the IP address, which is a part of the IP header, uniquely identifies each computer on the Internet. IP is a connectionless protocol. It provides the layers above with unreliable best-effort service. Note that IP could also support QoS. The QoS is provided with an Integrated Services (IntServ) framework using Resource Reservation Protocol (RSVP) [35] or with a

Differentiated Services (DiffServ) [24] framework by utilizing DiffServ codes (8-bit field in the IP header). When routed over the Internet, IP packets are aged by the routers so that they cannot indefinitely circle in the network. IP packets can be fragmented if their size is larger than the maximum allowed on a given link. Version 4 of IP does not provide security features, such as authentication, confidentially, and integrity. These features could be provided by a layer above IP.

IPSec is a protocol that provides security features to IP packets. In order to have a secure communication between two parties, a security association must be established first. The security association means that both users share a secret key that can be used for encryption and decryption of the messages exchanged. IPSec consists of two protocols: Authentication Header (AH), described in [26], and Encapsulating Security Payload (ESP), described in [27]. The IP AH enables authentication and integrity checking, while IP ESP adds confidentially.

The most significant limitation of IP version 4 is the small, 32-bit field address space. IP version 6 [36] offers 128-bit address fields and other improvements such as simplified header fields and built-in support for security; that is, IPSec is its integral part. The packet fragmentation is not allowed in IP version 6. Note that the address space limitation of IP version 4 can be to some degree mitigated with the use of private IP addresses and Network Address Translators (NAT).

2.6.3 Internet Key Exchange

The IKE [25] is an IETF protocol for secure key distribution. It is based on the Diffie-Hellman exchange procedure. The procedure assumes that two sides, denoted as α and β, use of a common large prime number, p, and the generator number $g < p$. The procedure consists of the following:

Exchanging the public key. Both sides generate random numbers that they keep private. From a generated value, each side computes a public number and transmits the number to the corresponding party. For example, side α generates a secret number x and then computes the public value as $X = g^x$ modulo p. Similarly, side β generates y and computes $Y = g^y$ modulo p. The public numbers are exchanged, and the private numbers are kept secret.

Computing the secret key. On reception of the public number from the corresponding party, each side computes the shared secret number, referred to as the shared secret key. Side α, with the private number x, computes the shared secret key, K, as $K = Y^x = (g^y)^x = g^{yx}$. Similarly, side β computes the same value as $K = X^y = (g^x)^y = g^{xy}$.

The Diffie-Hellman key exchange has a serious weakness. Even though it enables the establishment of the secret keys between the two parties, the procedure does not provide the means for users to authenticate each other. To address this issue, IKE mandates authentication in addition to the Diffie-Hellman key exchange. For this purpose, certified digital signatures, public key algorithms, and preshared secrets could be used.

2.6.4 Mobile Internet Protocol

Mobile IP version 4 is described in [13]. It is a network layer protocol that enables a node to change its point of attachment to the network without having to change its IP address. Mobile IP introduces a distinction between a home address and a care-of address (CoA). The home address is a permanent address of a node, and the care-of address is a temporary address that is topologically correct with the node's current location. Three types of entities are defined:

> **Mobile node (MN).** This is a node that can change its point of attachment to the Internet while maintaining the same IP address. In this book, a mobile node is the equivalent of a mobile station. In general, a mobile node does not have to be a wireless device.

> **Home agent (HA).** This entity is a router that interfaces to a mobile node's home link. The mobile node keeps the home agent informed of its current care-of address. The home agent intercepts packets addressed to the node's home address, and it forwards or tunnels the packets to the care-of address.

> **Foreign agent (FA).** This entity is a router that interfaces to a mobile node's foreign link. The care-of address typically corresponds to the foreign agent's own address.[9] In that case, the foreign agent detunnels packets arriving from the home agent and forwards them to the mobile node. In PCN, the function of the foreign agent is provided by PDSN.

In this section, we explain the mobile IP operation with corresponding extensions as described in [3]. On successful attachment to a new link, the mobile node listens for mobile IP agent advertisement messages that are periodically broadcast over that link. In PCN, the link between the mobile node and the foreign agent (PDSN) is established through PPP. The authentication and authorization does not have to be performed during PPP negotiation, because it must be performed during mobile IP registration. For authentication purposes, the agent advertisement message sent by the foreign agent contains the MN-FA challenge extension [33]. This extension allows a mobile node to authenticate itself to the foreign agent. After receiving the agent advertisement message, the mobile node responds with the mobile IP registration request message. For authentication and authorization purposes, the registration message contains an MN-NAI extension (contains user name) [34], an MN-FA challenge extension, and an MN-AAA authentication extension [33]. The mobile node is then authorized for service by the AAA server.

The main purpose of the mobile IP registration is for the mobile node to register its care-of address with the home agent. After authorizing the user to access the network, the foreign agent relays the registration message to the home agent. The mobile mode always authenticates itself to the home agent using an MN-HA authentication extension. The MN-HA authentication extension is a mandatory part of the mobile IP registration messages and, for this purpose, the mobile

---

9.   The care-of address may also be collocated with the mobile node. This method is not attractive for wireless links because of the extra overhead associated with tunneling.

node and home agent use a statically configured secret key. If the foreign agent and the home agent have established security association, the mobile IP registration messages could also contain an HA-FA authentication extension. After a successful registration, a tunnel is created between the foreign agent and the home agent. A brief overview of the tunneling protocols is given in Section 2.6.5. When the home agent and the foreign agent have established a secure association, the tunnel between the two could provide secure communications through IPSec.

The routing in mobile IP is not simple, unicast IP routing. The IP packets originating at the corresponding node and destined to the mobile node are first intercepted by the home agent. The home agent keeps track of the location of the mobile node and if the mobile node is in the home network, the packets are simply forwarded to it. However, if the mobile node is in a foreign network, the packets are tunneled to the mobiles node's care-of address, commonly collocated with the foreign agent. The foreign agent detunnels the packets and forwards them to the mobile node. If the mobile node wants to send data, it simply forwards packets to the foreign agent, which now acts as a default router. The foreign agent forwards the packets to the destination, using unicast IP routing. This routing scheme is referred to as *triangular routing for mobile IP* and is illustrated in Figure 2.24. Note, however, that the IP packets could also be reverse-tunneled to the home agent, de-tunneled by the home agent and only then forwarded to the server by unicast IP routing. The purpose of reverse tunneling [37] is to prevent topologically incorrect source IP addresses that could be filtered by some ISPs.

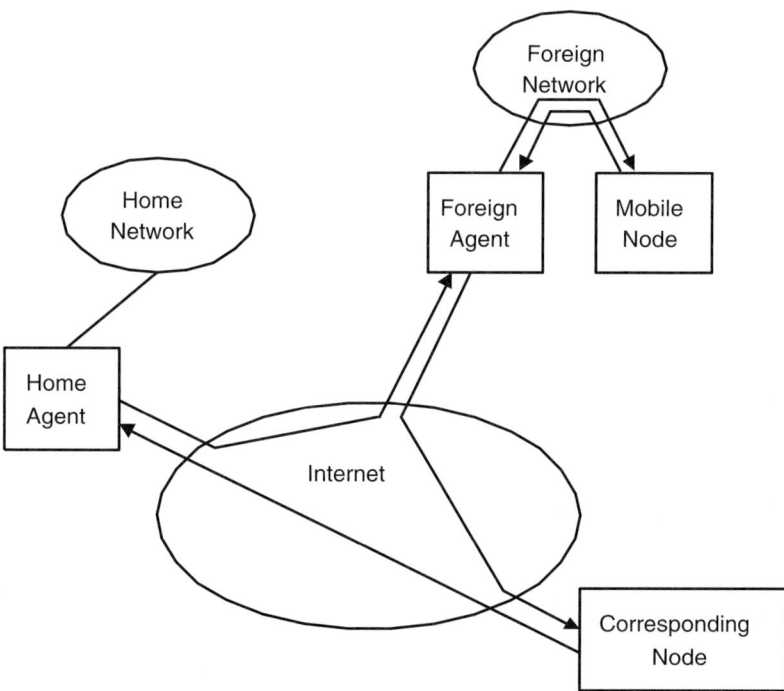

Figure 2.24 Triangular routing for mobile IP.

2.6.5 Tunneling Protocols

A tunnel is a virtual link that connects two points, referred to as the tunnel entry and the tunnel exit points. Packets enter the tunnel at the entry point and leave the exit point unchanged. In mobile IP, when the packets are sent by the corresponding node toward the mobile node, the tunnel entry point is the home agent and the tunnel exit point is the foreign agent. For reverse tunneling, the tunnel entry point is the foreign agent and the tunnel exit point is the home agent. The tunneling is achieved with the encapsulation. Three encapsulation methods are defined: IP in IP encapsulation [38], minimal IP in IP encapsulation [39], and GRE [18].

2.6.5.1   IP in IP Encapsulation

The IP in IP encapsulation works by simply adding an outer header to the IP packets that need to be tunneled. In mobile IP, the outer IP header contains the addresses of the home agent and foreign agent, while the inner IP header contains the IP addresses of the mobile node and the corresponding node. IPSec may be used with IP in IP encapsulation. When used, it provides security features for the tunneled packets.

2.6.5.2   Minimal IP in IP Encapsulation

The minimal IP in IP encapsulation is similar to IP in IP encapsulation. The differences are that the inner IP header is compressed and it is typically 12 bytes long. The outer IP header is uncompressed and it consists of a full 20 bytes. The disadvantage of the minimal encapsulation is that it does not allow fragmentation of IP packets.

2.6.5.3   Generic Routing Encapsulation

The advantage of GRE is that it allows encapsulation of a protocol data unit (PDU) of a given protocol into the payload of another protocol, not necessarily IP into IP. An IP packet encapsulated by the GRE method consists of an outer IP header, a GRE header, and an inner IP header. Unlike the other encapsulation methods, GRE prevents recursive encapsulation that occurs when tunnels are improperly set and several tunnels form a loop.

2.6.6 User Datagram Protocol

The UDP [17] is an unreliable and connectionless transport layer protocol that provides only multiplexing and optionally error-checking service. The multiplexing service is based on port numbers. Each communication stream is uniquely identified with a pair of IP addresses and port numbers. A port number identifies an upper-layer protocol. A combination of an IP address and a port number is commonly called a socket. The UDP header contains both the source and destination port numbers. An optional error-check service protects the header and payload data.

2.6.7 Transmission Control Protocol

The TCP [16] is a reliable, positive acknowledgement-based, connection-oriented transport layer protocol. It provides in sequence delivery of byte-stream data, multiplexing, and mandatory error-checking service for the header and payload data.

TCP implements a flow-control mechanism to prevent buffer overflow at the receiving side and congestion in the network. The flow control employs a sliding window mechanism, where the number of unacknowledged bytes at the transmitter side is controlled by the receiver-advertised window. The amount of bytes the transmitter can send to the receiver is also bounded by the congestion window maintained at the transmitting side. The purpose of the congestion window is to avoid congestion on the network. The congestion control mechanism dynamically adjusts the window size based on perceived network state. The TCP operation is explained in more detail in Chapter 9.

2.6.8 Remote Authentication Dial In User Service

The RADIUS [28] is an application layer protocol that exchanges authentication, authorization, and configuration information between the NAS (in IS-835 PDSN/FA) and the HAAA server, possibly through the VAAA and BAAA. The NAS operates as a RADIUS client, while the HAAA server acts as a RADIUS server. The RADIUS servers, such as VAAA or BAAA, act as proxy clients to other RADIUS servers, such as HAAA.

In order to authenticate and authorize a user, RADIUS works in conjunction with PPP and, in the case of mobile IP, in conjunction with the associated extensions explained in Section 2.6.4. In particular, the user authentication and authorization is performed as follows: The NAS issues a challenge to the mobile station (mobile node) and after it gets the response with the authentication information, the NAS (PDSN/FA) composes the RADIUS access request message and sends it to the RADUIS server. The RADIUS server or HAAA authenticates and authorizes the user and responds with the RADIUS Access Accept Message. After this point, the user is authorized to access the network. The RADIUS protocol is extended in [29] to enable transfer of accounting information from the NAS to HAAA. Operation with IP version 6 is described in [40]. RADIUS protocol could also be used for key distribution.

REFERENCES

[1] "Internet Protocol," *IETF RFC 791*, September 1981.

[2] "TIA/EIA-IS-41-D Cellular Radio Communications Intersystem Operations," *Telecommunications Industry Association*, 1997.

[3] "TIA/EIA-IS-835-C cdma2000 Wireless Network Standard," *Telecommunications Industry Association*, 2003.

[4] "TIA/EIA-IS-2001 Inter-operability Specification (IOS) for cdma2000 Access Network Interfaces," *Telecommunications Industry Association*, 2002.

[5] "TIA/EIA-IS-2000.1 Introduction to cdma2000 Spread Spectrum Systems Release C," *Telecommunications Industry Association*, 2002.

[6] "TIA/EIA-IS-2000.2 Physical Layer Standard for cdma2000 Spread Spectrum Systems Release C," *Telecommunications Industry Association,* 2002.

[7] "TIA/EIA-IS-2000.3 Medium Access Control (MAC) Standard for cdma2000 Spread Spectrum Systems Release C," *Telecommunications Industry Association*, 2002.

[8] "TIA/EIA-IS-2000.4 Signaling Link Access Control (LAC) Standard for cdma2000 Spread Spectrum Systems Release C," *Telecommunications Industry Association*, 2002.

[9] "TIA/EIA-IS-2000.5 Upper Layer (Layer 3) Signaling Standard for cdma2000 Spread Spectrum Systems Release C," *Telecommunications Industry Association*, 2002.

[10] "TIA/EAI-IS-856 cdma2000 High Rate Packet Data Air Interface Specification," *Telecommunications Industry Association*, 2002.

[11] "TIA/EIA-IS-707-B Data Service Options for Spread Spectrum Systems," *Telecommunications Industry Association*, 2004.

[12] Romkey, J., "Serial Link IP," *IETF RFC 1055*, June 1998.

[13] Perkins, C., "IPv4 Mobility," *IETF RFC 2002*, May 1995.

[14] "TIA/EIA-IS-828-A BTS-BSC Inter-operability (A_{bis} Interface)," *Telecommunications Industry Association*, 2001.

[15] McDysan, D., and D. Spohn, *ATM Theory and Applications.* McGraw-Hill Series in Computer Communications, 1998.

[16] "Transmission Control Protocol," *IETF RFC 793*, September 1981.

[17] Postel, J., "User Datagram Protocol," *IETF RFC 768*, August 1980.

[18] Farinacci, D., S. Hanks, D. Meyer, and P. Traina, "Generic Routing Encapsulation (GRE)," *IETF RFC 2784*, March 1999.

[19] Russell, T., *Signaling System #7,* McGraw-Hill, 1998.

[20] Gallagher, Michael, and R. Snyder, *Mobile Telecommunications Networking,* McGraw-Hill, 1997.

[21] "TIA/EIA-IS-93-B Wireless Telecommunications: A_i – D_i Interfaces Standard," *Telecommunications Industry Association*, 2001.

[22] Simpson, W., "The Point-to-Point Protocol (PPP)," *IETF RFC 1661*, July 1994.

[23] Simpson, W., "PPP in HDLC-like Framing," *IETF RFC 1662*, July 1994.

[24] Blake, S., D. Black, M. Carlson, E. Davies, Z. Wang, and W. Weiss, "An Architecture for Differentiated Services," *IETF RFC 2460*, December 1998.

[25] Harkins, D. and D. Carrel, "The Internet Key Exchange (IKE)," *IETF RFC 2409*, November 1998.

[26] Kent, S., and R. Atkinson, "IP Authentication Header," *IETF RFC 2402*, November 1998.

[27] Kent, S., and R. Atkinson, "IP Encapsulating Security Payload," *IETF RFC 2406*, November 1998.

[28] Rigney, C., C. Willens, A. Rubens, and W. Simson, "Remote Dial In User Service (RADIUS)," *IETF RFC 2865*, June 2000.

[29] Rigney, C., "RADIUS Accounting," *IETF RFC 2866*, June 2000.

[30] Simpson, W., "PPP Challenge Handshake Authentication Protocol (CHAP)," *IETF RFC 1994,* August 1994.

[31] Adoba, B., and M. Beadles, "Network Access Identifier," *IETF RFC 2486*, January 1999.

[32] McGregor, G., "The PPP Internet Protocol Control Protocol," *IETF RFC 1332*, May 1992.

[33] Calhoun, P., and C. Perkins, "Mobile IPv4 Challenge/Response Extensions," *IETF RFC 3012*, November 2000.

[34] Calhoun, P., and C. Perkins, "Mobile IP Network Access Identifier Extension for IPv4," *IETF RFC 2794*, March 2000.

[35] Braden, R., L. Zhang, S. Berson, S. Herzog, and S. Jamin, "Resource Reservation Protocol (RSVP) Version 1 Functional Specification," *IETF RFC 2205*, September 1997.

[36] Deering, S., and N. Hinden, "Internet Protocol, Version 6 (IPv6) Specification," *IETF RFC 2460*, December 1998.

[37] Montenegro, G., "Reverse Tunneling for Mobile IP," *IETF RFC 3024*, January 2001.

[38] Perkins, C., "IP Encapsulation within IP," *IETF RFC 2003*, October 1995.

[39] Perkins, C., "Minimal Encapsulation within IP," *IETF RFC 2004*, October 1995.

[40] Adoba, B., G. Zorn, and D. Mitton, "RADIUS and IPv6," *IETF RFC 3162*, August 2001.

CHAPTER 3

Applications and Services

The cdma2000 wireless networks provide users with a wide range of applications, such as voice, Web browsing, and email. As illustrated in Chapter 2, "Network Architecture," different applications may require different services. For example, a user requests circuit-switched voice service by pressing the send button after dialing the number of the calling party. After end-to-end service is established, the caller and the calling party begin using the voice application. Activating a packet data service for a Web browsing application, however, requires activation of a different set of functions that are not invoked when circuit-switched voice service is established.

Due to the limitation of the available spectrum, it is important that the frequency bandwidth is efficiently utilized. It is therefore necessary that the provided service closely matches the needs of the application. For example, email application requires reliable radio connection, but it is delay-tolerant. Hence, the efficiency of the air interface can be maximized at the expense of delay. For that purpose, Radio Link Protocol (RLP)[1] and link adaptation techniques are commonly employed. Voice application is delay-intolerant, so RLP and link adaptation techniques are not acceptable. However, voice application is to some extent error-tolerant, and the air efficiency can be increased at the expense of higher error probability.

The availability of a particular application is not only what users care about. Quality of Service (QoS) is also important, because QoS directly impacts user perception about the usefulness of the application.

---

1. RLP is described in more detail in Chapter 5, "IS-2000 Layer 2 (Media and Signaling Link Access Control Layers)".

3.1 Applications

The cdma2000 wireless networks allow for a wide range of applications. The requirements that they impose to the service range from real-time to non-real time and from error-tolerant to error-intolerant. The service provided to an application takes into account its general traffic characteristics. To understand the rationale for the available services, in this section we describe the traffic characteristics of the most common applications.

3.1.1 Conversational Voice

Voice application requires real-time service because conversational voice cannot tolerate delay. If the voice traffic is delayed by more than 200 ms, it can produce an annoying effect of users speaking simultaneously.

Voice call arrivals can be modeled as a Poisson process. Denote the arrival rate with λ_v. The call duration fits well into an exponential distribution with mean holding time of $1/\mu_v$. For wireless cellular networks, a typical mean holding time is between 1 and 2 minutes.

Speech waveforms contain a high degree of redundancy. The cdma2000 networks exploit that redundancy and perform speech coding before transmitting the encoded signal over the air. The coding process significantly reduces the data rate necessary to carry speech signal, which directly increases system capacity or the number of users that can be simultaneously served over a given bandwidth. The rate at the output of the speech coder is variable, and it depends on many factors such as voice activity and nature of speech. Even after compression, voice traffic can tolerate occasional errors. Typically, a frame error rate (FER) of 1% to 3% is acceptable. The voice traffic characteristics are summarized in Table 3.1.

Table 3.1   Voice Traffic Characteristics

Arrival rate, λ_v	Call duration, $1/\mu_v$
Several calls per user per day	60–120 s

3.1.2 Internet Applications

Many applications use the Internet. Most were originally designed for high-bandwidth wired networks and each has different traffic characteristics. What is common for most of them is that they are much more susceptible to errors than are voice applications. However, they do not have as stringent delay requirement as voice.

The traffic characteristics differ from one application to another. In the text that follows, we provide the traffic models, adopted in [1], for a Web browsing session using HyperText Transfer Protocol (HTTP) [2], file download/upload using File Transfer Protocol (FTP) [3], Wireless Application Protocol (WAP) traffic [4], and video steaming [5]. These traffic models are extensively used in the industry for the air interface system-level simulations. They are relatively simple, yet they reasonably accurately represent the corresponding traffic. Note that quantitative values for some parameters used to represent the traffic depend on many factors. Some of them are user-dependent, and some are external network-dependent. Hence, the quantitative values reported in this section should be considered as exemplary.

3.1.2.1 Web Browsing

A packet trace of a typical Web browsing session is shown in Figure 3.1. As can be seen from the figure, the data arrival is bursty. The session can be divided into an interval in which the content of the Web page is being downloaded—referred to as packet data call period—and the user reading time. A typical Web page is written in eXtensible HyperText Markup Language (XHTML) [6] and consists of a main page and a bunch of embedded objects.

By typing a Universal Resource Location (URL) address and sending the request to the server, a user triggers a download of the main page. On reception of the main object, the client software, residing at the mobile station, parses the main page for additional references regarding the embedded image files. Each embedded object generates a new request. The HTTP server, on receiving each of the requests, sends the embedded objects to the client. Web browsing applications are to some degree sensitive to the delay. What matters is the time from the instant the user requests a page download until the page with all the embedded objects is downloaded. The download time varies depending on the radio conditions, system load, and the size of a page. It could be as low as few seconds, but it could take 10 seconds or more. In Figure 3.2 we illustrate a typical call flow between a mobile station (MS) and the HTTP server during a Web browsing packet data call.

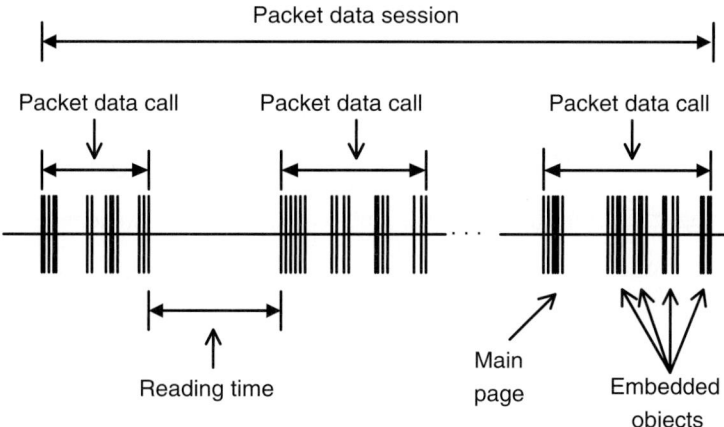

Figure 3.1 Packet trace of a typical Web browsing session.

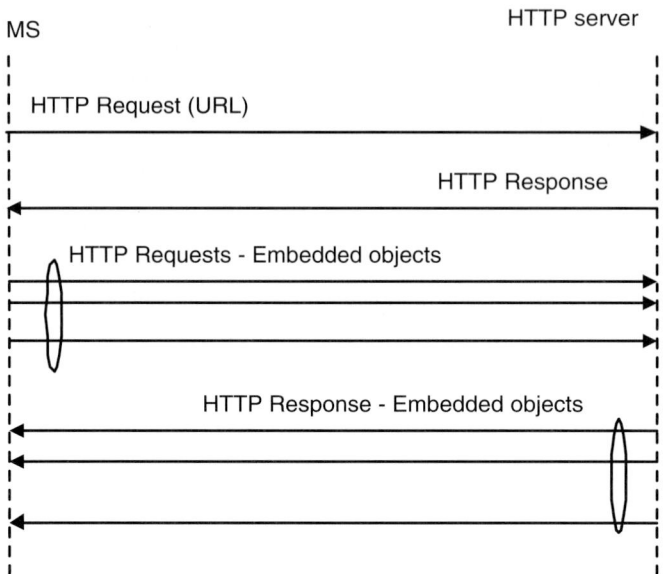

Figure 3.2   HTTP call flow.

The traffic model for the Web browsing session, as described in [1], can be characterized in terms of the following parameters:

- Size of the main object
- Size of the embedded objects in the Web page
- Number of embedded objects in the Web page
- Reading time
- Parsing time for the main page
- Size of the HTTP request

The main object size can be modeled as a truncated lognormal random variable, X, as

$$f_X(x) = \frac{1}{\sqrt{2\pi}\sigma x} \exp\left[\frac{-(\ln x - \mu)^2}{2\sigma^2}\right], x \geq 0 \qquad (3.1)$$

with the parameters $\sigma = \sigma_m = 1.37$, $\mu = \mu_m = 8.35$, the minimum size of the main page of 100 bytes, and the maximum size of 2 Mbytes. The size of the embedded objects can also be modeled as a truncated lognormal variable, with the parameters $\sigma = \sigma_e = 2.67$, $\mu = \mu_e = 6.17$, the minimum main page size of 50 bytes, and the maximum size of 2 Mbytes. The number of embedded objects per main page, N_e, can be modeled as follows. Generate a truncated Pareto random variable, Y:

$$f_Y(y) = \frac{\alpha k^{\alpha}}{y^{\alpha+1}}, k \le y < m$$

$$f_Y(y) = \left(k/m\right)^{\alpha}, y = m$$

(3.2)

with the parameters $\alpha = \alpha_{Ne} = 1.1$, $k = k_{Ne} = 2$, and the maximum $m = m_{Ne} = 55$. Subtract k from Y; that is, $N_e = Y - k$. The reading time and parsing time can both be modeled as exponentially distributed variable, Z:

$$f_Z(z) = \lambda e^{-\lambda z}, z \ge 0$$

(3.3)

For the reading time, that arrival rate is $\lambda = \lambda_r = 0.033$, while for the parsing time, it is $\lambda = \lambda_p = 7.69$. Note that the reading time is a heavily user-dependent parameter, and it can vary significantly. The HTTP requests can be modeled as fixed size, 350-byte packets. Table 3.2 summarizes the traffic model for a Web browsing session.

Table 3.2   HTTP Traffic Model Characteristics

Component	Distribution	Parameters
Main object size	Truncated lognormal	Mean = 10.71 kbytes Std. dev. = 25.032 kbytes
Embedded object size	Truncated lognormal	Mean = 7.758 kbytes Std. dev. = 126.168 kbytes
Number of embedded objects	Truncated Pareto	Mean = 5.64 Max. = 53
Reading time	Exponential	Mean = 30 s
Parsing time	Exponential	Mean = 0.13 s
Request size	Deterministic*	350 bytes

* In practice, HTTP request can significantly vary in size.

3.1.2.2 File Download/Upload

The traffic characteristics of a file download or upload are quite different from Web browsing traffic characteristics. A commonly used protocol for file transfer is FTP. The length of the file can vary greatly. The FTP applications are even less sensitive to delay than are Web browsing applications. A typical packet trace of an FTP session with multiple file downloads is shown in Figure 3.3.

The FTP download traffic model, as described in [1], can be mostly characterized in terms of the file size parameter. The file size is modeled as a truncated lognormal random variable, with the parameters $\sigma = \sigma_{fd} = 0.35$, $\mu = \mu_{fd} = 14.45$, and the maximum size of 5 Mbytes. The reading time could be modeled as an exponentially distributed random variable. Note that, as for the Web browsing session, the reading time is a heavily user-dependent parameter that can vary significantly. The arrival rate of $\lambda = \lambda_r = 0.006$ is suggested in [1], which corresponds to the mean reading time of 180 s. [1] summarizes the traffic model characteristics for an FTP download session.

Table 3.3 FTP Download Traffic Model Characteristics

Component	Distribution	Parameters
File size	Truncated lognormal	Mean = 2 Mbytes** Std. dev. = 0.722 Mbytes

** The file size is highly dependent on the application.

The FTP protocol is also commonly used for file upload. The FTP upload traffic model, as described in [1], can also be characterized in terms of the file size. The file size is modeled as a truncated lognormal random variable, with the parameters $\sigma = \sigma_{fu} = 2.09$, $\mu = \mu_{fu} = 0.93$, the minimum size of 500 bytes, and the maximum size of 500 kbytes. The arrival rate is typically lower for the file upload. The traffic model characteristics are summarized in Table 3.4.

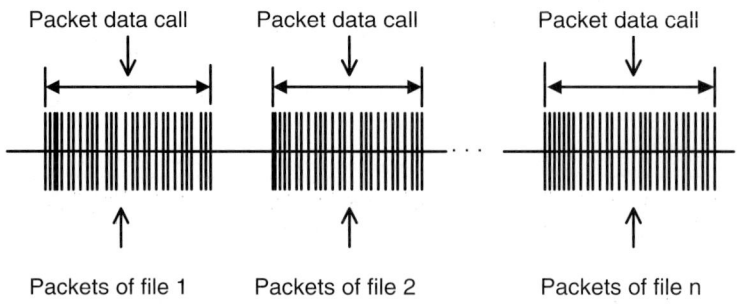

Figure 3.3 Packet trace of a typical FTP application session.

Table 3.4 FTP Upload Traffic Model Characteristics

Component	Distribution	Parameters
File size	Truncated lognormal	Mean = 22.7 kbytes Std. dev. = 200.3 kbytes

3.1.2.3 Wireless Application Protocol

Mobile wireless devices typically have small screens because the bandwidth in wireless networks is limited. To address this concern, WAP was developed. The initial idea was to develop a completely new protocol stack optimized for low-bandwidth, high-latency wireless links. However, the idea recently evolved to include the use of the existing Internet protocols with the wireless profile specifications that are fully interoperable with standard implementations.

WAP applications are written in XHTML Mobile Profile (XHTMLMP) markup language [7]. The XHTMLMP provides appropriate presentation for wireless devices. However, the documents written in core XHTML are completely operable on XHTMLMP browsers. The XHTML is designed to be extensible, permitting additional language elements to be added as needed. WAP does not require a WAP proxy, because the communications between the client and the server can be conducted using HTTP. However, deploying a WAP proxy can optimize the communications process and may offer mobile service enhancements. The concept behind the WAP proxy model is shown in Figure 3.4.

The data arrival is bursty. The WAP traffic model, as described in [1], can be characterized with the following parameters:

- Size of WAP request
- Size of WAP object

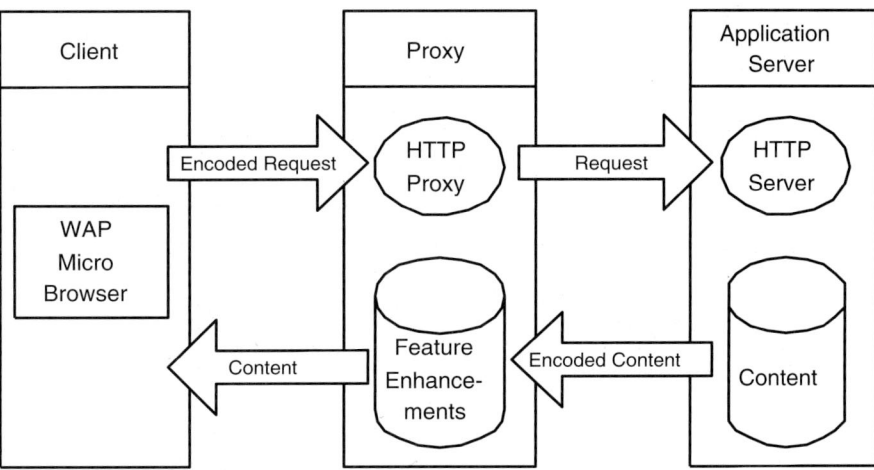

Figure 3.4 WAP proxy model.

- Number of objects per response
- Interarrival time between objects
- WAP gateway response time
- Reading time

A typical call flow for a WAP packet data call with a WAP proxy is illustrated in Figure 3.5. The WAP request can be modeled as deterministic in size, equal to 76 bytes. The size of each WAP object follows truncated Pareto distribution with parameters $\alpha = \alpha_{Ne} = 1.1$, $k = k_{Ne} = 71.1$, and the maximum object size $m = m_{Ne} = 1400$ bytes. The number of objects per request, N_e, can be modeled as a geometric random variable, G:

$$f_G(g) = p^g (1-p), \ p = 1/2. \tag{3.4}$$

incremented by 1; that is, $N_e = G + 1$. The interarrival time between objects, WAP gateway response time, and reading time are all modeled as exponential random variables with $\lambda = \lambda_i = 0.625$, $\lambda = \lambda_p = 0.4$, and $\lambda = \lambda_r = 0.182$, respectively. Note that WAP gateway response time is the time elapsed between the time instance the WAP gateway receives the request and the time instance the first object is transmitted to the mobile station. The reading time is the time elapsed between the time instance when the last object is downloaded and the new WAP request. Table 3.5 summarizes the WAP traffic characteristics.

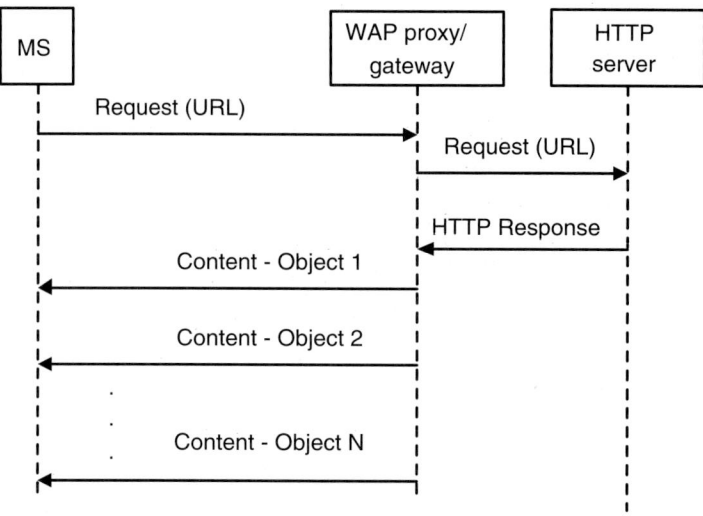

Figure 3.5   WAP call flow.

Table 3.5   WAP Traffic Model Characteristics

Component	Distribution	Parameters
Object size	Truncated Pareto	Mean = 256 bytes
Number of objects	Geometric plus 1	Mean = 2 plus offset of 1
Interarrival time	Exponential	Mean = 1.6 s
Gateway response time	Exponential	Mean = 2.5 s
Reading time	Exponential	Mean = 5.5 s
Request size	Deterministic	76 bytes

Note that WAP allows content to be sent or "pushed" to devices by the server-based application via a push proxy. This function allows servers to send information without requiring the devices to poll application servers for new information. This is important, because polling activities may consume considerable resources of wireless networks. The described traffic model does not apply to WAP push service. The WAP push service is used for multimedia messaging service (MMS) to deliver multimedia messages[2] to the mobile station.

3.1.2.4   *Video Streaming*

Video steaming applications enable full-motion video. For efficient transport of the air interface, the video stream is highly compressed [5]. Video streaming is a delay jitter-sensitive application. Packets must be played in order at the receiver. If packets arrive out of order, a de-jitter buffer must be implemented to reorder packets. When video streaming is used for two-way communications, the de-jitter buffer is limited in size, and more care must be taken to prevent delay and delay jitter in the network.

The video streaming session can be described with the model illustrated in Figure 3.6. The traffic model has been described in [1] with the following parameters:

- Interarrival time between frames
- Number of packets in the frame
- Interarrival time between packets
- Packet size

---

2.   MMS is an improvement over short message service (SMS), discussed later in this chapter. While SMS offers only alphanumerical text to be deliver to and from a mobile station, MMS offers rich multimedia content, including image files and short video clips.

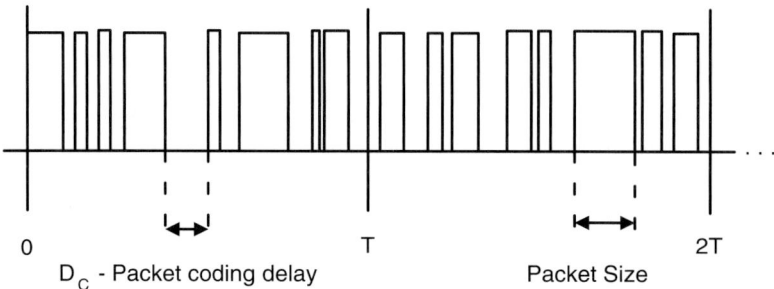

D_C - Packet coding delay   Packet Size

Figure 3.6   Video streaming traffic model.

Video traffic is divided into frames. Each frame arrives at a regular interval determined by the number of frames per seconds. For wireless devices, a good model is $T = 100$ ms, or 10 frames per second. Each frame is decomposed into a fixed number of packets. For wireless devices, a good example is 8 packets. The size of these packets is distributed as a truncated Pareto, with the parameters $\alpha = \alpha_s = 1.2$, $k = k_s = 20$ bytes, and the maximum packet size of $m = m_s = 125$ bytes. The video encoder introduces delay jitter between the packets of a given frame. These intervals are also modeled by a truncated Pareto distribution, with $\alpha = \alpha_i = 1.2$, $k = k_i = 2.5$ ms, and the maximum time between packets $m = m_s = 12.5$ ms. The average video streaming rate is 32 kbps. Table 3.6 summarizes the video streaming traffic model.

Table 3.6   Video Traffic Model

Component	Distribution	Parameters
Interarrival time between frames	Deterministic	T = 100 ms
Number of packets in the frame	Deterministic	8
Interarrival time between packets	Truncated Pareto	Mean = 6 ms
Packet size	Truncated Pareto	Mean = 50 bytes

3.2 Services

Before the application running on the mobile station can exchange user data with the corresponding node, an end-to-end service between the two points must be established. Setting up an end-to-end service can be divided into the following tasks. The first is the setup of the access bearer service. It consists of configuring necessary radio channels and the appropriate radio access network (RAN) and core network connections. After the access bearer service is set up, an external bearer service must be established. For example, for circuit-switched voice, setting up an external bearer service means reserving resources along the path of a voice call in the public switched telephone network (PSTN). However, for Web browsing, setting up an external bearer service means connecting to the Internet or configuring the mobile station with a proper Internet Protocol (IP) address. After access and external bearer service is established, end-to-end service is built on the top of them. Voice and packet data service setups are illustrated on exemplary call flow diagrams in Chapter 2.

The reference model for end-to-end and underlying services is shown in Figure 3.7. It shows two endpoints, access and external services and the access gateway. The access gateway is a point in the network where mapping between the access bearer service and the external bearer service takes place. Its purpose is to hide anything access-specific to the surrounding networks. For the circuit-switched voice service, the access gateway is the mobile switching center (MSC). Even though MSC is not usually referred to as the access gateway, it can conceptually be seen as the access gateway. In case of packet-switched data, the access gateway is the packet data serving node (PDSN).

As illustrated in Figure 3.7, the access bearer service can be further decomposed into radio bearer service and interoperability standard (IOS) service[3] [8]. The radio bearer service is between the mobile station and RAN, while IOS service is between RAN and the access gateway. In the following sections we do not discuss the IOS service. Rather, we focus on the radio bearer service and explain the access bearer service as a whole.

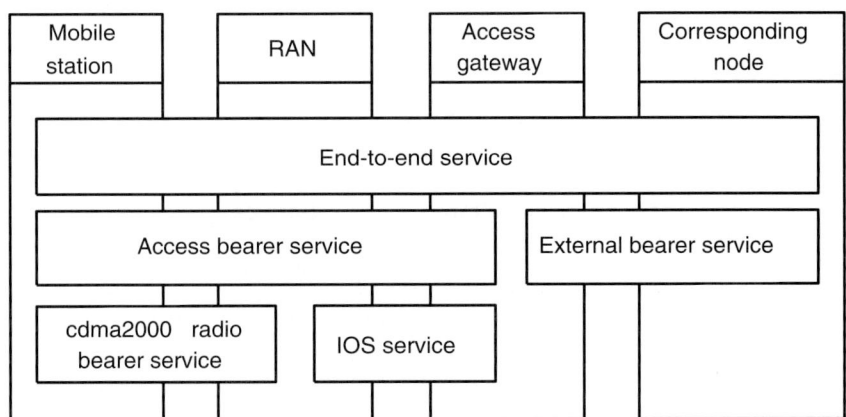

Figure 3.7 Reference model for services.

---

3.   Refers to service provided by the A interface (see Chapter 2 for a description of the A reference point).

3.2.1 Access Bearer Services

The access bearer services in cdma2000 networks connect a mobile station with the MSC or PDSN. For example, circuit-switched voice is switched though the MSC, but Voice over IP (VoIP) service and packet data service for Web browsing are switched through the PDSN. A mobile station can support multiple simultaneous services.

The services are activated through service options (SOs). The service option is a parameter that describes the type of service that is requested. Service details are characterized by the specific service option in use. The service option number identifies the treatment given to the corresponding traffic flow. For example, if the service option number indicates circuit-switched voice, the RAN must perform appropriate speech coding before it forwards the data to the mobile station. Each traffic flow is identified by a connection reference number. A service may be associated with a service reference identifier. The service reference identifier is transmitted over the air with the actual data and enables the mobile station and the base station to adequately multiplex and demultiplex traffic when multiple simultaneous services are supported. The service parameters may be negotiated when the service is established. This process is commonly called the *service negotiation*.

3.3 Circuit-Switched Service Options

The cdma2000 standard provides a large number of circuit-switched services. The most commonly employed are voice and short message service (SMS). The voice service is assigned multiple service option numbers. The numbers vary depending on the speech coder in use. The most spectrally efficient one is selectable mode vocoder (SMV), SO 56 [9]. The SMS is provided with two service options, depending on the data rate of the traffic channel. SO 6 [11] enables speeds of up to 9.6 kbps using Rate Set 1 traffic channel configuration, while SO 14 supports up to 14.4 kbps using Rate Set 2. Table 3.7 summarizes the service options for voice and SMS for Rate Set 1.

Table 3.7   Commonly Used Circuit Switched Service Options

Service Option Number	Service
56	Voice with SMV
6	SMS

3.3.1 Voice Service

The voice capacity of cdma2000 systems is inversely proportional to the average transmission data rate. Recognizing this fact, speech coders, commonly called vocoders, are used to compress voice traffic transmitted over the air. The stronger the compression, the higher the capacity of the system. Vocoders are specifically designed for voice signals, and they are not suitable for other analog signals. For circuit-switched voice service, voice compression is part of both mobile station and RAN service.

Voice arrives to MSC from the PSTN as pulse-coded modulated (PCM) traffic at the data rate of 64 kbps. The PCM system is capable of encoding an arbitrary random waveform whose maximum frequency component does not exceed one-half of the sampling rate [10]. However, an analysis of speech waveforms showed that there is significant redundancy in them. By exploiting the redundancy, it is possible to significantly reduce the data rate offered by PCM and only marginally reduce speech quality.

3.3.1.1   Voice Redundancies

Speech waveforms exhibit a high correlation from one sample to the next. Low amplitude values are more common than high. Moreover, at any particular time instant, certain sounds may be composed of only a few frequencies, in which case the waveform exhibits strong cycle-to-cycle correlation.

Human speech sounds are often categorized as *voiced* and *unvoiced*. Voiced sounds arise as a result of vibrations in the vocal cords. Each vibration allows a puff of air to flow from the lungs into the vocal tract. The interval between puffs of air exciting the vocal tract is referred as the pitch interval. Unvoiced sounds occur as a result of continuous air flowing from the lungs and the passing through a vocal tract constricted at some point to generate air turbulence. Generally speaking, voiced sounds arise in the generation of vowels. Unvoiced sounds correspond to consonants. A voiced sound exhibits a longer repetitive pattern corresponding to the duration of a pitch interval. A pitch interval lasts approximately from 2.5 to 20 ms. Usually, a single voice sound lasts for approximately 100 ms, and hence there may be as many as 40 pitch intervals in a single voice sound. Figure 3.8 illustrates typical time domain representations of voiced and unvoiced sounds. Note also that during telephone conversations, on average a party is usually active less than 40% of the call duration. Most inactivity however, occurs as a result of one person listening while the other is talking.

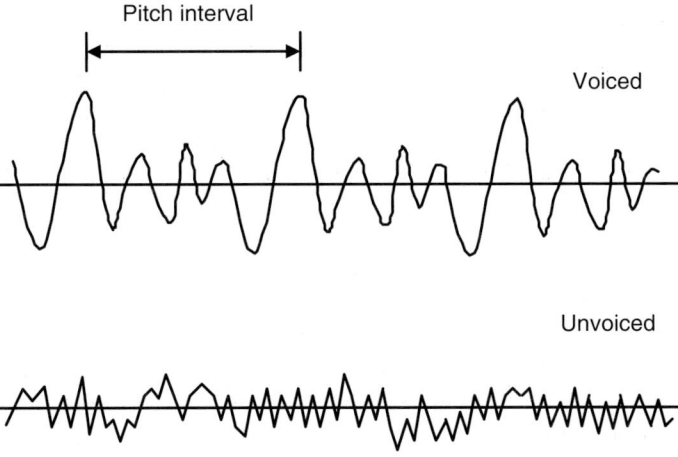

Figure 3.8   Voiced and unvoiced speech.

The above described redundancies focus on time domain representations of voice. In addition to time, there are frequency redundancies. They are, however, related. The vowels require most of the energy and occupy mainly the lower portion of the spectrum. The consonants carry much less energy and typically occupy higher frequencies. As a result, the higher portions of the voice spectrum have significantly reduced power levels. Consonants, however, carry most of the speech information and must not be neglected during the speech-coding process. Even over short periods of time, the spectral densities vary significantly. The frequency at which resonance occurs is called *formant*. Voiced speech typically contains three to four identifiable formants.

3.3.1.2   Selectable Mode Vocoder

The SMV is a variable rate vocoder. It can operate in several different modes, allowing the user to select the operating point that provides the desired quality versus average bit rate tradeoff. The SMV belongs to a class of code-excited linear predictive (CELP) coders. In the following paragraphs, we describe the features of a CELP coder.

The main component of a CELP coder is the linear predictive coder (LPC), which analyzes speech waveform in order to produce a time-varying model of the vocal tract excitation and the transfer function. A synthesizer in the receiver re-creates the speech by passing the specified excitation through a mathematical model of the vocal tract.

Rather than encoding the excitation on the sample-by-sample basis, CELP coders consider a block of samples at a time and use a codebook to generate inputs to the synthesizer filter. Each block is called *subframe*. The CELP coders use an analysis-by-synthesis search method to encode the signal. The vector of model parameters is obtained by synthesizing a signal for each of a set of such vectors and selecting a vector for which the synthesized signal most closely resembles a reference signal. The resemblance between the original and the trial signals is evaluated using perceptually relevant error criterion. The analysis-by-synthesis method is illustrated in Figure 3.9, and it is commonly called a closed-loop search. An open-loop search refers to the perceptually based decisions.

A pitch predictor is also an integral part of the CELP analysis-by-synthesis structure. The pitch predictor estimates the pitch delay and the pitch gain. For a conventional pitch predictor, the delay is constant within each frame and changes stepwise at the subframe boundary. The

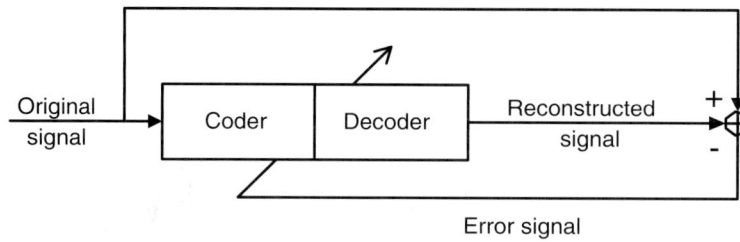

Figure 3.9   Analysis-by-synthesis method.

aggregate of the delay values at each subframe is called *delay contour*. Stepwise changes in the delay contour at the subframe boundaries can lead to a significant audible distortion. However, small, gradual changes in the pitch period are not disturbing. A method called generalized analysis-by-synthesis coding [12] provides for continuous delay contour. It consists of modifying the original signal by time warping and amplitude scaling, and it does not affect natural speech quality. The method is illustrated in Figure 3.10.

SMV Frame Structure. SMV produces one frame every 20 ms. A single frame may consist of several subframes. SMV is a variable rate vocoder with four distinct frame rates. The output rates are 8.55 kbps, 4 kbps, 2 kbps, and 0.8 kbps. The actual transmission rates over the physical channels are 9.6 kbps, 4.8 kbps, 2.4 kbps, and 1.2 kbps, respectively. The corresponding 20-ms frames are referred to as full-rate, half-rate, quarter-rate, and eighth-rate frames. The rate change is possible every 20 ms.

Figure 3.11 shows a block diagram of the SMV system. The preprocessing block includes high-pass filtering, noise suppression, and adaptive tilt compensation. The frame-processing block includes LPC analysis, open-loop pitch search, signal modification, and classification. Refer to [13] and [9] for more details on SMV. In this section, we provide only a brief overview.

SMV employs elaborate speech classification. Each frame is classified as either silence/background noise, stationary unvoiced, nonstationary unvoiced, onset, nonstationary voiced, and stationary voiced. Silence/background noise and stationary unvoiced frames can be represented using a spectrum and energy modulated noise. These frames might be coded with quarter- or eighth-rate coders. Half and full-rate frames can be of type 0 or 1. Frames with high-pitch prediction gain are declared type 1, while all others are declared type 0 frames.

Type 0 frames are divided into subframes. There are four subframes for full-rate and two for half-rate coding. A closed-loop pitch search is performed for each subframe, and the result of a pitch search is recorded in the adaptive codebook contribution. Pitch search is followed by excitation coding, or fixed-codebook search. The choice of each codebook vector and the quantization of the gains are completed for each subframe before the next subframe is processed.

Type 1 frames have stable pitch-prediction gains and pitch contour. Therefore, only a single pitch value is used for the whole type 1 frame. Since fewer bits are used for the representation of the pitch gains and lag, more bits are available for fixed-codebook coding. The subframe

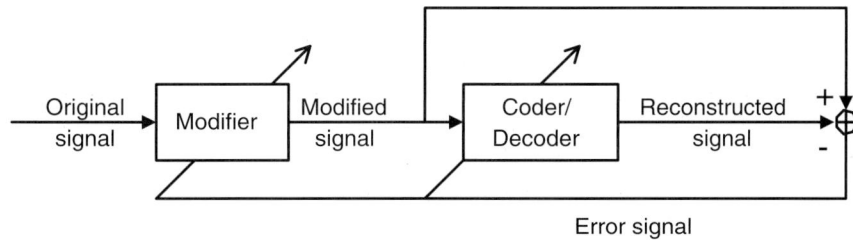

Figure 3.10 Generalized analysis-by-synthesis method.

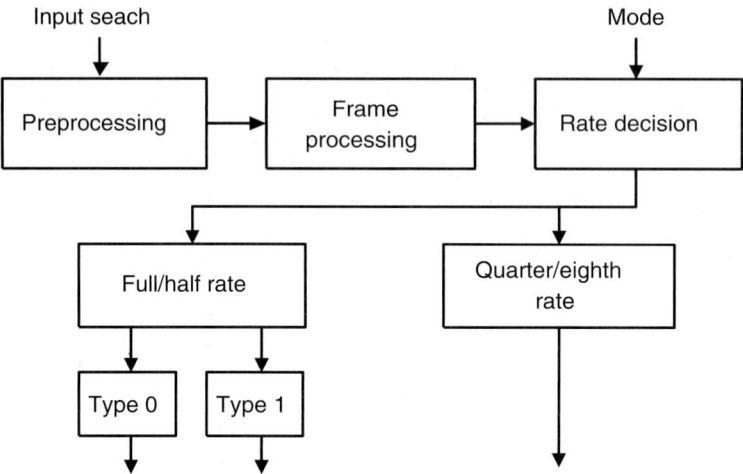

Figure 3.11 Block diagram of SMV system.

processing includes only the search for the fixed-codebook contribution. For full-rate type 1 frames, the number of subframes is four, and for half-rate frames, the number of subframes is three. Excitation coding for quarter-rate frames is also performed on the subframe basis. There are 10 subframes per frame. A pseudorandom noise is scaled appropriately and colored using a set of filters. Eighth-rate frames are not divided into subframes. A pseudorandom number-generator generates the excitation and a single gain value is applied to the entire frame.

Each coded frame also contains linear spectral frequencies (LSFs) codes. They are used to derive the coefficients of the LPC synthesis filter. On the transmitter side, at the beginning of the speech-encoding process, LSFs are derived from the LPC prediction coding coefficients. At the receiver side, the speech parameters are used to generate the speech signal. For full- and half-rate frames, the fixed-codebook and adaptive-codebook excitation is multiplied with the appropriate gains summed and passed through the LPC synthesis filter. Quarter- and eighth-rate frames are generated by passing a pseudorandom noise scaled by the appropriate gains though the LPC synthesis filter. Quarter-rate frame processing also includes a shaping filter. Pseudorandom noise is locally generated. Table 3.8 specifies the bit allocation for all frames produced by the SMV.

Table 3.8 Bit Allocation Table for SMV

Rate	Full		Half		Quarter	Eighth
Type	0	1	0	1		
LSFs	27	25	21	21	20	11
Energy					17	5
Shape					2	
Rate sanity bit	1	1			1	
Type bit	1	1	1	1		
Pitch	26	8	14	7		
Excitation	88	120	30	39		
Gains	28	16	14	12		
Total	171	171	80	80	40	16

Modes of Operation. There are three modes of operation defined for SMV: Mode 0, or premium mode; Mode 1, or standard mode; and Mode 2, or economy mode. Mode 0 gives highest average rate and voice quality. Mode 2 provides the lowest average rate, but the voice quality is the worst. Table 3.9 summarizes average rates and the voice quality, expressed in mean opinion score (MOS) for all three modes [14]. In addition to two-way conversational speech, the table includes all IS-2000 Layer 3 signaling messages and other functions such as users accessing voice mail service.

Table 3.9 Rate Probabilities and MOS Scores with FER = 1%

Mode	Average reverse-link rate	Average forward-link rate	MOS
0	4.4 kbps	5.2 kbps	3.86
1	3.5 kbps	3.6 kbps	3.84
2	2.9 kbps	3.0 kbps	3.69

SO 56 allows that a half-rate limitation be imposed to Mode 0 and Mode 1. In that case, SMV outputs only half-, quarter-, and eighth-rate frames. This feature allows for *dim-and-burst mode*, where 9.6 kbps capacity of traffic channel is shared with IS-2000 Layer 3 signaling messages. This mode comes at the expense of degraded voice quality.

On command, SMV can also produce a blank-rate frame. The blank-rate frame contains no bits. A blank-rate frame occurs when the transmitter uses the entire frame for IS-2000 Layer 3 signaling during *blank-and-burst mode* of operation.

3.3.2 Short Message Service

SMS allows short, alphanumeric messages to be sent to and from a mobile station. If the message is short enough, less than 256 bytes, it can be delivered to and from the mobile station without requiring a setup of SO 6. Message delivery can be done by encapsulating the SMS message into an IS-2000 Layer 3 signaling message. The delivery of the message is possible regardless whether the mobile station is in idle mode or in traffic mode. The message is transmitted to the mobile on a common channel and from the mobile on the access channel. If the message is more than 256 bytes, SO 6 (or SO 14) setup is necessary.

SMS messages require IS-2000 Layer 2 acknowledgments, regardless of whether SO 6 is set up or not. If only a part of an SMS message is not correctly received, the whole message must be repeated. An example of an SMS is a voice mail notification.

3.3.3 Quality of Service

For voice and SMS services, QoS over the radio access bearer can be described with the following parameters: data rate, delay, and FER. The radio traffic channels are configured so that they closely match the four data rates requested by the vocoder. The traffic channel can change the rate at every 20-ms frame boundary. For Rate Set 1, the maximum rate is 9.6 kbps. The receiver can blindly detect the rate so that explicit rate notification with signaling is not needed. In case of SMS service, rate matching is not necessary nor is it feasible, because the SMS message can be of any length. Hence, the data rate provided to the SMS service varies. As we already mentioned for Rate Set 1, the traffic channel rate is up to 9.6 kbps.

FER is a parameter configurable by the operator and/or the manufacturer. There are no parameters signaled with IS-2000 Layer 3 signaling that specify requested FER for voice or SMS service. For voice, mobile station and the base station negotiate the SMV mode, but not the FER on the traffic channel. For voice, random frame losses of 1% to 3% are considered acceptable because the current speech coders do not cause audible speech degradations. The speech quality rapidly decreases if FER is above 3%. If FER is more than 5%, the mobile is normally assumed to be out of the coverage area. An acceptable error rate for SMS service can be either lower or higher depending on the size of the SMS message.

The delay is not negotiated either. It is a function of the Media Access Control (MAC) layer to provide adequate priority for the traffic that requires expedited delivery. Voice traffic has stringent delay requirements. One-way delay over a cdm2000 network must not be larger than 100 ms. The manufacturers must design their systems so that the delay budget does not increase beyond unacceptable values. The delay components for voice traffic from RAN to mobile station or mobile station to RAN include buffering delay needed to accumulate enough voice samples so that speech coding can take place, processing delay, backhaul delay, transmission delay at the transmitting side, and forward-error correction (FEC) processing and speech-decoding time or conversion to PCM. Note that speech-coding processing and transmission over the air require at least 40 ms.

SMS is not a delay-sensitive application. The SMS messages have lower priority than voice or IS-2000 Layer 3 signaling. The MAC layer enforces adequate relative priority between SMS messages and voice.

3.4 Packet-Switched Service Options

In addition to circuit-switched service, cdma2000 networks provide packet-switched services. Two service options associated with the packet services are described in this chapter, summarized in Table 3.10. They are the high-speed packet data service, or SO 33 [15], and VoIP service with Link Layer-Assisted Robust Header Compression (LLA ROHC), or SO 61 [16].

Table 3.10 Packet-Switched Service Options

Service option number	Service
33	High-speed packet data
61	VoIP with LLA ROHC

The cdma2000 networks provide for differentiated levels of packet data service based upon the application needs. Applications can be classified into groups with similar needs. Four different traffic classes are recognized and are listed in Table 3.11.

Table 3.11 Traffice Classes and Their Characteristics

Traffic class	Priority	Characteristics
Background	Very low	The destination is not expecting the data within a certain period of time. The content of the packets must be transparently transferred. An example is background file download.
Interactive	Low	There is a request-response pattern of the end user. The entity at the destination is expecting the response within a certain time. The round-trip delay is important. The content of the packet must be transparently transferred. An example application is Web browsing.
Steaming	High	The time relations between information entities within a flow must be preserved. There is no requirement on very low transfer delay. An example application is audio and video streaming.
Conversational	Very high	The transfer time must be low, and the time relations between information entities within a flow must be preserved. An example application is VoIP.

To provide any kind of packet data service, the main service instance has to be established. The Point-to-Point Protocol (PPP) negotiation takes place over this service instance, which is why this service instance is commonly called a PPP service instance. This service instance is set up as an SO 33 or high-speed packet data service. Assuming that the always-on feature is enabled, this service instance remains active (or dormant) as long as the mobile station is powered on. Any other service instance is set up on a per-need basis, and it lives only throughout the time needed by the application requesting it. Certain applications may request a setup of an additional service instance, but some, like HTTP or FTP, may use the PPP service instance.

3.4.1 High-Speed Packet Data Service

SO 33 provides for high-speed packet data service. It can be used to provide service for all kinds of packet data applications other then VoIP. This service option is suitable for applications that do not have stringent delay requirements. However, SO 33 can also be employed for delay-sensitive traffic. Background, interactive, and streaming traffic classes could use SO 33. There are two modes of operation defined: nonassured and assured mode.

3.4.1.1   Service Option 33: Nonassured Mode

In the nonassured mode, SO 33 provides only relative priority. There are 14 different priority levels defined. The higher the level, the higher the priority for the service. QoS cannot be guaranteed. SO 33 commonly employs an air interface error-recovery mechanism using RLP with retransmissions and an HARQ type II mechanism. That means that the erasure rate, as seen by the application, is much smaller than the erasure rate on the physical channel. At the expense of delay, the air interface error-recovery mechanism provides low erasure rates, which are typically required for data services.

3.4.1.2   Service Option 33: Assured Mode

In the assured mode, QoS can be guaranteed to some degree. The error rate on the physical channel as well as the characteristics of the air interface error-recovery mechanisms are adjusted so that the desired QoS is provided to the layers above RLP. The QoS parameters that can be specified are:

- Priority, 14 levels
- Minimum data rate
- Maximum delay
- Data loss rate

Not all parameters need to be included for the assured mode. However, if a certain parameter is not included, only best effort service is provided. For example, if the data loss rate parameter is omitted, the data loss rate is not guaranteed. However, if for the same service the maximum delay is specified, the network attempts to meet the specified delay requirement.

Streaming traffic class is an example of a traffic class that should use the assured mode. Another example is interactive traffic requiring premium service.

3.4.2 VoIP Service

The air interface for voice traffic in cdma2000 networks is optimized for circuit-switched voice. As we discussed in Section 3.3.1.2, the rates supported over the traffic channel for Rate Set 1 closely match the rates generated by the SMV vocoder. VoIP frames contain a IP, User Datagram Protocol (UDP), and Real-Time Protocol (RTP) header. Clearly, sending the headers with SMV speech-coded frames would require a great deal of padding. Moreover, even if there were no associated padding, any header transmitted over the air would reduce the air interface efficiency. Note that for VoIP service, the associated uncompressed IP/UDP/RTP header is larger than the actual SMV compressed voice payload. Hence, in order to have air link–efficient VoIP service, the IP/UDP/RTP header must not be transmitted over the air with each voice frame. A service option that provides VoIP service is SO 61. It employs a header compression technique in order to reduce IP packet size transmitted over the air. In most cases, the header size is 0, and a full or compressed header is only occasionally transmitted. The header compression and decompression are performed by the mobile station and PDSN.

In addition to SO 61, VoIP service is also provided by SO 60 [16]. Unlike SO 61, SO 60 does not implement header compression. Instead, it utilizes header removal. In the forward direction, the header is removed prior to over-the-air transmission. In the reverse direction, the header is not even generated by the mobile station: It is generated by the PDSN. We do not discuss SO 60 in more detail in this book. We focus on SO 61.

The reason the header compression is feasible is that there is a significant redundancy between the header fields within the same packet and between the consecutive packets belonging to the same packet stream. The header can be significantly reduced for most packets by sending the static field information only initially and utilizing dependencies and predictability of other fields. The relevant information from past packets is maintained in the form of a *context*. The context represents a mapping function that is used to compress and decompress the header. The compressor and decompressor update their contexts when needed. Errors on the air link may lead to inconsistencies between the contexts of the compressor and decompressor.

SO 61 provides VoIP service using LLA ROHC [17], which is a header compression technique based on the ROHC IP/UDP/RTP profile [18]. In the following section we provide brief background information for this ROHC profile.

3.4.2.1 Robust Header Compression (ROHC): IP/UDP/RTP Profile

A good header compression technique, besides achieving strong compression, must be able to quickly identify context loss. Moreover, to minimize the number of lost data frames, it must also be able to promptly synchronize the compressor and decompressor. ROHC is a robust header compression technique that has means to reliably recover context information upon packet losses over the air.

Early header compressing schemes, such as Compressed Real Time Protocol (CRTP) [19], depend on the decompressor to signal to the compressor that context is out of sync and on the compressor to update the context. It would then take at least link one round-trip time until the context information is received by the decompressor. In the meantime, all packets received by the decompressor are discarded. Since in cellular links the round-trip delay can be significant,[4] it is not acceptable to wait for the context update and discard packets received in between. The decompressor can attempt recovery of the context information by guessing. However, there must be an indication to confirm successful context repair. For that purpose, the compressed ROHC header contains a cyclic redundant code (CRC) field computed over the uncompressed IP/UDP/RTP header.

Let us now examine in more detail the IP/UDP/RTP header. Most of the header fields are static. They can be inferred from other fields, or they are known. Only five fields are dynamic; these are summarized in Table 3.12.

Table 3.12 Dynamic IP/UDP/RTP Header Fields

Field	Length
IPv4 identification	16 bits
UDP checksum	16 bits
RTP marker	1 bit
RTP sequence number	16 bits
RTP timestamp	32 bits

Most of the time, the RTP timestamp and IPv4 identification can be inferred from the RTP sequence number, which is incremented by one for each packet transmitted by an RTP source. The RTP marker bit is usually unchanged and it can be communicated occasionally. The UDP checksum cannot be compressed, but it is an optional field and does not need be enabled. Therefore, most of the time, it is only necessary to update the RTP sequence numbers.

The ROHC IP/UDP/RTP profile first establishes relationships between the RTP sequence number and other fields. After the relationship has been established, ROHC reliably communicates only the RTP sequence number between the compressor and the decompressor. However, whenever the relationship changes, additional information is sent to update the context. The compressor and decompressor have three states each. Both machines start in the lowest compression state and gradually transit to higher compression states. The transitions do not have to be synchronized. During the normal operation, only the compressor temporarily transits back to

---

4.   The round-trip time between the mobile station and PDSN is on the order of 100 ms.

the lower compression states. The decompressor transits back to a lower compression state only when the context damage or loss has been detected.

Compressor States. The compressor states are initialization and refresh (IR), first order (FO), and second order (SO). The purpose of the IR state is to initialize the static parts of the context at the decompressor or to recover after a failure. In this state, the compressor sends complete header information, which includes all static and nonstatic fields in uncompressed form and some additional information. The compressor stays in the IR state until it is fairly confident that the decompressor has correctly received the static information.

When the compressor becomes fairly confident that the decompressor has correctly received the static information, it transits to the FO state. The compressor can also enter this state from the SO state, which happens whenever the headers of the packet stream do not conform to their previous pattern. The purpose of the FO state is to communicate irregularities in the packet stream. When operating in this state, the information about a few static fields and some (typically not all) all-dynamic fields is transmitted between the compressor and decompressor.

The SO state is the optimal state from the compression point of view. The compressor transitions to this state when the header is fully predictable from the RTP sequence number. The compressor must also be confident that the decompressor has acquired all the parameters needed to map the RTP sequence number to all other fields.

Decompressor States. There are three decompressor states: no context, static context, and full context. The decompressor starts in its lowest or no-context state and it gradually transits to higher compression states. In the no-context state, the decompressor has not yet successfully decompressed a packet header. Once the static and dynamic header fields are correctly recovered, the decompressor can transit all the way to the full-context state. On repeated failures to recover the context information, the decompressor goes back to the static-context state. When in the static-context state, a successful reception of any FO compressor state packets is sufficient to trigger the transition to the full-context state. If decompression of several packets sent in the FO state fails, the decompressor transits back to the no-context state. In the no-context state, the decompressor can process only packets that carry the static information field.

Modes of Operation. The ROHC scheme has three modes of operation: unidirectional (U), bidirectional optimistic (O), and bidirectional reliable (R).

In the U-mode of operation, packets are sent in one direction only: from the compressor to the decompressor. The transitions between compressor states are performed only because of periodic timeouts and irregularities in the header-field change patterns in the compressed packet stream. Due to a lack of feedback with periodic refreshes only, the loss propagation is more likely in this mode compared to bidirectional modes.

The O-mode is similar to the U-mode. The difference is that there exists a feedback channel. The feedback channel is used to send error-recovery requests and optionally acknowledgments of significant context updates. During the normal U/O mode of operation, the compressed header size is only a single byte. This header has three fields: packet type identification, sequence number (infers RTP sequence number), and CRC for the uncompressed header.

The difference between the O-mode and R-mode is that the R-mode uses the feedback channel more often. Also, a stricter logic at both the compressor and the decompressor reduces the probability of loss of context synchronization between the compressor and the decompressor. In practice, the loss of context happens only when the error rate on the air link is unusually high.

3.4.2.2 Service Option 61

SO 61 is based on LLA ROHC. The idea behind LLA ROHC is to substitute the functionality of the 1-byte ROHC header field that is sent during the normal operation of the U/O mode, with the link layer assistance. Relative to ROHC IP/UPD/RTP profile, LLA ROHC defines additional packet types: no-header packet (NHP), context synchronization packet (CSP), and context check packet (CCP).

NHP is a packet that consists of payload only. It is used when all headers fields follow the currently established change pattern. The condition for sending the NHP is that the decompressor is able to infer the proper sequencing.

If for some reason the decompressor context is invalidated, it might be beneficial to send a packet with the header information only and discard the payload. This case can be handled with the CSP, which is defined as one of the unused packet-type identifiers from the ROHC IP/UDP/RTP profile.

The CCP does not carry any payload, but only an optional CRC value in addition to the packet-type identifier. The purpose of CCP is to provide a useful packet that may be sent by a synchronized physical layer where data must be sent at fixed intervals, even if a compressed packet is not available. The CCP is defined by one of the unused packet-type identifiers from the ROHC IP/UDP/RTP profile.

As we have already mentioned, the main difference between the LLA ROHC and ROHC is that majority of 1-byte headers of ROHC RTP packets during normal U/O mode of operation are replaced by NHPs. The link layer substitutes all the functions (packet-type identification, sequence number, and the CRC check for the uncompressed header) that are provided by the 1-byte header.

All ROHC packets carry a packet-type identifier that indicates how the header should be interpreted. In the case of 0-byte header compression, the link layer provides the indication whether the packet has a 0-byte header. The purpose of the sequence numbers is to deal with packet reordering and packet loss. For the 0-byte header compression, the link layer guarantees in-order delivery and packet-loss indication. The CRC is used by the decompressor to verify that updated context is correct. In the U/O mode, this verification serves three purposes. The first is the detection of longer losses that cannot be detected by the sequence number transmitted over the air. The link layer already performs this function because it indicates all packet losses by utilizing physical layer frame CRC. The second is protection against failures caused by residual bit errors in the compressed headers. This is not an issue for 0-byte headers because the compressed header is not even transmitted. The third issue is protection against faulty implementations and other causes of errors. For this reason, it is recommended that periodic context verification be implemented.

Figure 3.12 shows the protocol stack diagram for the LLA ROHC operation. The header reduction (HR) layer consists of two parts: HR lower layer (HRL) and HR upper layer (HRU). The HRL refers to the control logic in the base station controller (BSC) and mobile station that interfaces directly to the IS-2000 multiplex sublayer. It provides the link-layer assistance function to HRU, which is the component performing compression and decompression based on ROHC IP/UDP/RTP profile. The termination points for HRU are mobile station and PDSN.

To cope with inability to detect context loss due to the absence of CRC in NHP packets, sending and receiving HRL applications enforce optimistic approach agreement. From the sender perspective, the agreement consists of sending the minimum agreed number of consecutive ROHC packets or CSP packets whenever an NHP packet cannot be sent. At the receiver, upon detection of the same agreed number of consecutive losses over the air link, if the context loss is suspected, HRU application invalidates the context and packets are not forwarded to the upper layers before the context has been synchronized.

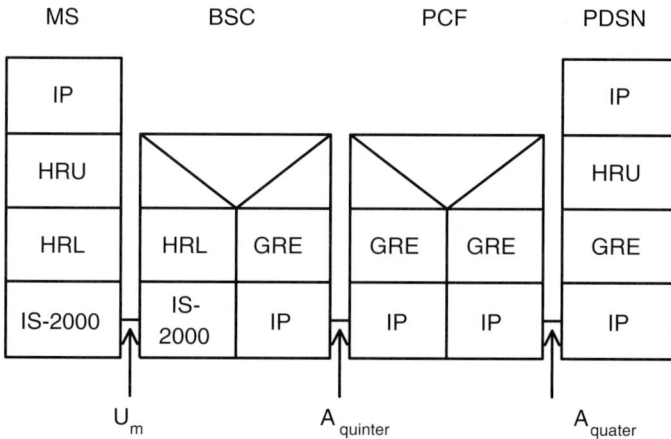

Figure 3.12 Protocol stack diagram for header compression operation (see Chapter 2 for the interface description).

REFERENCES

[1] Derryberry, R. T., "1xEV-DV Evaluation Methodology," *Contribution to the 3GPP2 standard committee TSG-C number 3GPP2-C30-20030616-043R1*, June 2003.

[2] Fielding R., J. Gettys, J. Mogul, H. Frystyk, L. Masinter, P. Leach, and T. Berners-Lee,. "Hypertext Transfer Protocol—HTTP/1.1," *IETF RFC 2616*, June 1999.

[3] Postel J., and J. Raynolds, "File Transfer Protocol (FTP)," *IETF RFC 959*, October 1985.

[4] "Wireless Application Protocol (WAP)," *WAP Forum*, available at http://www.wapforum.org.

[5] "ISO/IEC JTC1/SC29/WG11 N4668 Coding of Moving Pictures and Audio MPEG-4," *International Organization for Standardization,* March 2002.

[6] "XHTML 1.0: The Extensible Hypertext Markup Language: A Reformulation of HTML 4 in XML 1.0," *W3 Recommendation*, available at http://www.w3.org/TR/xtml1.

[7] "XHTML Mobile Profile, "*WAP Forum,* available at http://www1.wapforum.org/tech/documents/WAP-277-XHTMLMP-20011029-a.pdf.

[8] "TIA/EIA-IS-2001 Inter-operability Specification (IOS) for cdma2000 Access Network Interfaces," *Telecommunications Industry Association*, 2002.

[9] "TIA/EIA-IS-893 Selectable Mode Vocoder Service Option for Wideband Spread Spectrum Communication Systems," *Telecommunications Industry Association* (2001).

[10] Bellamy, John, *Digital Telephony*, Wiley, 1991.

[11] "TIA-EIA-IS-637-B Short Message Services for Wideband Spread Spectrum Systems," *Telecommunications Industry Association*, 2002.

[12] Kleijn, W. B., R. P. Ramachanrin, and P. Kroon, "Generalized Analysis-by-Synthesis Coding and its Application to Pitch Prediction," *In Proc. of ICASSP* 1992, vol. 1, pp. 337–340.

[13] Gao, Y., E. Shlomot, A. Benyassine, J. Thyssen, H. Su, and C. Murgia, "The SMV Algorithm Selected by TIA and 3GPP2 for CDMA Applications," *In proc. of ICASSP* 2001, pp. 709–712.

[14] DeJaco, A., "SMV Capacity Increase," *Contribution to the 3GPP2 standard committee TSG-C number 3GPP2-C11-20001016*, October 16, 2000.

[15] "TIA/EIA-IS-707-B Data Services Standard for Wideband Spread Spectrum Systems," *Telecommunications Industry Association*, 2003.

[16] "TIA-EIA-IS-923 Link Layer Assisted Service Options for Voice-over-IP: Header Removal (SO 60) and Robust Header Compression (SO 61)," *Telecommunications Industry Association*, 2003.

[17] Jonsson, L-E., and G. Pelletier, *Robust Header Compression (ROHC): A Link-Layer Assisted Profile for IP/UDP/RTP,* IEFT RFC 3242, April 2002.

[18] Bormann, C, C. Burmeister, M. Degermark, H. Fukushima, H. Hannu, L-E. Jonsson, R. Hakenberg, T. Koren, K. Le, Z. Liu, A. Martensson, Z Miyazaki, K. Svanbro, and K. Wiebke, "Robust Header Compression (ROHC): Framework and four profiles: RTP, UDP, ESP and uncompressed," *IEFT RFC 3095,* July 2001.

[19] Casner, C., and V. Jacobson, "Compressing IP/UDP/RTP Headers for Low-Speed Serial Links," *IEFT RFC 2508*, February 1999.

CHAPTER 4

IS-2000 Physical Layer

The physical layer protocol, specified in IS-2000 [1], represents the lowest layer of the cdma2000 protocol stack and defines the mobile station and base station interoperability procedures over the air interface or U_m reference point. The physical layer protocol specifies the physical channels baseband processing requirements for the following functions: power-control, encoding for error correction and detection, interleaving, modulation, scrambling, orthogonal spreading, and quadrate spreading with or without transmit diversity. The physical layer protocol also specifies the modulation pulse shape and its spectral characteristics, and the frequency bands over which cdma2000 can be operated. In the first part of this chapter, we describe the channels' structure and elaborate on their common modulation characteristics. In the second part, we describe in detail the channels' numerology and their individual properties.

cdma2000 allows for different spreading rates. With spreading rate 1 (SR1), both forward and reverse links use a single direct-sequence spread-spectrum carrier with spreading rate equal to 1.2288 Mcps. This configuration is commonly called a single carrier or 1×. With spreading rate 3 (SR3), three contiguous direct-sequence spread-spectrum carriers, each spread at 1.2288 Mcps, are bundled together on the forward link, while the reverse link operates on a single direct-sequence spread-spectrum carrier spread at 3.6864 Mcps. This configuration is commonly called a multicarrier or 3× and allows for high-speed services. Multicarrier systems can be deployed simultaneously supporting both mobile station cdma2000 single-carrier systems, as those are the ones largely deployed.

4.1 CDMA Channel Structure

The mobile station and base station communicate by means of several physical channels that are transmitted on a given frequency assignment. The physical channels can carry either user information or control information, or both. The aggregate set of all such channels is generically called a CDMA channel. A CDMA channel consists of a forward and a reverse CDMA channel, which in turn consist of multiple physical channels. To set the stage for the material that follows, we begin with a high-level description of the forward and reverse CDMA channels.

4.1.1 Forward CDMA Channel

The forward CDMA channel contains one or more spread-spectrum *code channels* that are transmitted on a given frequency assignment using a particular pilot offset. The frequency assignment is identified by a CDMA channel number, which corresponds to the carrier frequency. The pilot offset is a property of one of the code channels, the Forward Pilot Channel (F-PICH), which, as we discuss shortly, determines the timing of the forward CDMA channel transmissions. The forward CDMA channel structure with its constituent code channels is shown in Figure 4.1. The forward code channels can be used by one or multiple mobile stations to communicate simultaneously with the same base station. They are classified as dedicated, shared, or common.

Common channels carry information directed to one or more mobile stations at any given time. The Media Access Control (MAC) layer [2] controls access to the physical channel. There are a number of common channels that serve different purposes. One group of common channels is the pilot channels that allow the mobile station to acquire the forward CDMA channel, provide a phase reference for coherent demodulation, and provide a means for channel quality estimation. Four pilot channel types are allowed. The F-PICH is always present. If transmitter diversity is enabled, there is also a Forward Transmit Diversity Pilot Channel (F-TDPICH). Two additional pilot channels are intended for spot-beams applications directed to a group of mobile stations using smart antenna techniques. These pilot channels are called the Forward Auxiliary Pilot Channel (F-APICH) and the Forward Auxiliary Transmit Diversity Channel (F-ATD-PICH). The common channels also include the Forward Synchronization Channel (F-SYNCH) used by the mobile to acquire initial system information and a group of channels used to broadcast system information and to page the mobile station. The paging group includes the Forward Paging Channel (F-PCH).[1] The functions of the F-PCH can be replaced with the Forward Broadcast Control Channel (F-BCCH) and the Forward Common Control Channel (F-CCCH). The Forward Quick Paging Channel (F-QPCH) is optionally used in conjunction with the paging channels to reduce the amount of time needed to monitor the F-PCH or F-CCCH. The Forward Packet Data Control Channel (F-PDCCH) can also be classified as a common channel. The F-PDCCH provides signaling support for the F-PDCH. Finally, the common channels also include control channels called the Forward Common Assignment Channel (F-CACH) and the Forward Common Power Control Channel (F-CPCCH). These channels support a form of reverse-link access procedure, called reservation access mode. Note that the F-CPCCH can also be used to

---

1.   F-PCH is maintained for backward compatibility with legacy IS-95 mobile stations [3].

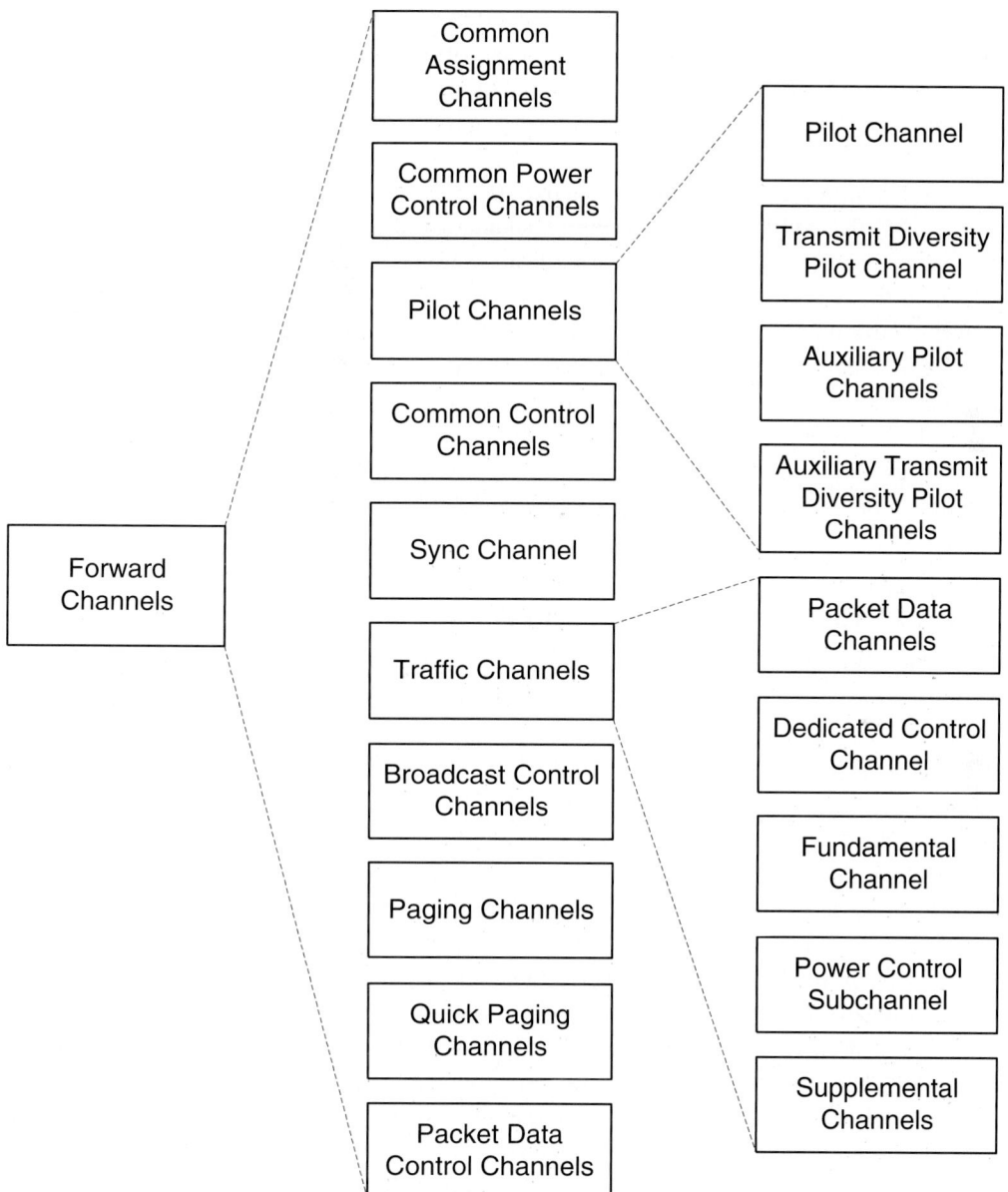

Figure 4.1 Forward link channel structure.

power control the Reverse Pilot Channel (R-PICH) of a mobile station when no dedicated channel carrying a forward power-control subchannel exists for that mobile station.

A dedicated channel is assigned to one and only one mobile station through IS-2000 Layer 3 signaling [6]. The dedicated channels can carry either user data or control information. They include the Forward Fundamental Channel (F-FCH), the Forward Supplemental Channel (F-SCH), the Forward Dedicated Control Channel (F-DCCH), and if directive antenna techniques are employed, the Forward Auxiliary Pilot Channel (F-APICH). The first three are also called traffic channels because they can carry user data. The F-FCH and the F-DCCH are the only traffic channels that can carry the power-control subchannel. The power-control subchannel is used to power control the R-PICH and is time-multiplexed onto the F-FCH or F-DCCH.

The Forward Packet Data Channel (F-PDCH), also a traffic channel, is used to carry bursty data mainly for high-speed nonreal-time services and is the only shared channel. It has many of the physical characteristics of a dedicated channel, and as such only a single mobile station decodes this channel at a time. However, unlike dedicated channels, the F-PDCH is assigned to a particular mobile station through IS-2000 Layer 2 [2] signaling. The assignment is very short, 1.25, 2.5, or 5 ms, and for that reason the channel is referred to as shared.

4.1.2 Reverse CDMA Channel

From a base station perspective, the reverse CDMA channel is the aggregate of all mobile station spread-spectrum transmissions on a CDMA frequency assignment. A base station is capable of processing simultaneous reverse CDMA channel transmissions from multiple mobile stations. The separation among the reverse CDMA channels transmitted by different mobiles is achieved by spreading the signal with sequences that are unique among mobile stations. The structure of the reverse CDMA channel is shown in Figure 4.2.

Reverse channels can be divided into dedicated and common. Dedicated channels are assigned to one and only one mobile station. They are spread by mobile-specific pseudonoise (PN) sequences so that they can be discriminated at the base station receiver. Dedicated reverse-link channels are the R-PICH, the Reverse Fundamental Channel (R-FCH), the Reverse Dedicated Control Channel (R-DCCH), and the Reverse Supplement Channel (R-SCH). The R-FCH, R-DCCH, and R-SCH are also called traffic channels. The R-FCH and R-DCCH carry signaling and/or user traffic, while the F-SCH carries only user traffic. There are also two channels in support of F-PDCH operation: the Reverse Channel Quality Indicator Channel (R-CQICH) and the Reverse Acknowledgment Channel (R-ACKCH).

Common channels employ a base station–specific PN sequence. All mobile stations active on the same reverse CDMA channel use the same PN sequence for a common channel, so explicit identification by means of upper-layer signaling is needed. As simultaneous transmissions by multiple mobile stations on the same channel cannot be discriminated by the base station receiver and leave the possibility for collisions, common channels are contention-based channels and use specialized random access protocols. Common reverse channels carry signaling in support of registration, authentication, call origination, and other procedures. They can

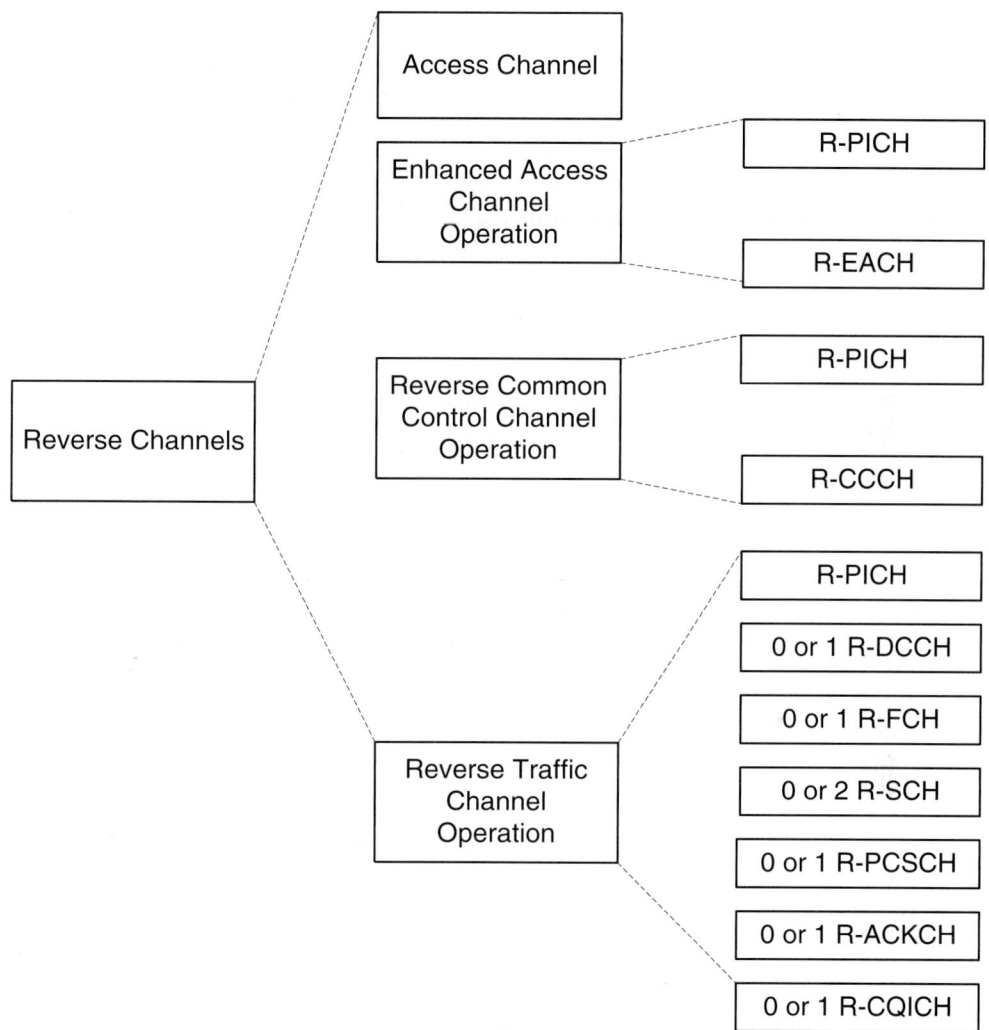

Figure 4.2   Reverse-link channel structure.

also carry small amounts of user data. The common reverse link channels are the Reverse Access Channel (R-ACH),[2] the Reverse Enhanced Access Channel (R-EACH), and the Reverse Common Control Channel (R-CCCH).

2.   R-ACH is maintained for backward compatibility with legacy IS-95 mobiles [3].

4.2 Modulation, Coding, and Spreading Characteristics

The modulation, coding, and spreading (MCS) characteristics of the forward and reverse channels vary greatly. One reason is that, as the standard evolved substantially since its first release, new and enhanced channels have been defined along with different physical layer protocol revisions. Differences are also intrinsic in the different uses of the channel. Channels that carry Layer 3 signaling and/or user data traffic, those that carry MAC layer signaling traffic, and those that support physical layer procedures, all require drastically different physical layer design. In the case of traffic channels, however, the fundamental functional building blocks are common to all channels, and it is therefore useful to describe those first.

4.2.1 Forward Traffic CDMA Channel

The functional building blocks of the forward traffic channels (F-FCH, F-DCCH, F-SCH, and F-PDCH) are shown in Figure 4.3. Source bits are first encoded, interleaved, modulated, and then spread before up-conversion and transmission on the radio channel. Although such functionality applies to all traffic channels, the specification details are different. That motivates the definition of radio configurations (RC) to describe the modulation characteristics that a traffic channel can have. The traffic channel RC defines the set of allowed bit rates, frame size, the type of encoder and its rate, and the type of modulation. There are 10 types of RC specified by the physical layer protocol. Table 4.1 summarizes the characteristics of the RCs applicable to SR1. The characteristics of IS-95 [3] legacy traffic channels (RC1 and RC2) are not shown. Note that other data rates, not listed in Table 4.1, are possible when flexible data rates are supported. However, we do not discuss flexible data rates operation in this book.

Table 4.1   Forward Traffic Channels RC Characteristics

RC	Data rate	Frame size	Code rate	Modulation
3	9.6 kbps	5 ms	1/4	QPSK
	1.5, 2.7, 4.8, 9.6, 19.2, 38.4, 76.8, and 153.6 kbps	20 ms		
	1.35, 2.4, 2.7, 4.8, 9.6, 19.2, 38.4, and 76.8 kbps	40 ms		
	1.2, 2.4, 4.8, 9.6, 19.2, and 38.4 kbps	80 ms		
4	9.6 kbps	5 ms	1/2	QPSK
	1.5, 2.7, 4.8, 9.6, 19.2, 38.4, 76.8, 153.6, and 307.2 kbps	20 ms		

Table 4.1   Forward Traffic Channels RC Characteristics (continued)

RC	Data rate	Frame size	Code rate	Modulation
	1.35, 2.4, 2.7, 4.8, 9.6, 19.2, 38.4, 76.8, and 153.6 kbps	40 ms		
	1.2, 2.4, 4.8, 9.6, 19.2, 38.4, and76.8 kbps	80 ms		
5	9.6 kbps	5 ms	1/4	QPSK
	1.8, 3.6, 7.2, 14.4, 28.8, 57.6, 115.2, and 203.4 kbps	20 ms	3/8*	
	1.8, 3.6, 7.2, 14.4, 28.8, 57.6, and 115.2 kbps	40 ms		
	1.8, 3.6, 7.2, 14.4, 28.8, and 57.6 kbps	80 ms		
10	from 81.6 to 3,091 kbps	1.25, 2.5, 5 ms	>1/5**	QPSK, 8-PSK, 16-QAM

* Obtained by puncturing rate 1/4 mother code.
** The mother code is rate 1/5; higher code rates are obtained with puncturing.

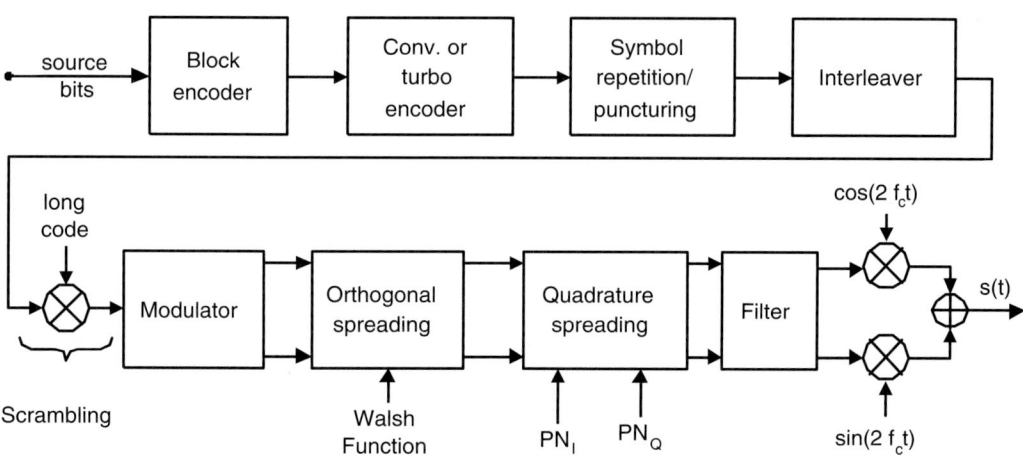

Figure 4.3 Forward traffic channels functional block diagram. Subscript I refers to I branch, subscript Q refers to Q branch, f_c is the carrier frequency, and s(t) is the output signal.

Not all RCs are applicable to a given traffic channel. For example, the F-PDCH can use only RC10, the F-FCH can use only RC3 through RC5, and the F-DCCH and F-SCH can use only RC3 through RC5. The set of data rates defined by RC3 and RC4 are called Rate Set 1. The set of rates defined for RC 5 are called Rate Set 2. Furthermore, a traffic channel may be able to use only a subset of the data rates or frame sizes specified for a valid RC. For example, the F-FCH (which carries either variable-rate voice traffic, low-speed data traffic, and/or signaling) can use neither data rates larger than 14.4 kbps nor frame sizes larger than 20 ms, both of which are instead applicable to the F-SCH (which carries high-speed data traffic). The frame size of 5 ms is applicable for F-FCH and F-DCCH only.

4.2.2 Reverse Traffic CDMA Channels

The functional building blocks of the reverse traffic channels (R-FCH, R-DCCH, and R-SCH) are shown in Figure 4.4. Source bits are first encoded, interleaved, modulated, and then spread before up-conversion and transmission on the radio channel. Just like the forward traffic channels, the reverse traffic channels are also specified in terms of their associated RC. There are six types of RC specified by the physical layer protocol. Table 4.2 summarizes the characteristics of the RC applicable to SR1. The characteristics of IS-95 [3] legacy traffic channels (RC1 and RC2) are not shown.

Table 4.2   Reverse Traffic Channels RC Characteristics

RC	Data rate	Frame size	Code rate	Modulation
3	9.6 kbps	5 ms	1/4	BPSK with pilot
	1.5, 2.7, 4.8, 9.6, 19.2, 38.4, 76.8, 153.6, and 307.2 kbps	20 ms	1/4 (except 1/2 for 307.2 kbps)	
	1.35, 2.4, 4.8, 9.6, 19.2, 38.4, 76.8, and 153.6 kbps	40 ms	1/4 (except 1/2 for 153.6 kbps)	
	1.2, 2.4, 4.8, 9.6, 19.2, 38.4, and 76.8 kbps	80 ms	1/4 (except 1/2 for 153.6 kbps)	
4	9.6 kbps	5 ms	1/4	BPSK with pilot
	1.8, 3.6, 7.2., 14.4, 28.8, 57.6, 115.2, and 230 kbps	20 ms		
	1.8, 3.6, 7.2., 14.4, 28.8, 57.6, and 115.2 kbps	40 ms		
	1.8, 3.6, 7.2., 14.4, 28.8, and 57.6 kbps	80 ms		

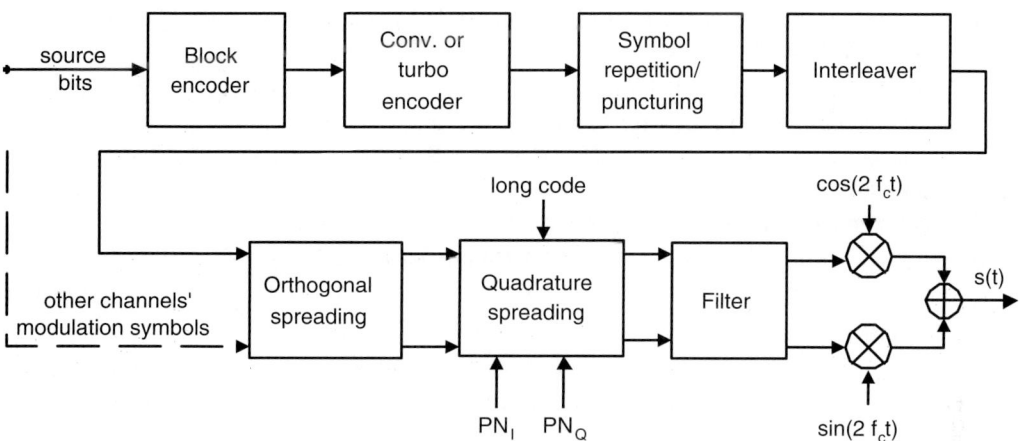

Figure 4.4   Reverse traffic channels functional block diagram.

RC3 and RC4 can be used by all traffic channels. The set of data rates defined by RC3 are called Rate Set 1. The set of rates defined for RC 4 are called Rate Set 2. Frame size larger than 20 ms and data rates larger than 14.4 Kbps with turbo coding are applicable to the R-SCH only. Frame size of 5 ms is applicable for R-FCH and R-DCCH only.

4.2.3 Error Detection

Data frames transmitted over the noisy wireless channel may be received in error. It is then necessary for the receiver to detect when such errors occur so that the data frame, unusable by the upper layers, can be discarded or retransmitted. In IS-2000, traffic channels' data frames are encoded by the transmitter using systematic linear block codes. With this technique, the block encoder computes a sequence of parity bits, also called frame quality indicator bits, which are appended to the data frame. If the data frame contains k information bits and the encoder output is a sequence of n code symbols, the code is referred to as a (n,k) linear code with code rate k/n. The receiver then uses a decoder that, on the basis of the frame quality bits, decides whether the frame is received error-free. With this error-detection scheme, erroneous data frames are delivered to the upper layers only if the decoder fails to detect the presence of errors. If the code for error detection is properly chosen, the probability that the receiver commits a decoding error can be quite small. For certain classes of (n,k) linear codes, for example, the probability of an undetected error is upper-bounded by $2^{-(n-k)}$. Clearly, if such codes are used with a sufficient number of frame quality bits, the probability of an undetected error can be made negligible.

In IS-2000, low-rate cyclic linear codes are used ($k/n \ll 1$). These codes are selected because their algebraic properties simplify the encoding and decoding implementation. The number of frame quality bits varies depending on the reliability that is required. That, in turn, depends on which of the following purposes error detection is serving: data rate determination,[3]

Hybrid Automatic Repeat Request (H-ARQ) support, or to prevent the receiver from acting upon incorrect control information.

As we have already discussed, the traffic channels carrying voice data (i.e., the F/R-FCH) are operated at a variable bit rate. The receiver does not know a priori at which bit rate data are being sent in the current frame period. Therefore, the receiver must decode the frame multiple times (in parallel), once for each of the possible bit rates in the rate set being used, and then select the frame for which no error is detected, if any. Except for the rare possibility of an undetected erroneous frame, at most one frame is selected and passed to the upper layers. Because the overhead must be kept to a minimum and the cost associated with passing to the vocoder an erroneous frame is relatively low, the error-detection performance requirement is not very stringent (e.g., compared to signaling message) and relatively few parity bits are used. Table 4.3 summarizes the use of parity bits, also called cyclic redundancy code (CRC) for Rate Set 1 and Rate Set 2. For example, for Rate Set 1, the F/R-FCH uses 6 parity bits for eighth- and quarter-rate frames, 8 bits for half-rate frames, and 12 bits for full-rate frames. Rate Set 2 are not further discussed in the book because there is nothing conceptually different from Rate Set 1 (only supported rates are different).

Table 4.3   Parity bits (CRC) for R/F-FCH

Rate Set	Data rate	CRC bits
1	1.5 kbps (1/8 rate)	6
	2.7 kbps (1/4 rate)	6
	4.8 kbps (1/2 rate)	8
	9.6 kbps (full rate)	12
2	1.8 kbps (1/8 rate)	6
	3.6 kbps (1/4 rate)	8
	7.2 bps (1/2 rate)	10
	14.4 kbps (full rate)	12

Even when the data rate is known prior, for example, because it was negotiated via upper-layer signaling, error detection is needed to support ARQ retransmission protocols used for error-insensitive applications, such as packet data. In the version of the ARQ protocol used in IS-2000 (see also Chapter 5), frames received in error must be discarded so that the receiver can

---

3.   Other rate determination measures are also used.

detect gaps in the sequence of successfully received frames and request retransmission of those that have been discarded. Then, the F/R-SCH frames are encoded with a 12-bit or 16-bit frame-quality indicator to enable error detection. The 16-bit indicator is used when the data rate is 19.2 kbps or higher.

Other miscellaneous uses of frame-error detection include power-control, message reassembly, and traffic channel supervision. Those are described in later chapters.

4.2.4 Forward Error Correction

Forward Error Correction (FEC) reduces the required signal-to-interference noise ratio (SINR) for the desired Frame Error Rate (FER) performance. The reduction in the required SINR is computed relative to uncoded systems and is commonly referred to as the coding gain.

In most cases, the physical layer employs two main types of FEC: convolutional and turbo coding.[4] Convolutional codes are suitable for short payloads, while turbo codes perform best with longer encoder packet sizes. The turbo encoder contains an embedded interleaver. Specifically, the coding gain is proportional to the turbo interleaver size, and the longer the encoder packet, the longer the interleaver and the higher the coding gain. The coding gain of turbo codes is in general larger than that of convolutional codes. However, if the payload is small, which is the case for voice frames, the convolutional codes perform as good as or better than turbo codes.

4.2.4.1   FEC with Convolutional Codes

The physical layer employs linear convolutional codes of constraint length $K = 9$. The encoder produces the code of rate one-half or one-quarter. A rate one-third code is also allowed for some legacy IS-95 channels. The rate one-quarter convolutional encoder is shown in Figure 4.5. The encoders for other rates are similar in structure. At the start of the encoding process, the encoder state in the linear shift register is all zeros. After the packet encoding is completed, the encoder is flushed with eight zeros. That is performed to ensure that after the packet is encoded, the encoder state is again all zeros. Having the all-zeros initial encoder state and flushing at the end improves the performance of the Viterbi decoder. The decoder now knows the state from which to begin the decoding and the state where the decoding process must end.

---

4.   In addition to convolutional and turbo codes, a block code is used to encode the R-CQICH. Note that the error-detection codes, discussed in Section 4.2.3, are also called block codes.

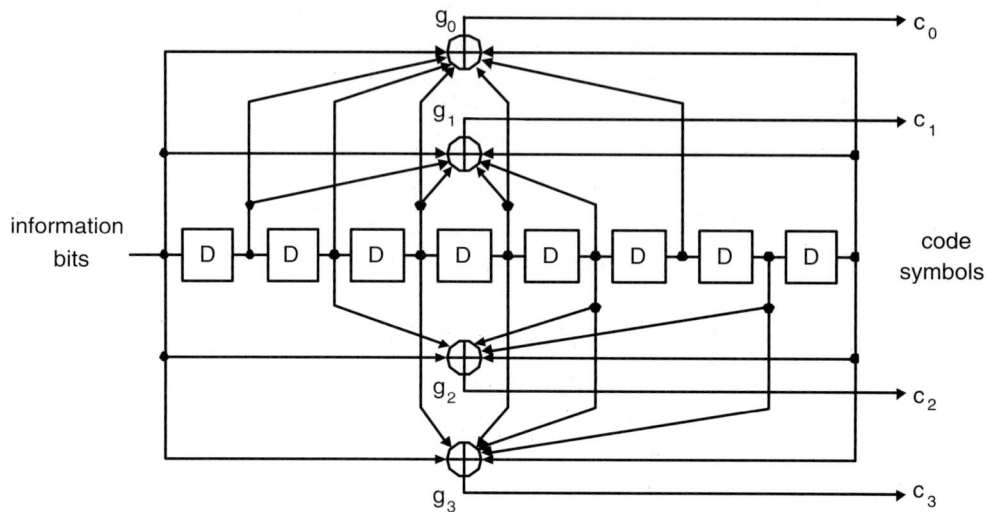

Figure 4.5   Convolutional encoder.

4.2.4.2   *FEC with Turbo Codes*

The IS-2000 turbo code encoder is shown in Figure 4.6. It consists of two constituent, recursive, systematic convolutional encoders of constraint length 4. At the beginning of the encoding process, the linear shift registers are initialized with zeros, and at the end of the packet encoding process, the constituent encoders are flushed so that the constituent encoders end in the all-zeros state. The flushing of the turbo encoder is different from the convolutional encoders because the turbo code consists of two recursive convolutional codes. Each recursive convolutional code is flushed by applying the feedback branch of the recursive encoder as input. The basic rate of the turbo code is one-fifth, which is achieved if the output consists of the information bit sequence and the parity bits from two constituent encoders. The interleaved information bits sequence, which is supplied as an input to the second constituent encoder, is punctured. Higher code rates are obtained through additional puncturing.

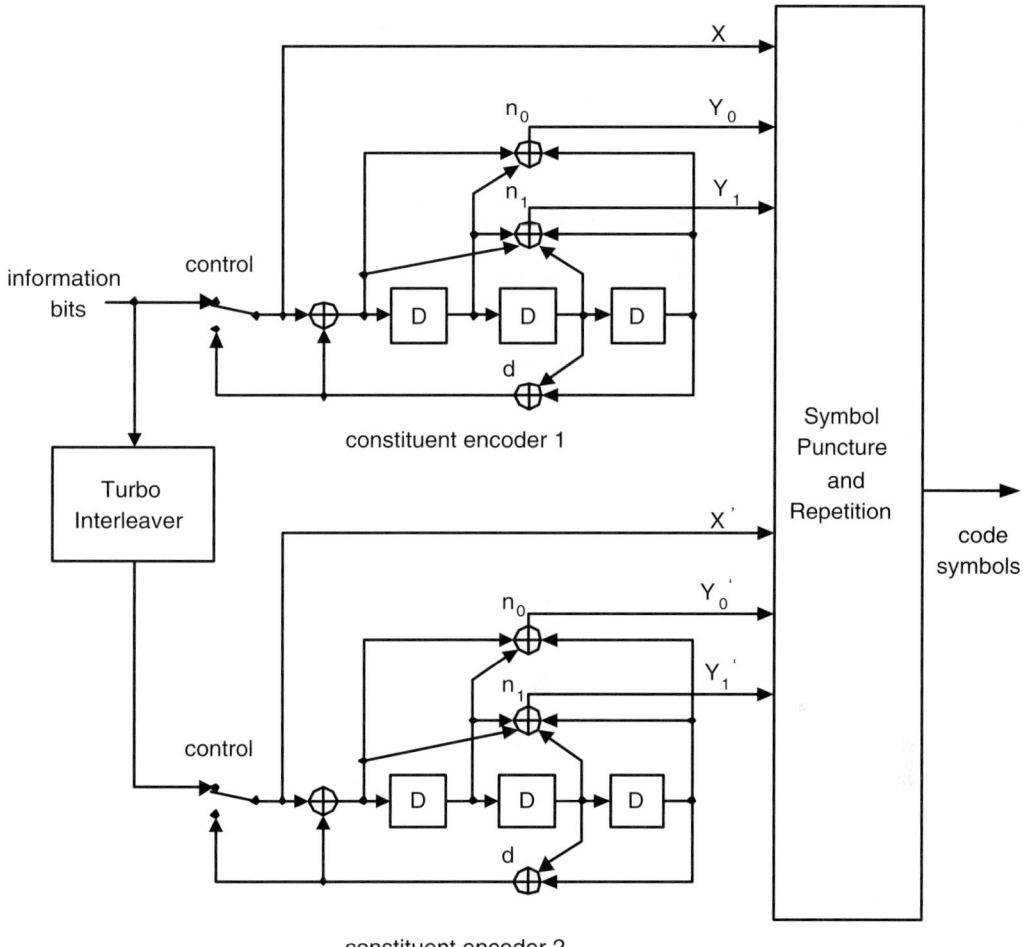

Figure 4.6 Turbo encoder.

An important factor for turbo code performance is the structure and the length of the turbo interleaver. The longer the interleaver, the better the FER performance. The interleaver structure is critical because the interleaver is embedded into the code structure and, effectively, different interleavers result in different codes. The interleaver shapes the weight distribution of the code, which ultimately controls its performance. The IS-2000 turbo interleaver is designed to optimize the performance. It can be described by the following:

- Write the bit sequence into a matrix with 2^5 rows and 2^N columns, row by row as shown in Figure 4.7.
- Shuffle the rows according to the bit-reversal rule.

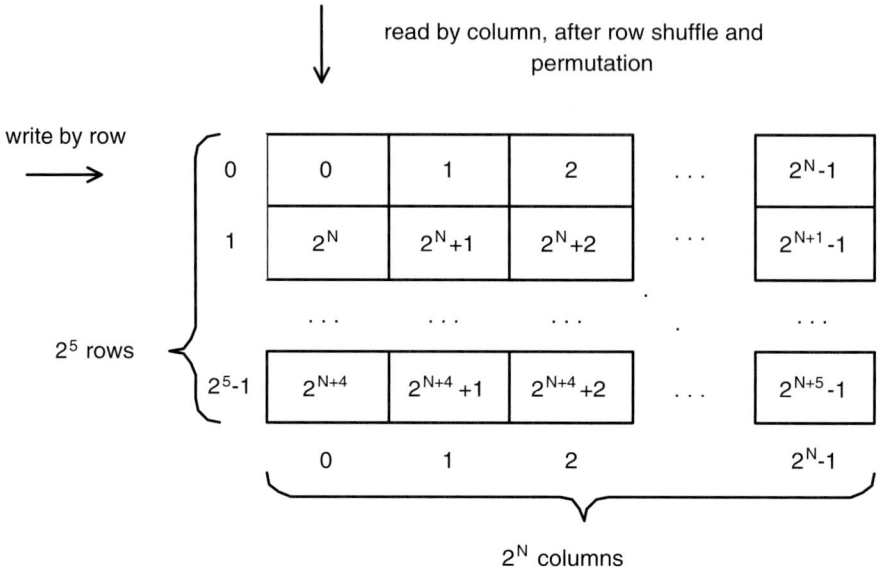

Figure 4.7 IS-2000 turbo interleaver structure.

- Permute the rows according to the following congruental rule: $x(i+1) = (x(i)+c) \bmod 2^N$, where i denotes the bit position in the row after permutation, $x(i)$ denotes the bit position before permutation, and c is a row-specific constant.

- Read the bits column by column.

Note that the size of the described interleaver structure assumes the sequence length is 2^{N+5}. However, the packet length, N_p, does not have to be equal to 2^{N+5}. The interleaver size is then chosen with the smallest N so that $N_p \leq 2^{N+5}$. If the congruental rule produces a bit location that falls out of the range of the packet, the bit is simply not transmitted. As an illustration, Figure 4.8 shows the I/O sequence for a packet length of 402 bits.

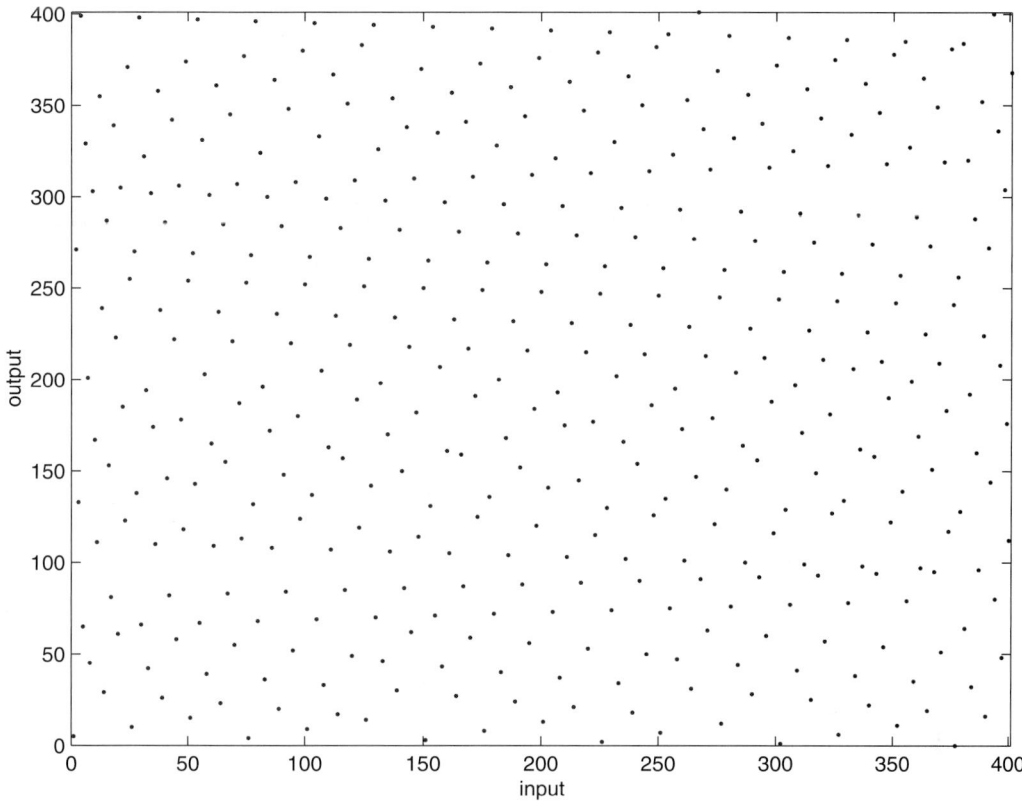

Figure 4.8 IS-2000 turbo interleaver I/O relationship.

4.2.5 Symbol Repetition and Puncturing

Traffic channels have different data rates and frame sizes, and are encoded with codes of different rates depending on the RC in use. However, following encoding, the coded symbol rate must be matched to that of one of a set of allowable modulation symbol rates. This process is called *rate matching*. Rate matching allows the functional blocks that follow in the transmitter chain to process the coded symbol in a manner that does not depend on the RC. In IS-2000, rate matching is done by means of symbol repetition and puncturing.

With symbol repetition, traffic channel-coded symbols of frames transmitted at a rate equal to or smaller than the full rate (9.6 kbps or 14.4 kbps) are sequentially repeated. The repetition factor is 8, 4, 2, and 1 times[5] for the eighth-, quarter-, half-, and full-rate frames of the forward traffic channels. The repetition factor is 16, 8, 4, and 2 times for the eighth-, quarter-, half-, and full-rate frames of the reverse traffic channels. Coded symbols of frames at a multiple of the full rate or multiple thereof are not repeated (or, using the standard terminology, they are repeated 1 time). Repeated coded symbols are then punctured according to an RC-dependent pattern, and the result is a sequence of modulation symbols. Puncturing patterns are defined in terms of a sequence of 0 and 1. Within a puncturing pattern, a 0 means that the repeated code symbol shall be deleted, and a 1 means that it shall be passed. The puncture pattern is repeated for all remaining symbols in the frame.

After repetition and puncturing, the sequence of coded symbols results in a sequence of modulation symbols. Modulation symbols are then fed to the interleaver. For 5-ms and 20-ms frames, the coded symbol rate of the forward traffic channels is matched to a modulation symbol rate that is a multiple 2^N of 19.2 ksps, with $N = 0,\ldots,5$. In case of reverse traffic channels, the modulation symbol rate is matched to a modulation rate that is a multiple 2^N of 76.8 ksps, with $N = 0,\ldots, 3$. Lower modulation symbol rates (proportionally reduced) are possible for 40-ms and 80-ms frames. For example, in the case of 40-ms frames, the modulation symbol rate can be as low as 9.6 ksps, while in case of 80-ms frames, the lowest modulation symbol rate is 4.8 ksps.

4.2.6 Channel Interleaving

To get the most gain from FEC in time-varying propagation environments, it is necessary to interleave the coded symbols so that deep fades do not impact adjacent coded symbols. On the transmitting side, interleaving follows the FEC block, while on the receiving side, the de-interleaving block precedes the FEC decoder.

The channel interleaver specified in the standard belongs to a class of block interleavers. There are two types defined: bit-reversal-order interleaver and forward-backwards bit-reversal interleaver. Common for both is that the input sequence is written column by column into a matrix with 2^M rows and J columns. The bits are then, as illustrated in Figure 4.9, read row by row. The row order is determined as explained in an example below. Consider the following example, with $M = 4$. The order in which rows are read is as follows:

- Label $2^4 = 16$ rows as 0, 1, 2, 3, 4, 5, 6, 7, 8, 9, 10, 11, 12, 13, 14, and 15.
- Express the numbers in binary format: 0000, 0001, 0010, … , 1111.
- Reverse the order of bits for each number: 0000, 1000, 0100, … , 1111.
- Convert back to decimal, so that the sequence becomes 0, 8, 4, 12, 2, 10, 6, 14, 1, 9, 5, 13, 3, 11, 7, and 15.
- Read the rows in the obtained order.

The forward-backward bit-reversal interleaver is similar to the bit-reversal interleaver. The even interleaved symbols are read in the permuted order, starting from the top row and leftmost column. The row sequence is therefore 0, 8, 4, …, 15 and the column sequence is 0, 1, 2, … , $J -$ 1. However, the odd interleaved symbols are read from the bottom and rightmost column. The

---

5.   The subrate traffic channel frames are transmitted at lower power to account for symbol repetition and thus decrease capacity consumption.

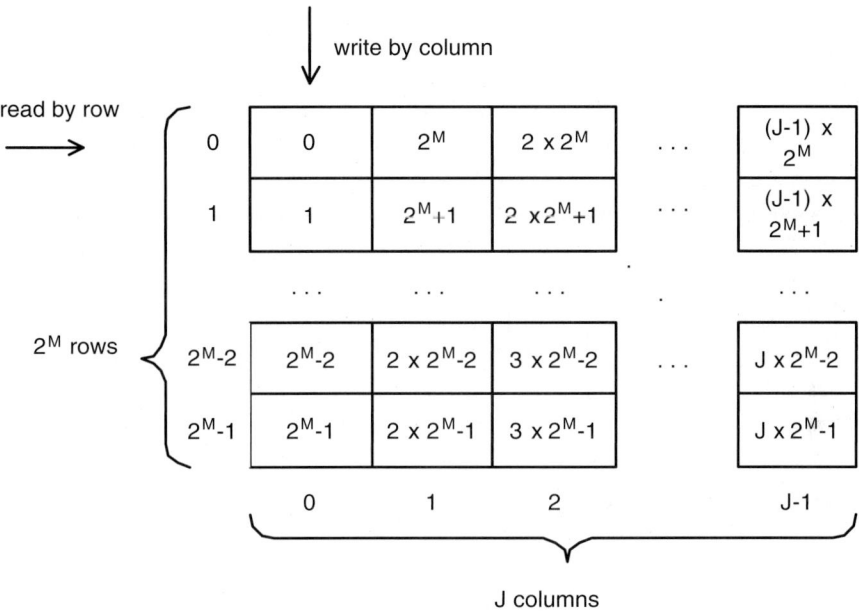

Figure 4.9 IS-2000 block interleaver structure.

row sequence is 15, 7, 11, ..., 0, and the column sequence is $J - 1, J - 2, J - 3, ..., 0$. The I/O sequences for the bit-reversal and the forward-backward bit-reversal interleaver, for a block length of 768 bits, are illustrated in Figure 4.10 and Figure 4.11 respectively. A sample of allowed combinations for M and J is shown in Table 4.4.

Table 4.4   Interleaver parameters

Interleaver block size	M	J
48	4	3
96	5	3
192	6	3
384	6	6
768	6	12
1,536	6	24
3,072	6	48
6,144	7	48
12,288	7	96

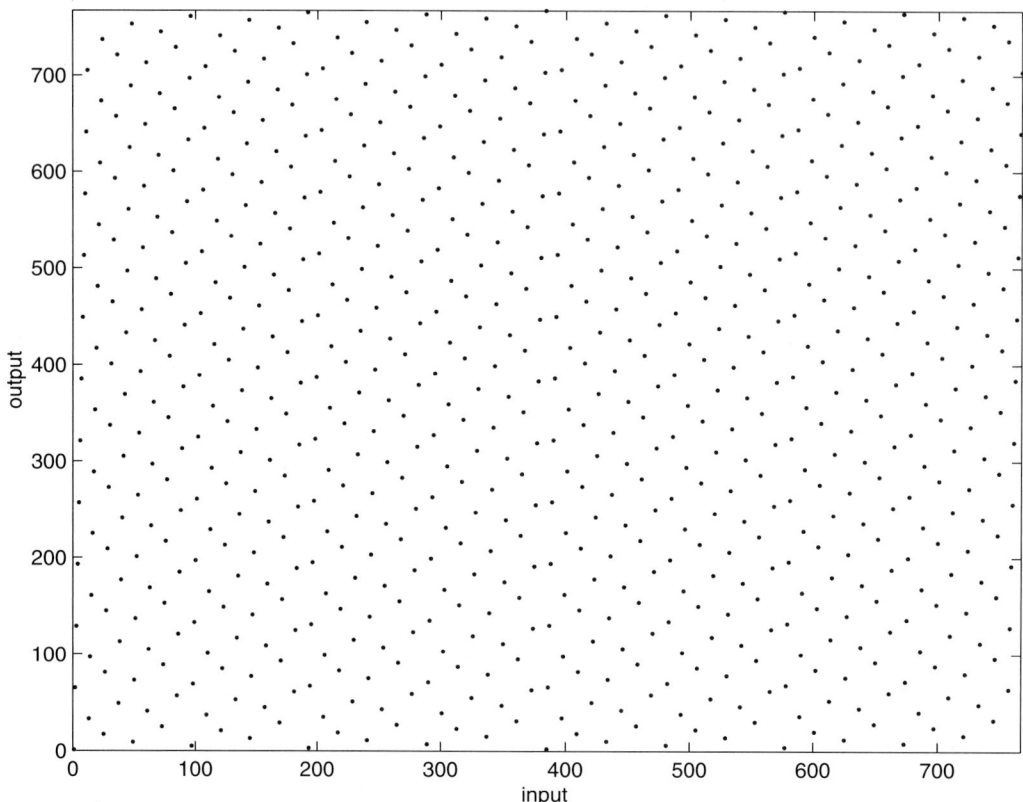

Figure 4.10 IS-2000 bit-reversal interleaver.

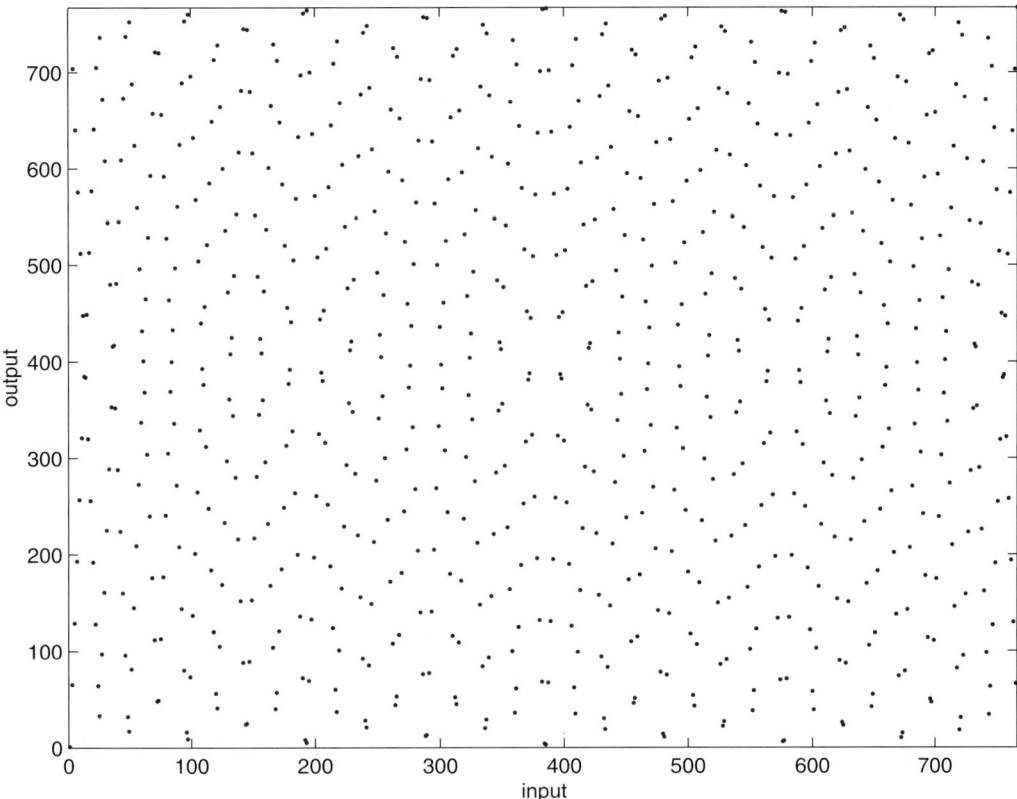

Figure 4.11 IS-2000 forward-backward bit-reversal interleaver.

4.2.6.1 *Channel Interleaver for Incremental Redundancy*

The channel interleaving for turbo-encoded packets can be done in two different ways. If incremental redundancy (IR) is not supported, channel interleaving is done as explained in Section 4.2.6, and it is not any different from the procedure specified for convolutional codes. The entire encoder packet is interleaved with a single interleaver. However, if IR is supported, interleaving is done differently, and it is integrated with the puncturing.

Figure 4.12 illustrates the interleaving process for a turbo-encoded packet that supports IR operation. The coded symbols of a turbo-encoded packet are grouped into five blocks. One of the blocks represents the systematic bits, and the remaining blocks consist of parity bits only. Each block corresponds to the output of a single generator polynomial of the constituent recursive convolutional codes. The bit-reversal interleaver, illustrated in Figure 4.10, interleaves each block; that is, each block is interleaved separately. The interleaver matrix may be larger than the length of the bit sequence. That does not cause any problem, because if a field in the interleaver

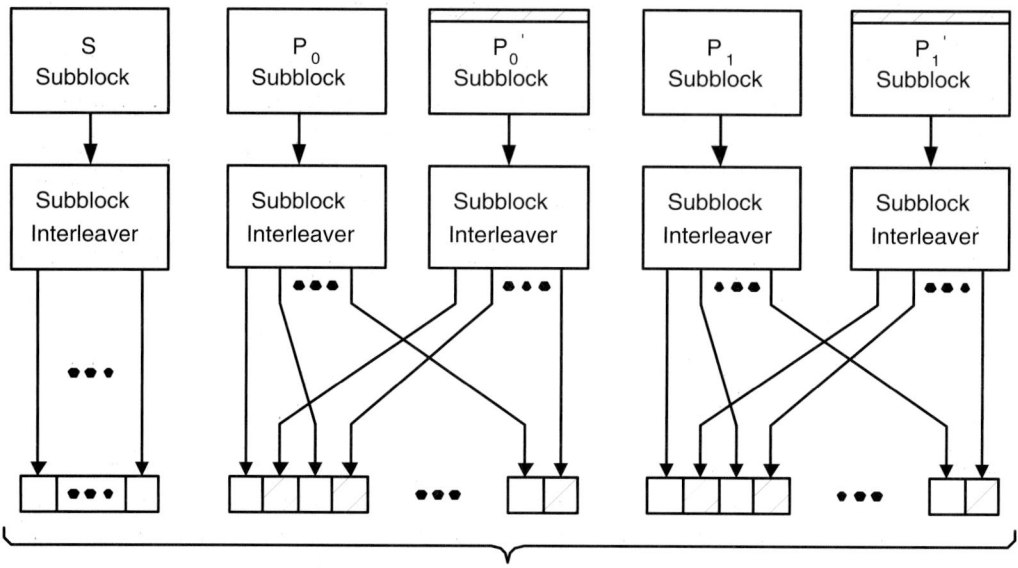

channel interleaver output sequence

Figure 4.12 IS-2000 channel interleaver structure for turbo codes with incremental redundancy.

matrix is empty, it is simply skipped. Table 4.5 shows the subblock interleaver parameters. The two blocks belonging to different constituent recursive convolutional encoders and equal generator polynomial branches are bit-by-bit multiplexed, forming three different groups of coded bits. The first group consists of systematic bits only, and the other two consist only of parity bits. The bits belonging to these three blocks are then concatenated and transmitted over the air. If the code rate is 1/5, all coded symbols are transmitted, but if the code rate is higher than one-fifth, for example one-third, only the first three-fifths of the coded bits are sent. The remaining coded bits are punctured.

Table 4.5   Subblock Interleaver Parameters

Packet size	M	J
408	7	4
792	8	4
1,560	9	4
2,328	10	3
3,096	10	4
3,864	11	2

4.2.7 Forward Link Data Scrambling

Data scrambling applies to all forward code channels except the pilot, F-SYNCH, F-QPCH, F-CPCCH, and F-PDCCH. Data scrambling provides privacy of communication. Communication privacy applies to the code channels dedicated to user's data transmission, that is, the F-FCH, F-DCCH, F-SCH, and F-PDCH. These channels are scrambled by modulo-2 addition of the modulation symbols at the interleaver output with a pseudorandom sequence of scrambling bits. The sequence of scrambling bits is unique among users. The scrambling sequence is obtained by decimating the bits of the so-called long PN sequence. As the long PN sequence is clocked at the chip rate, decimation is necessary to adapt to the rate of the modulation symbols at the interleaver output. The long PN sequence specified in cdma2000 belongs to a class of maximum-length linear-shift register sequences (MLLSRS). The generating polynomial has order 42, so the sequence period is $2^{42} - 1$ chips. Although the generating polynomial is identical for all code channels, a unique sequence is obtained by applying a user-specific (for dedicated channels) or code channel-specific (for common channels) mask to the registers' outputs. The masking process is illustrated in Figure 4.13. Each register's output is fed into an exclusive-OR (EOR) logic device together with the corresponding bit of the mask. The parallel outputs of the EOR devices are added together modulo-2, resulting in a single long PN code chip. The masking process effectively shifts the MLLSRS, resulting in a unique long PN sequence. If the mask is kept secret, an illegitimate receiver cannot descramble the traffic channel.

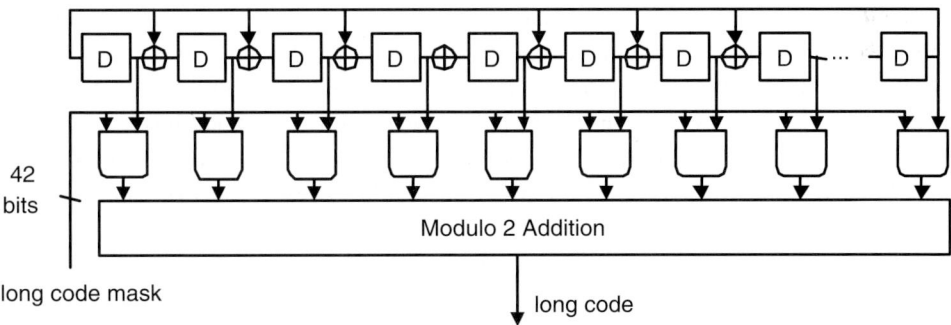

Figure 4.13 Long PN sequence masking.

4.2.8 Modulation

The forward and reverse modulations are different. As we see in the following section, the forward link is dimension-limited and, due to strong common pilot signal, can tolerate high-order modulation. The reverse link is not dimension-limited and, due to relatively week pilot signal, robust Binary Phase Shift Keying (BPSK) is used.

4.2.8.1   *Forward CDMA Channel Modulation*

After interleaving and, if applicable, data scrambling, the modulation symbols are fed into the modulator. A common modulation for forward code channels is Quadrature Phase Shift Keying (QPSK). The F-PDCH employs adaptive modulation, and in addition to QPSK, 8-PSK and 16-Quadriture Amplitude Modulation (16-QAM) are allowed.

The F-PICH carries all-zero symbols and is effectively an unmodulated signal. The legacy IS-95 channels, the F-SYNCH and the F-PCH, are BPSK-modulated. All other channels are QPSK-modulated except for the F-PDCH that employs adaptive modulation and is QPSK-, 8-PSK-, or 16-QAM-modulated. QPSK and higher order modulation schemes were introduced in IS-2000 to increase bandwidth efficiency. When QPSK is used for the F-FCH operated at 9.6 kbps, for example, the code rate can be reduced from half to quarter, leading to a considerable coding gain increase, while the complex modulation symbol duration is unchanged. Alternately, more Walsh channels are available if the code rate remains half. As the modulation symbol rate is unchanged, the spreading block that follows (see Section 4.1.1.6) achieves the same processing gain as that for BPSK without additional bandwidth expansion. At the receiver, however, the QPSK-coherent demodulator is more sensitive to inaccurate carrier-phase recovery relative to BPSK. For a given phase offset, the signal to SINR degradation of QPSK is larger than that of BSPK. The relative SINR degradation increases for increasing phase offset. The net effect of increased coding gain and SINR degradation due to phase-estimation error favors QPSK over BPSK, provided the phase-error variance is kept small. That is possible because carrier phase is estimated using the unmodulated F-PICH that is transmitted at relatively high power.

For the IS-2000 channels other than the F-PDCH, QPSK modulation consists of two stages. In the first stage, the logic binary symbols 0 and 1 are mapped to the scalars +1 and –1 respectively. In the second stage, the scalar modulation symbols are demultiplexed into two parallel streams, resulting in the in-phase (I) and in-quadrature (Q) branches of the complex symbol sequence. The demultiplexer behaves differently depending on whether the code channel supports transmit diversity. For ease of explanation, discussion of modulation and spreading when transmit diversity is enabled is deferred to Section 4.2.11. When transmit diversity is not enabled, the first scalar modulation symbol (represented by the symbol X in Figure 4.14) in each frame is mapped to the Y_I output, and the subsequent symbols are mapped to the Y_Q and Y_I in alternate fashion until the last symbol in the frame is mapped.

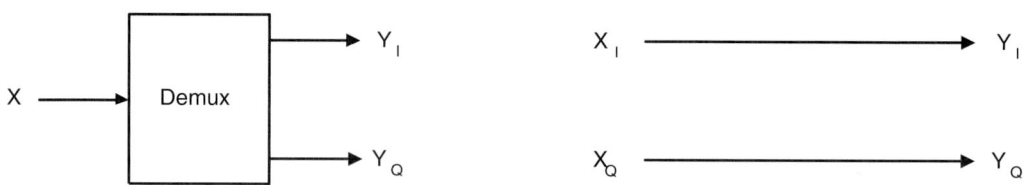

Figure 4.14 Demultiplexing when transmit diversity is not enabled.

In the case of the IS-95 legacy channels, the logic binary symbols 0 and 1 are mapped to the scalars +1 and −1, respectively. As these symbols are BPSK-modulated, the I and Q demultiplexer does not apply. However, it is convenient to model the BPSK modulator as one where the in-phase scalar modulation symbols (represented by the symbol X_I in Figure 4.14) are mapped to Y_I demultiplexer output, while the scalar in-quadrature modulation symbols are set to zero and mapped to the Y_Q demultiplexer output. That is, in Figure 4.14, $Y_Q = X_Q = 0$. This convention is useful because it allows specification of the following stages in the transmitter chain in an identical manner whether or not QPSK or BPSK is employed.

Modulation of the F-PDCH is radically different from all other code channels because it uses adaptive modulation with higher order modulation schemes with increased bandwidth efficiency. The modulation schemes allowed with the F-PDCH are QPSK, 8-PSK, and 16-QAM. The signal constellations and the coded symbols to complex modulation symbols mapping are shown in Figure 4.15. The coded symbol to modulation symbol mapping uses a form of Gray encoding.

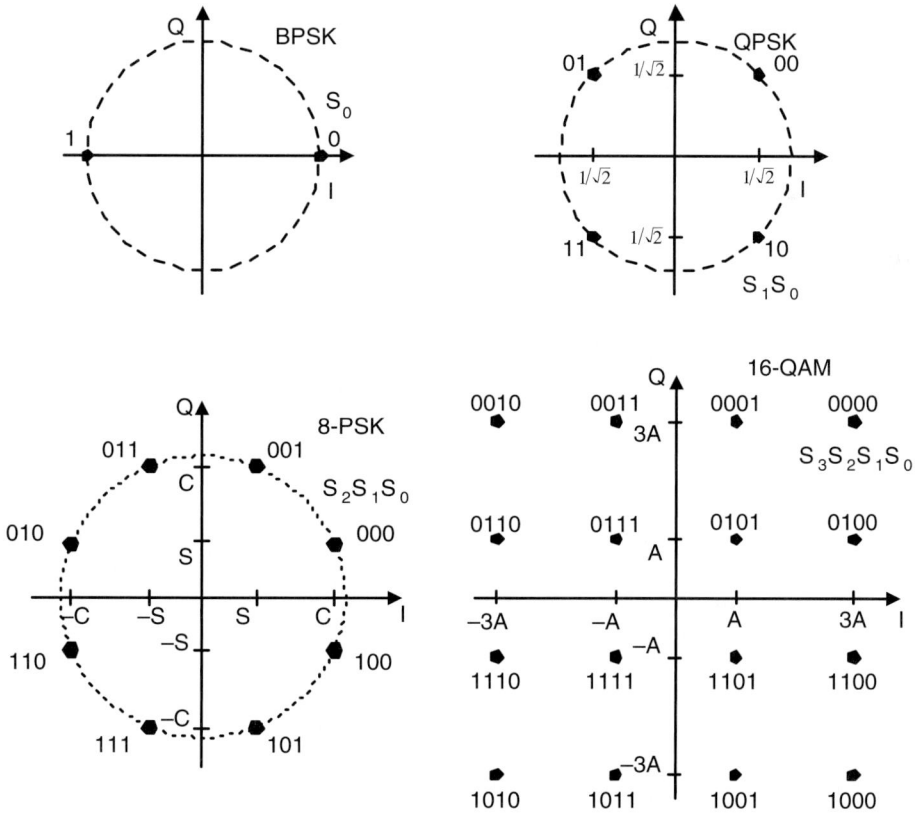

Figure 4.15 F-PDCH signal constellations. $C = \cos(\pi/8)$, $S = \sin(\pi/8)$, and $A = 1/\sqrt{10}$; S_j denotes the code symbol, j.

4.2.8.2   Reverse-Link Modulation

The legacy IS-95 R-FCH and R-ACH use a 64-ary orthogonal modulation that allows non-coherent demodulation at the base station receiver. The first release of IS-2000 specified enhanced traffic channels' RC that employ BPSK. With BPSK, the logic binary symbols 0 and 1 are mapped to the scalars +1 and −1, respectively.

As we shall see in the next section, the BPSK symbols are code-multiplexed with a pilot to allow for pilot-aided coherent demodulation at the base station receiver. Unlike the forward link that uses a common pilot for pilot-aided coherent transmit at relatively high power, on the reverse link the pilot is dedicated, and its transmit power must be limited to avoid excessive capacity consumption. Then, the carrier-phase estimation error is typically larger in the reverse direction than in the forward direction. That is one reason that BPSK, which is much less sensitive to relatively large carrier-phase errors than QPSK, is used with reverse channels while QPSK is used for forward channels. Another reason is that signal dimensionality is at premium on the forward CDMA channel but not on the reverse CDMA channel.

4.2.9 Orthogonal Spreading

In a CDMA system where multiple channels originate from the same transmitter, spectral efficiency can be significantly improved by using orthogonal sequences for signal spreading. When multiple channels are spread by an orthogonal chip sequence and the aggregate signal is transmitted synchronously over a nonfrequency selective channel, the receiver can separate the intended channel free of interference from all others by dispreading the signal with the same chip sequence used at the transmitter. The process of combining orthogonally spread channels is called code-multiplexing, and the channels themselves are called code channels. Code-multiplexing in a spread-spectrum system allows exploiting signal dimensionality in a more efficient way than simple QPSK modulation does.

Code-multiplexing applies to both the forward and the reverse channels. The orthogonal codes used in cdma2000 correspond to Walsh functions. Walsh functions form a closed set of normal orthogonal functions. Each function is defined over a normalized interval (0,1) and can take only the values +1 or −1. Normality implies that the result of integrating the product of any function and a copy of itself is unity. Orthogonality implies that the result of integrating the product of any two different functions is zero. Walsh functions are generated using Hadamard matrices. The Hadamard matrix is a square matrix whose elements take the values +1 and −1, and whose columns and rows are mutually orthogonal. We can use a binary logic representation of the Hadamard matrices by replacing the +1 value with 0 and the −1 value with 1. The N-th order $N \times N$ Hadamard matrix is generated recursively as

$$\mathbf{H}_1 = [0], \ \mathbf{H}_2 = \begin{bmatrix} 0 & 0 \\ 0 & 1 \end{bmatrix}, \mathbf{H}_{2N} = \begin{bmatrix} \mathbf{H}_N & \mathbf{H}_N \\ \mathbf{H}_N & \bar{\mathbf{H}}_N \end{bmatrix} \tag{4.1}$$

where N is a power of 2 and $\bar{\mathbf{H}}_N$ denotes the binary complement of $\mathbf{H}_N$. The Walsh function W_n^{N} of length N corresponds to the n-th row of the $N \times N$ Hadamard matrix, whose rows are indexed from 0 to $N - 1$. The elements of the Walsh function are referred to as Walsh chips. Orthogonal spreading by means of Walsh functions is done differently in forward and reverse channels, as detailed below.

4.2.9.1   *Forward-Link Orthogonal Spreading*

The forward code channels are associated with a Walsh function W_n^{N} of length 2^N Walsh chips, $N = 1, \ldots, 7$. The minimum and maximum length of a Walsh function is therefore 2 and 128 chips, respectively. Only for F-APICH and F-ATDPICH, the Walsh function length may be as high as 512. The I and Q values of the complex modulation symbols are multiplied, or spread, by the corresponding Walsh function at a fixed chip rate, R_c, equal to 1.2288 Mcps. After spreading, the code channels are summed together at chip level before entering the next stage of the transmitter chain. The Walsh functions used for spreading are time-aligned to ensure orthogonality among code channels. The first Walsh chip always starts at a time tick referenced to the base station timing.

The assignment of Walsh functions to the code channels is either predetermined or controlled by upper layers. The F-PICH is always assigned W_0^{64}. The F-SYNCH, if present, is always assigned W_{32}^{64}. The F-PCH, if one or more are present, is assigned W_1^{64} up to W_7^{64} consecutively. The remaining Walsh functions can be assigned to traffic channels. The assignment is done by the base station and is explicitly conveyed to the mobile station by means of Layer 3 signaling. The Walsh function length is selected so that it spans the duration of a complex modulation symbol.[6] That is required to ensure that, after dispreading, the symbol passed to the decoder is free of interference from the other code channels. For example, length-64 Walsh functions are used when the complex modulation symbol rate is 19.2 ksps, such as in the case of 9.6 kbps traffic channels using RC3. Higher data rates are associated with shorter Walsh functions. In general, the Walsh function length is $N = R_c / R_s$. For a given N, the Walsh function index is selected in such a way that each code channel is orthogonal to all other code channels in use. To understand how orthogonality among Walsh functions of variable length can be enforced, it is useful to organize the Walsh functions space in the form of the tree structure depicted in Figure 4.16. The tree structure is such that a Walsh function is orthogonal to all functions in the tree other than those corresponding to the branch they originate from. For example, W_2^8 is orthogonal to W_0^4, W_1^4, W_3^4, but it is not orthogonal to W_2^4. Then, once a Walsh function is allocated, all functions originating from that same branch are *blocked*. For example, if W_2^4 is in use, the length-8 Walsh functions W_2^8 and W_6^8 and all their children are blocked. The implications of Walsh functions management in the context of resource management are discussed in Chapter 6.

---

6.   Symbol repetition and puncturing (see Section 4.2.5) ensures that the complex modulation–symbol rate matches that of a Walsh function of appropriate length.

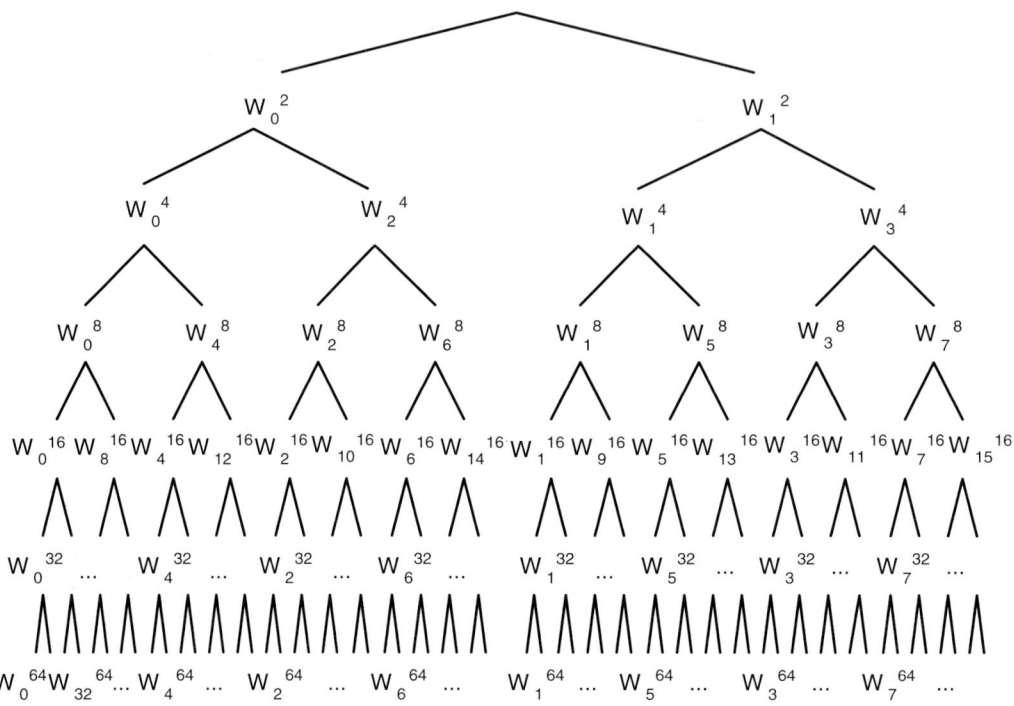

Figure 4.16 Walsh function tree example. Length-128 Walsh functions are not shown.

Traffic channels supporting voice calls at 9.6 kbps are generally assigned RC3, which requires a length-64 Walsh function and is the most power-efficient RC for voice calls. As a forward CDMA channel typically includes, besides the mandatory pilot, the F-SYNCH and one F-PCH, there are at most 61 available length-64 Walsh codes. If the number of simultaneous calls per CDMA channel exceeds such limit, one can increase the Walsh function space that can be assigned to 9.6 kbps traffic channels by using length-128 Walsh functions, corresponding to RC4. Note that with RC4, the FEC-code rate is doubled relative to RC3, from one quarter to one half, resulting in modulation symbol duration twice as long, which allows the use of the longer Walsh function for spreading. The drawback of RC4 is that the higher code rate means lower coding gain and higher capacity consumption for code channel. In general, the following relationships hold. The lower the code rate, the higher the coding gain, but also the lower the number of available Walsh functions, and vice versa.

An alternative method to expand the code space is to use spreading sequences that are not mutually orthogonal. In doing so, additional interference is created, but more users can be accommodated. To that aim, the cdma2000 physical layer provides for quasi-orthogonal functions (QOF) when the Walsh functions space is exhausted. The QOFs define three additional function sets. The Walsh functions within the set remain orthogonal, but the orthogonality between the

Walsh functions belonging to different sets is not maintained. QOFs are generated by multiplying the Walsh function by vectors called QOF masks. QOFs minimize the maximum correlation between functions that belong to different sets. The correlation between any Walsh function and the corresponding set of QOFs is independent of the Walsh function index and dependent only on its length. As we do not further discuss QOF in this book, refer to [4] for details.

4.2.9.2   *Reverse-Link Orthogonal Spreading*

On the reverse CDMA channel, orthogonal spreading is used to code-multiplex the R-PICH, R-FCH, R-DCCH, R-CQICH, R-ACKCH, R-CCCH, and R-EACH. When the R-CCCH or R-EACH is used, the only other active channel is the R-PICH. Orthogonal spreading does not apply to the legacy IS-95 channels.

The reverse code channels are associated with a Walsh function W_n^N of length 2^N Walsh chips, $N = 1,\ldots,6$. BPSK symbols are multiplied by the corresponding Walsh function at a fixed chip rate, R_c, equal to 1.2288 Mcps. The Walsh function is repeated a number of times such that the repeated Walsh function duration is equal to that of the BPSK symbol. After spreading, the gain of the code channels is set relative to that of the R-PICH by an amount specified by means of Layer 3 signaling. Then the code channels are mapped to either the I or Q branch of the complex multiplier that follows, depending on the channel type. The code channels mapped to the I branch are added together, and so are the ones mapped to the Q branch. The Walsh function assignment and code channel I/Q mapping are always predetermined. Assignment is devised in such a way that code channels within the I or Q code channels' set are orthogonal to one another. Walsh function assignment and I/Q mapping are depicted in Figure 4.17. Note that unlike the forward CDMA channel that is dimension-limited, as a large number of channels must be orthogonally code-multiplexed and, thus, uses QPSK, in the reverse CDMA channel the number of code channels is much smaller, so dimensionality is not an issue and the more robust BPSK can be used.

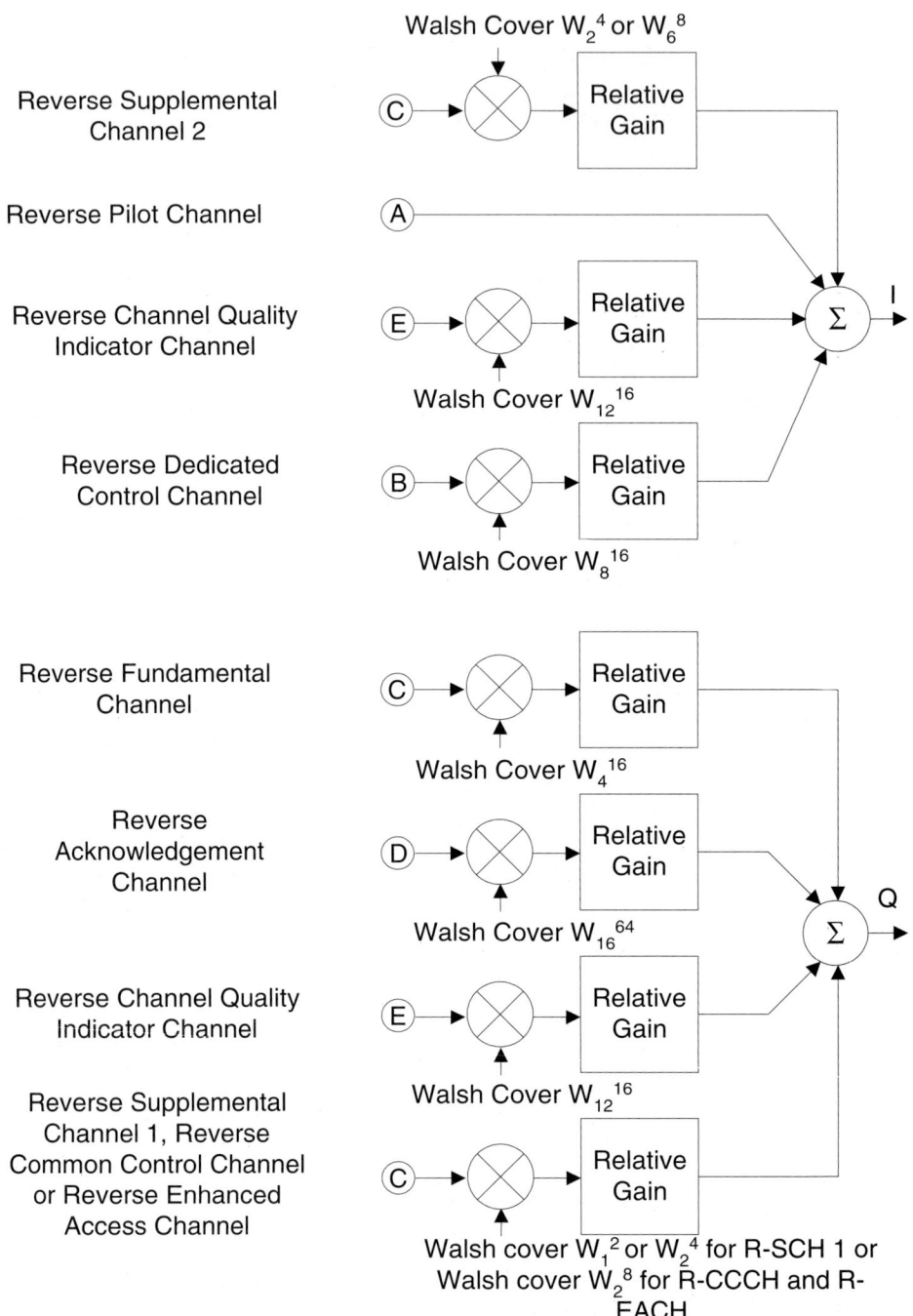

Figure 4.17 Reverse link I and Q mapping for IS-2000 channels.

4.2.10 Quadrature Spreading

Spreading is the process by which the information-bearing data sequence of relatively low rate is multiplied with a pseudorandom sequence of chips at a high rate, thus forming a broadband signal on the channel. At the receiver, the data sequence is recovered by dispreading, which consists of multiplying the received signal with a synchronized replica of the spreading sequence. After dispreading, the interference caused by other users' signals is reduced relative to the intended signal by an amount proportional to the chip rate to data rate ratio, also called processing gain. As the processing gain is typically large, the spreading/dispreading process effectively separate users' signals at the receiver.

4.2.10.1 Forward-Link Quadrature Spreading

In the forward direction, bandwidth expansion is achieved by orthogonal spreading at 1.2288 Mcps. However, with orthogonal spreading alone, signals of different base stations cannot be separated at the receiver, as they all use the same Walsh functions for orthogonal spreading. Moreover, when multipath is present, the mutual interference among paths would be excessive if orthogonal spreading alone was used, as Walsh functions have poor autocorrelation properties. Quadrature spreading is then used to ensure that both multipath and the forward CDMA channels belonging to different base stations can be separated at the receiver. With quadrature spreading, the sequence of Walsh chips on the I and Q branches are each modulo-2 added with one of two base station-specific sequences of pseudorandom chips. The pair of sequences are called short PN sequences because their length of the sequence is short, 2^{15} chips, relative to that used for data scrambling. The generator polynomials for I and Q PN sequences respectively are given by

$$PN_I(x) = x^{15} + x^{13} + x^9 + x^8 + x^7 + x^5 + 1 \tag{4.2}$$

$$PN_Q(x) = x^{15} + x^{12} + x^{11} + x^{10} + x^6 + x^5 + x^4 + x^3 + 1 \tag{4.3}$$

Note that, in general, a 15-bit long register generates a chip sequence of length $2^{15} - 1$. A length 2^{15} is obtained by inserting an extra zero after a run of 14 zeros. That means that each short PN sequence contains one run length of 15 zeros and no run length of 14 zeros. The run of 15 zeros is inserted for synchronization purposes so that the PN sequence is repeated exactly 75 times over a 2-second interval. Note that all base stations use identical short PN sequences. However, neighboring base stations use distinct transmission timing offsets relative to the same time reference so that the PN sequences, because of their autocorrelation properties, are effectively distinct and can be discriminated by the mobile station receiver.

Quadrature spreading by means of the short I and Q PN spreading codes for all channels other than the F-PDCH[7] and for the case when transmit diversity is not enabled is illustrated in Figure 4.18. After spreading, the chips are fed to a baseband filter for signal shaping and up-converted to a carrier frequency, as explained in the following section.

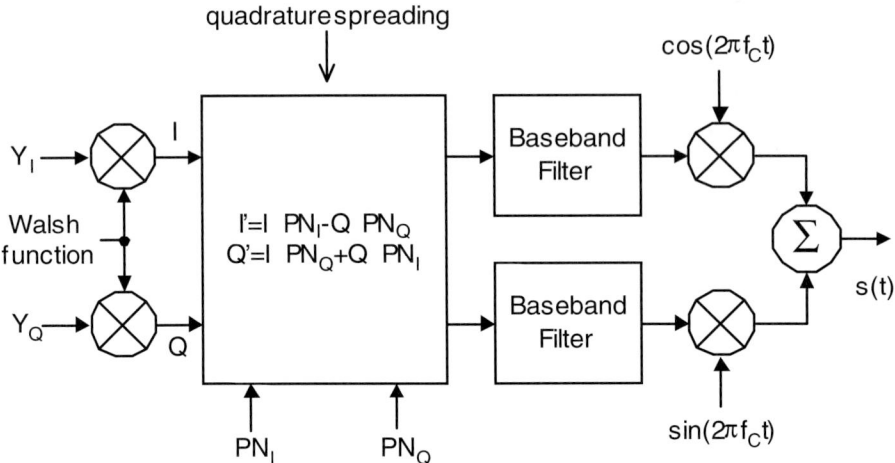

Figure 4.18 I and Q symbol mapping for nontransmit diversity mode.

4.2.10.2   Reverse-Link Quadrature Spreading

For all channels but the legacy IS-95 channels, the I-channel and Q-channel data are multiplied by a complex spreading sequence before baseband filtering, as depicted in Figure 4.19. The complex multiplier uses the same two short PN sequences used for forward-link spreading, and two long PN codes, one a 1-chip delayed replica of the other. The short PN sequences are the same for all mobile stations and are identical to those used on the forward CDMA channel. The long code is used to discriminate users' signals at the base station receiver. The long code is user-specific in case of dedicated channels and base station-specific in case of common channels. The long code can also be used for voice privacy, as only the legitimate user possessing the correct long code can despread the signal and decode the user's data.

The structure of the complex spreading device is designed to maintain the complex signal envelope nearly constant by avoiding zero-crossings between symbol transitions. That allows achieving higher efficiency of the mobile station power amplifier while still maintaining operation in the amplifier's linear region. If the peak-to-average was excessive, then the nominal operating point would have to be reduced to avoid saturating the amplifier, thus generating spectral regrowth that causes interference to the adjacent channels.

---

7.   Quadrature spreading is also used for F-PDCH. However, F-PDCH-specific quadrature spreading is illustrated in a separate figure.

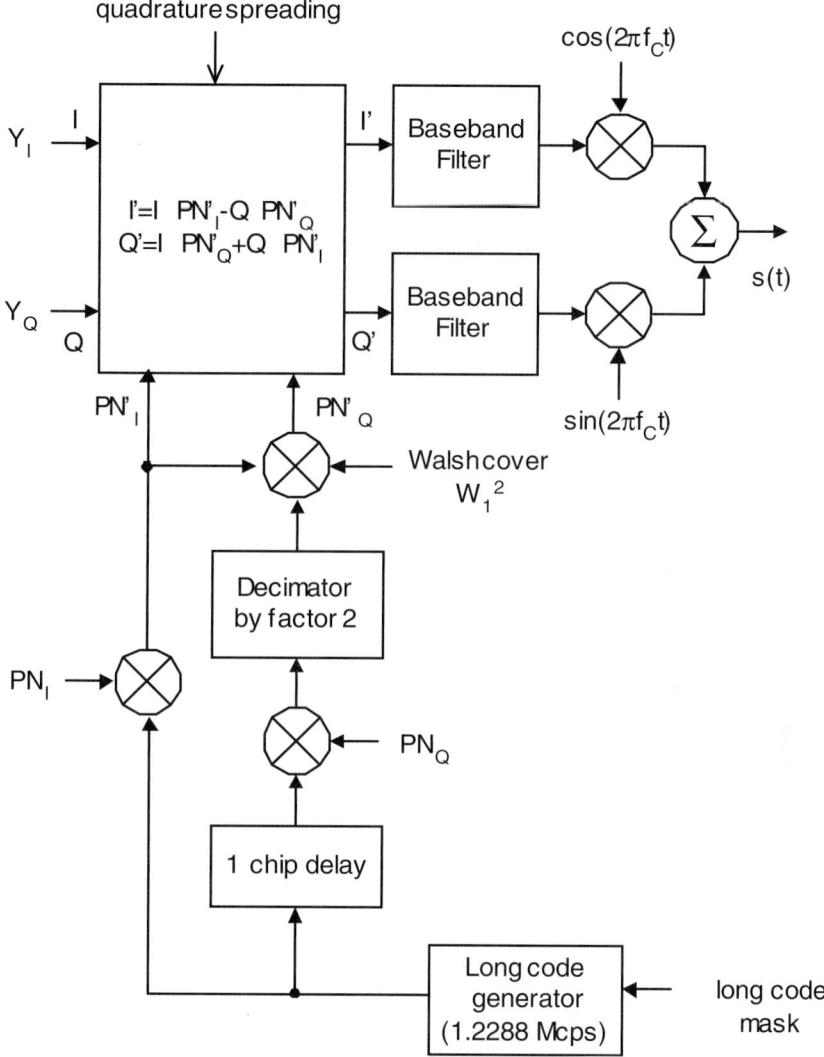

quadrature spreading

$$I' = I\ PN'_I - Q\ PN'_Q$$
$$Q' = I\ PN'_Q + Q\ PN'_I$$

$\cos(2\pi f_c t)$

$\sin(2\pi f_c t)$

Baseband Filter

Baseband Filter

Walsh cover W_1^2

Decimator by factor 2

PN_I

PN_Q

1 chip delay

Long code generator (1.2288 Mcps)

long code mask

s(t)

Figure 4.19 Reverse-link quadrature spreading.

4.2.11 Forward-Link Transmit Diversity

The cdma2000 physical layer supports transmit diversity operation. Transmit diversity, in which two antennas are used to transmit the same code channel, has the potential to achieve increased diversity gain relative to that provided by, for example, multipath and to improve forward-link performance. Two modes are allowed: Orthogonal Transmit Diversity (OTD) and

Space Time Spreading (STS). The I and Q mappings are different for OTD and STS. OTD simply multiplexes adjacent coded bits onto different antennas, while STS employs the orthogonal 2 × 2 space-time code proposed in [5].

Figure 4.20 shows the modulation symbol demultiplexing for OTD and STS, respectively. The demultiplexer, when transmit diversity is enabled, has four outputs, unlike when transmit diversity is not enabled in which only two outputs are present. For both OTD and STS, the demultiplexer maps the first symbol in each frame to the Y_{I1} output, and the subsequent symbols are mapped to the Y_{I2}, Y_{Q1} and Y_{Q2} outputs in alternate fashion until the last symbol in the frame. The symbol repetition depends on the transmit diversity mode that is employed.

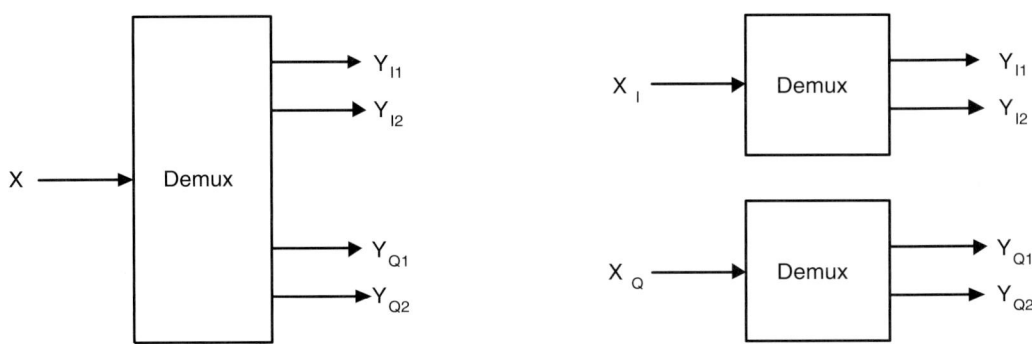

Figure 4.20 Demultiplexing for OTD and STS.

Figure 4.21 shows the symbol mapping when OTD is enabled. Each demultiplexed symbol output is repeated once to create two identical symbols for each input symbol. The first repeated symbol output on both the Y_{I1} and Y_{Q1} branches are not inverted, while subsequent outputs on the Y_{I2} and Y_{Q2} branches are alternately inverted.

When STS is enabled, each demultiplexed symbol output is repeated once to create two identical symbols for each input symbol. However, while in OTD mode symbols are mapped to only one of the two diversity antennas, in STS mode symbols are coded and sent on both diversity antennas. The coding mechanism is depicted in Figure 4.22.

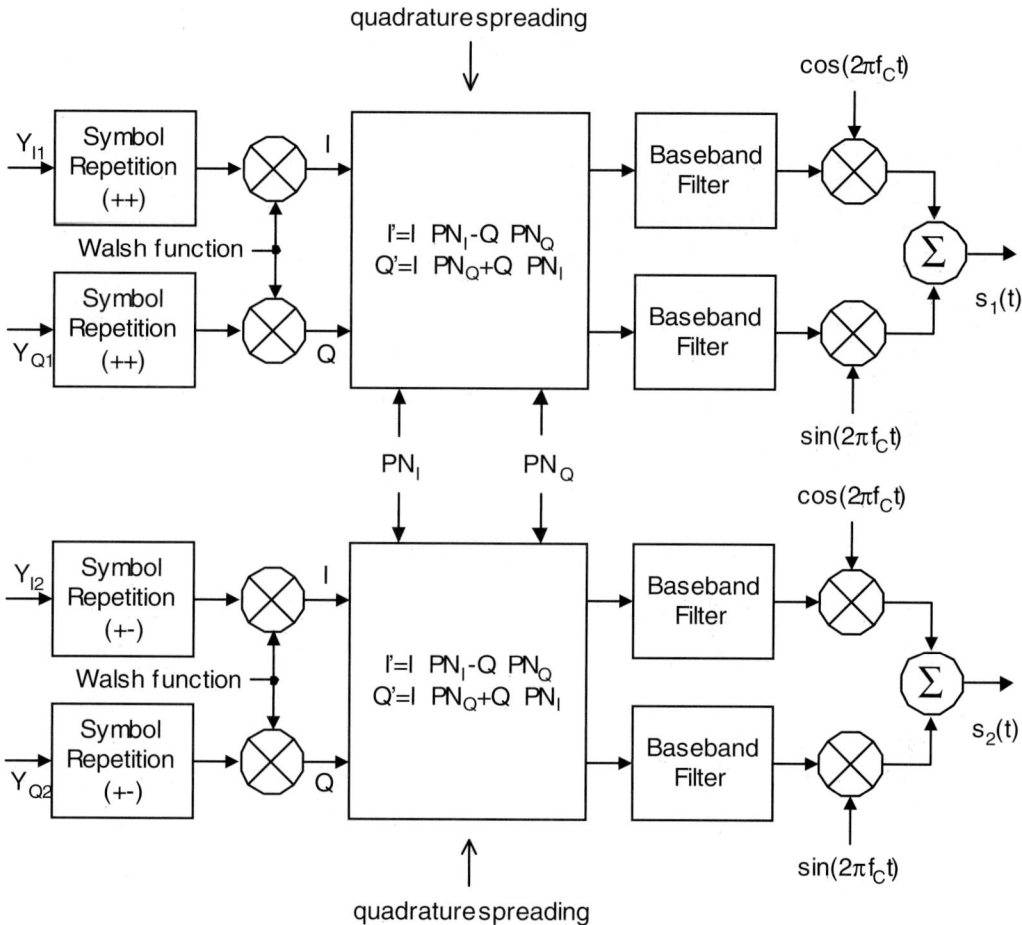

Figure 4.21 I and Q symbol mapping for the orthogonal transmit diversity mode.

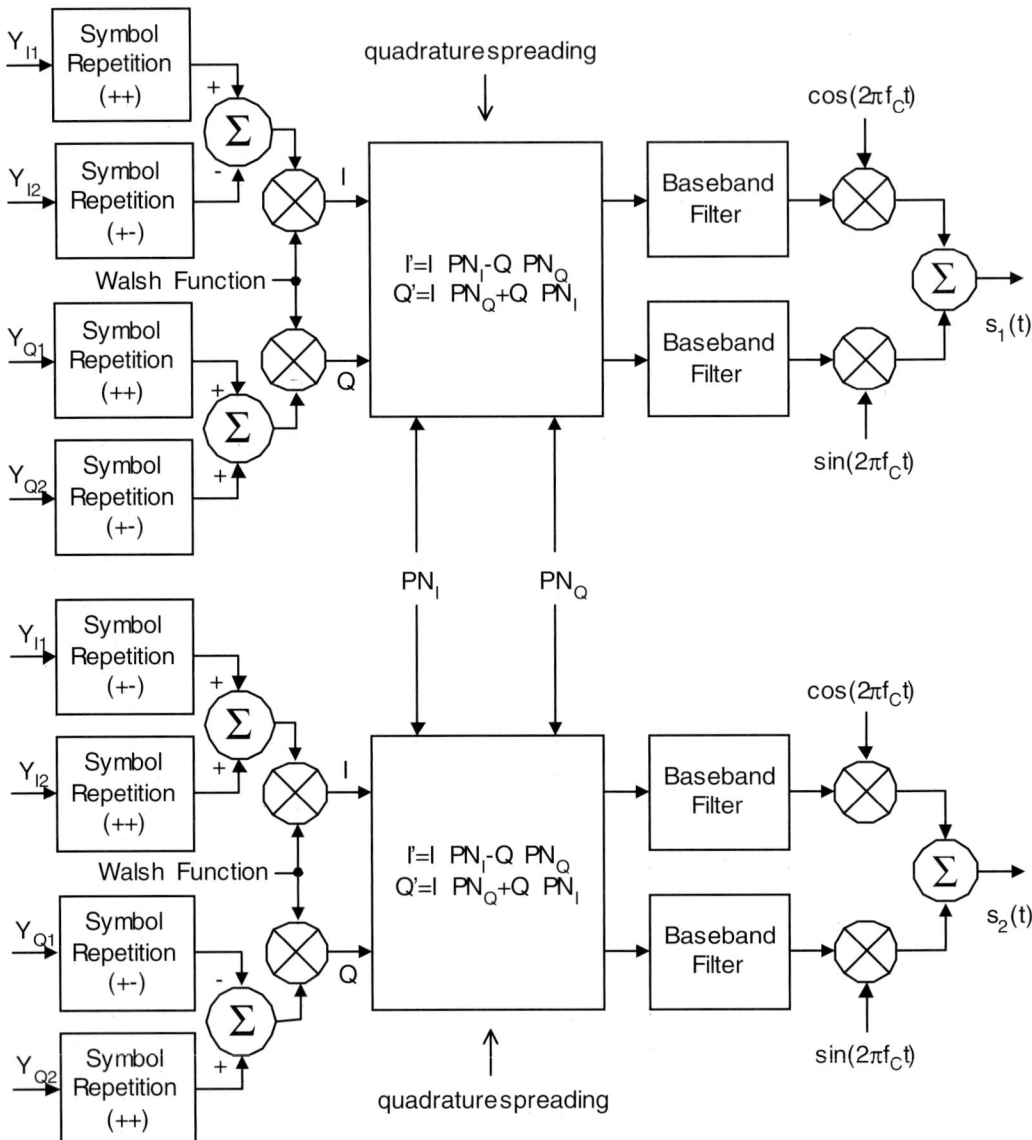

Figure 4.22 I and Q symbol mapping for the space-time spreading mode.

4.2.12 Baseband Filtering

Following spreading and before frequency up-conversion, the I and Q chips are applied to a low-pass filter in the form of a train of impulses. The filter impulse response determines the pulse shape of the signal transmitted over the air. The physical layer protocol specifies the nominal impulse response and frequency response of the filter. The nominal impulse response is specified in terms of a 48-tap finite impulse response (FIR) filter with a sampling rate equal to four times the chip rate, equivalent to 203.451-ns tap spacing. The filters used for baseband filtering of the forward and reverse CDMA channels are identical. In this section we discuss time-domain properties of the filter, while properties of its frequency response are discussed in the next section.

The pulse shape used in IS-2000 does not satisfy the Nyquist criteria for zero intersymbol (or interchip) interference. The filter is designed to achieve a high degree of spectral compactness by allowing some interchip interference (ICI) at the matched filter output. When sampling the matched filter impulse response at the optimum instant, the intended output, $s(0)$ is affected by the interference caused by chip-modulated contiguous impulses:

$$s(0) = \sum_{i=-\infty}^{\infty} c_i p(i \cdot T_c) = c_0 + \sum_{\substack{i=-\infty \\ i \neq 0}}^{\infty} c_i p(i \cdot T_c) \qquad (4.4)$$

where c_i represent the sequence of binary chips spanned by the impulse response and $p(t)$ denotes a pulse shape. The rightmost factor in Eq. (4.4) represents the ICI. If the chip sequence is pseudorandom and chips are assumed independent, the ICI is a zero-mean random variable with variance equal to $\sigma^2 = \sum_i p(i \cdot T_c)^2$. The matched-filter impulse response, considering 23 samples, is shown in Figure 4.23.

The ICI variance is equal to 0.0229, which corresponds to a signal-to-interference ratio (SIR) of 16.4 dB. This value is an upper bound on the maximum SINR achievable. ICI, however, only slightly affects demodulation performance because the actual SINR is dominated by multiple-access interference in the operating range of interest. The effect of the ICI in the exemplary case of forward code channels employing QPSK modulation can also be seen in Figure 4.24, which shows how the matched filter outputs are scattered around the signal constellation points.

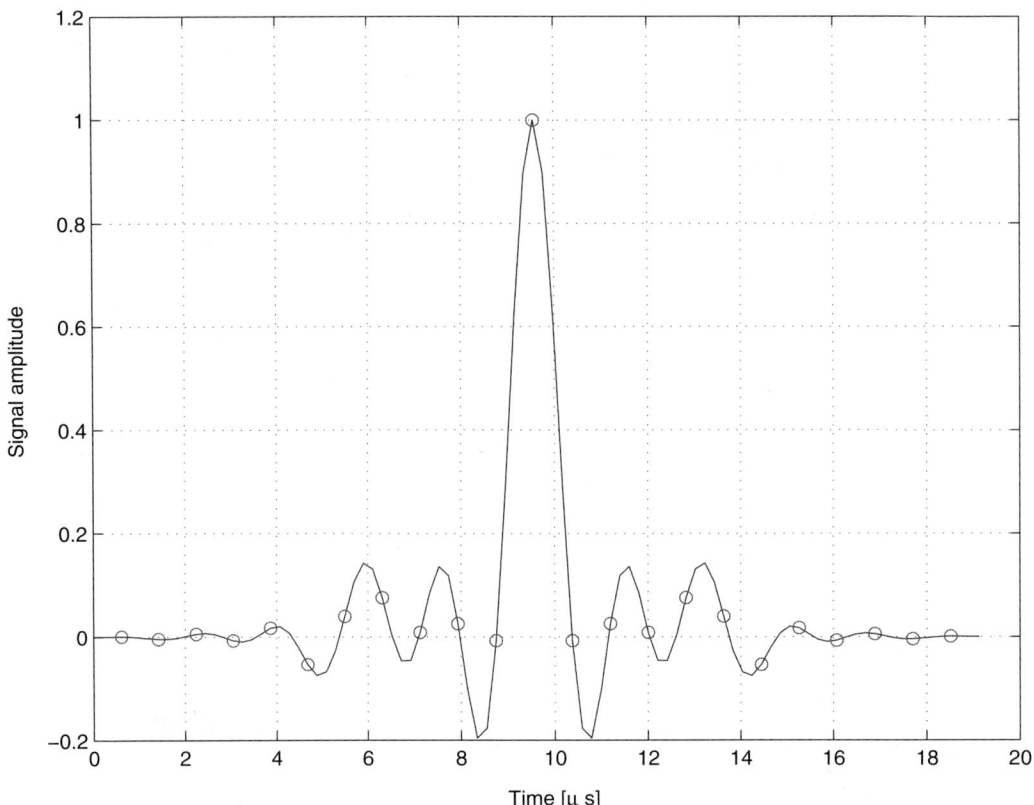

Figure 4.23 Sampled matched-filter impulse response.

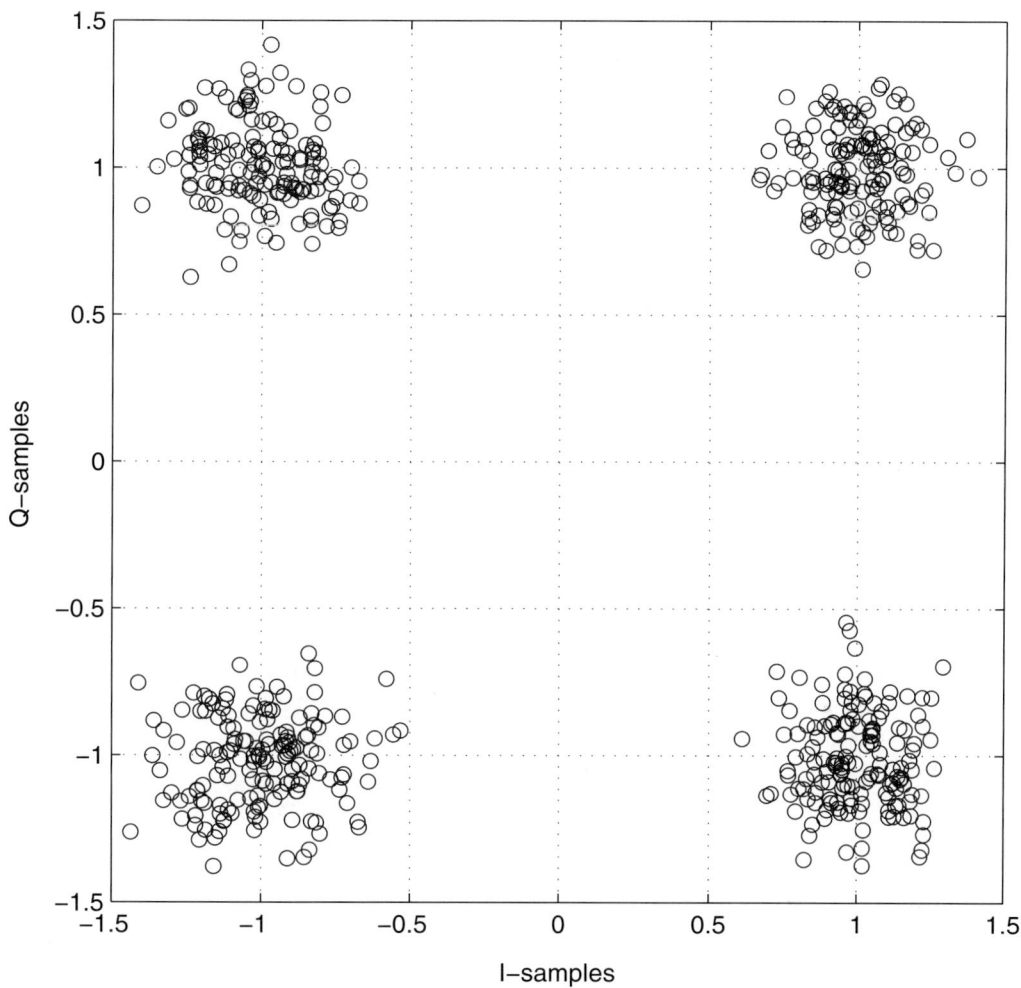

Figure 4.24 Scatter diagram of I-Q matched-filter outputs sampled at the optimum instant.

4.2.13 Band Packing

By using a non-Nyquist pulse shape, the interchip interference represents a tradeoff with spectral efficiency. The IS-2000 pulse shape achieves a high degree of efficiency that allows tight packing for contiguous CDMA carriers within the available band while maintaining adjacent channel interference at acceptable levels.

The pulse frequency characteristic is specified by the mask depicted in Figure 4.25. The mask's tolerances are expressed in terms of the maximum ripple in the passband ($\alpha_1 = \pm 1.5$ dB), the minimum passband frequency range ($2f_p = 2 \cdot 590$ kHz), the maximum stopband frequency range ($2f_s = 2 \cdot 740$ kHz), and the minimum rejection in the stopband ($\alpha_2 = 40$ dB).

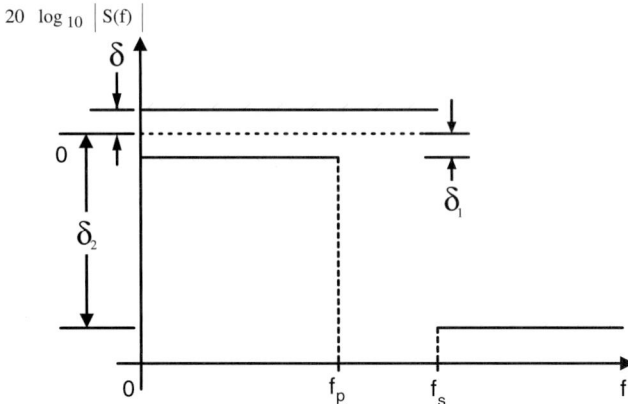

Figure 4.25 Baseband filter mask.

After filtering, the signal is up-converted by mixing with a local oscillator. The resulting passband signal is then fed to the transmit antenna. The local oscillator is tuned to the assigned carrier frequency. The assigned transmit-carrier frequency belongs to one of 13 different band classes currently specified in IS-2000. cdma2000 systems are most widely deployed in the bands specified in Table 4.6. The standard specifies the frequency assignments that are permissible within each band class. Frequency assignments are identified by a CDMA channel number. The permissible CDMA channels form a channel raster. The raster spacing depends on the band class. For example, Band Class 0 uses a 30 kHz raster because of the legacy channels used in first-generation cellular systems. Personal communications systems (PCS) systems, on the other hand, use a 50-kHz raster.

Table 4.6   Examples of Band Classes and Frequency Correspondence

Band class	MS Tx freq. band [MHz]	Fwd/Rev frequency separation	Channel raster spacing
0 (800 MHz, cellular)	824.025–848.985	45 MHz	30 kHz
1 (1900 MHz, U.S. PCS)	1,850–1,910	80 MHz	50 kHz
3 (800 MHz, JTACS)	887.0125–924.9875	55 MHz	12.5 kHz
4 (1700 MHz, Korean PCS)	1,750–1,780	90 MHz	50 kHz
6 (2 GHz, IMT2000)	1,920–1,979.95	190 MHz	50 kHz

As CDMA systems operate with full-duplex transmission, they must operate on a paired spectrum. That is, the forward and reverse channel carrier frequencies are separated by a fixed amount to avoid mutual interference. For the cellular, U.S. PCS, and Korean PCS bands, the separation is 45, 80, and 90 MHz, respectively. Typically, the mobile station operates at lower frequency than the base station. The reason is that, because path loss is proportional to the carrier frequency, the signal transmitted on the reverse direction is given a slight link budget advantage to compensate for the lower mobile station transmit power relative to the base station transmit power. A notable exception is Band Class 3 (JTACS) in which the mobile station transmit carrier frequency is 55 MHz higher than that of the base station.

Band classes are divided in blocks and the numbers of the CDMA channels in each block are specified. The definition of channel blocks within the same band stems from government regulations that allocate the blocks to different service providers, which therefore can coexists and offer competing services in the same geographical area and band class. Block sizes depend on the band class. For example, Band Class 4 use 10-MHz blocks, while Band Class 1 uses 5- or 15-MHz blocks.

The block size and the raster spacing are important because they determine the granularity with which CDMA carriers can be packed within a block. The minimum CDMA carrier spacing used for cellular systems is 1.23 MHz (equivalent to 41 channels with a 30-kHz raster). For PCS systems, the minimum CDMA carrier spacing is 1.25 MHz (equivalent to 25 channels with a 50-kHz raster). The CDMA carrier spacing in turn determines the amount of adjacent channel interference. It can be seen that the power captured within a 1.23- and a 1.25-MHz frequency band by the nominal filter frequency response is 95.2% and 96.5%, respectively. The interference caused by the adjacent channels puts an upper bound on the maximum achievable SIR. The maximum SIR in the presence of two-sided adjacent channel interference is equal to 13.2 dB and 14.6 dB for carrier spacing equal to 1.23 MHz and 1.25 MHz respectively, provided that the base stations operating on the adjacent channels are collocated. If base stations are not collocated, the effective SIR may be larger or smaller depending on whether the path loss between the mobile station and the serving base station is larger or smaller than that between the mobile station and the adjacent channel base station. In such a case, if spectrum availability allows for it, the carrier spacing can be increased to reduce the interference.

More stringent adjacent channel-interference requirements are imposed when systems operated on adjacent frequency blocks belong to different service providers. That is needed both because of local regulatory requirements and, more from a practical standpoint, because base stations of different operators are not collocated. As observed before, carrier spacing needs to be increased to account for the path-loss difference between the mobile station and the serving and adjacent channel base stations. That is achieved by using a guardband to further separate CDMA channels at the edge of the block. A sufficient guardband is 625 kHz. When using the 1.25 MHz carrier spacing, for example, three CDMA carriers can be packed in a 5-MHz block, leaving 625 kHz of guardband on each side of the block. In a 10-MHz block, seven carriers can be packed with 1.25 MHz spacing and still leave space for the 625-kHz guardband.

4.3 Forward Code Channels

We have already explained the spreading and modulation procedures that are to a large extent shared among most of the channels. In this section, we describe the channel-specific encoding processes for all the forward channels.

4.3.1 Forward Pilot Channel (F-PICH)

The first channel every mobile station attempts to acquire when it powers up is the F-PICH. The F-PICH is used for initial acquisition by the mobile station searcher, the time tracking loops, and for estimation of the radio propagation channel.

The F-PICH encoding process is shown in Figure 4.26. It consist of the all-zeros sequence transmitted on the I branch. Walsh spreading is performed with Walsh code W_0^{64}. The PN offset of the short PN scrambling code uniquely identifies the cell to a mobile station.

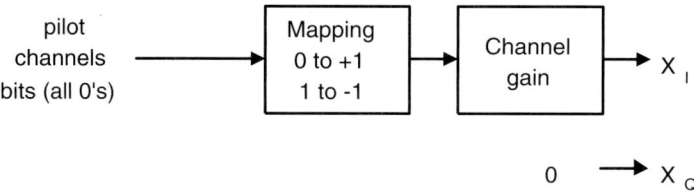

Figure 4.26 F-PICH encoding.

4.3.2 Forward Synchronization Channel (F-SYNCH)

Once the F-PICH is successfully acquired, mobile stations need to decode the F-SYNCH. The F-SYNCH identifies to the mobile the system information as well as the base station and which pilot has been acquired. The F-SYNCH frame length is equal to the short PN code sequence length, 26.66 ms, and it is transmitted synchronously with the 15 zeros run length. When a mobile station detects 15 consecutive zeros in the short PN code, it begins decoding the F-SYNCH.

The F-SYNCH encoding is shown in Figure 4.27. Each frame consists of 32 information bits, encoded with a rate of one-half convolutional code of constraint length 9 and transmitted on the I branch. Walsh code W_{32}^{64} is reserved for the F-SYNCH. The frame is not protected by a cyclic redundancy check (CRC) code. Rather, an entire F-SYNCH message that spans several frames contains its own CRC.

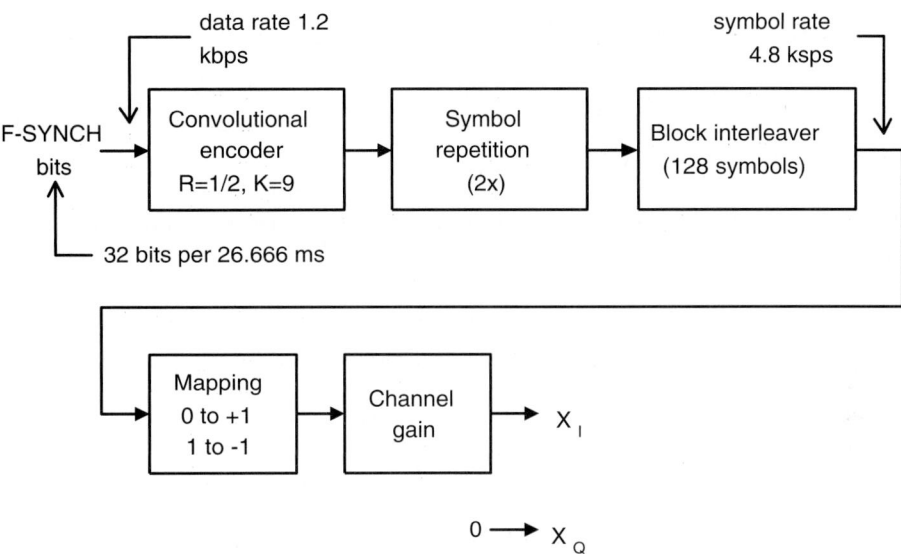

Figure 4.27 F-SYNCH encoding.

4.3.3 Forward Paging Channel (F-PCH)

After decoding the F-SYNCH message, the mobile station is instructed to decode the F-PCH in order to obtain more detailed system information. Besides carrying the overhead system information, the F-PCH can also carry mobile-specific messages. An example of the mobile-specific message is the IS-2000 Layer 3 [6] *General Page Message* that alerts the mobile station of an incoming call.

F-PCH encoding is shown in Figure 4.28. The channel can operate at a rate of 4.8 kbps or 9.6 kbps. The frame length is 20 ms, and the corresponding number of bits per frame is 96 and 192 bits, respectively. Each frame is divided into two half frames, and eight half frames make a paging slot. The paging slot duration is constant, 80 ms, but the number of paging slots is configurable, ranging from 16 to 2048. One paging slot is assigned to every mobile in the system, and multiple mobiles can share a slot. The long PN code sequence is used to scramble the coded bits. One primary paging channel is mandatory, and additional paging channels, up to six, are optional. Additional F-PCHs are implemented if a single F-PCH cannot handle the signaling load. A single F-PCH is, however, common implementation. Regardless of the supported rate, the F-PCH employs the Walsh code of length 64. The Walsh code space reserved for the primary paging channels is W_1^{64}, while secondary F-PCHs consume Walsh codes from W_2^{64} to W_7^{64}. However, if fewer then seven paging channels are supported by the system, the remaining Walsh codes can be reused for traffic channels. Each paging channel employs a unique 42-bit long code offset mask. The F-PCH frame is not protected by a CRC, but each paging message, which can span a number of half frames, is protected by its own CRC code.

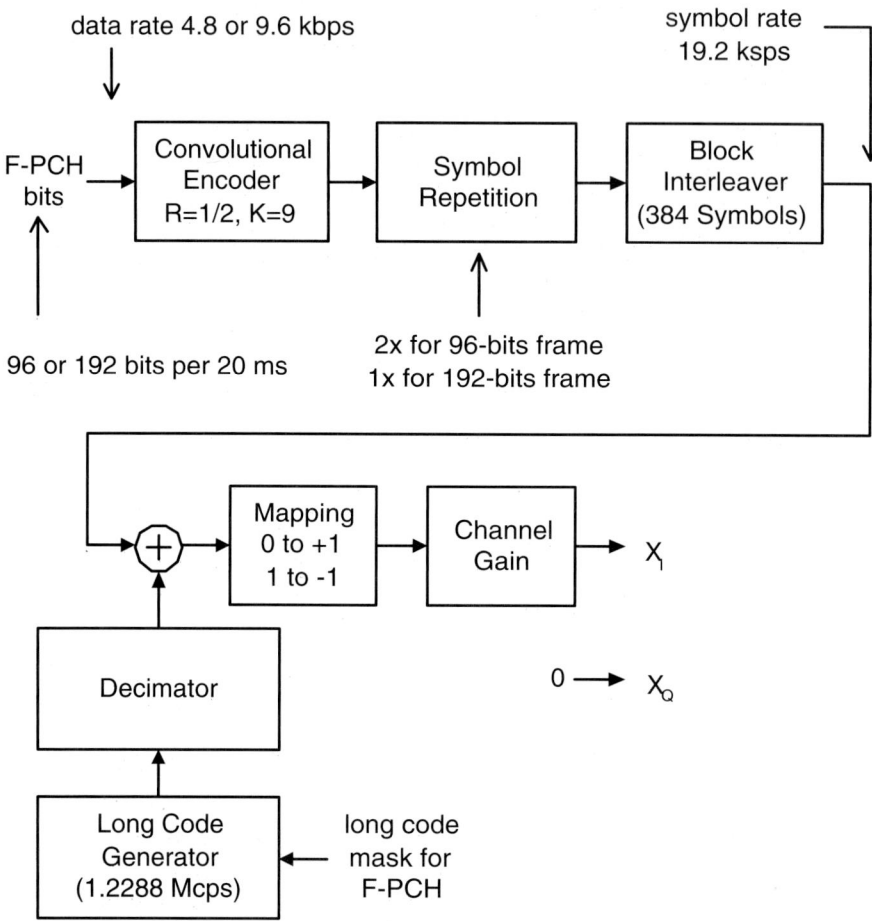

Figure 4.28 F-PCH encoding.

4.3.4 Forward Broadcast Control Channel (F-BCCH) and Forward Common Control Channel (F-CCCH)

The functions of the F-PCH can be substituted with two other channels, the F-BCCH and F-CCCH. The F-BCCH serves to broadcast system-specific and cell-specific overhead information, while the F-CCCH is used for mobile station-directed messages. The separation of the F-PCH into two physical channels allows for optimization of the channel structure design and optimization of system resources. For example, the F-BCCH and F-CCCH do not have to operate at the same rates and the same power level. The frame lengths can also be different.

The F-BCCH encoding process is shown in Figure 4.29. The channel frame length is 40 ms, and the message size is 744 bits followed by the CRC and encoder tail bits. The F-BCCH can operate at three different rates: 4.8, 9.6, and 19.2 kbps. When operating at 19.2 kbps, each frame is transmitted only once. However, when operating at 9.6 and 4.8 kbps, after the initial transmission, the message is repeated once and three times, respectively. When the frame is repeated, the receiver can soft-combine a newly received frame with the previous transmission of the same frame. This mechanism allows receivers to stop decoding as soon as the frame is correctly received. The code rate for this channel can be either half or quarter. With code rate half, the Walsh code length is 64, while with the code rate quarter, the Walsh code length is 32. The specific Walsh code for the primary F-BCCH is indicated in the F-SYNCH message. The secondary F-BCCHs are indicated on the primary F-BCCH through overhead messages. The maximum number of channels is eight. Each F-BCCH is scrambled with the long PN code with a unique offset mask.

The F-CCCH structure is shown in Figure 4.30 and Figure 4.31. The channel is divided into 80-ms slots, where each mobile has one assigned slot. As is the case for the F-PCH, the minimum number of slots is 16, and the maximum number of slots is 2,048. Each slot consists of a number of frames, with variable length. The allowed values are 5, 10, and 20 ms. The rate of

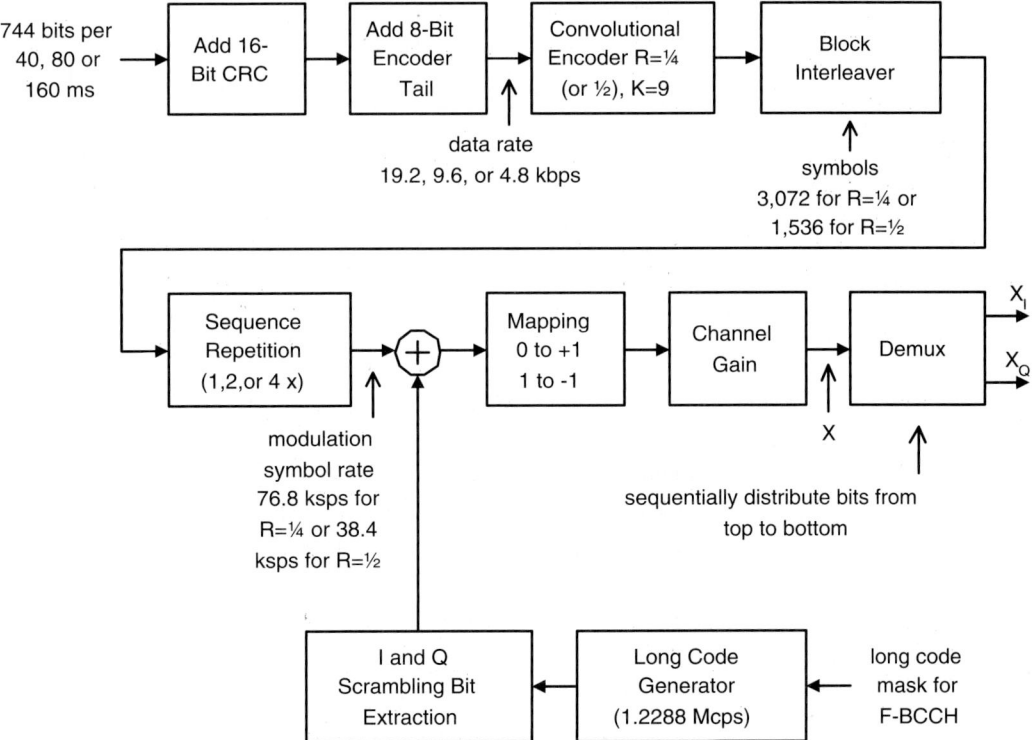

Figure 4.29 F-BCCH encoding.

the channel is also variable; however, the rate for a specific frame is predetermined and known by the mobile station the frame is directed to. The supported rates are 9.6, 19.2, and 38.4 kbps. With respect to the code rate, the channel operates in two modes: half- and quarter-code rate modes. The Walsh code length used for this channel depends on the data rate as well as on the code rate. For the highest data rate and the lowest code rate, the Walsh code length is 16, while for the lowest data rate and the highest code rate, the Walsh code length is 128. Overhead information on the F-BCCH provides F-CCCH Walsh code information. The F-CCCH can operate in the discontinuous mode, that is, the channel is not used if there is no message to transmit.

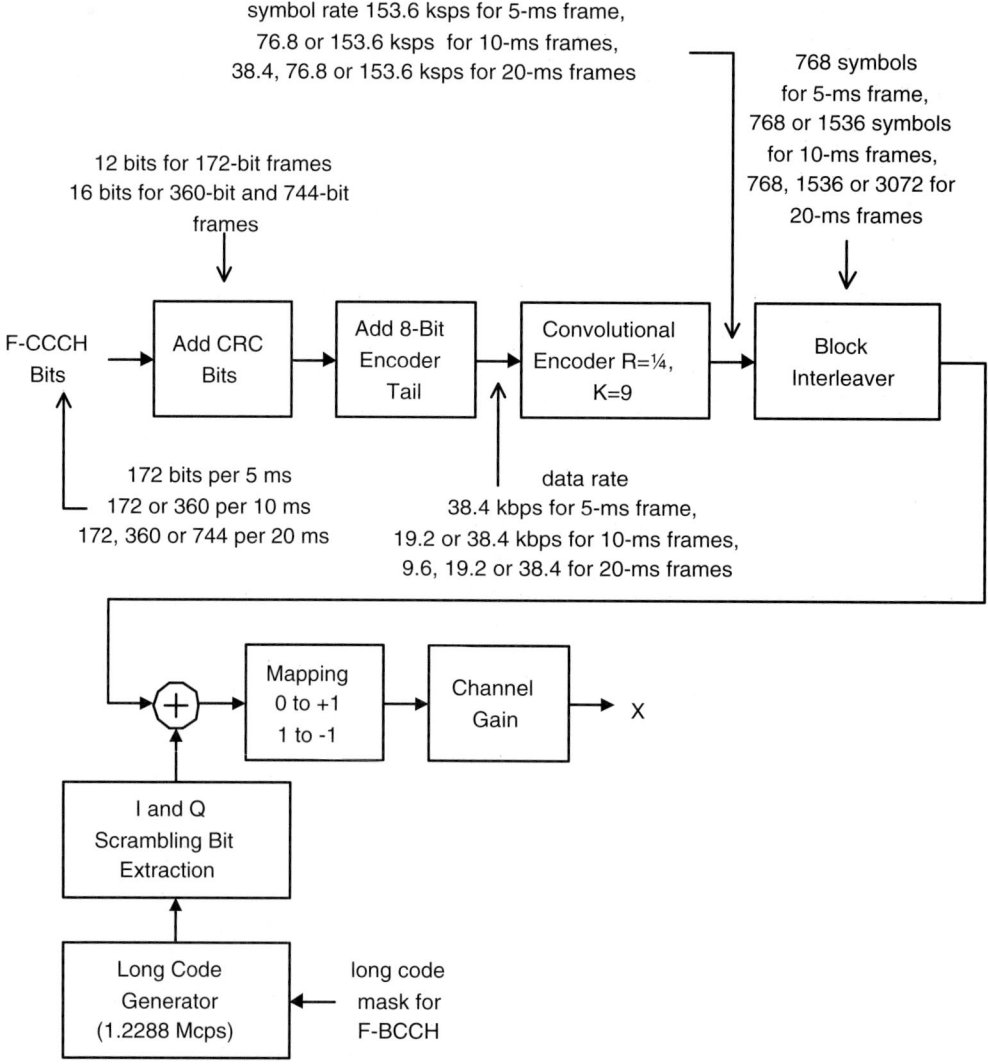

Figure 4.30 F-CCCH encoding: FEC rate R =1/4.

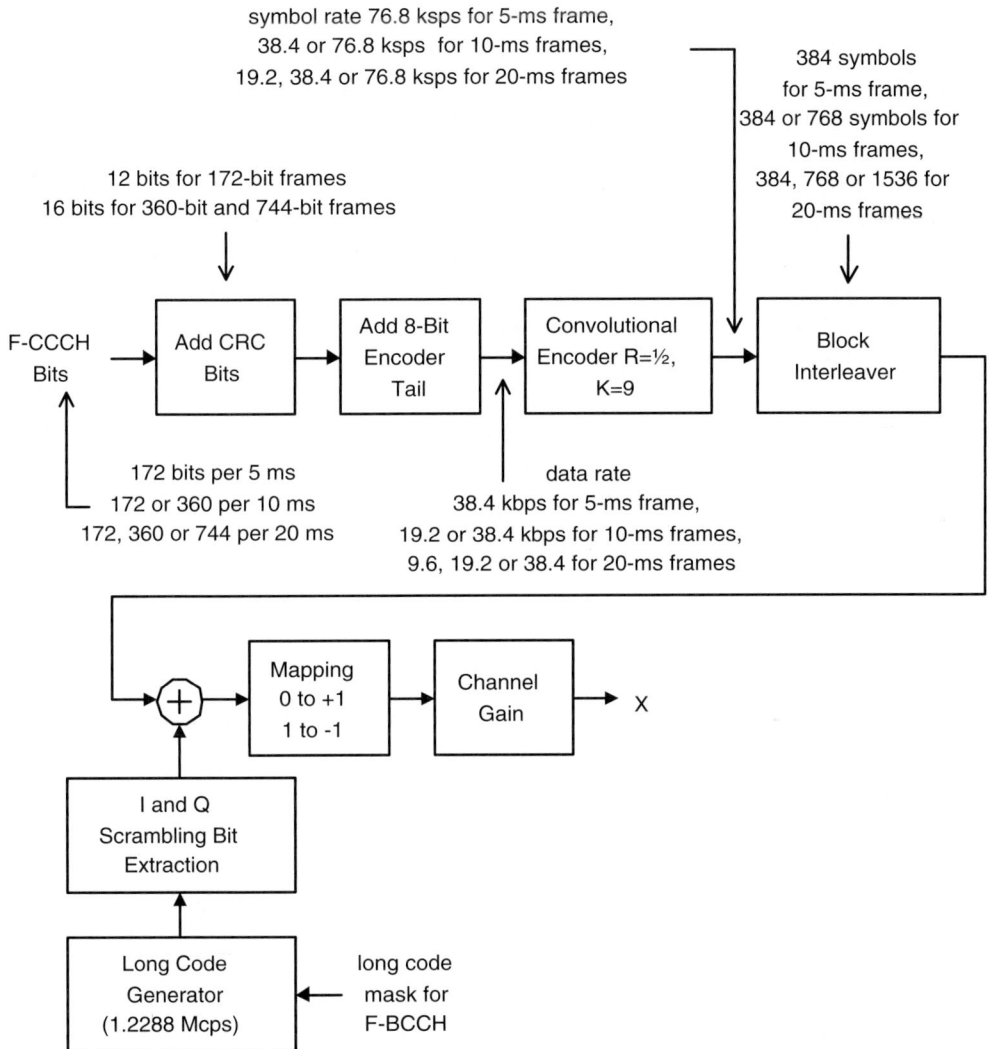

Figure 4.31 F-CCCH encoding: FEC rate R =1/2.

4.3.5 Forward Quick Paging Channel (F-QPCH)

The idea behind the F-QPCH is to decrease the time a mobile station needs to monitor the F-PCH or F-CCCH. Recall that each mobile is assigned an 80-ms slot, which it must periodically decode to receive page messages. The period at which the mobile station must decode paging channels (F-PCH or F-CCCH) can be as short as 1.28 ms. This means that the mobile has to "wake up" every 1.28 ms and decode the entire 80-ms slot on the F-PCH or F-CCCH. For a typical mobile that receives only several pages a day, this time significantly contributes to the battery-power consumption. The F-QPCH increases the battery life of the mobile because the F-PCH or F-CCCH is decoded only if the mobile receives an indication on the F-QPCH. The channel structure is quite simple, and decoding the F-QPCH is not a processing-intensive operation. It consists of a pair of indicator bits, the location of which are mobile station-dependent.

The channel structure is shown in Figure 4.32. The size of the F-QPCH slot is 80 ms. The channel data rates are 2.4 or 4.8 kbps, while the actual paging indicator rates are 4.8 and 9.6 ksps, respectively. The total number of paging indicators per slot is 384 for the lower and 768 for the higher data rate. There are two paging indicators per mobile and two mobiles can share a paging indicator. The actual mapping is determined by a hashing function. Up to three F-QPCHs are allowed. The reserved Walsh codes are W_{80}^{128}, W_{48}^{128}, and W_{112}^{128}. The information on which particular code or codes are used is contained in the overhead information transmitted over the F-PCH or F-BCCH. The modulation on the F-QPCH differs from other channels. That is because the F-QPCH employs on-off keying (OOK) modulation. However, Figure 4.18 shows that the I and Q branch mapping is still valid. When the paging indicator is present, the modulation symbol equals +1, but when the paging indicator is not present, the modulation symbol equals 0 and no signal is transmitted over the air. The detection performance of OOK is worse than that of pure QPSK, but the use of the indicator slot is so rare that OOK consumes much less forward-link capacity than QPSK would.

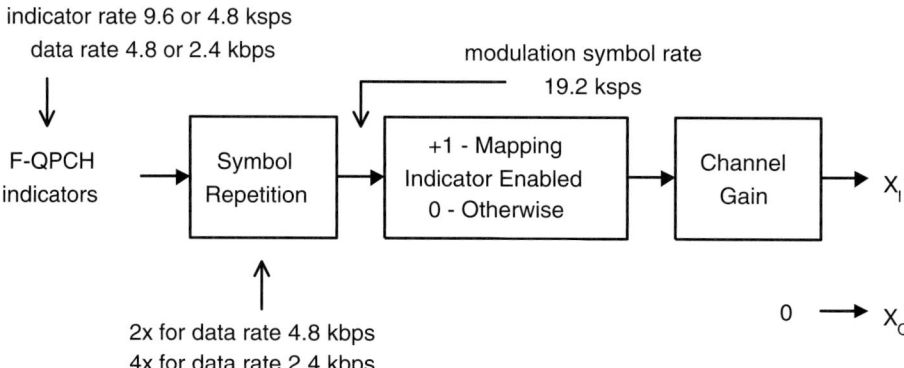

Figure 4.32 F-QPCH encoding.

4.3.6 Forward Common Assignment Channel (F-CACH)

The F-CACH is introduced to support the reservation access mode on the R-EACH. The message that assigns the R-CCCH and F-CPCCH subchannel after the reservation access probe is transmitted on the R-EACH is transmitted on the F-CACH.

The encoding process is illustrated in Figure 4.33. The frame length of this channel is constant, 5 ms. Depending on the mode of operation, the FEC code rate can be half or quarter. Up to seven such channels are allowed. The Walsh code length is 64 or 128 depending on the code rate employed. The F-CACH Walsh code information is broadcast over the F-PCH or F-BCCH.

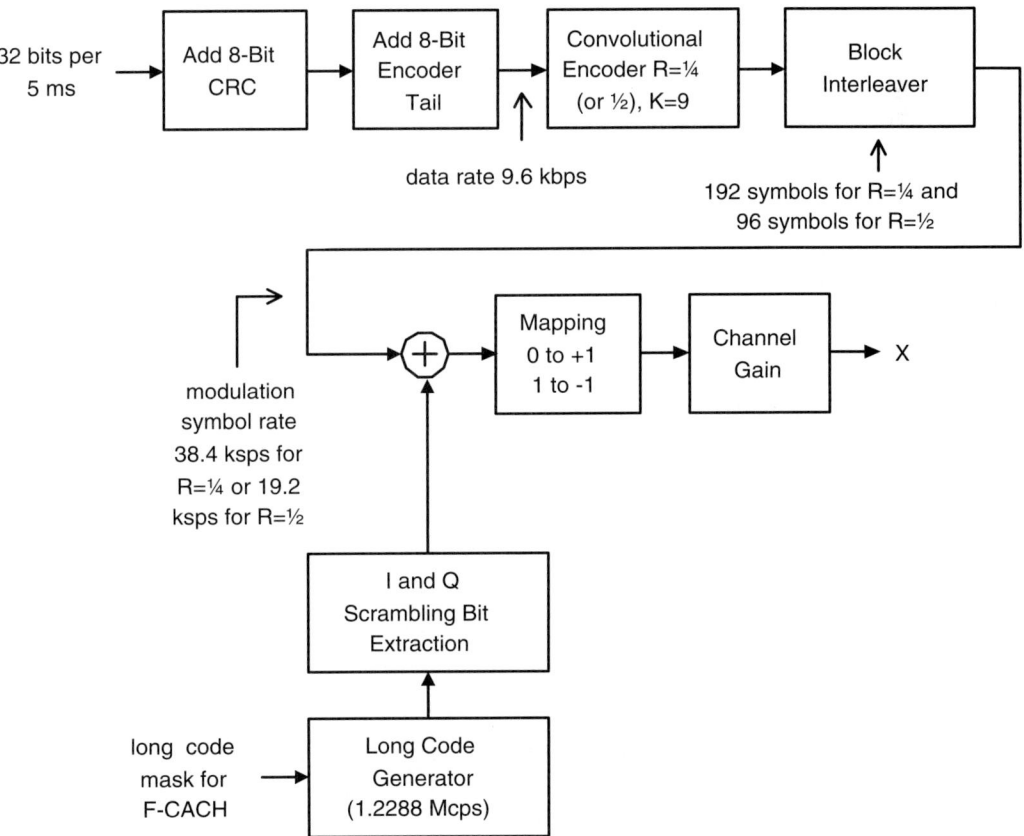

Figure 4.33 F-CACH encoding.

4.3.7 Forward Common Power-Control Channel (F-CPCCH)

The purpose of the F-CPCCH is twofold. One is to allow power control of the R-CCCH and R-PICH during the reservation-access mode-origination procedure, and the other is to power control the R-PICH when the mobile station is in the traffic state (see Chapter 6) and the F-FCH and F-DCCH are not assigned.

The F-CPCCH is the uncoded channel because it carries only up and down power-control commands. The standard allows for up to 15 channels. The Walsh code length is 128 in nontransmit diversity mode and 64 when any form of transmit diversity is employed. Each F-CPCCH is divided into subchannels. At an 800-bps power-control rate, there are 24 subchannels. If the power-control rate is reduced, the number of subchannels is proportionally increased. At the power-control rate of 200 bps, each F-CPCCH can support 96 subchannels. Each subchannel is uniquely associated with the corresponding R-CCCH and/or R-PICH. The F-CPCCH structure is shown in Figure 4.34. Subchannels are equally split between I and Q branches. Therefore, referring to Figure 4.34, there are *2N* subchannels per power-control group. The power-control group duration is 1.25 ms for 800-bps power-control rate, 2.5 ms for 400-bps, and 5 ms for 200-bps. The reservation access mode supports all power-control rates, while R-PICH, when the mobile station is in the traffic state, can only be power controlled at 800 bps. For transmit diversity techniques, the power-control bits are repeated so that after I and Q symbol mapping, shown in Figure 4.18, each power-control bit is transmitted over both I and Q branches. The F-CPCCH Walsh code information is signaled to the mobile with IS-2000 Layer 2 or MAC layer messages on the F-CACH.

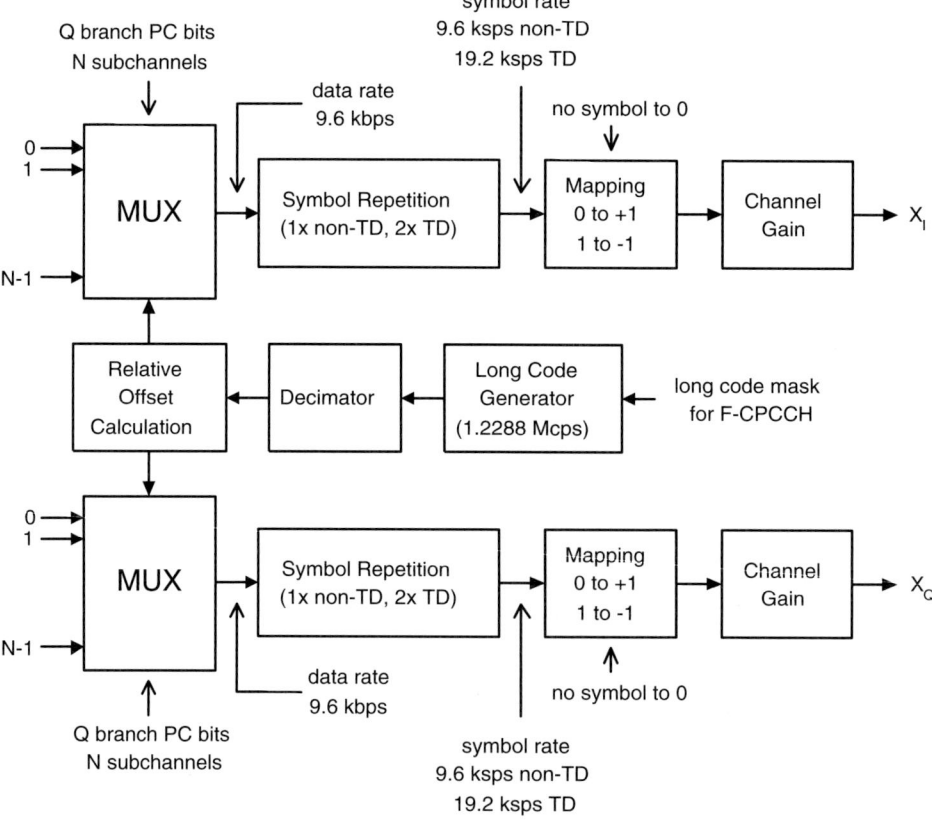

Figure 4.34 F-CPCCH encoding.

4.3.8   Forward Packet Data Channel (F-PDCH)

The F-PDCH is a shared packet data channel that supports high-speed operation traffic that can tolerate some degree of jitter. Access to the channel is handled through MAC layer scheduling. This channel is different from all other channels because of adaptive modulation and coding. Also, variable Walsh code space is utilized. Namely, modulation and coding can change from frame to frame as directed by the MAC layer based on the feedback information from the mobile station. The feedback information is contained in the R-CQICH—which reports the pilot chip energy, E_c to total noise density, N_t ratio, E_c/N_t of the strongest received F-PICH—and in the R-ACKCH, which indicates whether the frame reception was successful. The selected modulation and coding also depend on the available Walsh codes. Unlike all other channels, the F-PDCH utilizes only leftover resources at the base station. This means that the power as well as the Walsh codes space consumed by the F-PDCH can change from frame to frame.

The F-PDCH Walsh code length is 32, and the channel may utilize up to 28 of these codes. The frame duration is variable, and as already indicated in Section 4.2.1, permitted lengths are 1.25, 2.5, and 5 ms. The encoder packet size can take six different values ranging from 386 to 3842 bits. The F-PDCH utilizes only turbo codes, and a wide range of code rates obtained by puncturing is allowed. The code rate can be as low as $\approx 1/5$ and as high as $\approx 3/4$. The channel structure is shown in Figure 4.35, while I and Q channel mappings, unique for F-PDCH, are presented in Figure 4.36. The standard allows for parallel operation of up to two F-PDCHs.

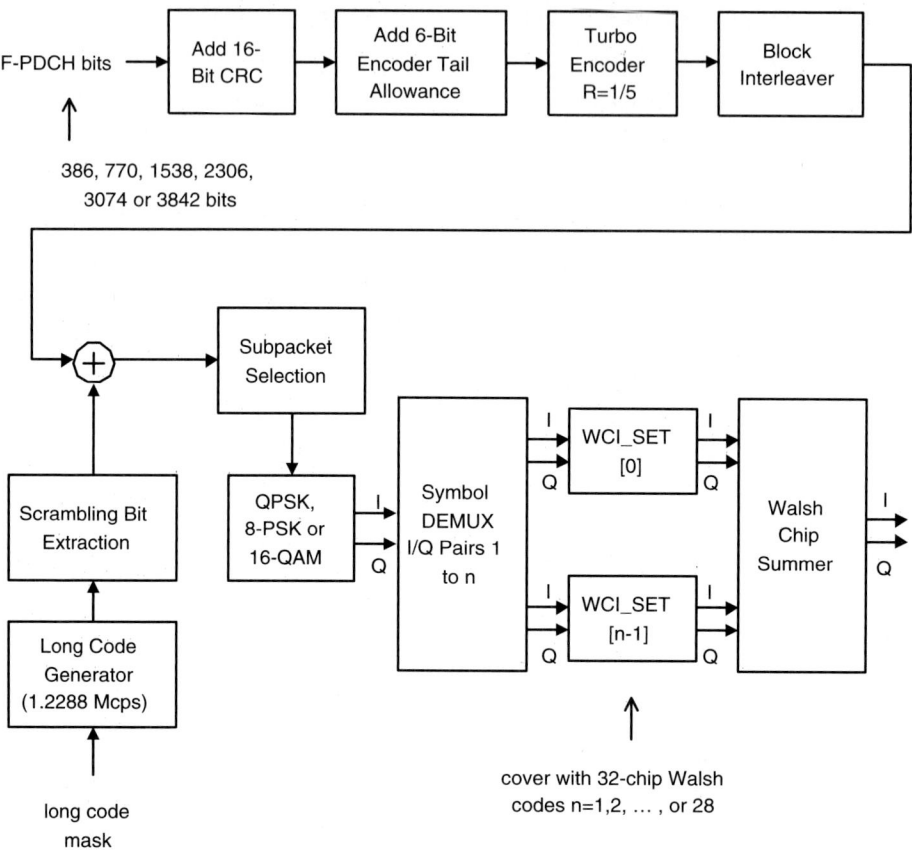

Figure 4.35 F-PDCH encoding.

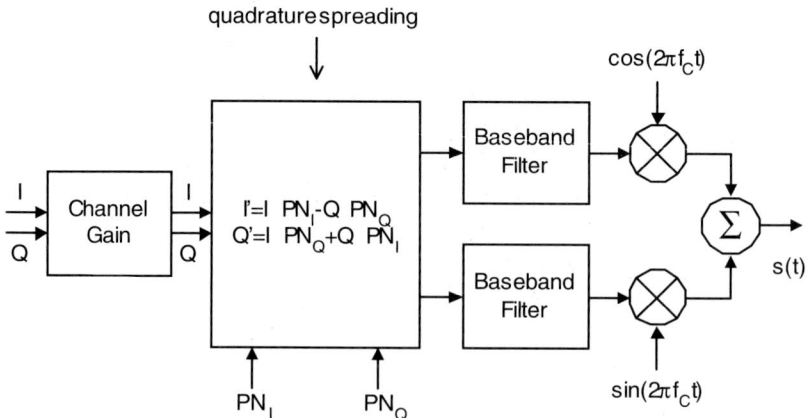

Figure 4.36 F-PDCH I and Q mapping.

4.3.9 Forward Packet Data Control Channel (F-PDCCH)

The data transmission on the F-PDCH is always accompanied by Layer 2 control information transmitted in parallel over the F-PDCCH. The control information is necessary for correct demodulation and decoding of the associated F-PDCH frame. The number of information bits is 13. The frame is protected by two CRC codes, outer and inner. The inner CRC code is generated from the F-PDCCH data bits, while the outer CRC code is generated from the data and inner CRC bits after a mobile station-specific identification number is EOR'd with the inner CRC bits.

The frame length is variable, 1.25, 2.5, and 5 ms, matching the frame sizes on the F-PDCH. Since the number of information bits is constant, the data rate is variable. The highest resulting rate naturally corresponds to the shortest frame, and the rate is equal to 29.6 kbps. For longer frames, the rate is proportionally lower. The F-PDCCH employs a convolutional encoder with a variable code rate, half for the shortest frame length and quarter for the other two frame sizes. However, the effective code rate can be either lower or higher after coded bits are punctured or repeated. Walsh code information is signaled to the mobile station through system overhead information transmitted over the F-PCH or F-BCCH. The F-PDCCH uses one Walsh code of length 32. The channel structure is illustrated in Figure 4.37.

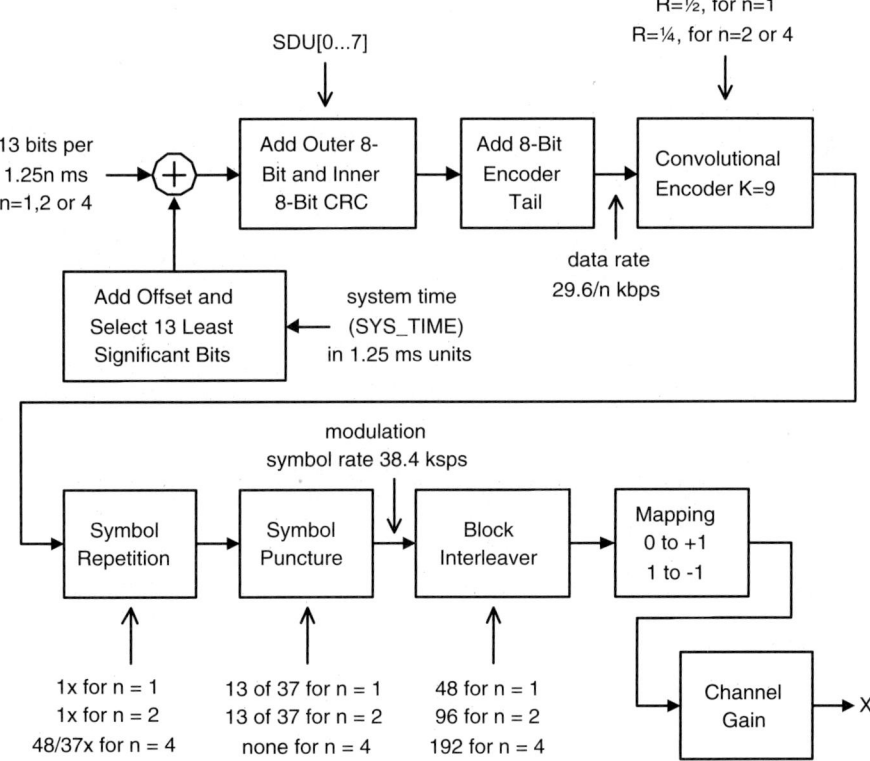

Figure 4.37 F-PDCCH encoding.

4.3.10 Forward Fundamental Channel (F-FCH)

The F-FCH is a dedicated power-controlled traffic channel, typically power-controlled at 800 bps. It is also a variable rate channel with four distinct rates. The actual rates depend on the rate set, and there are two sets defined. Rate Set 1 allows for the following rates: 9.6, 4.8, 2.7, and 1.5 kbps.[8] These rates are often referred to as full rate, half rate, quarter rate, and eighth rate, respectively. Rate Set 2 permits the following rates 14.4, 7.2, 3.6, and 1.8 kbps. These rates are also commonly referred to as full rate, half rate, quarter rate, and eighth rate, respectively. Therefore, when referring to a full-rate frame, for example, a reference to the rate set should also be given. However, we show only the structure for Rate Set 1 and in particular RC3. It is understood that the standard also allows for other RCs (and some of them support Rate Set 2). As we mentioned in Section 4.2.1, the IS-2000 standard allows for flexible rates. In this scenario, the rate on any dedicated channel, including F-FCH, can be configured with Layer 3 signaling and can take any value between the minimum and the maximum rate for a given RC.

The F-FCH structure is presented in Figure 4.38. The frame length is either 5 or 20 ms. The shorter frames are used for transferring delay-sensitive IS-2000 Layer 3 control information, while longer frames are primarily used for voice traffic but also for IS-2000 Layer 3 signaling. The number of bits per frame varies depending on the data rate, and it is optimized to fit the output of the vocoder. Even though it is optimized for voice traffic, the F-FCH can also carry data traffic. Discontinuous transmission is not allowed on the F-FCH; that is, after the channel is set up, a frame must be transmitted in each 20-ms interval. This restriction on the F-FCH operation is handled by the SMV, because the vocoder produces a frame every 20 ms. However, when the F-FCH carries only data traffic, the MAC layer ensures that a frame is supplied to the physical layer every 20 ms. The Walsh code length used for the F-FCH and RC3 is 64, and RC3 is illustrated in Figure 4.38. The channel structure for other RCs is similar. The difference is in the set of supported rates, employed code rate, and Walsh code length. For example, RC4 enables the use of a Walsh code of length 128. RC5 supports Rate Set 2. The power-control bits that power control R-PICH are punctured onto the F-FCH at the rate of 800 bps. This means that one power-control bit is punctured every 1.25 ms. The length of the punctured power-control bit is equal to one-twelfth of 1.25 ms. This means that for RC3, a power-control bit replaces four coded symbols.

Figure 4.38 does not show scrambling and possible puncturing of power-control bits onto a power-control subchannel. This procedure is illustrated in Figure 4.39, and it is common for the F-FCH and the F-DCCH.

---

8.   Rate Set 1 also refers to the set of rates provided with legacy IS-95 channel configuration, RC1. In case of RC1, the lower two rates are slightly different. R1 is not discussed in this book.

16 bits for 5-ms frame
6, 6, 8 or 12 bits for 20-ms frames

9.6 kbps for 5-ms frame
1.5, 2.7 4.8 or 9.6 kbps
for 20-ms frames

1x for 5-ms frame
8, 4, 2, or 1x for 20-ms frame

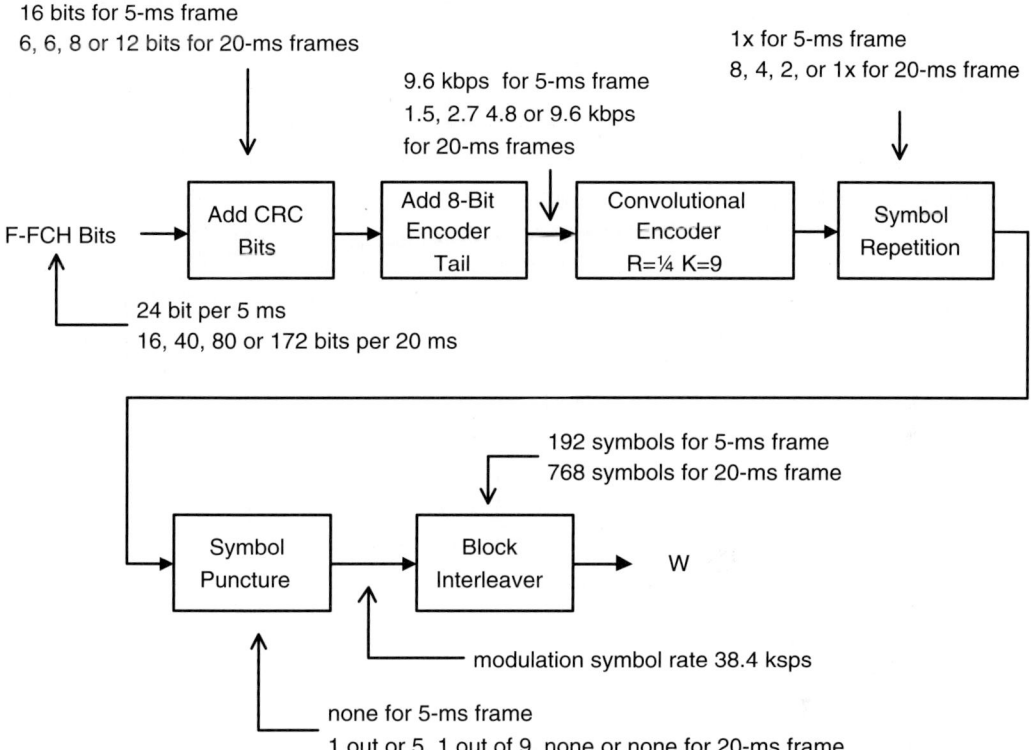

24 bit per 5 ms
16, 40, 80 or 172 bits per 20 ms

192 symbols for 5-ms frame
768 symbols for 20-ms frame

modulation symbol rate 38.4 ksps

none for 5-ms frame
1 out or 5, 1 out of 9, none or none for 20-ms frame

Figure 4.38 F-FCH encoding (RC3 only).

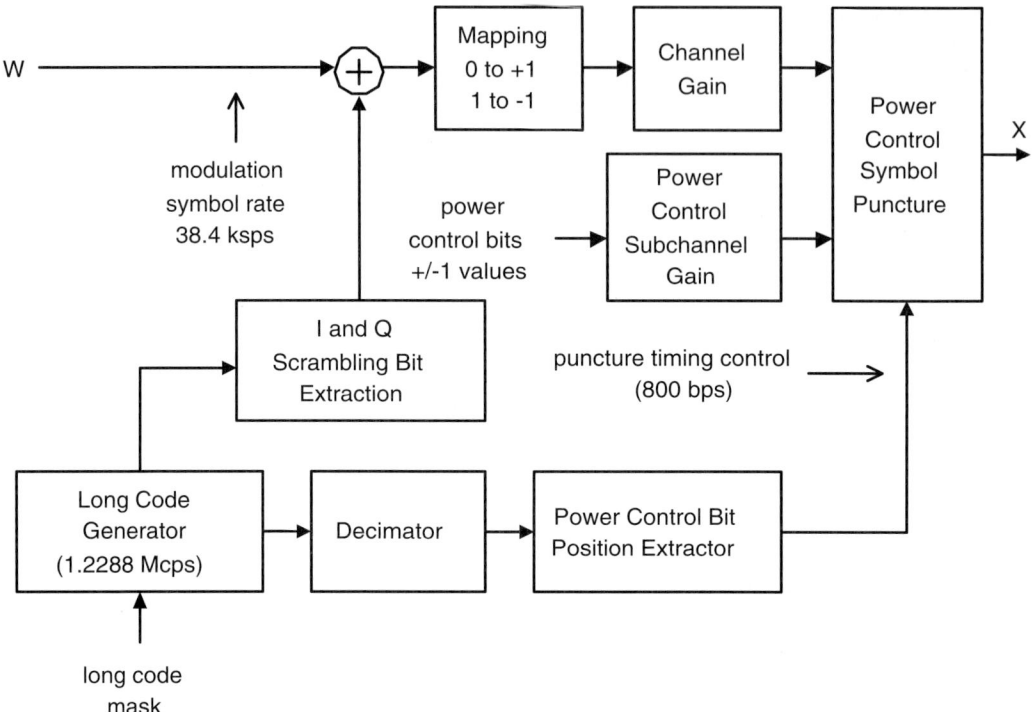

Figure 4.39 Scrambling and puncturing of power-control bits onto the power-control subchannel (RC3 only).

4.3.11 Forward Dedicated Control Channel (F-DCCH)

The F-DCCH provides a dedicated power-controlled channel for IS-2000 Layer 3 signaling. The channel name is, however, a misnomer because this channel can also transfer user data. It cannot carry voice traffic, because it does not support variable rate operation. The F-DCCH structure is shown in Figure 4.40 for RC 3. It is similar to the F-FCH, with a caveat that only full-rate or 9.6-kbps operation is allowed. It is true, however, that if flexible-rate operation is supported, the F-DCCH can take any rate negotiated with IS-2000 Layer 3 signaling. The other difference relative to the F-FCH is that discontinuous transmission on the F-DCCH is permitted. As is the case for the F-FCH, power-control symbols are punctured on the F-DCCH, but only if they are not already punctured on the F-FCH. The procedure, illustrated in Figure 4.39 applies to F-DCCH, except that power-control bits may not be punctured. If punctured onto the F-DCCH, power-control commands are transmitted even if the F-DCCH has temporarily discontinued its transmission. The frame size is the same as for F-FCH, 5 ms and 20 ms, and the coding is convolutional only. The code rate for RC3 is one-quarter, and the corresponding Walsh code length is 64. The channel structure for other RCs is similar. The difference is in the supported rate, employed code rate, and Walsh code length. For example, RC4, supports the same data rate, but it employs rate, $R = 1/2$ code, which enables use of a Walsh code of length 128. RC5 supports 14.4-kbps operation. Relative to RC3, the code rate is increased, $R = 3/8$, but the Walsh code utilization remains unchanged. The Walsh code length is 64.

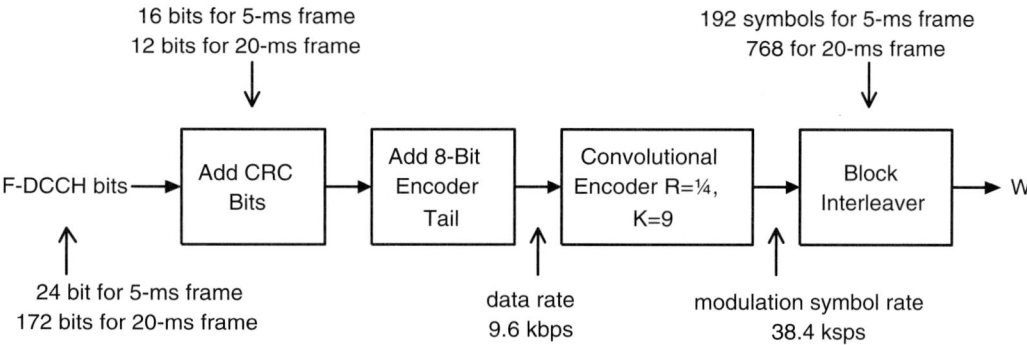

Figure 4.40 F-DCCH encoding (RC3 only).

4.3.12 Forward Supplemental Channel (F-SCH)

When the F-FCH does not provide enough throughput for the data traffic, the F-SCH may be set up. However, the F-SCH cannot operate alone: The F-FCH or the F-DCCH must be active at the same time. The F-SCH is a dedicated channel that is power controlled at a rate of up to 800 bps. The power may simply follow, for example, the F-FCH (see Chapter 9 for power-control modes). Another option is that the F-SCH follows the F-DCCH. In both cases, the power-control rate is 800 bps. The alternative to these two options is to power control the F-SCH separately from the primary power control channel, such as the F-FCH. In this case, part of the power-control bits control the F-FCH, and the remaining bits control the F-SCH. The permitted rates are 200 bps for the F-FCH or F-DCCH and 600 bps for the F-SCH, or 400 bps for the F-FCH or F-DCCH and 400 bps for the F-SCH.

The channel structure is presented in Figure 4.41, for RC 3 only. This channel allows for both convolutional and turbo coding. The frame length is not restricted to 20 ms. It can also use 40 ms and 80 ms (5-ms frame length is not supported). Longer frame size allows for a longer interleaver that improves channel interleaving and the coding gain for turbo code. For RC3, the code rate is one-quarter and the maximum supported rate is 153.6 kbps. The Walsh code length for the highest supported data rate is only 4. Power-control bits are not punctured onto the F-SCH. Note that Figure 4.41 does not show all the rates supported by the R-SCH. Lower rates are also possible. The encoder structure is similar to the one shown in Figure 4.41. In addition to what is shown on the block diagram in Figure 4.41, the lower rates require symbol repetition. Note that without specifically mentioning F-SCH, we have already illustrated all the supported

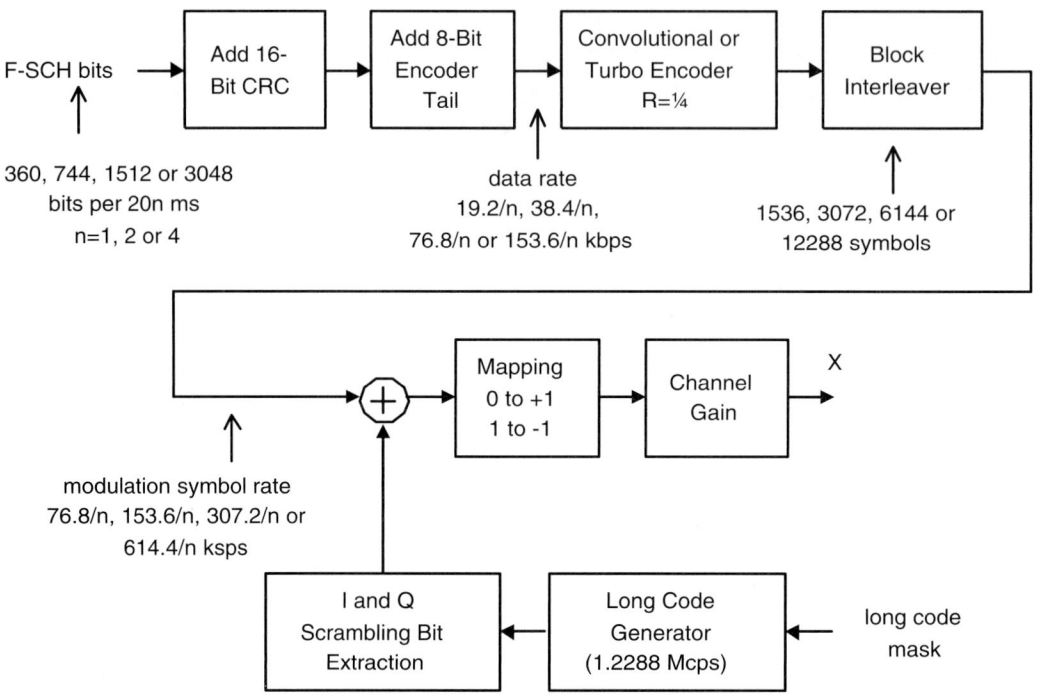

Figure 4.41 F-SCH encoding (RC3 only; lower rates are not shown).

rates in Table 4.1. The rates illustrated in Figure 4.41, however, are commonly used in practice, while other rates are not widely implemented.

For RC4, the code rate is increased to $R = 1/2$, and the maximum supportable rate is 307.2 kbps. RC5 employs code rate, $R = 3/8$ and supports Rate Set 2.

4.4 Reverse Channels

In the previous sections we explained the coding and spreading for all the reverse channels introduced in IS-2000. In this section we describe the channel-specific encoding processes for all the reverse channels.

4.4.1 Reverse Pilot Channel (R-PICH)

All reverse-link channels introduced by the IS-2000 standard contain the R-PICH, which consists of all-zeros symbols. During reverse-access channel operation, the R-PICH enables initial acquisition by the base station searcher. For that purpose, the R-PICH is transmitted as a preamble in discontinuous mode, at higher power than normal, in order to enable faster acquisition. After the preamble phase, the R-PICH is transmitted at a lower power level, and it is used by the time-tracking loop and for the phase and amplitude estimation. Also, the base station searcher uses the R-PICH to continuously search for new paths.

When one or more forward power-controlled dedicated channels are active, for example the F-FCH, the R-PICH is punctured with power-control bits. These power-control bits are used for fast power control of those channels. In Figure 4.42, we illustrate how the pilot signal is multiplexed with power-control bits. The power-control bits are transmitted with the same power level as the pilot signal and occupy one-quarter of the 1.25-ms slot, commonly called the power-control group. Unless it is transmitted during the reverse-access procedure, the R-PICH itself is fast-power-controlled at the rate of 800 bps. The power-control commands are transmitted either on a corresponding F-CPCCH subchannel or punctured on the associated F-FCH or F-DCCH. A mobile station contains only a single R-PICH.

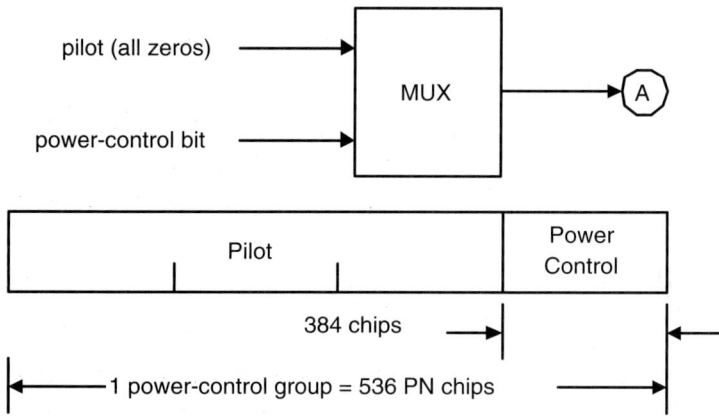

Figure 4.42 R-PICH structure.

4.4.2 Reverse Access Channel (R-ACH)

The R-ACH was introduced in IS-95, but it is commonly supported by cdma2000 systems for backward-compatibility reasons. It is a common reverse channel that is used by all mobile stations to establish the link with the base station associated with the F-PCH the mobile station is monitoring. The R-ACH does not follow the spreading and modulation, as depicted in Figure 4.17. The complete channel structure, including modulation, spreading, and coding is illustrated in Figure 4.43.

The length of the R-ACH frame is 20 ms, and each frame contains 88 information bits and 8 tail bits. The resulting data rate is 4.8 kbps. The R-ACH employs a constraint length 9 convolutional code of rate one-third, and 64-ary orthogonal modulation. The pilot is not present because the receiver noncoherently decodes the information. After long and short PN code

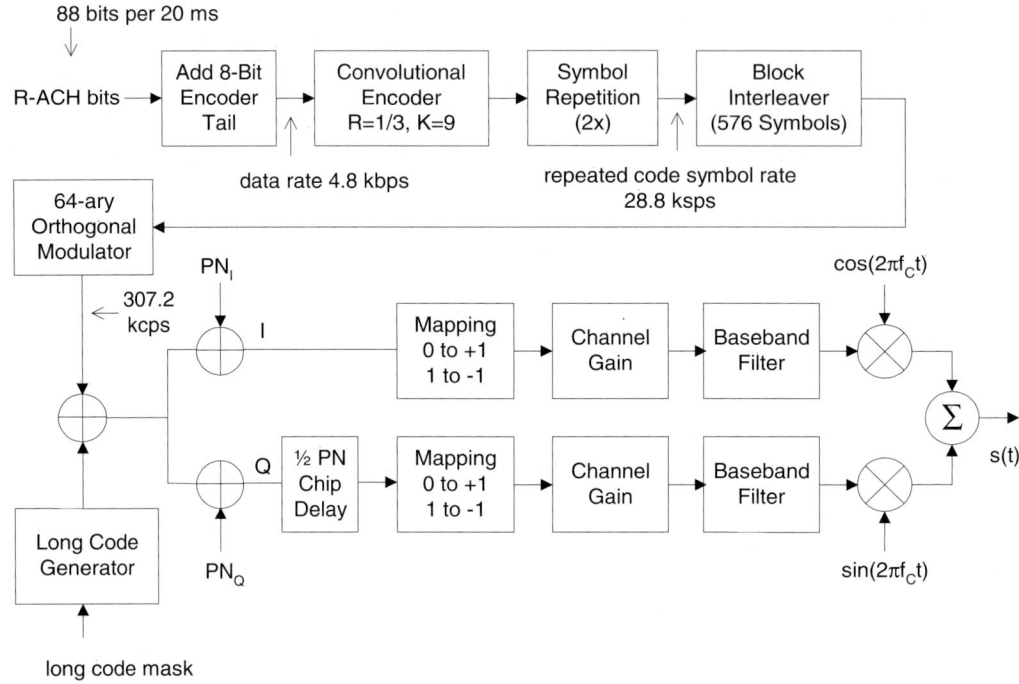

Figure 4.43 R-ACH structure, I and Q mapping.

spreading, the Q branch is delayed by a half-chip to produce Offset QPSK (OQPSK) chip modulation. OQPSK produces fewer harmonics than QPSK because zero-crossing is avoided. Figure 4.44 illustrates the OQPSK signal constellation (I and Q branch after PN_I and PN_Q spreading) and the permitted transitions.

The R-ACH access probe can be divided into the preamble and the message. The preamble consists of all-zeros modulation symbols. The power level is estimated based on the received F-

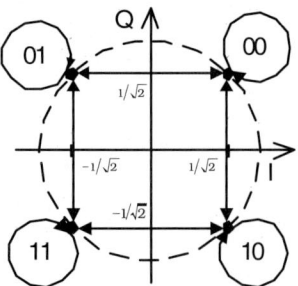

Figure 4.44 OQPSK signal constellation.

PICH power, and it is not changed during the entire probe. This power-control mechanism is called an open-loop power-control method. Access on the channel adheres to a slotted ALOHA protocol.

Figure 4.45 shows the access probe transmission on the R-ACH. The typical length of the access probe slot is on the order of several hundred milliseconds. There is only one R-ACH per mobile station, but there can be up to 32 R-ACHs associated with a single F-PCH. The R-ACH PN spreading code mask is transmitted as overhead information over the F-PCH. Each mobile station randomly chooses the R-ACH on which it attempts to access the system.

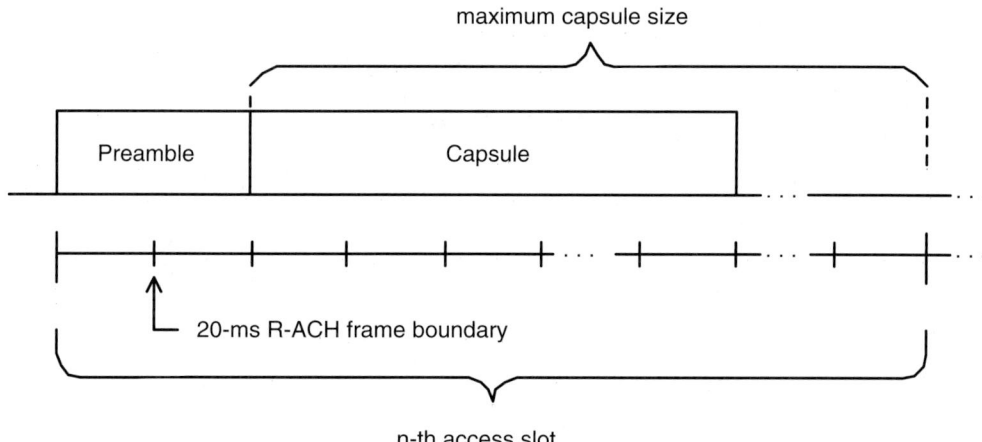

Figure 4.45 Access probe transmission on the R-ACH.

4.4.3 Reverse Enhanced Access Channel (R-EACH)

The R-EACH was introduced in IS-2000 as an enhanced access procedure relative to that of the R-ACH. The R-EACH structure differs depending on whether the header or information bits are sent. Figure 4.46 illustrates the R-EACH structure when information bits are sent. This is structure is used in the basic access mode. The frame length can be 5, 10, or 20 ms. The number of information bits varies from 172 to 744. The resulting data rate is between 9.6 and 38.4 kbps. The preamble consists of the pilot signal. The FEC is realized with a convolution code of rate one-quarter and constraint length 9.

Figure 4.47 shows the R-EACH structure when the header is sent. This structure is used during the reservation access mode. In this mode, the frame length is 5 ms and the data rate is 9.6 kbps. Also in this case, the preamble consists of the pilot signal, and the FEC is realized with a convolution code of rate one-quarter and constraint length 9.

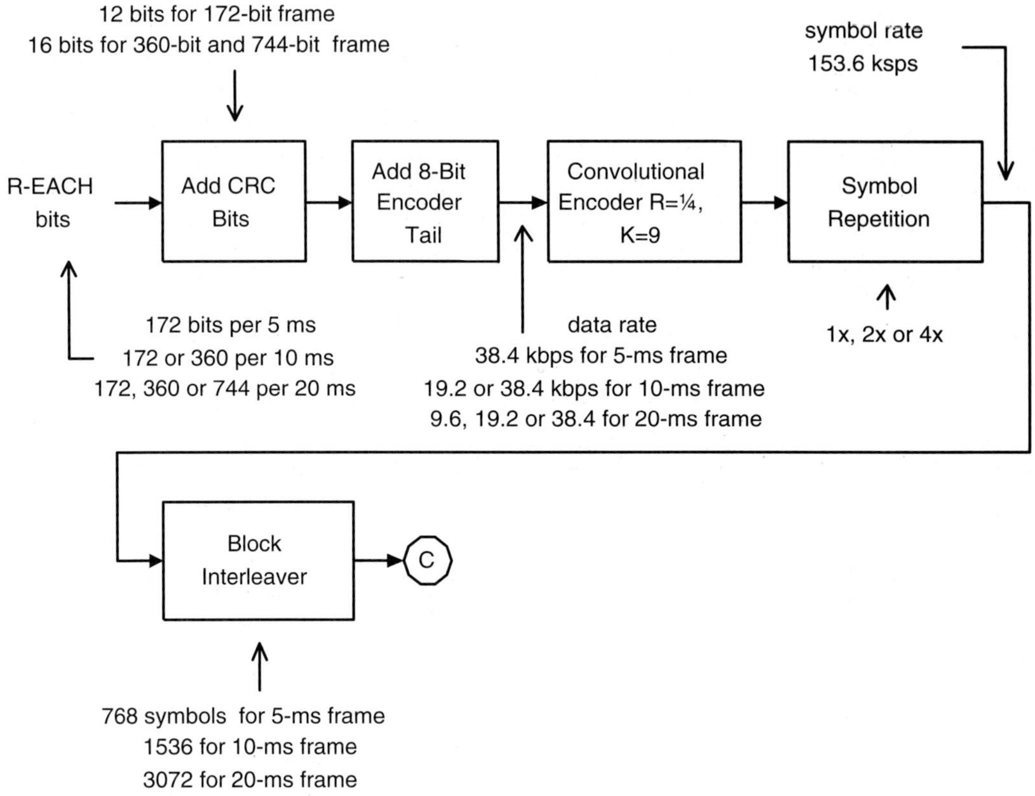

12 bits for 172-bit frame
16 bits for 360-bit and 744-bit frame

symbol rate
153.6 ksps

R-EACH
bits
→ Add CRC Bits → Add 8-Bit Encoder Tail → Convolutional Encoder R=¼, K=9 → Symbol Repetition →

172 bits per 5 ms
172 or 360 per 10 ms
172, 360 or 744 per 20 ms

data rate
38.4 kbps for 5-ms frame
19.2 or 38.4 kbps for 10-ms frame
9.6, 19.2 or 38.4 for 20-ms frame

1x, 2x or 4x

→ Block Interleaver → Ⓒ

768 symbols for 5-ms frame
1536 for 10-ms frame
3072 for 20-ms frame

Figure 4.46 R-EACH encoding for information-bits transmission.

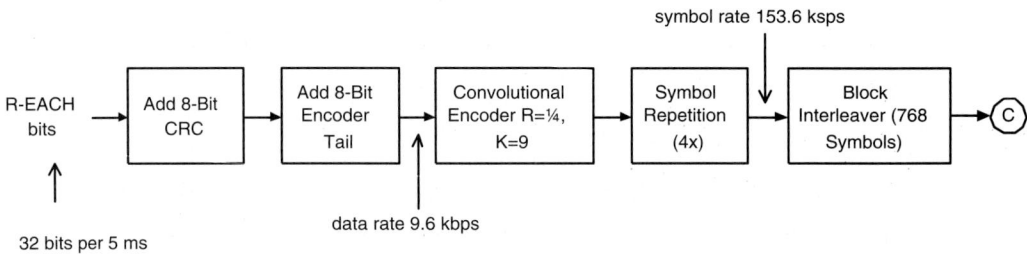

symbol rate 153.6 ksps

R-EACH bits → Add 8-Bit CRC → Add 8-Bit Encoder Tail → Convolutional Encoder R=¼, K=9 → Symbol Repetition (4x) → Block Interleaver (768 Symbols) → Ⓒ

32 bits per 5 ms

data rate 9.6 kbps

Figure 4.47 R-EACH encoding for header transmission.

There can be up to 32 R-EACHs associated with a single F-CCCH. The R-EACH PN spreading code mask is transmitted as overhead information over the F-BCCH. Each mobile station randomly selects the R-EACH when accessing the system.

As mentioned earlier, the R-EACH supports two distinct access modes. The first mode the basic access mode. In this case, the purpose of the R-EACH is similar to that of the R-ACH. The main improvement of R-ACH is that mobile stations can attempt access at R-EACH slot boundaries. The slot size is a configurable number, and it can be as low as 1.25 ms. This is an advantage over the R-ACH because the mobile station does not have to wait before transmitting the access probe. In Figure 4.48 we show the access probe in the basic mode. The preamble consists of a sequence of fractional preambles. Referring to Figure 4.48, the following parameters are used to define the preamble. The number of fractional preambles is denoted as N. The fractional preamble length is defined by the parameter P, which is a multiple of 1.25 ms. The time between fractional preambles is denoted as B and it is also a multiple of 1.25 ms. Additional preamble is transmitted prior to R-EACH data. The length of the additional preamble is denoted as A and it is also a multiple of 1.25 ms.

In the second access mode, the reservation access mode, the R-EACH reserves transmission only on the R-CCCH. The R-EACH probe, which identifies the mobile and desired R-CCCH mode of operation, is only open-loop power controlled. However, the transmission on the R-CCCH is fast power controlled through the R-PICH at a rate of up to 800 bps. The F-CACH is used to signal to the mobile that the reservation was successful, to identify the granted R-CCCH and to indicate whether soft handoff was granted. There is a one-to-one correspondence between the F-CPCCH subchannel and the R-CCCH. Figure 4.49 illustrates the access probe in the reservation access mode. The preamble has exactly the same structure as for basic access mode.

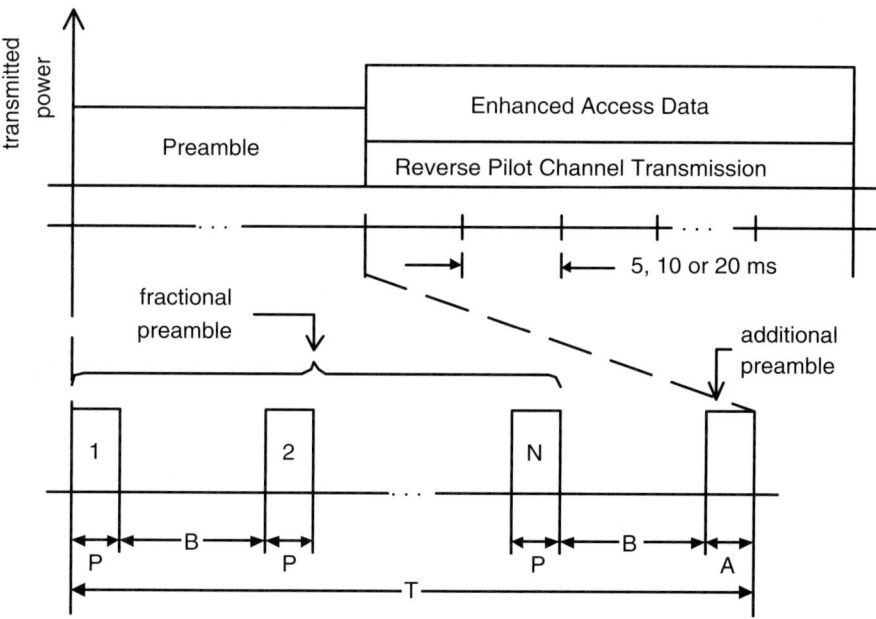

Figure 4.48 Basic access mode.

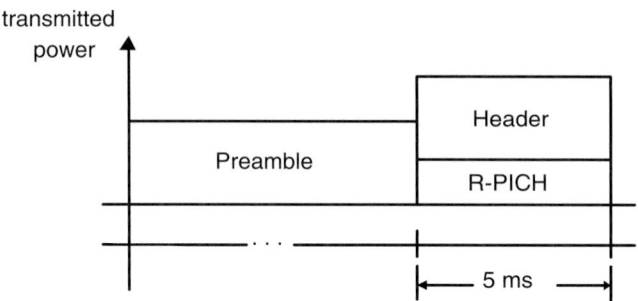

Figure 4.49 Reservation mode.

4.4.4 Reverse Common Control Channel (R-CCCH)

When operating in the reservation access mode, the actual data message is transmitted on the R-CCCH. This channel is fast power controlled with commands received on the corresponding F-CPCCH subchannel. The R-PICH gain follows the power-control commands, and the R-CCCH has a fixed offset relative to the R-PICH. The power-control rate can be 200, 400, or 800 bps. Since R-CCCH is power-controlled, access probes transmitted over this channel introduce less interference than access probes transmitted over the R-ACH or R-EACH. The fast-power-control reduces the error rate, which is important for efficient transport of long access probe messages.

Figure 4.50 shows the R-CCCH structure. The frame structure is identical to that of the R-EACH, but in this case the input bits belong to the R-CCCH message. Each mobile station can simultaneously transmit only over a single R-CCCH, but up to 32 R-CCCHs can be supported. The R-CCCH PN spreading code mask is transmitted as overhead information over the F-BCCH.

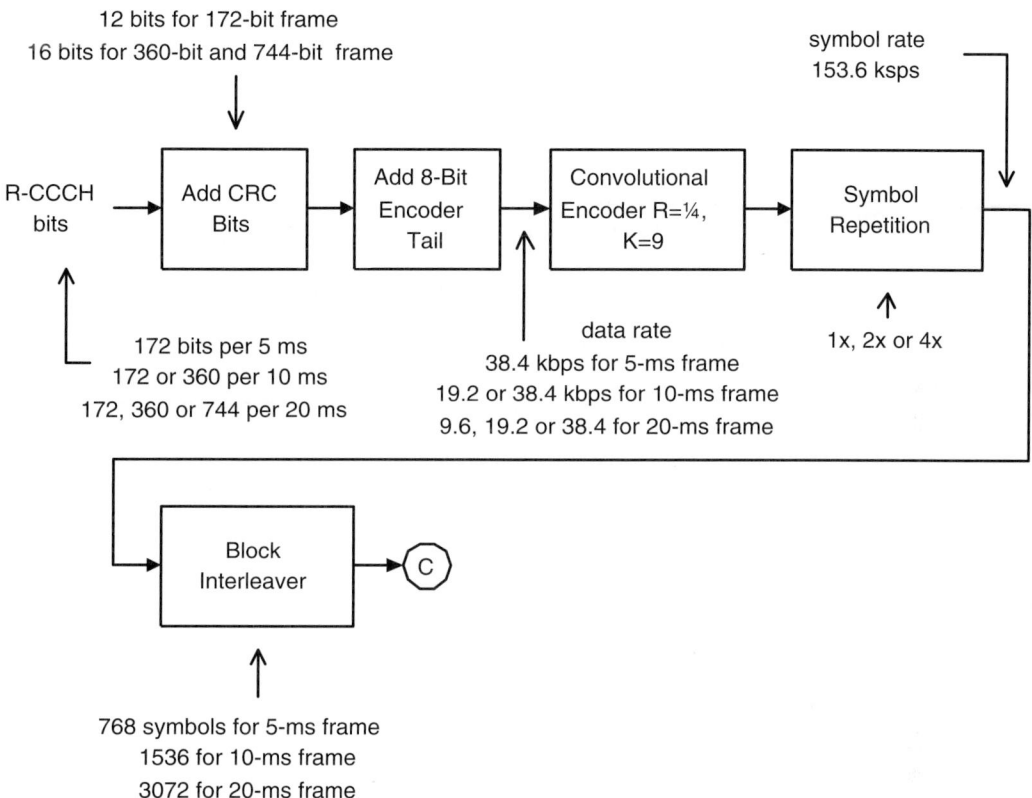

Figure 4.50 R-CCCH structure.

4.4.5 Reverse Fundamental Channel (R-FCH)

The R-FCH is a dedicated fast-power-controlled reverse-traffic channel. As in the case of the F-FCH, it allows for variable-rate operation with four rates. However, if flexible-rate operation is supported, the R-FCH can take any Layer 3 negotiated rate.

The R-FCH structure for RC3 is presented in Figure 4.51. Note that RC4 on the reverse link actually provides support for Rate Set 2 (similar to RC5 for the forward link). The frame size is 5 or 20 ms. The shorter frames are used for transferring delay-sensitive IS-2000 Layer 3 Signaling data, while longer frames are primarily used for voice traffic but also for data and IS-2000 Layer 3 signaling. The channel is fast-power-controlled through the R-PICH. The R-FCH gain offset relative to R-PICH remains constant for a given rate and, like F-FCH, R-FCH does not allow for discontinuous transmission. There is only a single R-FCH per mobile station.

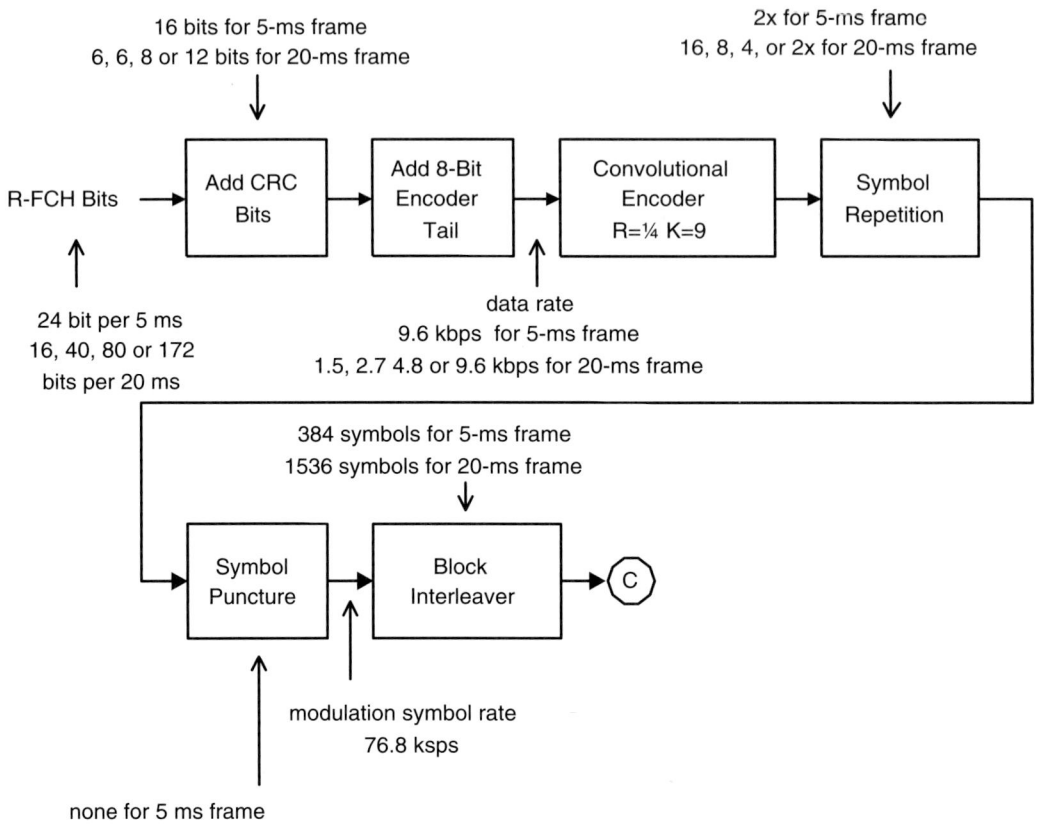

Figure 4.51 R-FCH encoding (RC3 only).

4.4.6 Reverse Dedicated Control Channel (R-DCCH)

The R-DCCH is a dedicated channel that carries IS-2000 Layer 3 signaling. The name is a misnomer, because R-DCCH can also carry user data. Voice traffic is not transported over this channel. The structure of the channel for RC3 is shown in Figure 4.52. It is similar to the R-FCH with the restriction that only full-rate or 9.6-kbps operation is allowed. However, if flexible-rate operation is supported, the R-DCCH can take any IS-2000 Layer 3 negotiated rate.

Discontinuous transmission on the R-DCCH is permitted. The channel is fast power controlled at 800 Hz through the R-PICH. The R-DCCH gain offset relative to the R-PICH is constant. The frame length can be 5 or 20 ms. RC4, in addition to 5-ms frame length at 9.6 kbps, provides support for 14.4-kbps operation. Relative to RC3, the code rate is increased, $R = 3/8$.

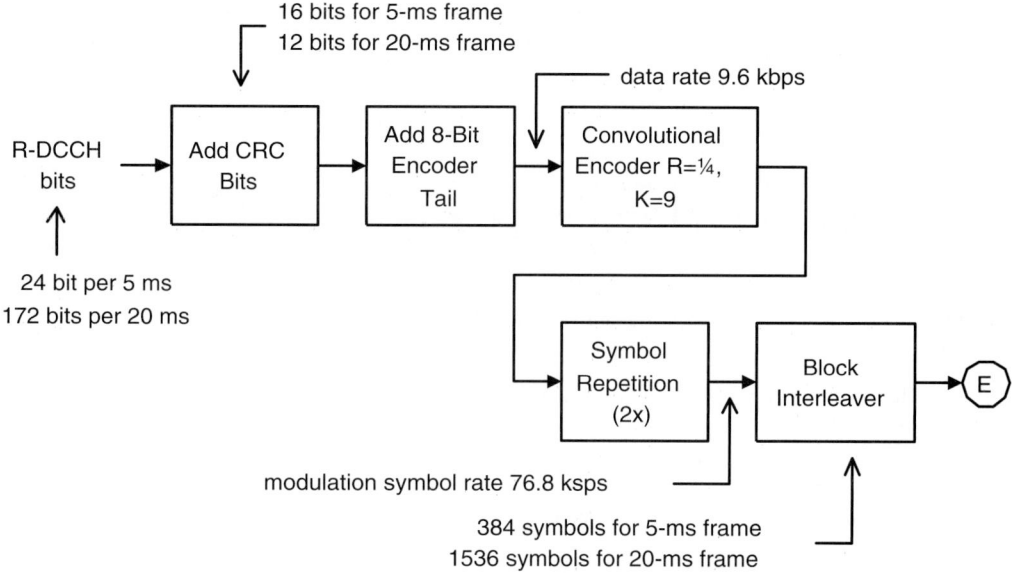

Figure 4.52 R-DCCH encoding (RC3 only).

4.4.7 Reverse Supplemental Channel (R-SCH)

The R-SCH supports high-rate transmission for data traffic when the R-FCH is not suffi-cient. When the R-SCH is present, the R-FCH or R-DCCH must also be active. The R-SCH is a fast-power-controlled dedicated channel because the R-SCH has a constant gain relative to the R-PICH, which is power controlled at 800 bps.

The R-SCH structure for RC3 is presented in Figure 4.53. This channel allows for both convolutional and turbo coding. In order to provide stronger interleaver gains, the frame length is not restricted to 20 ms. It can also be 40 and 80 ms (5-ms frame length is not allowed). For RC3, the R-SCH allows for data rates up to 153.6 or 307.2 kbps, depending whether quarter or half code rate is employed respectively. Note that Figure 4.53 does not show all the rates sup-ported by the R-SCH. Lower rates are also possible, but they are rarely used in practice. In addi-tion to what is shown on the block diagram in Figure 4.53, the lower rates require symbol repetition. Note that without specifically mentioning R-SCH, we have already illustrated all the supported rates in Table 4.2. RC4 provides support for Rate Set 2. The supported rates are also illustrated in Table 4.2.

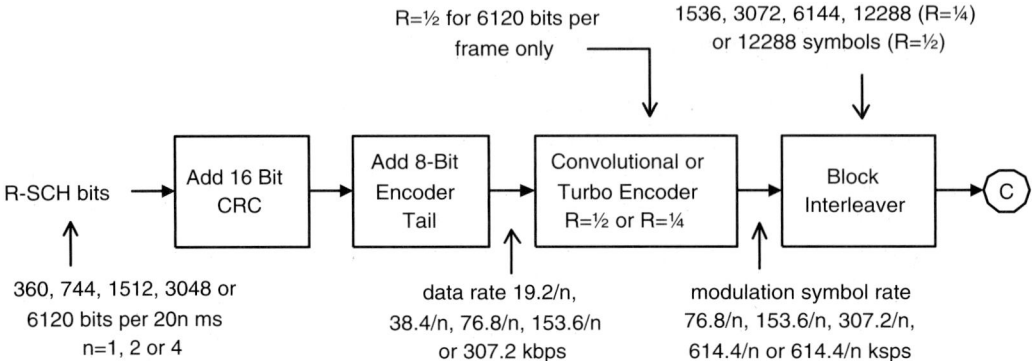

Figure 4.53 R-SCH encoding (for RC3 only; lower rates are not shown).

4.4.8 Reverse Channel Quality Indicator Channel (R-CQICH)

The R-CQICH is a support channel for adaptive coding and modulation over the F-PDCH. The channel is used to convey the F-PICH E_c/N_t to the serving base station. This information is critical for choosing the appropriate modulation and coding scheme. Moreover, a mobile and in some cases two mobiles that are selected for transmission over the F-PDCH are chosen based on the predetermined fairness metric and the R-CQICH value.

The information that is transported over the channel may be a full 4-bit coded value of the F-PICH E_c/N_t or an up/down indication relative to the previous accumulated value. The up or down indication is interpreted as a +0.5 or –0.5 dB change in F-PICH E_c/N_t, respectively. The R-CQICH is also used to indicate the base station of the reported F-PICH E_c/N_t. The Walsh cover identifies the base station. Figure 4.54 shows the R-CQICH channel structure. When a full 4-bit report is sent, it is encoded with a (12,4) block code that consists of the last 12 symbols of each of the length-16 Walsh codes.

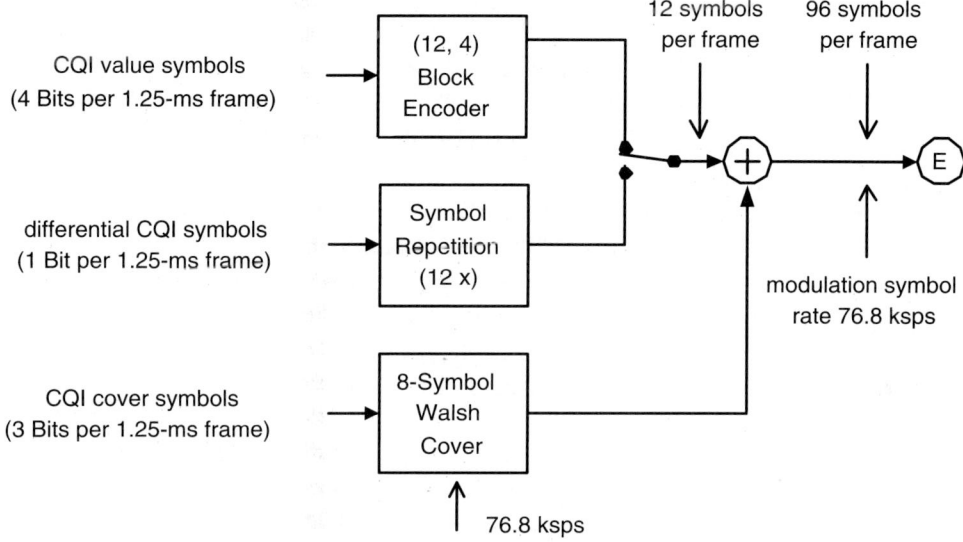

Figure 4.54 R-CQICH encoding.

4.4.9 Reverse Acknowledgment Channel (R-ACKCH)

The R-ACKCH supports H-ARQ type II operation over the F-PDCH. The purpose of this channel is to convey information from the mobile station to the base station whether the CRC check of the decoded packet has passed or failed. The R-ACKCH is illustrated in Figure 4.55. The R-ACKCH is a three-state channel. It allows positive acknowledgment or +1, negative acknowledgment or −1, and zero or null. The null is used if decoding of F-CPCCH fails.

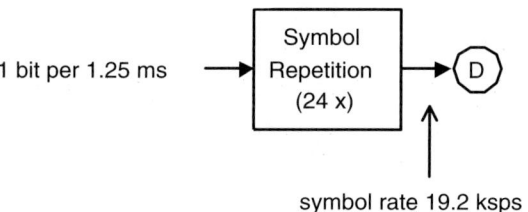

Figure 4.55 R-ACKCH encoding.

REFERENCES

[1] "TIA/EIA IS-2000.2 Physical Layer Standard for cdma2000 Spread Spectrum Systems, Release C," *Telecommunications Industry Association, 2002.*

[2] "TIA/EIA IS-2000.3 Medium Access Control (MAC) Standard for cdma2000 Spread Spectrum Systems, Release C," *Telecommunications Industry Association, 2002.*

[3] "TIA/EIA IS-95 A Mobile Station—Base Station Compatibility Standard for Dual Mode Wideband Spread Spectrum Cellular Systems," *Telecommunications Industry Association,* 1993.

[4] Shanbhag, A., and J. Holtzman, "Optimal QPSK Modulated Quasi-orthogonal Functions for IS-2000." *In Proc. Of IEEE Sixth Annual Symposium on Spread Spectrum Applications and Techniques,* Vol. 2, pp. 756–760, Parsippany, NJ, September 2000.

[5] Alamuti, S. "A Simple Transmit Diversity Technique for Wireless Communications," *IEEE JSAC,* 16(8), October 1998.

[6] "TIA/EIA IS-2000.5 Upper Layer (Layer 3) Signaling Standard for cdma2000 Spread Spectrum Systems, Release C," *Telecommunications Industry Association, 2002.*

CHAPTER 5

IS-2000 Layer 2 (Medium and Signaling Link Access Control Layers)

The IS-2000 Layer 2 protocol provides link-layer services over the U_m interface. Such services can be distinguished as those offered to the IS-2000 signaling layer, or Layer 3, and those offered to end-user applications and services. Layer 2 is split into two sublayers, the Medium Access Control (MAC) [1] and Signaling Link Access Control (Signaling LAC) [2].

The main function of the LAC layer is to provide transport to signaling messages originating from Layer 3. To achieve that, the LAC properly encapsulates the Layer 3 messages into LAC protocol data units (PDU) that are subject to fragmentation and reassembly to make them suitable for transport by the lower layers. The LAC can also provide Layer 3 with reliable message delivery. Data units are exchanged between Layer 3 and LAC on logical channels. The logical channels determine *what* data is transported rather than *how* it is transported. The how depends on the properties of the physical channel that the logical channel is mapped onto, but not on the logical channel itself. The logical channels hide the upper layers from the characteristics of the physical channels.

The MAC uses a generalized multimedia service model that allows a combination of services, such as voice, packet data, and circuit data, to operate concurrently. The MAC also provides a quality of service (QoS) control mechanism to mediate conflicting requests from competing services. It implements a multiplexing/demultiplexing function to allow access to the medium to both data units received from the LAC and those received from user applications. Reliable transmission of end-user application data may be optionally enabled, typically for packet data services, using the Radio Link Protocol (RLP).

In this chapter, we describe the MAC and LAC layers in terms of their communication protocols with the peer entities and the services offered to the layer above.

5.1 Medium Access Control Layer

The MAC provides services to both the signaling LAC layer and to the upper layers. Services and corresponding protocols are briefly listed below and discussed in detail in the following sections:

Multiplexing. The MAC layer exchanges data units with the LAC layer and with the upper layer over the logical channels. The MAC shields the upper layers from the physical layer with its radio-dependent characteristics. It is the responsibility of the MAC to provide the logical channels with access to the physical medium. Multiple logical channels may share the same physical channel to achieve greater efficiency. That requires the receiver to discriminate data according to the logical channels they belong to and transfer them to the upper layer through the appropriate service access point (SAP). Such service is provided by the multiplex sublayer protocol. We distinguish between the dedicated and the common multiplex sublayers. The former is responsible for dedicated channels—that is, those that are mapped to physical channels assigned to a single user. The latter is responsible for common channels—that is, those that are mapped to physical channels shared by multiple users. On dedicated channels, user traffic and signaling LAC protocol data units, or fragments thereof, are multiplexed into a data structure called multiplex protocol data units, or MuxPDU. The specification of the MuxPDU format represents the communication protocol between peer entities: multiplexer at the transmitter and demultiplexer at the receiving end. On common channels, signaling LAC protocol data units are delivered to the peer entity using a protocol called Short Radio Burst Protocol (SRBP).

Reliable transport of user-generated traffic. Such service is provided to the upper layer on the user plane to guarantee nearly error-free and ordered delivery of user traffic data. The service is made possible by the RLP, a specialized form of selective-repeat Automatic Repeat Request (ARQ) protocol.

QoS control. Signaling LAC data units, which encapsulate Layer 3 messages, and user data may have different QoS requirements. With concurrent services, multiple user data streams may have different QoS requirements depending on the service or service instance they correspond to. The multiplex sublayer provides the QoS control required to mediate between requests from competing services, which at a basic level may simply entail establishing relative priority.

5.1.1 Dedicated Channels Multiplex Sublayer

The dedicated channels multiplex sublayer is responsible for multiplexing Layer 2 signaling data units (SDU) from multiple sources, or logical channels, onto the set of physical channels assigned to the call in the progress according to the defined priority and QoS criteria.

Consider the example given in Figure 5.1, which depicts the multiplexing function performed at the base station when a data service is connected. User data bits are delivered in blocks to the multiplex layer for transmission on the Forward Supplemental Channel (F-SCH) and Forward Fundamental Channel (F-FCH). The F-FCH is shared between signaling and user traffic. The F-SCH carries user data only. To assemble the F-SCH SDU (or F-SCH physical layer payload), the multiplex sublayer appends a header to each data block to be sent on the F-SCH. Two data blocks are concatenated to form one F-SCH SDU. Depending on the data rate and the PDU type, as we see later, up to eight PDU can be fit in a single F-SCH frame. User data can also be transmitted on the F-FCH. If during a frame interval, Layer 3 signaling also delivers a data block, then the multiplex sublayer may multiplex the signaling and traffic data blocks onto the same F-FCH frame payload, or can transmit them one at a time, typically giving priority to the signaling data.

The operation of the dedicated channels multiplex sublayer is described in terms of *multiplex options* and multiplex layer PDU formats, or MuxPDU types. These are detailed in the following sections.

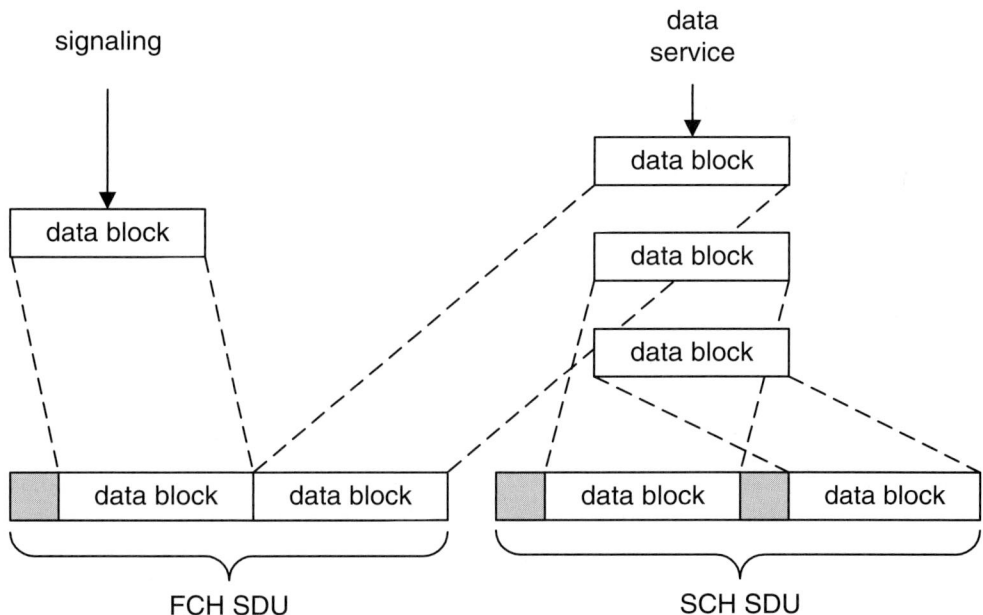

Figure 5.1 Multiplex layer transmitting function (shaded blocks represent the Mux header).

5.1.1.1   Multiplex Options

The multiplex option is one of the service configuration parameters negotiated between mobile station and base station when the service is connected (see also Chapter 6). It specifies the way in which information bits transmitted and received on the traffic channels in either forward or reverse direction are divided into various types of traffic, such as signaling and user data.

The multiplex option used with the F/R-FCH can be of only two types: multiplex option 1, which uses Rate Set 1, and multiplex option 2, which uses Rate Set 2. Both options allow multiplexing of primary traffic, secondary traffic, and signaling on the same physical layer frame. Similarly, the Forward/Reverse Dedicated Control Channel (F/R-DCCH) supports multiplex options 1 and 2 only. The distinction with respect to F/R-FCH is that F/R-DCCH supports only full-rate frames.

The multiplex options that can be used with the F/R-SCH, however, are much more numerous because the F/R-SCH can be operated at various data rates and is typically used to carry packet data traffic, possibly from multiple services connected concurrently and, therefore, a great deal of flexibility is required. The F/R-SCH multiplex option characterizes the service configuration in terms of data rate; maximum number of multiplex layer PDU that can be included in the physical layer payload; size of the multiplex layer PDU; type or format of the multiplex layer PDU. The 16-bit multiplex option number itself is obtained by concatenating the binary encoded values of the service configuration parameters listed above, as shown in Table 5.1. Consider for example the case when multiplex option 0x0911 has been negotiated. Its binary equivalent is $(0000100100010001)_b$, where, starting from the 5th most significant digit, 10 = MuxPDU type 3; 01 = double-size PDU; and 000100 = up to four PDU per F/R-SCH frame; 01 = rate set 1. All the information required for the multiplex sublayer at the receiver end to interpret the received F/R-SCH payload is contained in the 16 bits above.

There is only one multiplex option applicable for the Forward Packet Data Channel (F-PDCH). The specified multiplex option, however, still allows flexible multiplexing of all user data traffic and signaling onto F-PDCH. The multiplex option code is $(0000111100000000)_b$, where 11 = MuxPDU type 5; 11 = variable size PDU; 000000 = no restriction on the number of PDUs per F-PDCH frame; 00 = the rate set is not applicable.

Table 5.1   Multiplex Options Descriptors

	Bits	Value
Rate set	2*	01 = rate set 1, 10 = rate set 2
Max. MuxPDUs included in an L1 SDU	6	000000 = no restriction, 000001 to 001000 = 1 to 8 Mux PDU per L1 SDU
MuxPDU size	2	00 = single, 01 = double, 11 = variable
MuxPDU type	2	00 = type 1, 2, or 4, 10 = type 3, 01 = type 4 or 6, 11 = type 5
Format descriptor	4†	0000 = format 1

* least significant bits; † most significant bits.

5.1.1.2   Mux Layer Protocol Data Units and Their Formats

The MuxPDU type fully describes the frame structure—that is, header bit mapping, payload length, and payload allocation to one or multiple traffic types (e.g., signaling or user data). Although a one-to-one mapping between physical channels and MuxPDU types does not exist (see Table 5.2), the various types are tailored around the needs of specific services, and they account for the constraints of the physical channels used as transport for such services. Within such constraints, they are designed to optimize multiplexing efficiency and keep protocol overhead to a minimum. For example, MuxPDU type 1 is tailored for use with 8-kbps vocoders but also to provide the capability to transport simultaneous signaling as well as secondary user data (for user applications other than voice). MuxPDU type 5, on the other hand, was designed to carry user data exclusively. Salient aspects of the most commonly used MuxPDU types are discussed below.

Table 5.2   MuxPDU Types Supported for Physical Channel Type

Physical channel	Supported MuxPDU types
F/R-FCH	1, 2, 4 and 6
F/R-DCCH	1, 2 ,4 and 6
F/R-SCH	1, 2, 3 and 5
F-PDCH*	5

* The multiplex sublayer actually does not interface with the F-PDCH
directly. It interfaces with the PDCH Control Function (PDCHCF).

MuxPDU type 1 format is shown in Figure 5.2. Of all PDU types, it is the only one supported by all physical channel types other than F-PDCH.[1] It corresponds to Rate Set 1 transmission and is tailored to support 8 kbps vocoders such as Selectable Mode Vocoder (SMV). The MuxPDU type 1 can be either 171, 80, 40, or 16 bits long, corresponding to full-, half-, quarter-, and eighth-rate vocoder frames.

Data blocks from the signaling layer can only be encapsulated in the PDU corresponding to the fundamental data rate, 9,600 bps. Its format includes a variable length header that indicates whether the PDU carries user data, or signaling, or both. When carrying a mix of user data and signaling, the PDU payload allocation to user data is 80, 40, or 16 bits in order to accommodate half-, quarter-, and eighth-rate vocoder frames. The remainder of the bit space is allocated to signaling traffic. This type of signaling transmission is called *dim-and-burst*. When the PDU carries only signaling traffic, the transmission is called *blank-and-burst*. On receipt of a signal-

---

1.   MuxPDU type 1 can be encapsulated into MuxPDU type 5, which is supported on F-PDCH.

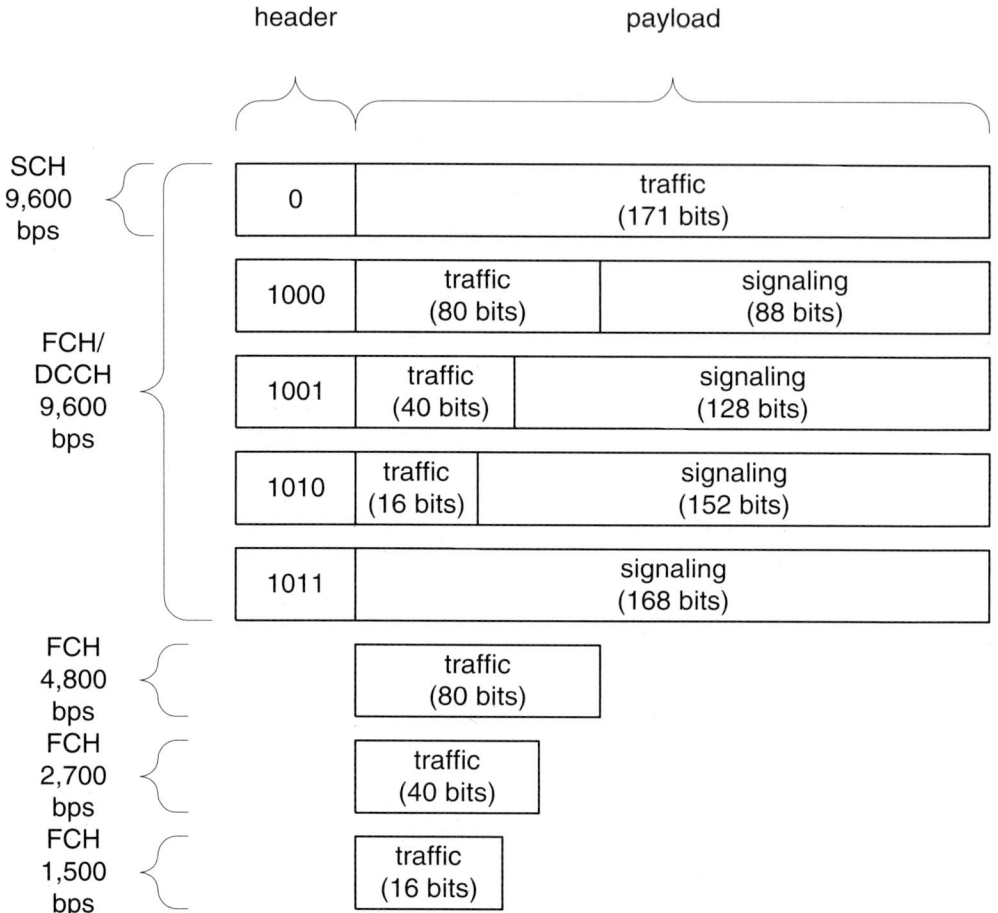

Figure 5.2 Mux PDU type 1 format with primary traffic and/or signaling (format of PDU carrying secondary traffic not shown).

ing PDU from the signaling layer, the multiplex sublayer must decide whether to use dim-and-burst or blank-and-burst. The choice is implementation-dependent. Dim-and-burst is preferable if the user-data frame pending for transmission corresponds to 4,800 bps or lower, since both signaling message and user data can be inserted in the MuxPDU. Blank-and-burst, by contrast, would require discarding the user-data frame, therefore affecting voice quality. However, blank-and-burst is preferable when the message length is such that it would span multiple frames if the dim-and-burst were used (therefore increasing the probability of message erasure)[2] or if the

2.   Probability of message erasure is proportional to the message transmission period.

user-data frame pending for transmission corresponds to a full-rate frame and the message has a stringent latency requirement.

Note that PDUs corresponding to physical channel transmission at rate 4,800, 2,700, or 1,500 bps do not contain the header. That does not create ambiguity at the receiving end of the multiplex sublayer. The physical layer performs parallel decoding of the received symbols under all possible data-rate hypotheses and then selects the data rate for which the decoded bits successfully pass the cyclic redundancy code (CRC) check.[3] Hence, no header is needed for rate information or to indicate the presence of signaling traffic, which is not allowed when the data rate is lower than 9,600 bps.

MuxPDU type 3 is used for the F/R-SCH when the assigned data rate is 19.2 kbps or larger. Its format is shown in Figure 5.3. The header contains the service reference identifier (sr_id), a tag that uniquely identifies one of the possibly many logical channels mapped onto the F/R-SCH. This PDU type can be either a single-size or double-size PDU. Multiple PDUs are serially concatenated to fill the F/R-SCH frame payload. Double-size PDU are more efficient, since the overhead is 6/350 ~ 1.7%, while for single-size PDU the overhead is ~3.4%.

MuxPDU type 5 was originally introduced to support more efficient F/R-SCH high-speed packet data operation than its predecessor, type 3. The frame format is shown in Figure 5.4. The improved efficiency comes with the variable-length format, which results in lower overhead for larger payloads. A field in the header indicates the PDU length. Like type 3, the type 5 header also contains the service reference identifier, a tag that uniquely identifies one of possibly many logical channels mapped onto the physical channel. This MuxPDU type can also be used with F-PDCH, and it is the only type mapped onto that physical channel. When mapped onto the F-PDCH, PDU type 5 can carry signaling, which is indicated with sr_id = 000. MuxPDU type 5 can also encapsulate MuxPDU types 1, 2, and 4. The presence of encapsulated MuxPDU is signaled with the extension indicator. The 2-bit length-indicator field follows the extension indicator and points out how payload size is computed. The encoding of

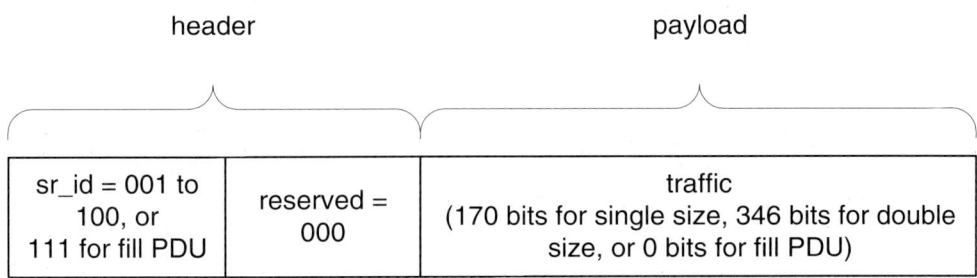

Figure 5.3 MuxPDU type 3 format.

---

3.   Either only one hypothesis passes the CRC check or none does, because the probability of error misdetection is negligibly small.

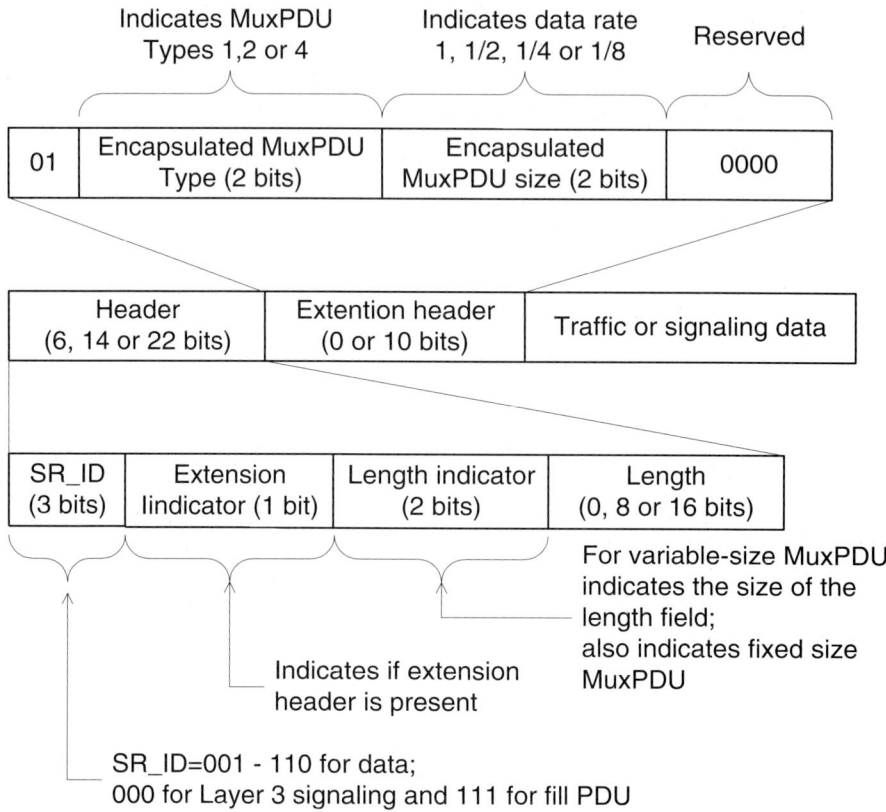

Figure 5.4   MuxPDU type 5 format.

the field is shown in Table 5.3. The payload size is signaled with the length field or provided by the physical layer. It can also be fixed and optimized for F-PDCH operation. The fixed, optimized payload size for F-PDCH is 378 bits.[4]

Table 5.3   Length Indicator Encoding

Length Indicator	Payload size
00	Indicated by physical layer
01	Indicated by 8-bit length field
10	Indicated by 16-bit length field
11	Fixed size, optimized for F-PDCH

---

4.   The payload size is 378 bits if an extension header is absent and 368 if an extension header is present.

MuxPDU 2 type is similar to MuxPDU type 1. The only difference is that MuxPDU type 2 supports Rate Set 2. MuxPDU type 4 is tailored for 5-ms frames, which are used to carry Layer 3 signaling 'mini-messages' with short transmission time and fast turnaround. There is no associated header with the MuxPDU type 4. MuxPDU type 6 allows for a configurable header format. The header format is specified in a partition table.

5.1.2 Packet Data Channel Control Function

The Packet Data Channel Control Function (PDCHCF) is the entity that terminates all the physical channels associated with F-PDCH. The PDCHCF handles link adaptation and hybrid ARQ (HARQ) operation over four independent ARQ channels. An overview of the PDCHCF is given below. More details are provided in Chapter 9 when we discuss packet data operation.

At the base station side, the PDCHCF selects an appropriate modulation and coding scheme for the F-PDCH based on the information received on the Reverse Channel Quality Indicator Channel (R-CQICH). The PDCHCF also assembles the control signaling information necessary to appropriately configure the mobile station receiver to be able to decode the F-PDCH packet. Such information is transferred to the mobile station on the Forward Packet Data Control Channel (F-PDCCH). Since F-PDCCH is a common channel, the messages must be addressed to the intended mobile station. For that purpose, each mobile station is assigned an 8-bit identifier referred to as MAC_ID. In addition to information related to the modulation and coding, Walsh codes, and HARQ instance, MAC_ID is included in the F-PDCCH message.

At the mobile station side, the PDCHCF evaluates the PDCCH message and takes appropriate action. For example, the PDCHCF configures the mobile station decoder so that the received packets can be successfully decoded. If the packet is decoded error free, the PDCHCF sends a positive acknowledgment on the Reverse Acknowledgment Channel (R-ACKCH). The mobile station receiver supports HARQ. The erroneously decoded packets are not discarded; they are rather soft-combined with the retransmissions (see Chapter 9). In addition to R-ACKCH, the PDCHCF handles R-CQICH operation. Besides channel quality indication, R-CQICH is also used to signal cell or sector selection and reselection, referred to as *switching*. The R-CQICH may contain a full 4-bit channel quality report or a 1-bit differential update. Cell selection is signaled using the corresponding Walsh cover.

5.1.3 Radio Link Protocol

Most data applications are sensitive to data loss and require reliable transmission. Despite the use of forward error correction, and possibly HARQ, the physical layer protocol may not be able to guarantee reliable data transfer over the radio channel. If the data application is delay tolerant, however, the residual errors on the physical channels (after HARQ, if applicable) can be corrected with RLP. RLP is a negative acknowledgment (NAK)-based ARQ protocol. It reduces the physical layer frame error rate (FER) to values more suitable for the packet data service running above. That is achieved by retransmission of erased frames upon request of the receiver. RLP, although functionally part of the MAC layer, is specified separately in the IS-707 standard [3].

5.1.3.1   *Parameter Negotiation and SYNC Exchange Procedure*

RLP data transmission and retransmission procedures are regulated by a set of parameters that are negotiated when the service is connected. RLP parameters are contained in the so-called RLP_BLOB, where BLOB stands for 'block of bits.' The RLP_BLOB itself is one of the fields in the service configuration record carried by signaling messages during service negotiation (see Chapter 6). The RLP parameters include the number of NAKs per round, the number of rounds, and possibly the estimated round-trip delay between the base station and the mobile station. As explained in the following section, the number of NAKs per round and number of rounds regulate the NACK retransmission mechanism, while the round-trip delay regulate its timing. If the F-PDCH is supported, the RLP_BLOB may contain a delay-detection window parameter, DELAY_DETECTION_WINDOW. As discussed in Chapter 9, the delayed detection of the missing frames is necessary to prevent unnecessary retransmissions caused by asynchronous nature of HARQ type II and parallel channels operation.

Correct setting of the round-trip delay is necessary to ensure efficient operation of RLP. When it is not known a priori, the round-trip delay may be estimated during the SYNC exchange procedure, which is a three-way handshake protocol. The procedure is illustrated in Figure 5.5.

The procedure starts with one RLP entity, e.g., at the base station, sending a continuous stream of SYNC control frames. On receipt of the first SYNC frame, the mobile station RLP sends a continuous stream of SYNC/ACK frames till it receives the first ACK frame. At the

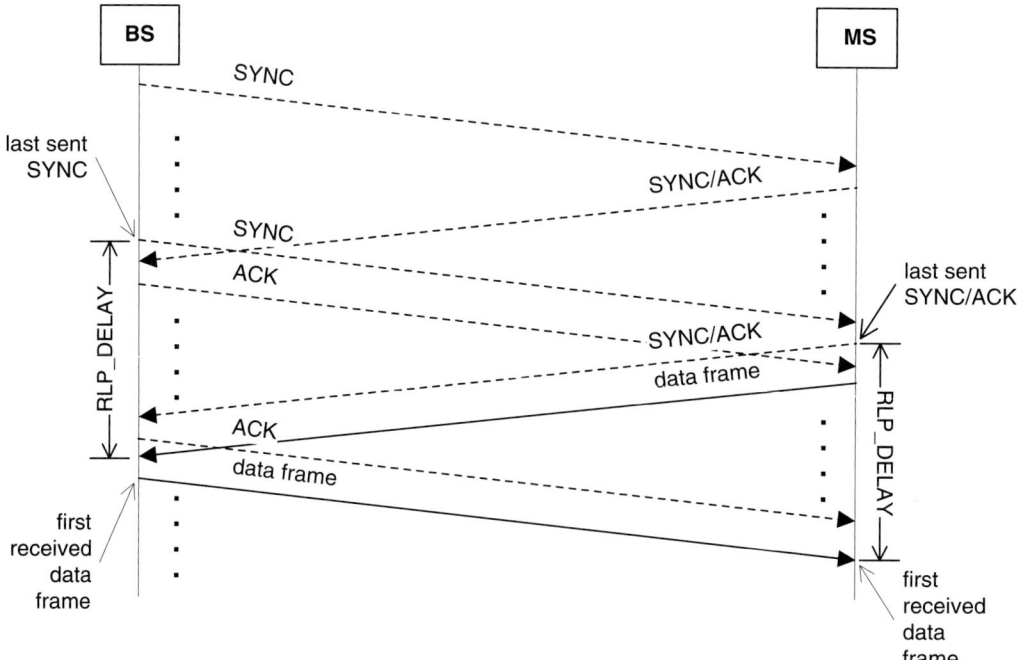

Figure 5.5   RLP SYNC exchange procedure for round trip delay estimation.

same time, the base station RLP stops sending SYNC frames and sends instead ACK frames when it receives the first SYNC/ACK. When the mobile station RLP receives the ACK, it starts sending data frames. When the base station receives the first data frame, it stops sending ACK frames and sends instead data frames. At this time the sync exchange procedure is terminated. The mobile station computes the round-trip delay (RLP_DELAY) as the time elapsed between transmission of the last SYNC/ACK frame till receipt of the first data frame. The base station, on the other hand, estimates the RLP_DELAY as the time elapsed between transmission of the last SYNC frame till receipt of the first data frame. As illustrated in Figure 5.5, the mobile station and base station will obtain identical estimates of the round-trip delay.

5.1.3.2   Data Transfer

Once RLP parameters are initialized, data transfer can start. On the transmitting side, RLP maintains a 12-bit sequence number, L_V(S), which is the sequence number of the next data frame that will be supplied to the multiplex sublayer. After the multiplex sublayer adds an appropriate header, it passes down the frame to the physical layer for encoding and transmission over the air.

On the receiving side, the receiver maintains two 12-bit sequence numbers, L_V(R) and L_V(N). The L_V(R) contains the expected value of the RLP sequence number field in the next data frame to be received. The L_V(N) contains the sequence number of the next data frame that is needed for sequential delivery to the higher layers. RLP transmit and receive sequence number variables are illustrated in Figure 5.6.

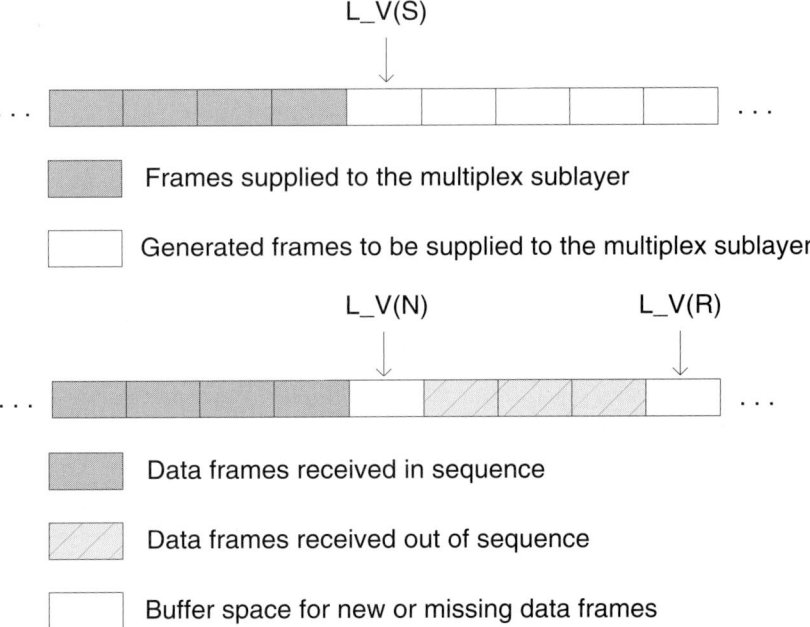

Figure 5.6   RLP transmit and receive sequence number variables.

The figure illustrates how RLP detects missing frames. The RLP waits for out-of-order frames so that it can conclude that a frame is missing. When a missing frame is detected, one or more RLP NAK control frames are sent back to the transmitter, indicating the sequence number of the missing frame or frames. The NAK frame can be transmitted as soon as the missing frame is detected or after the negotiated delay-detection window has elapsed (in the case of F-PDCH). The receiver could send more than one NAK when a missing frame is detected so that the transmitter retransmits multiple copies of the missing frame. After sending the NAK frame, the receiver starts the retransmit timer, REXMIT_TIMER. The timer value is based on the estimated round-trip delay (RLP_DELAY), which is obtained using the SYNC exchange procedure explained above. On reception of a NAK frame, the transmitter retransmits the missing frame or frames. If the missing frame is still not received when the retransmit timer expires, and if the maximum number of NAK rounds is not exhausted, the receiver sends another round of NAK frames to the transmitter. Note that correct setting of the timer is important to ensure efficient operation of the retransmission scheme. If the timer is too short, NAKs would be sent prematurely, thus causing unnecessary retransmissions and decreasing throughput. If the timer is too long, NAKs would be delayed and the average time it takes to recover from physical layer errors increases. RLP abandons the error-recovery procedure only after the number of NAK rounds reaches its negotiated limit. Aborting retransmission necessitates recovery from the layers operating above the RLP (i.e., TCP). The number of NACKs per round and the number of rounds are usually denoted by ($a1$, $a2$, $a3$, ...), where $a1$ is the number of NAKs sent in the first round, $a2$ is the number of NAKs sent in the second round, and so on. Thus, a (1,2,3) RLP scheme implies the following procedure. One NAK is sent in the first round, two NAKs are sent in the second round, three NAKs are sent in the third and final round, at which point transmission is aborted. Each round corresponds to the event of not receiving the missing data packet in sequence after the round-trip time. Thus in total, 6 NAKs would be sent before declaring an abort. Increasing the number of NAKs in each round decreases the residual probability of error seen by the upper layers, but at the expense of increasing the delay variance and overhead for transmitting the additional NAKs. Consider for example the (2,3) and the (1,2,3) RLP schemes. When using the (2,3) scheme it is less likely that a second round of NAKs will be required because a packet is retransmitted twice during the very first round.[5] However, retransmitting a packet at least twice for every lost RLP packet causes overhead at the expense of throughout. When using the (1,2,3) scheme, on the other hand, unnecessary retransmissions can be avoided since the first single retransmission may suffice. However, this scheme can result in higher delay. The scheme that results in the best delay/throughput depends on the physical layer FER and the characteristics of the channel, e.g., distribution of length of the frame error bursts. In [4] it was shown by means of simulation that the (1,2,3) scheme is quite robust and achieves good performance in most channel conditions.

---

5.   However, when frame errors are correlated and occur in bursts, the benefit of multiple NAKs per round is diminished.

When the transmitter data buffer is empty, the transmitter sends RLP idle frames. Recall that in order to detect a missing frame, the receiver must receive an out-of-order frame. The RLP idle frames are short control messages that enable the error-recovery mechanism for the last transmitted data frame. An RLP instance may correspond to primary or secondary traffic. In the case of an RLP instance that carries primary traffic, idle frames are transmitted until a dedicated channel, such as F/R-FCH, is released. In this case, the purpose of idle frames is not only for error recovery, but also to supply data for the traffic channel. Remember that discontinuous transmission is not allowed over F/R-FCH. However, if RLP traffic corresponds to secondary traffic, that is, if there is another primary RLP instance or voice mapped to the same physical channel, RLP idle frames are only transmitted for a period of time that corresponds to round-trip delay. In this case, their purpose is purely for error recovery.

Besides an idle frame, another important RLP control frame type is a fill frame. The fill frames are short signaling messages, used by the RLP receiver to inform the RLP transmitter about the sequence number of the next data frame needed for sequential data delivery to the layer above RLP. This sequence number is called L_V(N). The fill frame implicitly positively acknowledges the reception of all RLP frames with a sequence number less than L_V(N).

In this section we described the data transfer of unsegmented RLP frames. Segmentation of RLP frames is permitted, but the operation, even though different in detail, is the same in principle and is not discussed in this book.

5.1.3.3 Performance

In this section we estimate the performance of RLP, using the same approach as in [5]. We consider a simplistic scenario that assumes independent RLP frame errors and denotes the frame error probability as p. This assumption may not be entirely correct in practice because, especially in fading channels, frame errors tend to occur in bursts, but it certainly gives an insight into the protocol performance. Denote the probability of error of NAK control frames with q. Assume that both p and q are time invariant. The probability that in the j-th round the frame is not received correctly can then be computed as

$$P_j = \left[1 - (1-p)(1-q) \right]^{n(j)} \tag{5.1}$$

where $n(j)$ is the number of NAKs in the round j. The residual error rate after the J-th round can be written as a product of the probability that the initial transmission had been in error and that after J rounds of NAKs, the frame was still not correctly decoded:

$$P_J = p \prod_{j=1}^{J} \left[1 - (1-p)(1-q) \right]^{n(j)} \tag{5.2}$$

Of interest now is to compute the probability that the frame is correctly received at the receiver after the i-th NAK in the j-th round. Let us denote that probability as $P_c(i,j)$. The probability can be evaluated as a product of the probability that up to the $J-1$-th round, the frame was in error, P_{J-1}; the probability that the first $i-1$ NAKs in the J-th round did not result in an error-free frame; and the probability of successful frame decoding after the i-th NAK. Therefore, we can write

$$P_c(i,j) = p(1-p)(1-q)\left[1-(1-p)(1-q)\right]^{i-1} \prod_{j=1}^{J-1}\left[1-(1-p)(1-q)\right]^{n(j)} \qquad (5.3)$$

To estimate the frame delay after RLP, we must make an assumption for the one-way delay between the RLP termination points, T, and the interframe time, τ. If we assume that a mobile station is assigned a channel configuration with dedicated traffic channels, the assumption of constant T and τ is reasonable. Assume also that the retransmit timer value is equal to $2T$. We can now compute the average frame delay, $E[D]$, between the RLP termination points (base station and mobile station) as

$$E[D] = T(1-p) + \sum_{j=1}^{J_{MAX}}\sum_{i=1}^{n(j)} P_c(i,j)(2jT+2(i-1)\tau) \qquad (5.4)$$

In Table 5.4 we provide numerical examples for RLP residual error, delay and the average number of transmissions per frame when using RLP scheme (1,2,3). We consider a scenario in which a mobile station downloads data over F-SCH (with the probability of error, p, ranging from 1% to 10%). T is set to 100 ms and τ is equal to 20 ms. The error rate on the reverse-traffic channel is assumed to be much lower, $q = 1\%$, because RLP control frames are carried by the R-FCH. Note that the RLP residual error rate can be made negligibly small, even in case of high F-SCH FER, at the expense of increased delay and deceased throughput.

Table 5.4   RLP Performance with Uncorrelated Errors (R-FCH error rate, q = 1%)

F-SCH error rate	Residual RLP error rate	Average delay (ms)	Average number of transmissions per frame
1%	$<10^{-6}$	101.0	1.01
5%	$<10^{-6}$	105.6	1.05
10%	$<10^{-6}$	112.2	1.11

5.1.3.4   RLP Data Frame Formats

In this subsection we illustrate the RLP data frame formats on a several examples. A format A frame is used for F/R-FCH or F/R-DCCH. Table 5.5 shows the frame format for the primary traffic over a full-rate, Rate Set 1, F/R-FCH. This is called a Rate 1 RLP frame. The RLP frame consists of the information (data or control) and type field. The size of data or signaling is 168 bits (in this example). The size of primary RLP traffic can also take values that correspond to half-, quarter-, and eighth-rate F/R-FCH frames. These frames are called Rate 1/2, Rate 1/4, and Rate 1/8 RLP frames, respectively. The number of information bits is 80, 40, and 16, respectively. The secondary traffic takes values that are complements (relative to Rate 1) to the primary rates. The actual signaling and data frame size does not have to exactly match the size of the RLP frame. Padding is used to match the payload to the RLP frame size.

Table 5.6 illustrates the data frame format for all the primary and secondary RLP frame rates other than Rate 1/8. The Rate 1/8 RLP frame is for signaling, such as idle or fill frames. For example, in the idle control frame, 12 bits convey L_V(S) and 4 bits determines the control frame type.

All RLP control signaling uses format A, which is mapped to MuxPDU type 1 or 2 and F/R-FCH or F/R-DCCH. If dedicated channels are not present (as in the case of the traffic channel configuration with F-PDCH only on the forward link), MuxPDU type 1 or 2 is encapsulated into MuxPDU type 5. The encapsulation process was described in Section 5.1.4 and illustrated in Figure 5.4.

Table 5.5   Rate 1 RLP Frame Format A

Field	Length	Description
Information	168	Control or data frame
Type	3	Frame type, set to 001

Table 5.6   Data Frame

Field	Length	Description
SEQ	8	Sequence number, 8 least significant bits
CTL	1	Set to 0 for unsegmented data frames
REXMIT	1	Indicates retransmissions
LEN	6	Data length in octets
Data	8 x LEN	Data octets
Padding	Variable	Padding bits

The alternative to format A for data traffic only is format B. Format B is more efficient than format A due to a smaller header. However, as shown in Table 5.7, padding is not allowed.

Table 5.7   RLP Frame Format B

Field	Length	Description
SEQ	8	Sequence number, 8 least significant bits
Data	160	Data octets
Type	3	Frame type, also distinguish new data from retransmissions

F/R-SCH and F-PDCH require frame formats other than A or B. For these channels, format C and format D are defined. Format C is intended for traffic with negotiated frame size, and it is illustrated in Table 5.8. Format C actually supports variable data blocks, but the block size must be indicated by the multiplex sublayer. This frame format has small overhead. The disadvantage is that the sequence number field cannot be larger than 8 bits. For the high-rate channels, the small sequence space can create frame number ambiguity. The ambiguity occurs due to sequence numbers wraparound. This is more likely to happen with 8-bit sequence numbers than with 12-bit space, which is provided with format D. However, as shown in Table 5.9, format D has larger overhead than format C. Fill control frames can be used to determine whether or not there is an ambiguity with 8-bit sequence numbers.

Table 5.8   RLP Frame Format C

Field	Length	Description
Type	2	Frame type, also distinguish new data from retransmissions
SEQ	8	Sequence number, 8 least significant bits
Data	Variable	Data octets

Table 5.9   RLP Frame Format D

Field	Length	Description
Type	2	Frame type, also distinguish new data from retransmissions
SEQ	8	Sequence number, 8 least significant bits
SQI	1	Indicates whether SEQ_HI is present
...		

Table 5.9   RLP Frame Format D (continued)

Field	Length	Description
REXMIT	1	Indicates retransmissions
LEN	0 or 8	Data length in octets, multiplex sublayer does not indicate
SEQ_HI	0 or 4	Sequence number, 4 most significant bits
…		
Padding_1	Variable	Padding bits to provide octet alignment for the data field
Data	8 x LEN	Data octets
Padding_2	Variable	Padding bits

5.1.4 Burst Transmission on the Reverse Access Channel

The Reverse Access Channel[6] (R-ACH) is used for communication from the mobile station to the base station when no dedicated traffic channel connection exists. As discussed in Chapter 6, the R-ACH is used by the mobile station while in the system-access state to transmit signaling messages such as origination, page response, and short data messages. The R-ACH is a random multiple-access channel in the sense that mobile stations access this channel without explicit authorization from the base station, and neither their identities nor their number is known by the base station a priori. It is also a contention channel, which means that mobile stations contend against each other for base station resources. The contention process may result in collisions. The R-ACH is a slotted channel; that is, message transmission can be initiated only at the slot boundary, which is synchronized to system time. Therefore, the R-ACH has some characteristics similar to a slotted ALOHA channel, but many of its characteristics are changed by the properties of CDMA. Such characteristics motivate the control procedures adopted by the access channel protocol, which we explain below.

The parameters controlling the access channel protocol are contained in the *Access Parameters Message* sent on the Forward Paging Channel (F-PCH). They are summarized in Table 5.10 for ease of reference in the discussion that follows.

---

6.   Some aspects of the R-ACH operation presented hereafter are outside the scope of the MAC layer but are included here for ease of discussion.

Table 5.10 R-ACH Protocol Parameters

Field	Description
NOM_PWR	Correction factor for open-loop power control
INIT_PWR	Correction factor for open-loop power control used at the beginning of the access attempt.
PWR_STEP	Power increment between successive access probes
NUM_STEP	Determines the maximum number of probes within an access sequence
MAX_CAP_SZ	Determines the maximum message size allowed on the R-ACH
PAM_SZ	Determines the length of the access probe preamble
PSIST(n)	Persistence values used for flow control of request messages, one for each of 16 different mobile station's classes
MSG_PSIST	Persistence modifier, relative to origination messages, used for data burst messages
REG_PSIST	Persistence modifier, relative to origination messages, used for registration messages
PROBE_PN_RAN	Used to generate the pseudorandom PN delay
ACC_TMO	Timeout to receive base station acknowledgment to a probe transmission
PROBE_BKOFF	Controls the random probe backoff
BKOFF	Controls the random sequence backoff
MAX_REQ_SEQ	Maximum number of access probe sequences for a request message
MAX_RSP_SEQ	Maximum number of access probe sequences for a response message

Note: Not all parameters are shown.

5.1.4.1   R-ACH Structure and Properties

The R-ACH is a message-oriented, connectionless channel and therefore can be operated at a fixed rate, unlike the connection-oriented traffic channel that must be operated at a variable rate to increase capacity. The relatively low data rate of the R-ACH, 4800 bps, was originally selected in IS-95 [6] because the channel use was envisioned for short messages. In such scenarios, message transmission time is small relative to the overall access network processing delay, so a higher data rate would provide no advantage.

The R-ACH has a fixed-size preamble consisting of multiple frames. The preamble frame consists of all zero symbols in order to facilitate preamble acquisition at the base station receiver. The number of preamble frames is decided by the access network and depends on the rate at which the base station demodulator can search the pseudonoise (PN) space, the cell radius, and the channel multipath spread. The following are the reasons for these dependencies. The access channel is slotted, and all mobile stations will start transmission at the slot boundary. Hence, the base station knows when to start acquiring the preamble within the uncertainty caused by the propagation delay. The propagation delay depends on mobile station distance relative to the base station and the presence of reflectors causing certain paths to arrive at the base station demodulator with further delay relative to that of the line-of-sight component. Moreover, when testing the hypothesis of a preamble signal being present at a given PN offset from the slot boundary, the energy-detector dwell time at the base station receiver must be sufficiently large to minimize the probability of false alarm while achieving a high probability of detection. On the other hand, as for a conventional slotted ALOHA protocol, increasing the preamble size requires a larger slot size for a given message size, which in turn decreases the capacity of the channel.[7] It is therefore apparent that the preamble length setting is a trade-off between improved acquisition performance (which increases for increasing preamble size) and access channel capacity (which decreases for increasing preamble size).

Unlike the traffic channel, the R-ACH does not need a frame quality indicator because its data rate is fixed[8] and because a CRC is needed on the entire message itself. A frame in error would be detected by the message CRC, thus making the frame CRC redundant. A potential advantage of a frame CRC would be early detection of message transmission failure, but that would not allow the base station receiver resources to be reassigned to another message reception due to the slotted nature of the channel.

The R-ACH slot structure is depicted in Figure 5.7. During one access channel slot, one access probe can be transmitted. An access probe consists of the access probe preamble of size PAM_SIZE + 1 frames and the message capsule of size MAX_CAP_SIZE + 3 frames. The message capsule contains the message CRC bits, for message error detection, and a variable amount of padding bits to fill out the entire slot. As is explained in Section 5.1.4.3, the actual probe transmission is delayed by a pseudorandom time between 0 and 2 ^ (PROBE_PN_RAN) – 1, in units of one chip duration.

7.   Although capacity increases using a slotted protocol, the main reason for using slotted transmission is to simplify the acquisition process at the base station by providing it with a time reference at which acquisition can start.
8.   Recall that the CRC is used on traffic channel frames to determine both frame quality and data rate, since the data rate is variable and not known a priori by the base station.

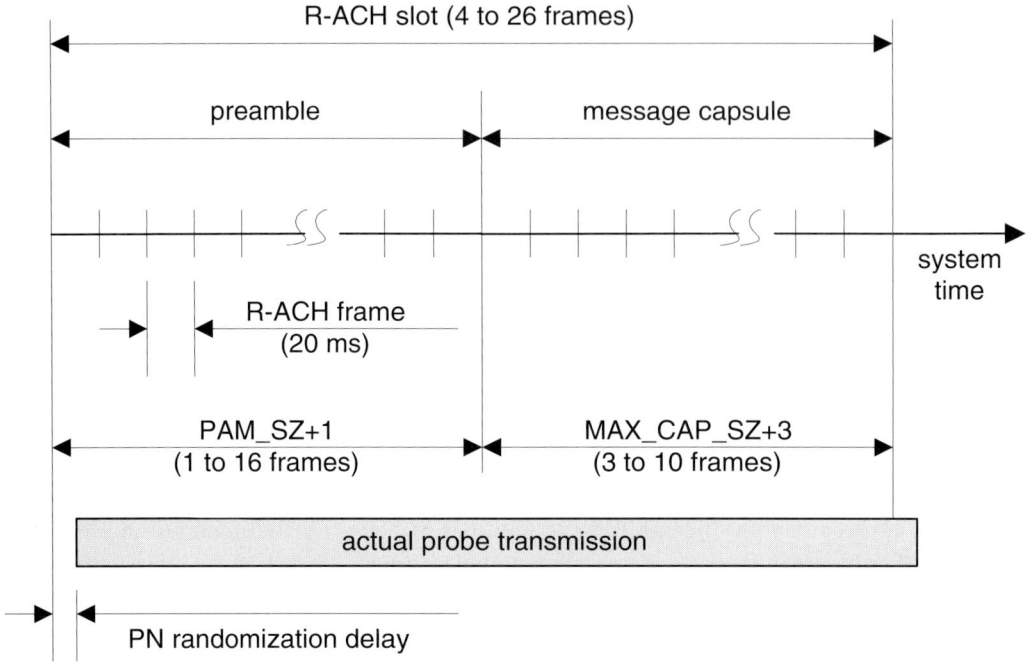

Figure 5.7   R-ACH slot structure.

5.1.4.2   R-ACH Power Control

The R-ACH uses the same open-loop power control method used by the traffic channel. The mobile station sets the transmit power to a value inversely proportional to the total received forward-link power to compensate for path loss, adjusted by an amount dictated by the base station. As depicted in Figure 5.9, the initial power, *IP*, is set to −73 minus the mean received power in dBm plus 0.5 × NOM_PWR+INIT_PWR.

The R-ACH also uses a particular form of closed-loop power control called access probing. On transmission of a message, the mobile station waits for an acknowledgment from the base station. If that is not received within a predefined amount of time, the message transmission is repeated at higher power relative to the open-loop value. The power increment, *PI*, is equal to 0.5 × PWR_STEP. With this form of probing, the mobile station adjust its transmit power relative to the open-loop estimate to compensate for forward/reverse link–path loss differences. The adjustment must be done gradually to avoid jamming the base station receiver, so typically PWR_STEP is set to 1 or 2 dB. Of course, access probing is a rather slow and inefficient closed-loop power-control method, but it was deemed sufficient in the IS-95 standard due to the relatively small access channel message size and utilization expected of the early CDMA systems.

5.1.4.3 Handling Access Channel Collisions

The R-ACH is a slotted, multiple-access protocol with random slot selection by multiple mobile stations contending access to the channel. Therefore, when two or more mobile stations start message transmission at the same slot boundary, a collision may occur. A collision does not happen at all times even when the same slot is being contended because of two reasons.

First, the signal of one mobile station may be received at much higher power than the others due to the combined effect of open-loop power control inaccuracy and/or access probing. That is called the *capture effect*.

Second, if the mobile stations' distance from the base station differ by more than ~122 m, equivalent to one chip differential round-trip delay,[9] the multiple transmissions can be distinguished, allowing the base station to acquire at least[10] one of them. However, if the cell size is quite small, or if the mobile stations are likely to originate a message in the same geographical area,[11] the differential delay may not suffice to allow distinguishing message transmissions and the collision probability greatly increases. That is the reason for *PN randomization*, a randomization method in addition to that used for slot selection that randomizes the transmission time relative to the beginning of the slot by an amount up to 511 chips. The PN randomization delay is determined by a hashing algorithm keyed by the mobile station electronic serial number (ESN). This random delay effectively enlarges the cell diameter.

5.1.4.4 R-ACH Control Procedures

Let us now consider the procedure for transmission of a message on the R-ACH. The entire process for message transmission is called *access attempt* (see Figure 5.8). An access attempt consists of up to MAX_REQ_SEQ *access probe sequences*. An access probe sequence consists of up to NUM_STEP + 1 access probes (see Figure 5.9).

The access protocol has three main characteristics: stop-and-wait message transmission, message retransmission random backoff, and flow control.[12]

From a LAC layer perspective, the R-ACH is a stop-and-wait protocol. On transmission of a message, the mobile station waits for the base station acknowledgment on the F-PCH. If the acknowledgment is not received within a period of ACC_TMO × 80 ms, the mobile station may retransmit the probe at the next access slot. The ACC_TMO is set by the base station and must

---

9.   Since R-ACH transmission is referenced to the forward link pilot timing, round-trip delay must be considered.
10. Multiple-access transmissions in the same slot can be acquired if parallel searching and multiple demodulators are employed at the base station receiver.
11. That is the case, for example, when zone-based registration is used, because mobile stations entering a new registration zone are likely to start message transmission at the cell edge, i.e., approximately at the same distance from the base station. See Chapter 6 for details.
12. Flow control is used only for request messages—unsolicited messages such as the *Origination Message, Data Burst,* or *Registration Message*. Response messages—those sent in response to a base station order, such as the *Page Response Message,* do not necessitate explicit flow control since the access network can regulate the rate of those by regulating the rate at which it sends page messages.

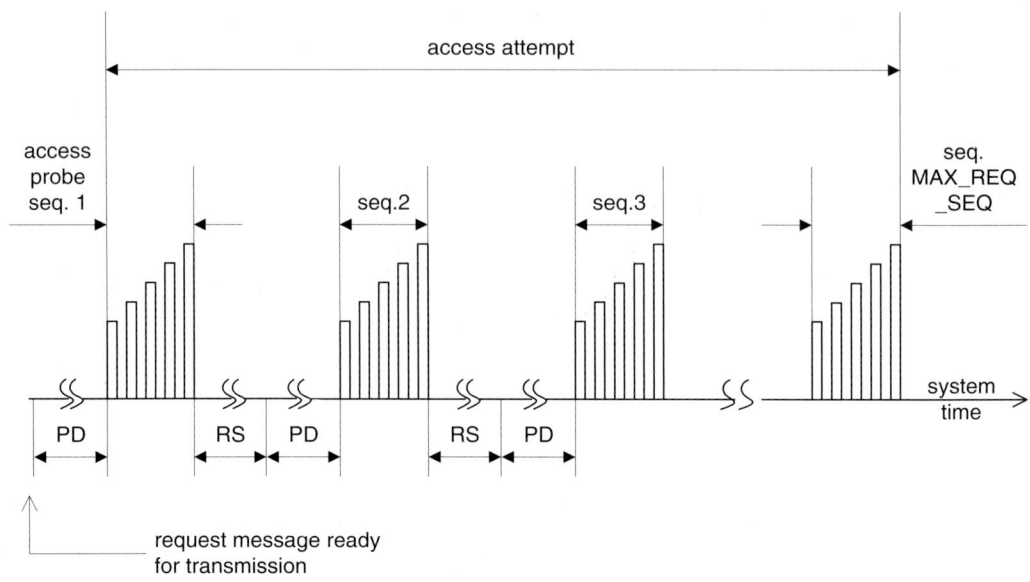

Figure 5.8   Request message transmission attempt.

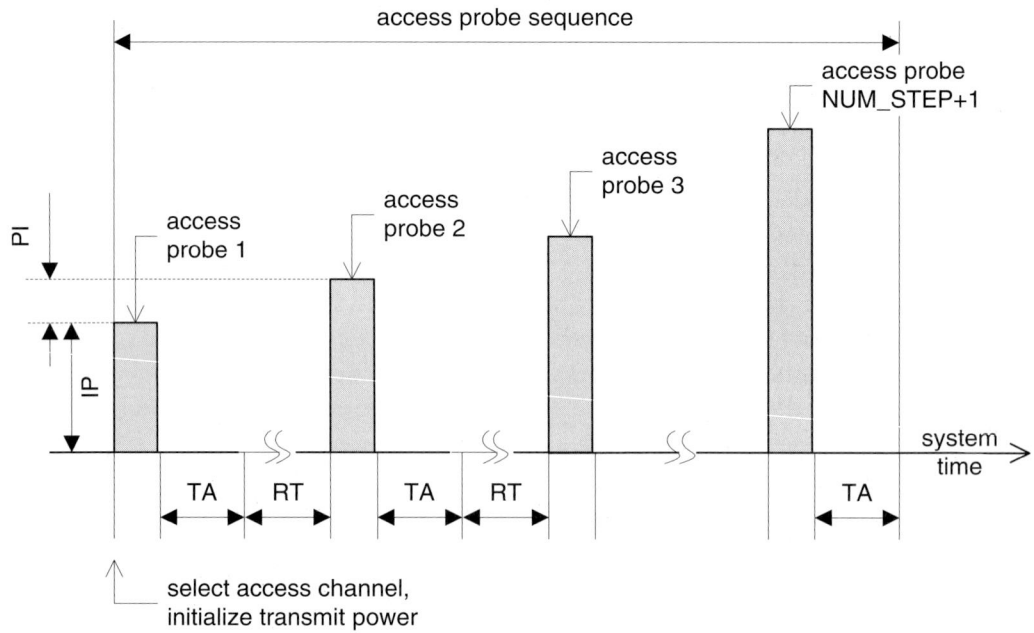

Figure 5.9   Access probe sequence.

be larger than the message decoding delay, plus the internal processing delays, plus the acknowledgment message queuing and transmission delay on the F-PCH.

A random backoff, in units of slot time, is used to randomize the time of retransmissions and minimize the probability that retransmissions will collide again. The random backoff is applied both between probe sequences and between probes within a sequence. The sequence backoff, RS, is a random integer between 0 and BKOFF + 1. The probe backoff, RT, is a random integer between 0 and PROBE_BKOFF + 1.

Flow control is used to throttle the rate at which request messages are sent. When the mobile station has a request message to send, it has to run a persistence test before it can start the access attempt. The mobile station generates a random number as if it were tossing a biased coin. If the outcome is favorable, it starts the access attempt at the slot boundary; otherwise it waits for the next slot and repeats the test. The persistence test is also run before transmission of each probe sequence within the access attempt. The coin bias, P, is equal to the probability of favorable outcome and depends on the parameter PSIST:

$$ P = \begin{cases} 2^{-\text{PSIST}/4} & \text{if PSIST} \neq 63 \\ 0 & \text{otherwise} \end{cases} \tag{5.5} $$

Note that the PSIST parameter is set differently for different mobile stations depending on the access overload class they belong to. The access network controls setting of the PSIST parameters, which can be changed dynamically depending on loading and availability of radio resources. In an extreme situation, for example, when the network is overloaded, the overload class for emergency and law-enforcement mobile stations can be given PSIST = 0, while all other classes can be given PSIST = 63. Then, no one but emergency callers can access the network. Finally, the last detail: Since, for the access network, it is desirable to throttle delay-insensitive messages more heavily than origination messages, the coin bias P is further multiplied by 2 ^ (–REG_PSIST) and 2 ^ (–MSG_PSIST) for registrations and data burst messages, respectively.

5.1.5 Burst Transmission on the Enhanced Reverse Access Channel

The R-ACH specified in the early IS-95 standard was meant for transmission of short messages. When the message size exceeds several frames, the R-ACH channel structure with its protocol mimicking that of a slotted ALOHA system performs poorly. We begin with listing the R-ACH shortcomings, and in the next subsections we introduce enhanced channels and protocols designed to remedy them.

Long slot cycle and access probing time. The R-ACH slot is designed to accommodate both preamble and message. When the message size increases, so does the slot size, with negative effect on throughput and delay. The mobile station has no way to assess whether message transmission has collided other than to wait for expiration of the acknowledgment timer. By that time, the mobile station has unnecessarily wasted power and added to the reverse-link interference.

Long acquisition time. The R-ACH preamble length is typically set to 1 to 3 frames to allow for reliable acquisition. This corresponds to 25% to 75% of the typical *maximum* message capsule length, which is a significant overhead.

Low data rate and fixed message size. The R-ACH operates at 4800 bps only, and the data burst is formatted to fit into a fixed (quite long) access slot. This design cannot efficiently handle packet data applications, since any fixed slot size would be sometimes too small and sometimes too large to accommodate the variable-size packet. If a random access channel is to be used for data transmission rather than for control messages only, more flexibility is needed in terms of data rates and frame sizes for an efficient utilization of channel resources.

Open-loop power control without soft handoff. The transmitted power on the R-ACH is determined based on an approximate estimation of path loss and is slowly adjusted with each probe transmission. With increasing message size, closed-power control with fast feedback is required to achieve low message error rate and power efficiency. Soft handoff is also required, since the likelihood that the cell selected at the time of access remains that the best serving cell throughout message transmission decreases with increasing message transmission time.

In light of the shortcomings above, two alternative reverse-link access channels and associated protocols can be used: the Reverse Enhanced Access Channel (R-EACH) *in basic access mode* and the R-EACH and Reverse Common Control Channel (R-CCCH) in *reservation access mode*.

5.1.5.1   R-EACH in Basic Access Mode

The R-EACH has several physical layer improvements relative to the R-ACH already mentioned in Chapter 4. But the main difference enabling significant Layer 2 performance improvement is the time-dependent long-code mask used for signal spreading.

Just like the RACH, the R-EACH uses a slotted random access protocol with transmission beginning only at the start of the slot. However, unlike the RACH, the R-EACH slot is not designed to accommodate the entire message. Rather, a message transmission may overlap multiple slots. That is possible because the R-EACH long code mask changes with every slot, but once message transmission is initiated, the mask remains unchanged for its entire duration. The impact on performance is profound. The vulnerability period, that is, the period of channel contention when simultaneous users' transmissions may collide,[13] is limited to a (mini) slot period,

---

13.   Moreover, transmissions started within the same mini-slot are not guaranteed to collide because of the capture phenomena discussed in Section 5.1.4.3.

not to the entire message transmission. Once a particular users' signal has been detected in a R-EACH mini-slot, the remainder of the transmission is unaffected (except for the increase in background noise) by that of incoming users because their spreading sequences are guaranteed to be different and, therefore, signals can be distinguished at the receiver after despreading. The advantage of using access slots much shorter than the message duration is also seen on the average transmission delay. In case of acknowledgment timeout, the message retransmission delay, or probe backoff, must be a random number of mini-slots (a multiple of 1.25 ms) rather than message slots (a multiple of the message duration, ~100-200 ms).

The R-EACH slot size (EACH_SLOT) is a multiple of 1.25 ms. The long-code mask time dependency rests in its least significant 9 bits, which identify one of 512 possible slot offsets and therefore one of 512 possible signature sequences. The beginning of the slot cycle coincides with the time instant when system time is an integer multiple[14] of $512 \times$ EACH_SLOT $\times 1.25$ ms. Although not mandated by the standard, the slot size should be set equal to the preamble length (also a multiple integer of 1.25 ms) because there is typically only one hardware element at the base station allocated to acquisition and demodulation of a R-EACH channel. Hence, a preamble spanning multiple slots would have no chance of being acquired (and therefore be unnecessarily long) once the first slot has elapsed, since its spreading code remains unchanged, while the base station searcher element switches to a new long-code mask at each slot boundary.

Another important consequence of the time-dependent long-code mask is worth mentioning. The base station may well choose to perform maximum ratio combining of all received paths instead of performing path selection, as in the case of the R-ACH. That is because collisions are limited to the first mini-slot rather than to the entire message, and trading a (small) decrease in the probability of capture for a (substantial) improvement in physical layer performance is advantageous.

The R-EACH supports multiple data rates and frame sizes. This allows the mobile station to select the optimum data rate and frame size on the basis on the required transmit power and amount of data queued for transmission. The R-EACH also has better preamble design that allows for discontinuous transmission, which is more power-efficient while achieving nearly the same acquisition performance as that of continuous preamble transmission.

Unlike the R-ACH, the variable rate R-EACH uses a frame quality indicator to aid the receiver in determining the data rate. The frame CRC also allows early detection of message transmission failure so that the base station receiver resources may be reassigned to demodulate another incoming message (message transmissions on the R-EACH may overlap, while on the R-ACH they do not because of the slotted nature of the channel).

With the exceptions above, the R-EACH protocol is also a stop-and-wait protocol with flow control conceptually similar to that of the R-ACH. The difference is in the specification details, which we omit for brevity. Section 5.1.5.2 offers some insight on the protocol performance.

---

14. The R-EACH slot is further offset from the slot cycle boundary by a multiple of 1.25 ms, so that parallel transmission attempts on separate R-EACH channels are staggered in time.

5.1.5.2   R-EACH Throughput and Delay Analysis

We characterize the traffic as follows. Each packet transmitted on the R-EACH is of constant length, requiring M seconds for transmission that we assume to be a multiple integer of the R-EACH slot size, T. The average number of packets generated for transmission time, also called the steady-state channel throughput or utilization, is $S = \lambda M$. The maximum achievable utilization is called capacity of the channel. Since packets that collide need to be retransmitted, the offered traffic rate is G (packets per transmission time) with $G \geq S$. The interarrival times of packet transmissions as well retransmissions are assumed independent and exponentially distributed. In the following, we wish to solve for the R-EACH capacity when using the basic access mode protocol.

We assume that when multiple transmissions are initiated at the same slot boundary, no one signal can be resolved from the others (zero probability of capture in case of collision). Furthermore, we neglect the probability of packet erasure due to multiple access interference, since we are interested in the R-EACH capacity from a random access protocol standpoint. Therefore, once a packet transmission is detected and is being serviced, packet transmissions initiated at subsequent slot boundaries do not affect the one being serviced. We consider a single-server system, which is of practical interest, since there is one and only one demodulating element allocated to each R-EACH channel.

Consider Figure 5.10 for an exemplary sequence of access attempts. A user becomes ready when a packet is queued for transmission. The first ready user (users becoming ready are represented by vertical arrows) triggers packet transmission at the slot boundary. Since the server is idle and no other user becomes ready in the same slot, the packet is successfully transmitted. Users number 2 and 3 become ready in the same slot just prior to completion of the transmission in progress. Their packet transmissions are initiated simultaneously and are detected, but go unresolved, ending up in a collision. User number 4 goes undetected, since it becomes ready and starts transmission during a busy period. User number 5 finds the server idle and its transmission is successfully completed.

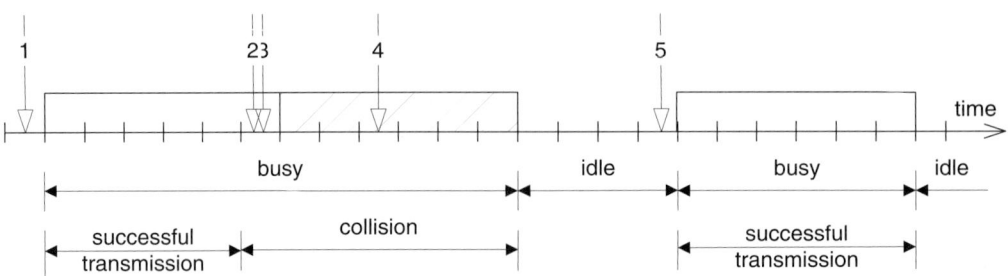

Figure 5.10 Example of access attempts on R-EACH.

Using renewal theory arguments [7], the average channel utilization can be expressed as

$$S = \frac{\bar{U}}{\bar{B} + \bar{I}} \tag{5.6}$$

where $\bar{B}$ and $\bar{I}$ are the average busy period and idle period, respectively. Their sum represents a cycle. Then, $\bar{U}$ represents the average time during a cycle that the channel is used without conflicts. Denote with K the random (Poisson-distributed) number of packet arrivals corresponding to both first transmissions and retransmission attempts in a slot period. The normalized values $\bar{B}$ and $\bar{I}$ —in units of message transmission time—can be obtained from their probability mass functions as follows.

$$
\begin{aligned}
\bar{I} &= a\left(1 + \sum_{i=0}^{\infty} i \Pr[I = i]\right) = a\left(1 + \sum_{i=0}^{\infty} i \Pr[K \geq 1] \cdot \Pr[K = 0]^i\right) \\
&= a\left[1 + \left(1 - e^{-aG}\right)\sum_{i=0}^{\infty} i\left(e^{-aG}\right)^i\right] = \frac{a}{1 - e^{-aG}}
\end{aligned} \tag{5.7}
$$

$$
\begin{aligned}
\bar{B} &= 1 + \sum_{i=0}^{\infty} i \Pr[B = i] = 1 + \sum_{i=0}^{\infty} i \Pr[K \geq 1]^i \cdot \Pr[K = 0] \\
&= 1 + e^{-aG}\sum_{i=0}^{\infty}\left(1 - e^{-aG}\right)^i = \frac{1}{e^{-aG}}
\end{aligned} \tag{5.8}
$$

where $a = T/M$ is the slot size in units of message transmission time. $\bar{U}$ can be obtained as that fraction of $\bar{B}$ proportional to the probability that the transmission is successful (conditioned on the channel being busy).

$$\bar{U} = \Pr[K = 1 | K > 0] \cdot \bar{B} = \frac{aGe^{-aG}}{1 - e^{-aG}} \cdot \bar{B} = \frac{aG}{1 - e^{-aG}} \tag{5.9}$$

Using Eqs. (5.6) to (5.9), we can show that the average channel utilization is given by

$$S = \frac{aGe^{-aG}}{1 - e^{-aG} + ae^{-aG}} \tag{5.10}$$

When $a = 1$, the R-EACH protocol is identical to that of the R-ACH (and its throughput equivalent to that of slotted ALOHA, $S = Ge^{-G}$). Note that $\lim_{a \to 0} S = G/(1 - G)$. This shows that when $a = 0$, a throughput of 1 can be achieved when the offered load approaches infinity. S versus G for different values of a is plotted in Figure 5.11.

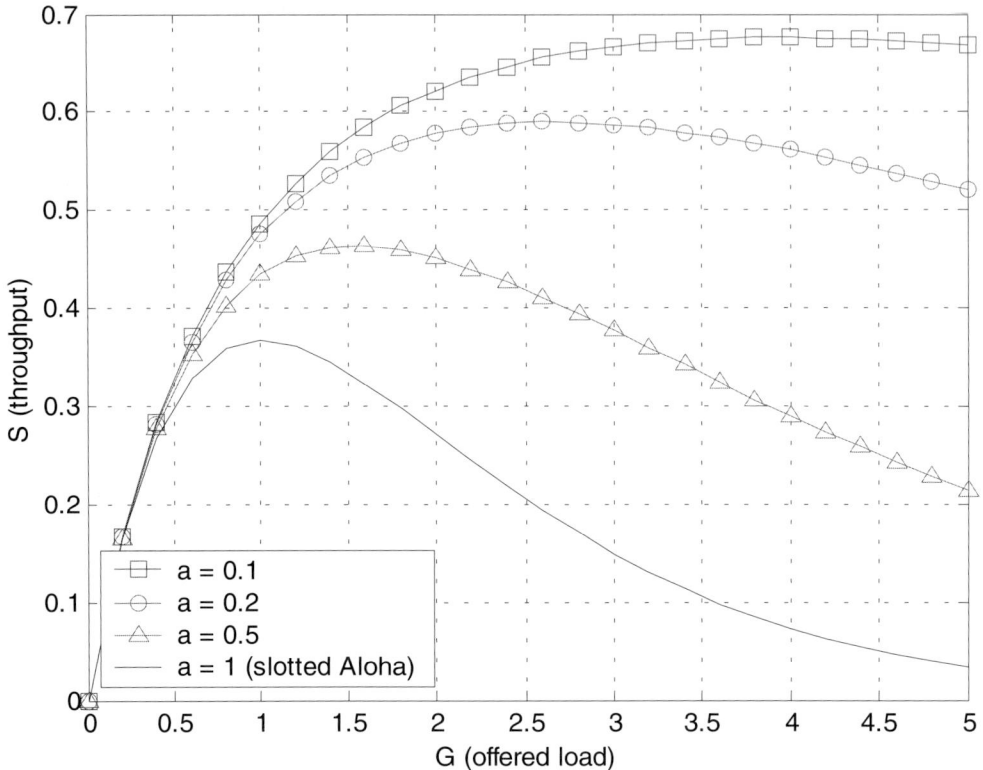

Figure 5.11 R-EACH normalized throughput versus load.

We can use Figure 5.11 for a practical example. Assume the R-ACH and R-EACH slot sizes to be 160 ms (60 ms preamble plus 100 ms message capsule) and 10 ms (preamble only), respectively. The R-EACH is operated at 9600 bps (twice that of the R-ACH), and the message transmission period is equal to 50 ms. The R-ACH capacity is equal to $1/e = 0.38$ packets per 160 ms, or 2.37 packets/s. The R-EACH capacity is 0.59 packets per 60 ms, or 9.83 packets/s.

Any given packet transmission has a probability of successful delivery equal to $p = S/G$. To obtain the average delay, we observe that the number of transmissions of a given packet, N, is geometrically distributed, and its average is $\bar{N} = 1/p = G/S$ (see Figure 5.12). The average R-EACH packet transmission delay, $\bar{D}$, is then

$$\bar{D} \approx \frac{T}{2} + \left(M + \tau_{ACK} + \tau_{backoff} \right)\bar{N} \tag{5.11}$$

where τ_{ACK} is the Layer 2 acknowledgment timeout and $\tau_{backoff}$ is the *average* probe backoff. The approximation is because we neglect the sequence backoff for simplicity.

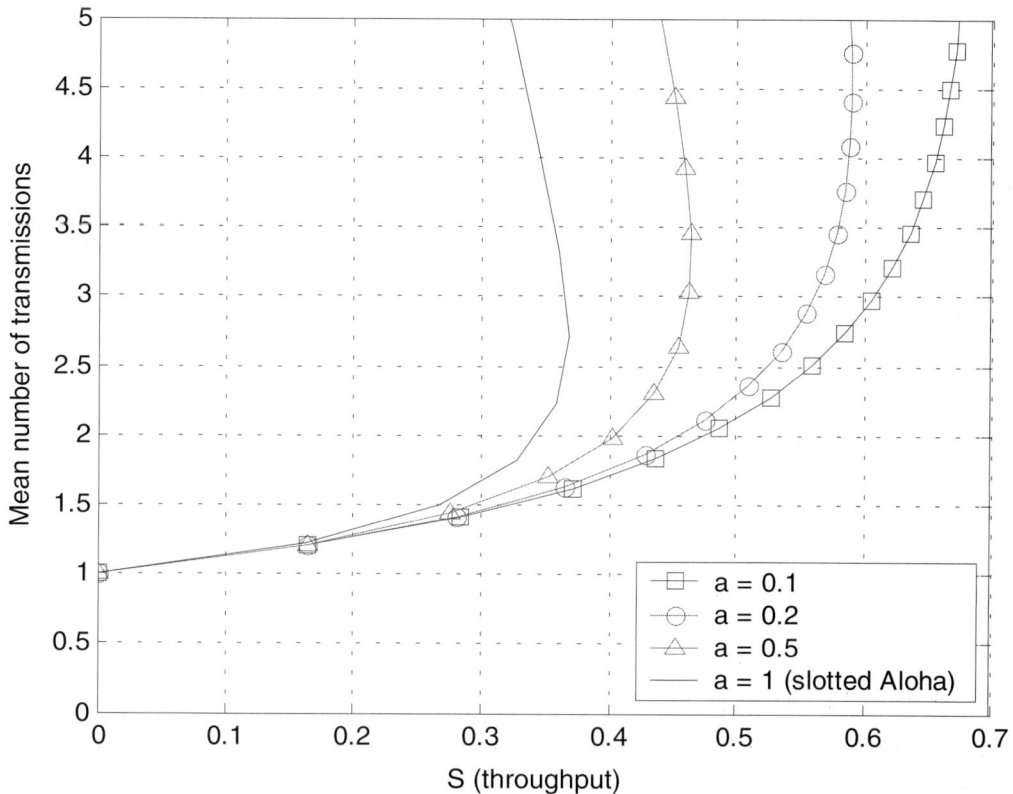

Figure 5.12 G/S versus throughput.

5.1.5.3   R-CCCH in Reservation Access Mode

The R-EACH is efficient for messages of relatively short duration (less than ~100–150 ms), but when the data burst is quite long, it still suffers from many of the drawbacks we discussed above. For extremely long messages, system access using the R-EACH/R-CCCH combination in reservation access mode is the method of choice. The main characteristics of the reservation access mode are fast-capture feedback, closed-loop power control, and soft handoff [8].

The fast-capture feedback mechanism works as follows. The mobile station sends a short message, or access header, containing a short, randomly selected request identifier and information about the access attempts it intends to make. If the base station captures the access header, it sends a short acknowledgment (which is a MAC-level message rather than a LAC acknowledgment) by echoing the request identifier used by the mobile station. Mobile stations that do not receive the quick feedback soon realize that their attempt has failed and can initiate another request following a random backoff time. Thus, unnecessary interference and delay associated with entire message transmissions of users that have not been captured is avoided.

Let us examine protocol procedures in reservation access mode by means of the simple (without soft handoff) example in Figure 5.13. Access on the R-EACH/R-CCCH is subject to flow control. When a data burst is ready for transmission, the mobile station runs a persistency test controlled by parameter broadcast by the base station parameter, just like in the case of the R-ACH protocol (see Section 5.1.4.4). Once the test passes, the mobile station can initiate transmission of the R-EACH preamble at the slot boundary, followed by the R-EACH header. The header is 5 ms long and contains, among other fields, a randomly selected 16-bit HASH_ID and a rate indicator specifying the requested data rate. When header transmission is complete, the mobile station sets a timer, T_{ECAM} and waits for the Early Acknowledgment Channel Assignment Message (or EACAM, a MAC-level message) to be received on a predetermined Forward Channel Assignment Channel (F-CACH). At the base station side, once the header is captured, the base station reserves resources on an R-CCCH and sends the mobile station the EACAM. The EACAM contains the HASH_ID that was contained in the R-EACH header as well as the identity of the reserved R-CCCH and associated CPCCH. A mobile station that receives the EACAM with matching HASH_ID knows that it has been captured[15] and proceeds further in the transmission attempt. Any mobile station that fails to receive the EACAM with matching HASH_ID prior to timer expiration aborts the attempt and reinitiates a new one after a random backoff. Note that the timer duration is a system parameter. Its setting is controlled by the base station, since it must be larger than the signaling delay and the latency in allocation/initializing radio resources, both of which are base station implementation–dependent. The captured mobile station initiates transmission of the R-CCCH preamble at the slot boundary, followed by the R-CCCH message part. At the time of preamble transmission, the mobile station also sets a power-

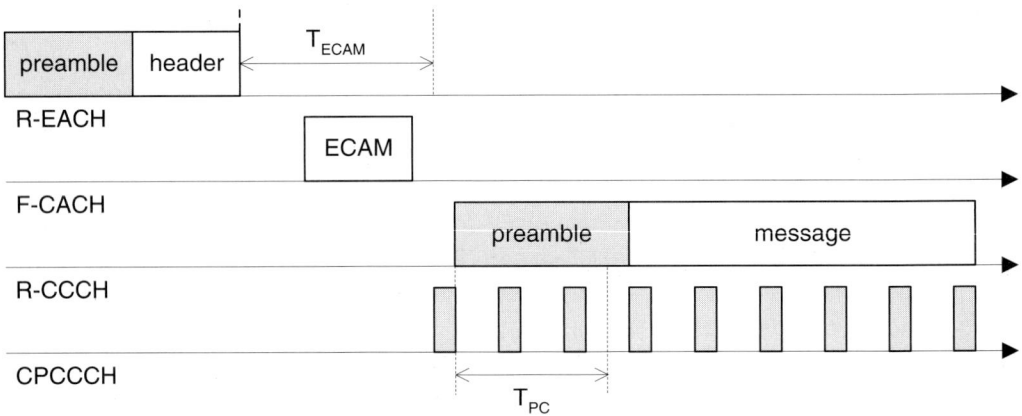

Figure 5.13 Example of reservation access mode (without soft handoff).

---

15.   The random 16-bit HASH_ID is long enough to make the probability of *false capture*, that is, the probability that
      two or more terminals in the process of a simultaneous access attempt choose the same ID, negligibly small.

control timer, T_{PC}. As soon as it sends the EACAM, the base station attempts to acquire the preamble and performs power control. Power control consists of periodically estimating the received signal energy and sending the mobile station a command to either increase or decrease the transit power if the estimated energy falls below or above a preset threshold. The power-control commands are sent as bits on the Forward Common Power Control Channel (F-CPCCH) subchannel assigned to the mobile station. Note that the mobile station does not immediately react to the power-control commands. Rather, it waits until the expiration of the T_{PC} timer before doing so to ensure that the base station has captured the preamble and has sent valid power-control bits. When the message transmission is completed, the mobile station monitors the Forward Common Control Channel (F-CCCH) for the Layer 2 acknowledgment message.

If R-CCCH in soft handoff is supported by the base station and if the second strongest pilot is of sufficient quality, the mobile station can request to access the R-CCCH in soft handoff. The request is carried by the R-EACH header. If handoff is granted, the base station sends both the ECAM and a second MAC-level message, the Power Control Channel Assignment Message (PCCAM), the latter containing information about the CPCCH subchannel corresponding to the additional R-CCCH that has been reserved. We omit details for brevity.

5.2 Signaling Link Access Control Layer

The LAC layer provides services to Layer 3 for transmission of signaling messages. The LAC protocol consists of independent functional entities that process upper-layer PDUs in a sequential manner. Therefore, protocol architecture is better described in terms of protocol sublayers, each corresponding to a specific functionality and service provided to the upper layer. The sublayers and their protocol functions are the following.

The **authentication** sublayer provides access control and authentication of subscriber identity.

The **ARQ** sublayer provides reliable delivery of messages (on Layer 3 request) using a selective-repeat ARQ protocol and duplicate message detection.

The **addressing** sublayer provides address control for mobile station-directed messages.

The **integrity** sublayer prevents an illegitimate message from being inserted into the channel and a legitimate message from being modified while on its way to the intended recipient

The **utility** sublayer tags the Layer 3 PDU according to the message type and further assembles and validates PDUs.

The **segmentation and reassembly (SAR)** sublayer provides segmentation of Layer 3 messages into fragments of size suitable for transport by the MAC layer and corresponding message reassembly functionality.

As a generated or received data unit traverses the protocol stack, it is processed by various protocol sublayers in sequence. Each sublayer processes only specific fields of the data unit that are associated with the sublayer functionality. Figure 5.14 shows the data encapsulation process throughout the LAC sublayers.

The LAC and Layer 3 exchange signaling messages on *logical channels*. The concept of logical channels was introduced with the first release of the IS-2000 standard (there was no mention of logical channels in its predecessor, the IS-95 standard) because of the need to make specification of signaling procedures insensitive to the characteristics of the physical channels that can be selected to carry signaling messages. From a signaling perspective, given the LAC services provided by its sublayers, the only characteristics that matter are whether the message is destined to a specific mobile station or to all mobile stations camping in a cell, and in case of the

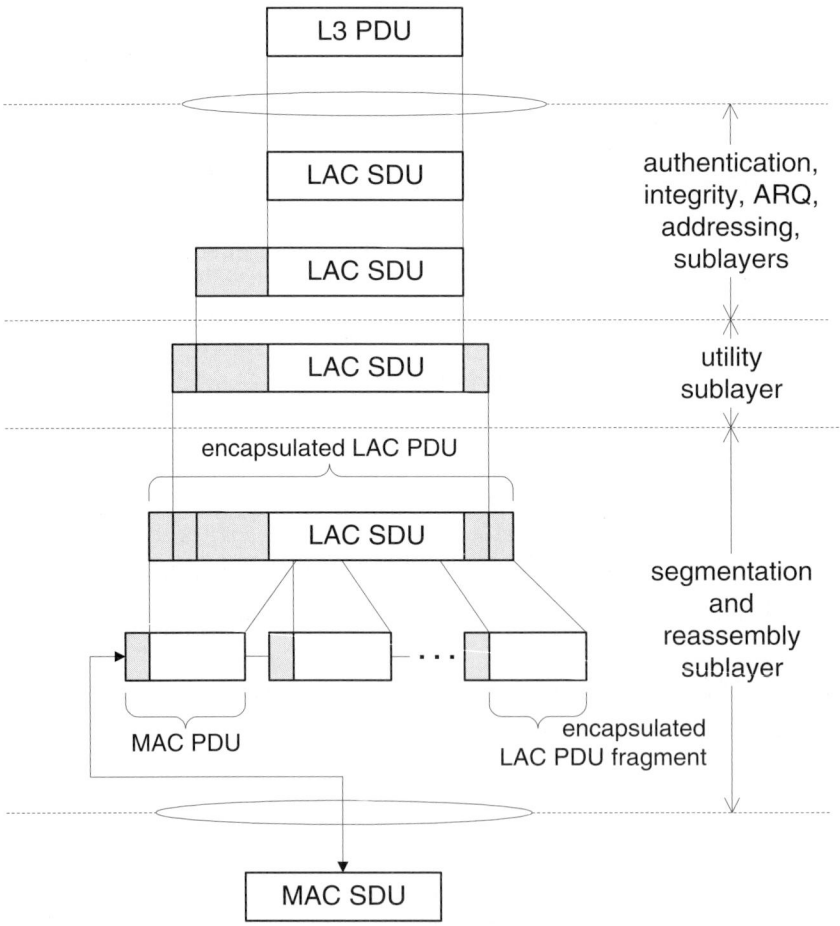

Figure 5.14 Data encapsulation throughout the LAC sublayers.

former whether or not the mobile station has an already established connection with the radio access network. Once the above is established, for example, a message must be sent to a mobile station in the traffic channel state, from a signaling perspective it does not matter whether the message is sent over the F/R-FCH or F/R-DCCH. Hence, logical channels eased protocol specification work and facilitated understanding, notwithstanding the additional acronyms that we are about to endure.

Logical channels are classified on the basis of whether they carry information destined to a single or multiple users, whether the information is signaling or user traffic, and whether the direction of transfer is base station to mobile station, or vice versa. Then, a logical channel can be one of six types: forward or reverse common signaling channel (*f-csch* or *r-csch*); forward or reverse dedicated signaling channel (*f-dsch* and *r-dsch*); and forward or reverse dedicated traffic channel (*f-dtch* or *r-dtch*). Logical channels on forward and reverse direction are depicted in Figure 5.15 and Figure 5.16.

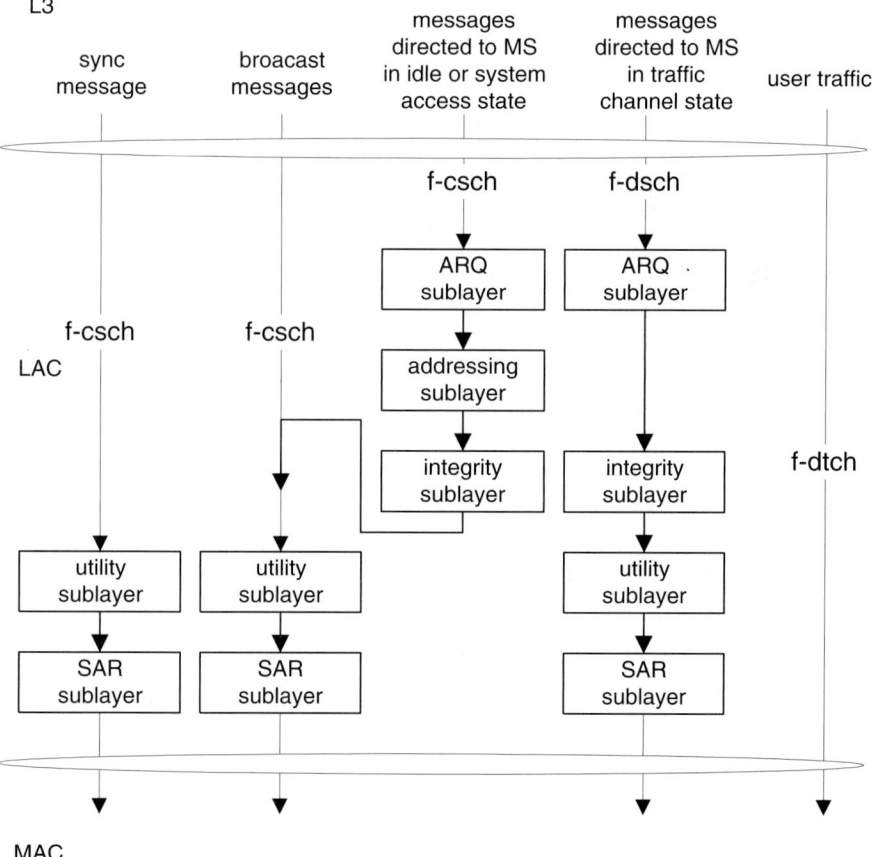

Figure 5.15 Forward-link logical channels.

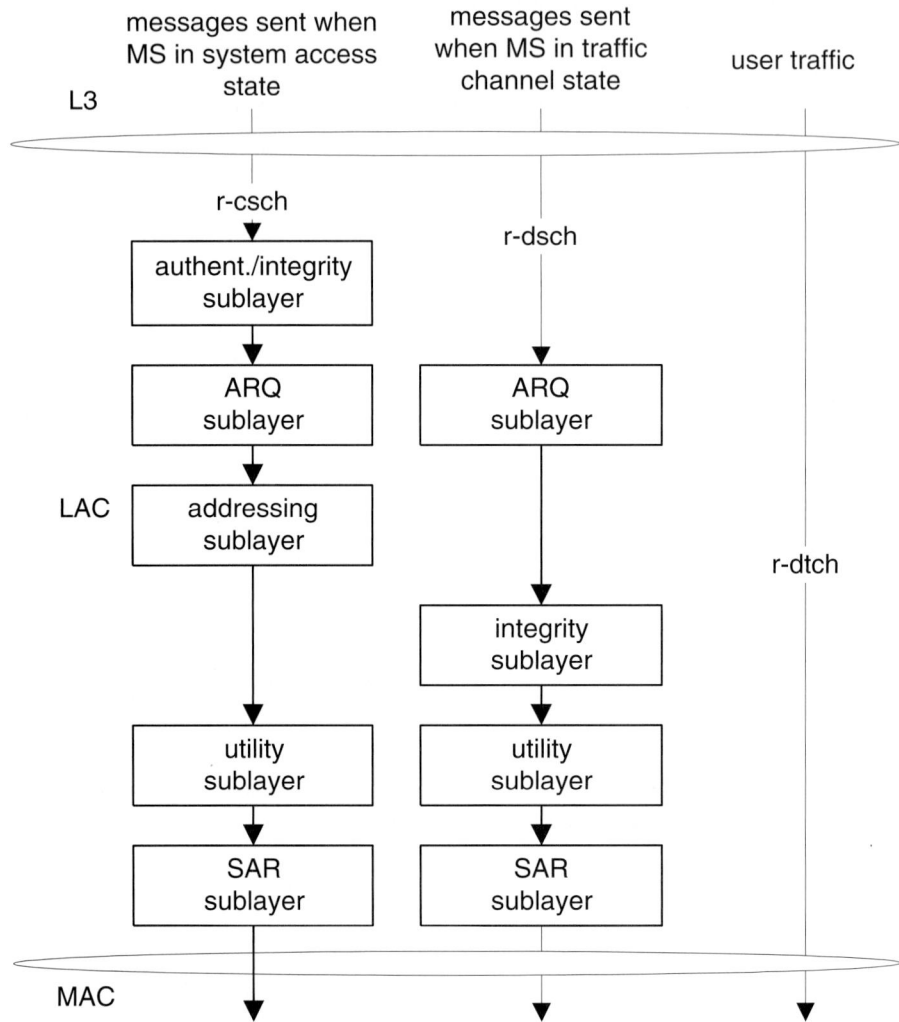

Figure 5.16 Reverse link logical channels.

   Common channels, being shared, require in-band signaling for user identification. Dedicated channels, conversely, are assigned to one and only one mobile station and do not require in-band signaling for addressing purposes; rather, they are uniquely identified by their physical characteristics, such as spreading codes and carrier frequency. That explains why the addressing sublayer of dedicated channels is empty.

   Note that the LAC protocol acts solely on Layer 3 signaling messages. The LAC layer is transparent to end-user traffic, such as vocoded data flowing directly between the vocoder and the MAC layer on the *r/f-dtch*. Mapping between logical channels and physical channels is summarized in Table 5.11.

Table 5.11 Mapping Between Logical and Physical Channels

Logical channel	Physical channel
f-csch	SYNC, PCH, BCCH, F-CCCH
r-csch	RACH. REACH, R-CCCH
f-dsch	F-FCH, F-DCCH
r-dsch	R-FCH, R-DCCH
f-dtch	F-FCH, F-DCCH, F-SCH, F-PDCH
r-dtch	R-FCH, R-DCCH, R-SCH

Note: Some physical channels, such as QPCH, F-CPCCH, and F-CACH, are not associated with any logical channel.

5.2.1 Authentication Sublayer

Authentication is the process by which information is exchanged between a mobile station and a base station for the purpose of confirming the identity of the mobile station. A successful outcome of the authentication process occurs only when it can be demonstrated that the mobile station and base station possess identical sets of shared secret data. The authentication sublayer operates only on the common channels (see Figure 5.15 and Figure 5.16). Some aspects of authentication are related to Layer 3 functions, while some reside in Layer 2. The *global challenge* type of authentication is placed in the LAC specification because it can be viewed as related to access. We nonetheless defer discussion of authentication procedure to Chapter 6, where security features of IS-2000, including authentication, are discussed in detail.

5.2.2 Message Integrity Sublayer

Message integrity is the process by which information is exchanged between a mobile station and a base station for the purpose of validating the content of signaling message. Message integrity prevents illegitimate messages from being inserted into the channel or legitimate messages from being modified. Message integrity is protected with a 32-bit Message Authentication Code (MAC-I). MAC-I is computed using an integrity algorithm based on the message itself, an integrity key, and a crypto-sync. MAC-I, the message, and crypto-sync are transmitted over the air. The crypto-sync makes sure that the message integrity algorithms at the mobile station and the base station are synchronized. If the message is altered during the transmission, the MAC-I will not check. The receiver discards a message, which MAC-I does not check. The integrity key could be set up using a procedure similar to the one explained in Chapter 6. That procedure is called second-generation authentication. The key can also be set up using third-generation authentication, also called Authentication and Key Agreement (AKA). AKA is discussed in Chapter 6.

5.2.3 ARQ Sublayer

The ARQ sublayer has the important task of providing reliable transmission of signaling messages on both common and dedicated signaling channels and on both forward and reverse directions. When transmitting a signaling message in assured mode, it uses a selective-repeat transmission protocol in which messages are retransmitted if no acknowledgment is received after the expiration of a timer. Ordered delivery of messages is not guaranteed by the ARQ sublayer that, unlike RLP, does not have a resequencing buffer. Unrelated signaling messages (e.g., destined to different service instances or corresponding to different protocol procedures) may be transmitted and successfully received in different order and still be processed by the upper layers without ambiguity. The ARQ protocol also allows for detection of duplicate message transmission. Duplicate messages are discarded by the ARQ sublayer.

We now describe the ARQ protocol employed on the dedicated channels only, since that used on the common channels is conceptually equivalent and differs in the implementation details only. Both mobile station and base station implement an AQR sublayer transmitter and receiver function. However, we do not distinguish between procedures carried out by the mobile station and base station, since they are equivalent with few exceptions, which will be noted. The ARQ-related fields that are appended onto dedicated signaling messages are listed in Table 5.12 and will be referenced hereafter.

Table 5.12   ARQ Fields in Dedicated Signaling Messages

Field	Description
ACK_SEQ	Acknowledgment sequence number
MSG_SEQ	Message sequence number
ACK_REQ	Acknowledgment required indicator

When transmitting a message requiring acknowledgment (as dictated by the signaling layer originating the message), the ACK_REQ field is set to 1. The 3-bit MSG_SEQ is incremented by one (modulo 8) relative to that used for the last transmitted message (excluding retransmissions) requiring acknowledgment. After transmitting a message requiring acknowledgment, the transmitter starts the retransmission timer (with expiration time $T_{1m} = 0.4s$) and stores the message until a valid acknowledgment is received for that message, which is one whose ACK_SEQ number matches the MSG_SEQ number of the message awaiting acknowledgment. If no acknowledgment is received prior to the expiration of the timer, the message is retransmitted using the same MSG_SEQ. Up to 13 transmissions are allowed for each message sent by the mobile station on the *r-dsch*, while for the base station such limit is implementation dependant.

The receiver, on receipt of a message sent in assured mode, arranges for transmission of the acknowledgment. The acknowledgment is either carried by an *Acknowledgment Order Mes-*

sage specifically built for that purpose or piggy-backed by a signaling message already queued for transmission. Piggy-backing is more efficient and typically used when the signaling layer protocol requires transmission of a message in response to a message just received (e.g., receipt of a *Handoff Direction Message* triggers transmission of the *Handoff Completion Message*, the latter carrying the Layer 2 acknowledgment for the former). The receiver shall send acknowledgment using either one of the methods above within 0.2 sec after receipt of the message requiring acknowledgment. Considering that the retransmission timer is 0.4 sec, that leaves 0.4 – 0.2 = 0.2 sec (ample time, given the speed at which processors operate) for transmission and processing delay before a retransmission is unnecessarily triggered.

Although the MSG_SEQ field is 3 bits long, which allows for unique tagging of up to eight messages, only up to four messages can be sent before receiving acknowledgment for one of the transmitted messages. Such constraint is misleadingly akin to that used by a sliding-window flow-control protocol. Its motivation is not to prevent congestion, as in the case of flow control. Rather, it is to detect duplicate messages at the receiver. To understand that, let us see how duplicate detection works. The receiver maintains a received status indicator for each possible MSG_SEQ number with logical value equal to NO or YES for new or old messages, respectively. When receiving a message with a MSG_SEQ whose corresponding status indicator is set to NO, the message is acknowledged, treated as new, and forwarded to the signaling layer. The received status indicator corresponding to MSG_SEQ is toggled to YES, while that corresponding to (MSG_SEQ + 4) modulo 8 is toggled to NO. When receiving a message with MSG_SEQ whose corresponding status indicator is set to YES, the message is considered a duplicate and discarded (it is nevertheless acknowledged once again, since duplicate transmission indicates that the acknowledgment previously sent had been erased).

An example of assured-mode message delivery from the base station (here considered to be the transmitter) to the mobile station and management of ARQ state variables is described next (refer to Figure 5.17).

1. No message acknowledgment is currently pending at the base station, and the last message sent that required acknowledgment had MSG_SEQ = 7. The ARQ sublayer receives a message to be delivered reliably from the signaling layer. The BS transmits the signaling message with MSG_SEQ = 0 with acknowledgment required.
2. The mobile station receives the message that now requires acknowledgment. Its corresponding received status indicator, MSG_SEQ_RCVD[0], is currently set to NO, so the mobile station treats the message as a new one. Then, it sets MSG_SEQ_RCVD[0] = YES and MSG_SEQ_RCVD[4] = NO. Finally, it sends the *Acknowledgment Order Message* with ACK_SEQ = 0, which, however, is erased due to poor reverse-link conditions.
3. While acknowledgment to the first message is still pending, the base station increments MSG_SEQ from 0 to 1 and sends a second message requiring acknowledgment to the mobile station. That message is erased due to poor forward-link conditions. At this time the base station and mobile station state variables are as depicted in Figure 5.18.

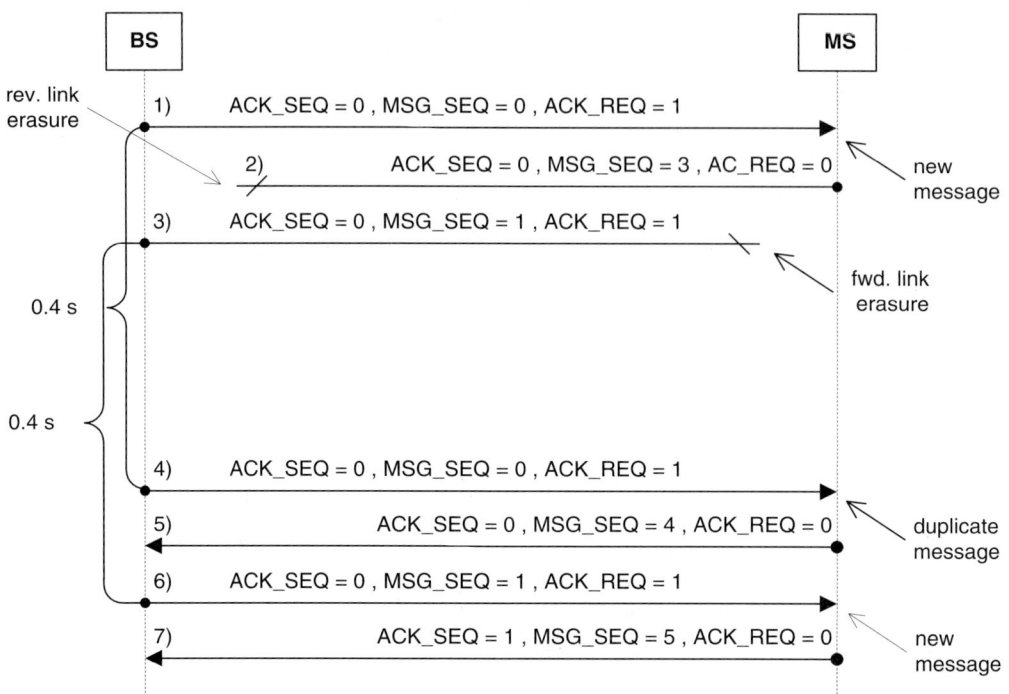

Figure 5.17 Example of assured-mode message delivery on the forward link: Arrows represent either Layer 3 signaling PDUs (forward direction) or *Layer 2 Acknowledgment Order Messages* (reverse direction).

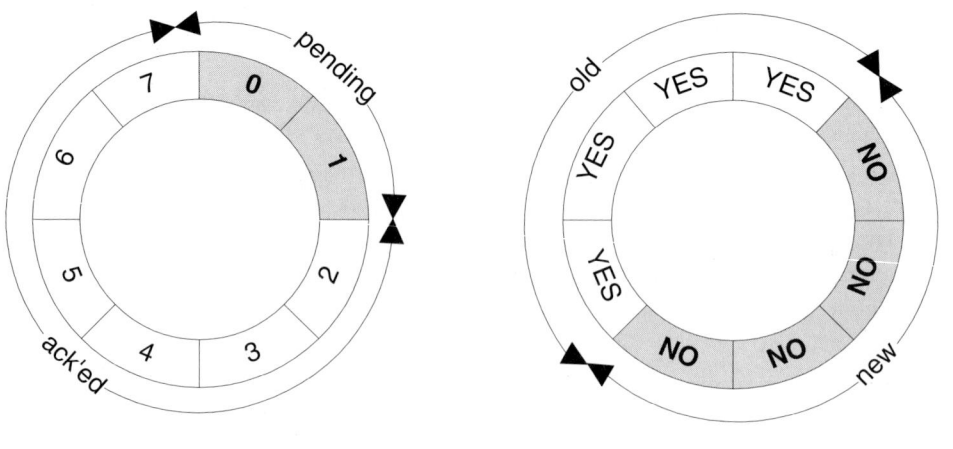

BS ARQ transmitter   MS ARQ receiver

Figure 5.18 BS and MS ARQ state variables management (see step 3 of exemplary signaling flow in Figure 5.17).

4. The retransmission timer ($T_{1m} = 0.4s$) for the first message (MSG_SEQ = 0) expires, and the base station retransmits with identical MSG_SEQ value.

5. The mobile station receives the message with MSG_SEQ = 0. Since currently MSG_SEQ_RCVD[0] = YES, it treats it as a duplicate message. The message is discarded but nonetheless acknowledged. The *Acknowledgment Order Message* is received at the base station, which clears the corresponding retransmission timer.

6. The retransmission timer ($T_{1m} = 0.4s$) for the second message (MSG_SEQ = 1) expires, and the base station retransmits with identical MSG_SEQ value.

7. The mobile station receives the message with MSG_SEQ = 1. Since currently MSG_SEQ_RCVD[1] = NO, it treats the message as a new one. Then, it sets MSG_SEQ_RCVD[1] = YES and MSG_SEQ_RCVD[5] = NO. Finally, it sends the *Acknowledgment Order Message* with ACK_SEQ = 1. The order is received at the base station, which clears the corresponding retransmission timer.

Signaling messages can also be sent in nonassured mode. Such messages do not require acknowledgment, and the ACK_REQ field is set to 0. With each transmission of a new message not requiring acknowledgment, the 3-bit MSG_SEQ field is incremented by one (modulo 8). Note that the sequence of messages not requiring acknowledgment is different and independent of that of messages that do require acknowledgment. Unlike those sent in assured mode, those sent in nonassured mode can be repeated multiple times in rapid succession by the transmitter. That is useful, for example, when a high degree of transmission reliability is required and yet the transmitter cannot tolerate the delay incurred with retransmission triggered by acknowledgment timer expiration, as in the case of messages sent in assured mode. This method of sending multiple copies of the same frame in rapid succession is called *quick repeat*. To increase the probability that at least one message of the repeat sequence is successfully decoded, retransmissions are interleaved by one or two data frames so that radio channel conditions change randomly from one transmission to the next. That is particularly useful in case of slowly moving terminals, since frame erasure events for successive frames have some degree of correlation. However, to allow for duplicate detection of nonassured messages, all quick repeats must occur within $T_{3m} = 0.32s$ from the instant of the first transmission. Considering, for example, transmission of a message of length 2 frames with 2 frames interleaved between repeats, the quick-repeat sequence can be up to $T_{3m}/(0.02 \cdot 4) = 4$ messages long. Clearly, transmission of messages in quick repeat has the disadvantage that more user-data frames than needed are, on average, blanked by the signaling message, hence degrading voice quality or data throughput. In the case of many signaling procedures, the standard dictates whether to use assure or nonassured mode transmission. For all other cases, guidelines for selecting the type of message transmission are as follows. Messages that are noncritical but require low latency and occur frequently throughout the duration of the call, such as periodic measurement report messages, should be sent in nonassured mode without repeats. Messages that are critical and require low latency but that occur infrequently throughout the duration of the call, such as messages related to hard handoff (see Chapter 7), should be sent in nonassured mode with quick repeats. All other messages should be sent in assured mode without quick repeats.

5.2.4 Addressing Sublayer

The addressing sublayer is responsible for selecting and formatting the mobile station's address-related fields. Explicit addressing is required on all common channels. The forward common channels are one-to-many channels, so addressing is used by the base station to indicate the intended recipient of the message. The reverse common channels are many-to-one channels, so addressing is used by the mobile station to identify itself. Addressing on dedicated channels is implicit in that the long code used for signal spreading on the forward and reverse links uniquely identifies one and one only mobile station. Therefore, the dedicated channels' addressing sublayer is empty.

On the *r-csch*, the addressing scheme used by the mobile station can be one of several combinations of the mobile identification number (MIN), ESN, international mobile station identification (IMSI), and temporary mobile station identification (TMSI) addresses (see Chapter 6). The format of such addresses is specified in the Layer 3 protocol, and so we defer discussion of their use (and significance) to Chapter 6. In the context of the LAC, the addressing sublayer selects the addressing scheme based on base station preferences and mobile station status. The base station's expressed preference is carried by the PREF_MSID_TYPE field of the *Extended System Parameters Message* that is broadcast by the base station on the F-PCH. Mobile station roaming and provisioning status also influences the address selection. For example, the mobile station may not have been provisioned yet with an IMSI, in which case the ESN is the only address information included in messages sent on the *r-csch*. Once the address type is selected, the addressing sublayer is responsible for proper formatting of the addressing fields of the LAC layer PDU.

On the *f-csch*, the addressing scheme used by the base station can also be one of several combinations of the MIN, ESN, IMSI, or TMSI addresses. Address selection depends on the address type the mobile station has registered with. Once the address type is selected, the addressing sublayer is responsible for proper formatting of the addressing fields of the LAC layer PDU. The base station informs the mobile station of the type of address that is used within each page record by properly encoding the PAGE_CLASS field of the *General Page Message* sent on the F-PCH.

5.2.5 Utility Sublayer

The utility sublayer is a thin protocol mainly responsible, at the transmitter, for assembling PDUs with the information received from the upper layers and, at the receiver, classifying and validating the received PDUs. A specific function allocated to the utility sublayer (mainly because of lack of a better fit) on the *r-csch* is to populate the PDU with the radio environment report fields. Such fields carry information about received pilot strength that is useful for many Layer 3 functions such as access handoff (see Chapter 7) or forward-link open-loop power control (see Chapter 8).

5.2.6 Segmentation and Re-assembly Sublayer

Signaling messages originated at the signaling layer are of variable length and typically do not fit precisely in the portion of the multiplex sublayer PDU allocated to transport signaling traffic. Recall from Section 5.1.4 that the signaling traffic transport blocks can only be 88, 128, 152, or 168 bits long. If the message length including LAC overhead is larger than 168 bits, the message must be split across multiple multiplex layer PDUs and their transmission will span multiple frames. A protocol needs to be devised to segment and reassemble messages at the transmitter and receiver, respectively. That is precisely the role of the segmentation and reassembly (SAR) sublayer. The SAR protocol parameters and procedures differ slightly depending on whether the message is transmitted on common or dedicated signaling channels. In the case of common channels, the physical channel type also dictates the use of specific SAR parameters. However, protocol procedures are conceptually identical, and therefore we focus on the dedicated channels only.

Consider, for example, signaling transmission on the *r-dsch*, as in Figure 5.19 (procedures on the *f-dsch* are exactly symmetrical). When a message is received from the signaling layer, the SAR appends at the beginning of the PDU the EXT_MSG_SEQ and MSG_SEQ fields to indicate message length. Then, it computes the 16-bit message CRC and appends it to the trailing end of the PDU. The so obtained encapsulated PDU is divided in multiple fragments that will fit the MUX layer payload (with the exception of the last fragment, which may be shorter and will be padded by the lower layer). At the beginning of each fragment, the SAR appends the 2-bit segment indicator (SI) field to indicate whether the current segment is the first of a sequence or is a trailing segment. Note that only one bit would have sufficed for the SI. The reason that two

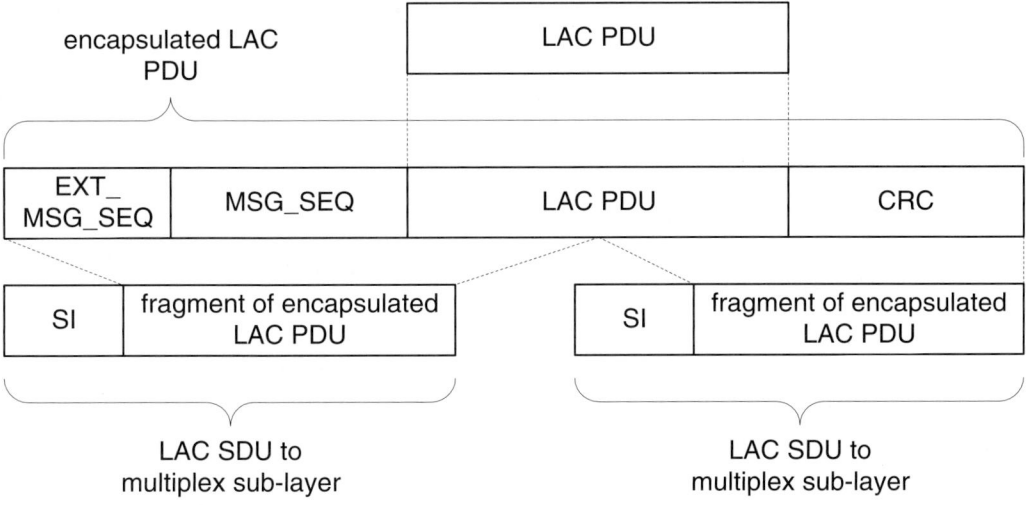

Figure 5.19 Segmentation procedure for message transmission on *r-dsch* and *f-dsch*.

bits are allocated to the SI is for future extension of the ARQ protocol to allow selective repeat transmission of message fragments instead of entire messages.

At the receiver side, the SAR protocol reassembly procedure is depicted in Figure 5.20. On receipt of a message fragment from the multiplex sublayer, the SAR sublayer strips out the SI field and checks its setting.

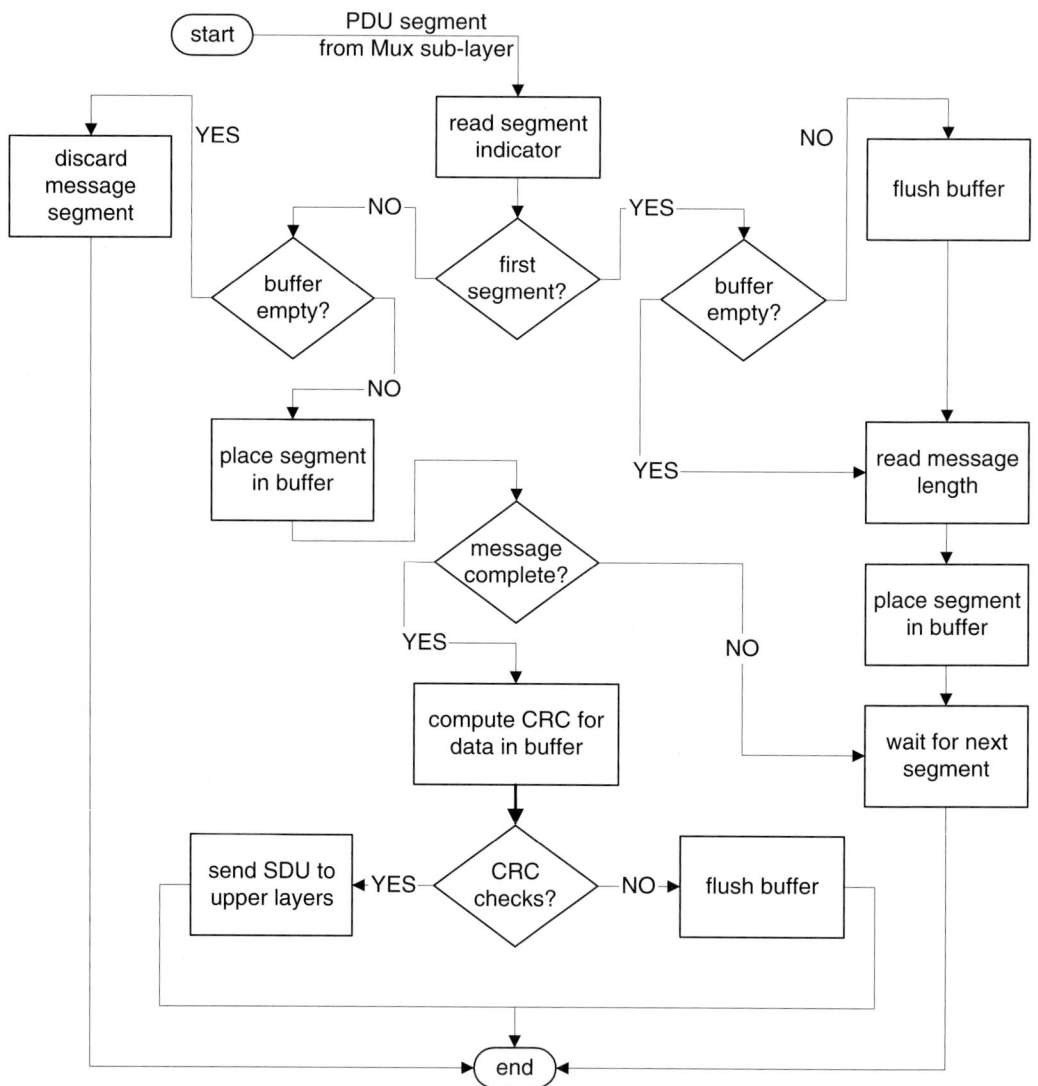

Figure 5.20 Reassembly procedure for message reception on *r-dsch* and *f-dsch*.

If the SI field indicates that the fragment is the first of a sequence, it determines the message length, strips out the EXT_MSG_LENGTH and MSG_LENGTH fields, and places the remainder of the fragment in the reassembly buffer. If just prior to that the buffer was not empty (for example, because of stale fragments from a previous, unsuccessful, message transmission), the buffer is flushed before insertion of the first fragment.

If the SI field indicates that the fragment is a trailing fragment, it appends the fragment to the content of the reassembly buffer. If just prior to that the buffer was empty, the fragment is discarded because the empty buffer indicates that the first message fragment was lost (for example, because the physical channel frame carrying the first fragment was erased), and message reassembly cannot be completed.

Once the buffer reassembly depth has reached the expected message length, the SAR layer computes the CRC for the data in the buffer. If the computed CRC matches that appended to the last fragment, the reassembled message is delivered to the signaling layer, otherwise it is discarded. The message-level CRC (in addition to physical layer frame CRC) prevents fragments belonging to different messages from being erroneously reassembled together (which may happen if, for example, a trailing fragment of a message is erased and the first fragment of the subsequent message is also erased).

REFERENCES

[1] "TIA/EIA-IS-2000.3 Medium Access Control (MAC) Standard for cdma2000 Spread Spectrum Systems Release C," *Telecommunications Industry Association,* 2002.

[2] "TIA/EIA-IS-2000.4 Signaling Link Access Control (LAC) Standard for cdma2000 Spread Spectrum Systems Release C," *Telecommunications Industry Association,* 2002.

[3] "TIA/EIA-IS-707-B Data Services Standard for Wideband Spread Spectrum Systems," *Telecommunications Industry Association,* 2004.

[4] Chaponniere, Ethienne, S. Kandukuri, and W. Hamdy, "Impact of TCP/RLP Parameters on the Performance of cdma2000," in the proceedings of VTC 2003.

[5] Bao, Gang, "Performance evaluation of TCP/RLP Protocol Stack over CDMA Wireless Link," *Wireless Networks*, 2, 1996.

[6] "TIA/EIA IS-95 A Mobile Station—Base Station Compatibility Standard for Dual Mode Wideband Spread Spectrum Cellular Systems," *Telecommunications Industry Association,* 1993.

[7] Kleinrock, Leonard, and F. Tobagi, "Packet Switching in Radio Channels," *IEEE Trans. On Communications*, 23(12), December 1975.

[8] Etemad, Kamran, "Enhanced Random Access and Reservation Scheme in CDMA2000," *IEEE Personal Communications*, April 2001.

CHAPTER 6

IS-2000 Layer 3 Protocol

The IS-2000 Layer 3 [1], also called the signaling layer, comprises base station and mobile station interoperability procedures and associated signaling that allow the end user to receive service. From a functional standpoint, much of Layer 3 is about call processing, that is, the procedures used to set up, maintain, and release a connection between mobile station and base station. But Layer 3 is also responsible for other key interoperability procedures in the area of mobility management, identification, and security, to list a few. From an architecture standpoint, the signaling layer is modeled in terms of a protocol layer, service access points, and the control and data planes. The protocol layer on the transmitting end generates Layer 3 protocol data units (PDUs) for transmission to the Link Access Control (LAC) layer [2], while on the receiving end it decapsulates lower layer PDUs before processing the resulting service data unit (SDU). The service access points are used for exchange of communications primitives with the lower layers. The control plane is used for protocol supervision, and the data plane is used to carry both user traffic and Layer 3 signaling.

As we attempt to distill in one chapter the voluminous IS-2000 Layer 3 standard, omissions and simplifications are inevitable. We concentrate on the description of the data plane only, where the PDUs are generated and processed as dictated by the various call-processing functions, since all other building blocks are implementation-dependent and do not affect interoperability between mobile station and base station. Also note that discussion of handoff and power-control functions, although within the scope of Layer 3, are deferred to Chapters 7 and 8, as they are better treated independently and from an algorithmic, rather than protocol-oriented, point of view.

6.1 Identification

Mobile station identification comprises the set of information elements and related signaling procedures used by the mobile station and the base station when exchanging information on common channels. When messages are sent on a reverse common channel, the mobile station must explicitly reveal its identity. Identification also allows the base station to address messages sent on a forward common channel to the intended mobile station. We already mentioned identification and addressing in the context of Layer 2 when describing the LAC addressing sublayer. The LAC protocol deals only with message formatting. It is a Layer 3 responsibility to define the identification-related information elements and their use.

6.1.1 International Mobile Station Identity (IMSI)

Every mobile station is assigned an identifier that uniquely distinguishes it all over the globe. This identifier is called international mobile subscriber identity (IMSI) and is standardized by ITU in the ITU-T recommendation E.212 [3]. To appreciate the significance of the international standardization of the IMSI, we must go back to the early days of cellular when the mobile identification number (MIN) was the only identifier used by the IS-41 core network. The MIN was originally designed to contain a 10-digit number formatted as per the North American Numbering Plan (NANP). According to the NANP, the first three digits corresponding to the area code are followed by three digits corresponding to the local office, or exchange, and four digits corresponding to the subscriber. Moreover, the MIN was typically set equal to the subscriber dialable directory number. That created two types of problems. First, it meant that a subscriber could not maintain his or her directory number when changing service provider. Second, as the MIN follows a national numbering format and contains no country identifier, problems surfaced both with registration of international subscribers roaming in North American networks and with call delivery to North American subscribers roaming internationally. Unlike the MIN, the IMSI is separated from the subscriber directory number, thus enabling number portability and seamless international roaming.

The IMSI consists of up to 15 numerical characters (0 through 9). The first three digits represent the mobile country code (MCC), followed by the mobile network code (MNC), which are either two or three digits long, and the mobile station identification number (MSIN). The combination of MNC and MSIN is also called the national mobile station identity (NMSI), which is up to 12 digits in length. The IMSI format is depicted in Figure 6.1. Identifiers obtained by shortening the IMSI are used operationally. Although those are not discussed any further, suffice it to say that they are used to improve paging channel and access channel[1] efficiency.

---

1.   Throughout this chapter we generically refer to the Forward Paging Channel (F-PCH) or the Forward Broadcast Control Channel (F-BCCH) and Forward Common Control Channel (F-CCCH) as the paging channel. Similarly, access channel may refer to either the Reverse Access Channel (R-ACH) or Reverse Enhanced Access Channel (R-EACH). A distinction is made only when necessary, depending on the context.

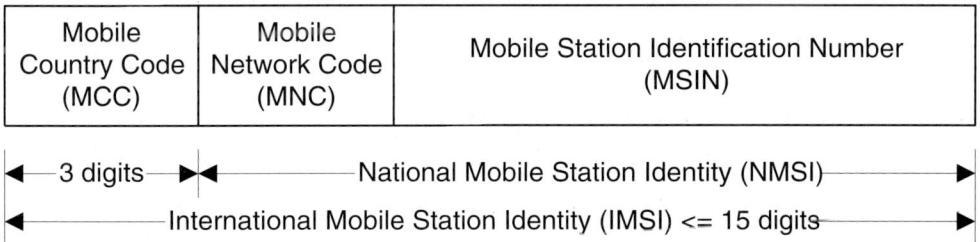

Figure 6.1   IMSI format.

6.1.1.1   *International Roaming MIN (IRM)*

Some North American operators have not yet fully deployed E.212 IMSI-compliant networks and still rely on the MIN instead. To address this situation, the International Forum on ANSI-41 Standards Technologies[2] (IFAST) has assumed an active role by creating and administering International Roaming MIN (IRM).

As previously discussed, the MIN is associated with the 10-digit directory number and formatted according to the NANP. According to the NANP, a directory number cannot start with either a 0 or a 1 digit and, therefore, any such number cannot be a legitimate MIN. These unutilized numbers can then be used as IRMs. IRMs perform, internationally, the same identification function as the MIN but, unlike MINs, their assignment is coordinated internationally to avoid reuse across national boundaries. IRMs have the following format: 0-XXX+6D or 1-XXX+6D. The first four-digit code, beginning with a 0 or 1, represents the IRM network identifier. The IRM network identifier defines an IRM block, a group of 1 million unique IRMs, which all have the same IRM network identifier. There are 2,000 IRM network identifiers. The IRM, when transmitted by a roaming terminal to a visited network, is utilized by the visitor location register (VLR) of that network to identify the home network of the roaming subscriber to query that network to determine the validity of the subscriber and to ascertain billing and services information. The digits following the IRM network identifier uniquely identify a subscriber of the network identified by the IRM and are in the format appropriate for the domestic numbering plan of the home network.

Note that from a Layer 3 perspective, the IRM is undistinguishable from a MIN. Also, use of IRM is only a temporary solution for those network operators migrating to IMSIs.

6.1.2 Electronic Serial Number (ESN)

The electronic serial number (ESN) is a 32-bit binary number that uniquely identifies the mobile station. It is assigned by the mobile station vendor rather than by the network operator. To

2.   Additional information may be found on the IFAST Web site: http://www.ifast.org.

coordinate the assignment of ESN, vendors are assigned blocks of ESNs for the mobile stations they manufacture. The ESN is important because it is used operationally in many procedures, both at physical layer and Layer 3 level. The ESN is used, for example, to derive the public long-code mask that is used for spreading of the physical channel assigned to that mobile station while in a call. The ESN is also used as input to pseudorandom hashing procedures needed, for example, to select the random channel to be use for access or the paging channel slot.

Since the publication of IS-2000 revision C, the ESN is no longer necessary to compute the public long mask. Instead, the base station has the option to select the bits forming the public long-code mask itself. This option is useful when the ESN is not available at the base station, so that call setup can proceed without the delay incurred with querying the ESN from the core network. The base station-assigned long code was also introduced because the industry predicts that the ESN will be exhausted in the near future, and a new equipment identifier will be needed as a replacement. If a base station-assigned long code is selected, the base station has to convey the public long-code mask to the mobile station during the call-setup procedure, prior to establishment of the traffic channel.

6.1.3 Temporary Mobile Station Identity (TMSI)

The TMSI is a temporary, locally assigned number used for addressing the mobile station. The TMSI as a number has no association with the mobile station permanent identifications such as the IMSI, ESN, or MIN. Rather, it is assigned by the serving VLR with the only restriction of being unique within one administrative zone (called the TMSI zone) and may be reused across zones.

The main reason for using the TMSI is user anonymity. User anonymity is achieved in that the user identity is not transmitted over the air in the clear. The TMSI only is used for addressing the mobile station once it has been assigned. Since the TMSI has no association with the mobile station permanent identification numbers, user identity is concealed. A second reason is to improve paging channel efficiency. The TMSI is much shorter than the IMSI. Use of the TMSI reduces considerably the length of the page messages, since the main content of the page is the mobile station identity. The TMSI can be either 2, 3, or 4 octets long. The shorter the TMSI, the smaller the number of mobile stations that can be uniquely addressed within a TMSI zone. That in turn increases the traffic associated with TMSI assignment, since a mobile station moving from one zone to another needs to be reassigned a TMSI. The variable length TMSI, then, allows a trade-off of paging efficiency with TMSI assignment traffic. The addressing sublayer of the LAC only specifies the format of the TMSI. The signaling procedures for TMSI assignment are specified by the Layer 3 protocol.

Mobile stations' battery life was also a motivation for using TMSI when it was first proposed for IS-95 systems. Standby improvement comes with the smaller page message that allows the mobile station to power down earlier in a paging slot (see Section 6.2.2.1). That is no longer a significant benefit in IS-2000 systems that use the Forward Quick Paging Channel (F-QPCH) to improve standby time. In such systems, the mobile station does not have to monitor the entire paging channel slot even when TMSI is not used (see Section 6.2.2.2).

Use of TMSI has not gained much popularity among network operators because of the cost of implementing, administering, and maintaining a TMSI database at the VLR.

6.2 Layer 3 Processing

Layer 3 processing in IS-2000 is specified in terms of a state machine, with states and substates. From a mobile station perspective,[3] the call-processing states and the events causing state transitions (see Figure 6.2) are the following:

Initialization State. After power-up, the mobile station enters the initialization state to select and acquire a system. On acquisition of system timing, the mobile station enters the idle state.

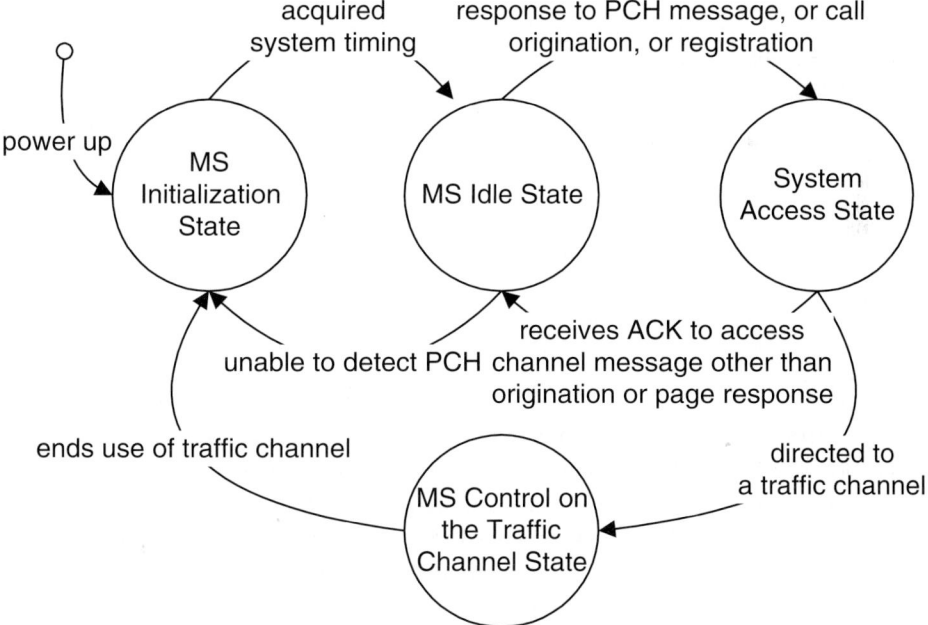

Figure 6.2 Mobile Station (MS) Layer 3 processing states.[4]

---

3.   The base station's call-processing state machine is similar in concept to that of the mobile station. For brevity, in this chapter we describe Layer 3 procedures mainly from a mobile station perspective.
4.   As of IS-2000 Rel.B, the mobile station may enter the idle state directly from the traffic channel state.

Idle State. On entering the idle state, the mobile station attempts to demodulate the paging channel. If it fails in doing so, it reenters the initialization state to reselect another system; otherwise it monitors the Forward Paging Channel (PCH) or Forward Broadcast Control Channel (F-BCCH) messages and acquires information that will be needed for call processing in other states.

System Access State. If the mobile station registers with the base station, or receives a page message, or originates a call, it enters the system access state. In the case of a registration, it simply waits for acknowledgment from the base station and then reenters the idle state. In the case of call setup, the mobile station is directed to a traffic channel and enters the mobile station control of the traffic channel state.

Mobile Station Control on the Traffic Channel State. In this state the mobile station performs various procedures to establish and maintain two-way communication with the base station until the call is released, at which point it reenters the initialization state.

In the sections that follow we describe the signaling layer protocol following the same state machine, but unlike the description presented in IS-2000, we often refer to exemplary call flows and explain the rationale behind protocol design (as we did not in the description presented in IS-2000).

6.2.1 Initialization State

When entering the initialization state after power-up, the mobile station selects which system to use, acquires the Forward Pilot Channel (F-PICH) of a cdma2000 system, obtains system configuration and timing information from the Forward Synchronization Channel (F-SYNCH), and synchronizes its timing to that of the acquired cdma2000 system.

6.2.1.1 System Determination

The algorithm for system determination, although not specified by the IS-2000 standard, typically makes use of a *preferred roaming list*. The preferred roaming list is used in such a way that the mobile station, upon power-up, tries to acquire the home system, that is, the one on which it is subscribed or one belonging to a network operator that has a roaming agreement with that of the home system. Each entry of the roaming list contains the system identifier (SID) and the network identifier (NID) of the system to acquire, the block of CDMA channels numbers, and the corresponding roaming flag. The roaming flag, if turned on, makes the roaming indicator of the mobile station's display flash once service is provided on this entry. The algorithm also accounts for the previously visited systems to speed up acquisition. To that purpose the mobile station may also maintain a table containing the channels on which service was most recently provided.

The system determination procedure then simply consists of building an ordered scan list. At the top of the list are placed the entries from the most recently visited channels, followed by those in the preferred roaming list. The most recently visited channels have priority, since it is likely that once the mobile station is powered up, it is located in the same geographical region in which it recently received service. Once the scan list is built, the mobile station sequentially attempts to acquire each channel in the scan list using the procedures described in the following sections.

6.2.1.2 *Pilot and Synchronization Channel Acquisition*

After determining which system to use, the mobile station searches for the F-PICH (search procedures are described in Chapter 7). If the mobile station fails to acquire the F-PICH within 20 s, it reenters the system determination substate; otherwise it demodulates the F-SYNCH. On the F-SYNCH, it receives the *Sync Channel Message* (SCHM) that carries timing and system configuration information. Before we detail the timing adjustment procedure, seemingly complex, we must understand motivations and constraints. First, recall that the F-PICH pseudonoise (PN) sequence period is equal to 2^{15} (or 32,768) chips, corresponding to 26.667 ms, while the traffic channel frame duration is 20 ms. The implication is that the mobile station upon acquisition of the F-PICH can adjust its timing to that of the revolution period of the F-PICH PN sequence, that is, to the 26.667-ms mark. But since such mark is offset from that of the traffic channel frame boundaries by a time-varying amount, the mobile station is unable to autonomously acquire traffic channel timing unless additional information is provided to it.[5] The timing information in the SCHM serves exactly that purpose. Let us now see how all that works.

The mobile station, upon acquiring the F-PICH, is able to acquire the 26.667-ms mark and start decoding the F-SYNCH frames at each 26.667-ms interval. The most significant bit of the frame is the SOM bit (see Figure 6.3), which is set to 1 if and only if transmission of a new SCHM starts at this frame boundary. Furthermore, the start of the message is restricted to the start of an F-SYNCH channel superframe, which comprises three F-SYNCH frames and therefore is $3 \times 26.667 = 80$ ms in duration. That allows the mobile station to acquire the 80-ms mark of the superframe. At that point the mobile station decodes the entire message. Note that the message is protected by a Layer 3 cyclic redundant code (CRC), while the physical frame is not, so it is only at this point that the mobile station can assess if the SOM bit was incorrectly decoded. Assuming that the message passes the CRC check, the mobile station obtains the following timing information: the pilot PN sequence offset index (PILOT_PN), the long-code state (LC_STATE), and the system time (SYSTEM_TIME) fields. Let us see why such information is relevant and how it is used. The 80-ms mark acquired so far is offset from the traffic channel 20-

---

5.   The difference between a F-PICH PN sequence period and traffic channel frame duration is the result of different design constraints. Choice of F-PICH PN sequence length is dictated by the need for an adequate number of pilots with different offsets. Traffic channel frame duration is a trade-off between opposing needs to maximize interleaver depth and minimize frame decoding (and hence speech) delay.

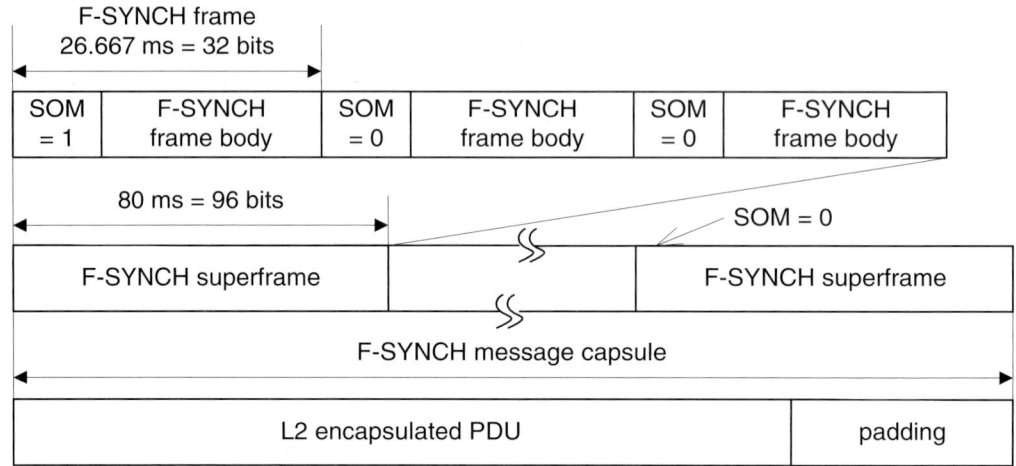

Figure 6.3   F-SYNCH structure.

ms mark by an amount that depends on the PICH PN sequence offset index. Specifically, such offset is equal to PILOT_PN × 64 chips. At this point the traffic channel frames boundary is acquired, and the mobile station proceeds with adjusting the state of the long code. To do that, the mobile station uses the LC_STATE field of the SHCM, which represents the state of the code valid at 320 ms minus PILOT_PN × 64 chips after the end of the message. Finally, the mobile station adjust its system time to that of the SYSTEM_TIME field, which is also valid at 320 ms minus PILOT_PN × 64 chips after the end of the message. System time plays an important role in many Layer 3 procedures.

The SCHM also contains system configuration such as the SID and NID. The mobile station uses the SID to determine its roaming status. The roaming indicator (usually the letter R) flashes on the mobile station's display whenever the received SID differs from that provisioned into the mobile station and corresponding to the home system on which it is subscribed. SID and NID are also used to control registration procedures (see Section 6.6).

6.2.2 Idle State

In the idle state the mobile station monitors the F-PCH or F-QPCH. In this state the mobile station can receive messages, receive an incoming call, initiate registration procedures, or initiate message transmission. The procedures performed while in the idle state are F-PCH/F-QPCH monitoring, response to overhead information, idle handoff, and registration. Description of idle handoffs procedures is deferred to Chapter 7.

6.2.2.1   Paging Channel Monitoring Procedure

The F-PCH is divided into 80-ms slots called F-PCH slots. The F-PCH protocol provides for scheduling transmission of messages destined to a specific mobile station in certain assigned slots, thus allowing the mobile station to monitor the F-PCH periodically. The mobile station is said to operate in *slotted mode*, since it monitors only its assigned slots.[6] When monitoring the assigned slots (the *wake-up* state), the mobile station's circuitry is fully enabled, therefore draining current and reducing battery life. At all other times the mobile station is said to be *asleep*. When asleep, the mobile station's circuitry is disabled. Slotted-mode operation is then important, since it prolongs the mobile station's standby time.

To determine its assigned slots, the mobile station must first determine the length of the slot cycle, the slot cycle start time, and the offset of its assigned slot within each cycle from the cycle start time. The slot cycle length is equal to $2^{\text{SLOT\_CYCLE\_INDEX}}$ in units of 1.28 second (16 slots). The slot cycle start time, in 80-ms slots, is that for which system time, t, in units of 20-ms frames, is such that

$$\left(t/4 - \text{PGSLOT}\right) \bmod \left(16 \cdot 2^{\text{SLOT\_CYCLE\_INDEX}}\right) = 0 \tag{6.1}$$

where the pseudorandom number PGSLOT is an integer uniformly distributed[7] between 0 and 2047. PGSLOT is determined based on a hashing function keyed by the mobile station IMSI. The base station uses the mobile stations' IMSI and SLOT_CYCLE_INDEX when computing hashing function[8] (the same used by the mobile station) to determine PGSLOT and the slot in which to send a message directed to the mobile station.

We omit discussion of the F-BCCH monitoring procedure because they are similar to those used for the F-PCH.

6.2.2.2   Quick Paging Channel Procedure

To prolong the standby time, it is desirable to keep the rate of occurrence and duration of the wake-up state to a minimum. We can reduce the rate of occurrence by increasing the paging channel slot cycle index. However, such approach would increase latency of messages directed to mobile stations in slotted mode, since the base station must wait for the next available slot assigned to the mobile station before it can send a message on the F-PCH or F-CCCH. The increased latency impacts negatively the call-setup delay in the case of mobile station-termi-

---

6.   Although an optional feature, all commercial mobile stations support slotted-mode operation. Therefore, we omit discussion of the non-slotted mode.

7.   The maximum F-PCH slot cycle length is 2048 slots, or 163.84 s, corresponding to SLOT_CYCLE_INDEX = 8. PGSLOT is uniformly distributed so that, on average, the number of mobile stations assigned to a slot is identical for all slots and the F-PCH capacity is maximized.

8.   The mobile station's SLOT_CYCLE_INDEX is included in the registration message. Hence, the base station knows the indexes of all registered mobile stations.

nated calls. Therefore, to increase the standby time, we may want to decrease the duration of the wake-up state rather than its rate of occurrence. The F-QPCH serves exactly that purpose. F-QPCH support is optional for both mobile station and base station.[9] The base station may support up to three F-QPCH per CDMA channel, operated at either 4,800 or 9,600 bps. We now analyze in details the F-QPCH protocol.

The F-QPCH protocol provides for scheduling the transmission of paging indicators in the F-QPCH slots assigned to the mobile station to alert it of an incoming message on the F-PCH (or F-CCCH). The F-QPCH assigned slots, each of 80-ms duration, are offset from the assigned F-PCH slot by 100 ms. The mobile station is assigned two paging indicators. If time t^* represents the assigned F-PCH slot boundary, the F-QPCH quarter slots marked as 1 to 4 in Figure 6.4 are defined as those for which $t \in (t^* - 100, t^* - 80)$, $(t^* - 80, t^* - 60)$, $(t^* - 60, t^* - 40)$, and $(t^* - 40, t^* - 20)$, respectively. The exact position of the paging indicators assigned to the mobile station is determined by a pseudorandom hashing algorithm devised in such a way that the indicators will be either in the first and third quarter slots or in the second and fourth quarter slots. The rationale for the F-QPCH timing is twofold. It ensures that the paging indicators are separated by at least 20 ms so that the channel fade experienced by each indicator is nearly evenly uncorrelated, thus allowing for a high degree of receiver time diversity. It also ensures that the last indicator is offset from the assigned PCH slot by at least 20 ms, which allows the mobile station to power up its circuitry and settle the demodulator tracking loops in time for receipt of the F-PCH message. When the mobile station detects that at least one indicator is set to OFF, it goes back to sleep until the next assigned F-QPCH; otherwise it wakes up to receive the messages in the next assigned F-PCH slot.

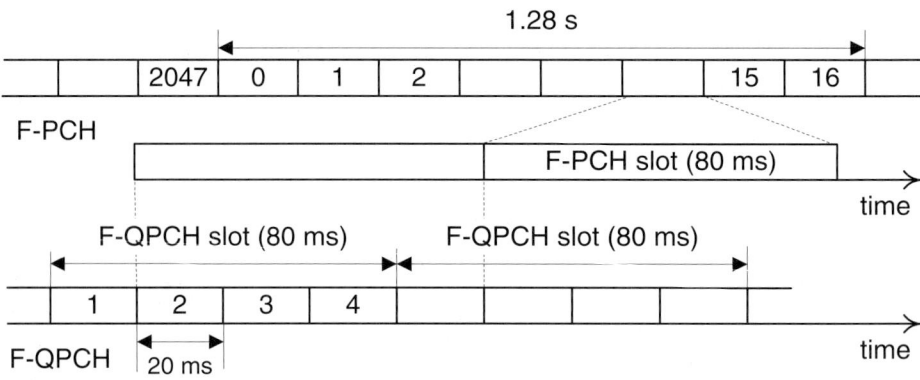

Figure 6.4   F-QPCH timeline.

9.   The mobile station informs the network of its F-QPCH capability by means of the QPCH_SUPPORTED field in the *Registration Message*. F-QPCH capability is then stored in the mobile station profile at the home location register (HLR). At call setup, the base station is informed by the network whether the F-QPCH shall be used to page the mobile station.

The base station protocol mirrors that of the mobile station. The base station determines the position of the mobile station's assigned paging indicators using the same hash function used by the mobile station. When a message has to be directed to a mobile station that is known to support the F-QPCH, the base station turns ON both indicators in the F-QPCH slot immediately before the assigned F-PCH slot.

6.2.2.3   Standby Time Improvement When Using the F-QPCH

We now estimate the mobile station standby time improvement achievable by the F-QPCH following a similar approach to that used in [4]. Denote with I_A and I_S the current drawn when the mobile station is awake and asleep, respectively, and with C the battery capacity. Then, if f_A and f_B are the fraction of time the mobile station is awake and asleep, respectively, the standby time, ST, is simply $ST = C/(f_A I_A + f_B I_B)$. The improvement of the standby time with the F-QPCH, ST_2, relative to that without the F-QPCH, ST_1, is then given by

$$\frac{ST_2 - ST_1}{ST_1} = \frac{1 + f_{A,1}\left(I_A / I_S - 1\right)}{1 + f_{A,2}\left(I_A / I_S - 1\right)} - 1 \tag{6.2}$$

In case of operation without the F-QPCH, prior to each F-PCH the mobile station must enable its circuitry and reacquire pilot timing and carrier phase. We call such period the warm-up time (T_w). Then, with each slot, the mobile station must demodulate the F-PCH messages until it determines that no more page messages are pending in such slot, and only at that time it can reenter the sleep state. We call this period the F-PCH slot monitoring cost (T_{PCH}). Assuming the slot cycle index equal to zero, the fraction $f_{A,1}$ is equal to $(T_w + T_{PCH})/1280 = (10 + 50)/1280 = 4.7\%$.

With the F-QPCH, $f_{A,2}$ is also affected by the same warm-up period, but the F-QPCH monitoring cost is negligible, as the paging indicator duration is very small. Ideally, $f_{A,2}$ can then be as small as 0.8%. However, there are two events that negatively affect F-QPCH standby time improvement because they cause an increase of the effective time the mobile station is awake: false alarms and collisions.

A false alarm occurs when the mobile station's detector mistakes an OFF indication for an ON. The probability of false alarm depends on channel conditions and on how much power is allocated by the base station to the F-QPCH. The probability of false alarm is higher for mobile stations located at the fringe of the cell. Because of false alarms, the mobile station occasionally incurs the cost of monitoring the F-PCH even when the F-QPCH is used.

F-QPCH collisions are the second event type affecting standby time improvement. Recall that the number of paging indicators available per F-QPCH slot are 188 and 94 when the F-QPCH is operated at 9,600 and 4,800 bps, respectively. It may happen that multiple mobile stations are assigned, pseudorandomly, one or both paging indicators with the same time offset. When one such mobile station is being paged, its paging indicators may cause multiple mobile stations to wake-up beside the intended one. We refer to this phenomenon as F-QPCH *collision*. Collisions add on to false alarms to make the F-QPCH less effective and we are therefore interested in estimating their rate of occurrence. The probability of collision can be estimated following the method used in [4]. The collision probability increases with increasing paging rate. However, it turns out that the collision rate is negligibly small relative to the false alarm rate even when the cell is loaded to its maximum capacity. We can then approximate $f_{A,2}$ accounting for the false alarms only and, assuming again the slot cycle index to be equal to zero, obtain $f_{A,2}$ $\approx (P_f * T_{PCH} + T_w)/1280$. Figure 6.5 depicts the standby time increase as a function of the ratio I_A / I_S and for different false alarm probability. Since for many commercially available terminals the ratio I_A / I_S is 100 or larger, we conclude that the standby time is at least doubled when using the F-QPCH, except for those mobile stations that experience unfavorable channel conditions.

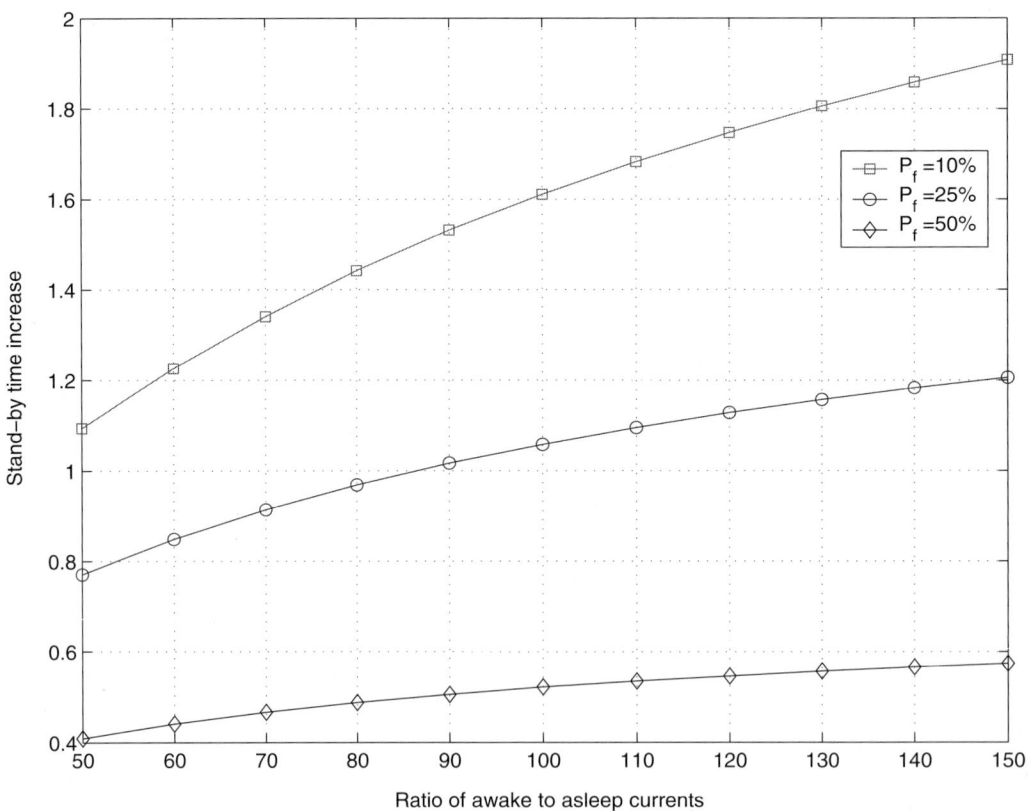

Figure 6.5   Standby time increase with F-QPCH.

6.2.2.4   *Response to Overhead Information*

While in the idle state the mobile station monitors its assigned F-PCH or F-BCCH slots to receive overhead messages. Overhead messages are those containing configuration information that the mobile station needs to interoperate with the base station. Several are the overhead messages that the base station may transmit, and their format and content differs depending on whether the F-BCCH or the legacy F-PCH is used. Some messages are optionally sent by the base station depending on its configuration and the feature set it supports. Other are mandatory and must be transmitted periodically, at least once every 1.28 s. The constraint is in place to put a limit to the delay that a mobile station has to experience whenever it must update its stored configuration information, for example, at the time of call attempt. Rather than listing all messages and describing their format, in the following we briefly describe the various categories of configuration parameters and their use.

The first category of configuration parameters is the one including generic system parameters, which regulate procedures such as registration, identification, and security. Radio resource parameters include those related to power-control procedures and handoff procedures. Some other parameters carry information about the physical characteristics of other common channels, such as Walsh code, data rate, and power levels. The access parameters are those needed to control operation on the reverse common channels. That includes Reverse Access Channel (R-ACH), Reverse Enhanced Access Channel (R-EACH), and Reverse Common Control Channel (R-CCCH) protocol parameters. The neighbor base stations' parameters convey information about the configuration of neighboring base stations that are likely candidates for idle handoff. If the base station supports multiple CDMA channels on different frequency assignments (simply called a multicarrier base station), the overhead messages carry information regarding the configuration of such channels. Using such information, the mobile station performs a hashing procedure to determine which CDMA channel it should use. The hashing algorithm is such that the mobile station is pseudorandomly assigned to one of the CDMA channels so that the common channels' load is uniformly distributed across all available CDMA channels. Other overhead information that may be carried by the overhead messages is that related to optional features such as authentication or services such as CDMA Tiered services (see Section 6.8).

When the mobile station receives a new overhead message, it processes it and stores the configuration parameters in volatile memory. The mobile station can distinguish new messages from old ones because of the configuration message sequence number. A configuration message sequence number (CONFIG_MSG_SEQ) is associated with each message sent on the F-PCH or F-BCCH. The base station increments setting of the sequence number whenever the message content changes. The mobile station stores the most recently received CONFIG_MSG_SEQ for each message. When a new message is received, its CONFIG_MSG_SEQ is compared against the one stored to determine whether the stored configuration parameters are current. If the comparison results in a match, the mobile station ignores the message, or else it updates the configuration parameters with the newly received ones.

6.2.2.5 Directed Message Processing

Beside overhead messages, a mobile station in the idle state is able to receive messages intended for that mobile station only, that is, *directed* messages. All directed messages contain addressing information that allows the mobile station to perform the address-matching operation. The base station may use either IMSI, TMSI, or ESN to address the mobile station for all directed messages except the *General Page Message* (GPM) for which ESN addressing is not allowed.

Directed messages include those used for paging, for registration and authentication procedures, to transport teleservices information such as Short Message Service (SMS), or to support features such as voice mail notification. The Layer 3 protocol specifies how the mobile station is to process the information carried by mobile station–directed messages and also specifies the actions the mobile station has to perform on receipt of any such message of which it is the intended recipient.

6.2.3 System Access State

In the system access state the mobile station exchanges signaling messages with the base station in response to messages requiring acknowledgment, or in order to register with the base station, or because it has been paged, or because of user intervention, for example, at call origination or SMS transmission. The system access state (see Figure 6.6) is entered from the idle state and consists of the following substates.[10]

> **Update Overhead Information Substate**. In this substate the mobile station monitors the F-PCH or F-BCCH and updates its configuration parameters until it has received a current set of overhead messages. That is necessary because the configuration messages' content may have changed since the last time the mobile station demodulated its assigned PCH or BCCH slot. When it enters this substate, the mobile station sets the 4-second system access timer. The timer is cleared once the mobile station determines that its configuration is current and enters one of the following substates depending on the cause that triggered the access attempt. The mobile station aborts the access attempt and reenters the idle state if it is unable to process the overhead messages by the time the access timer expires (e.g., because of poor channel conditions).
>
> **Mobile Station Origination Attempt Substate**. In this substate the mobile station attempting call setup sends the *Origination Message* (ORM) on the access channel. If the mobile station is attempting to reconnect a dormant packet data session, it may transmit the *Reconnect Message* (RCNM) instead. Once message acknowledgment

---

10.   The priority access call assignment (PACA) cancel substate is not listed here, since PACA is not described in this book.

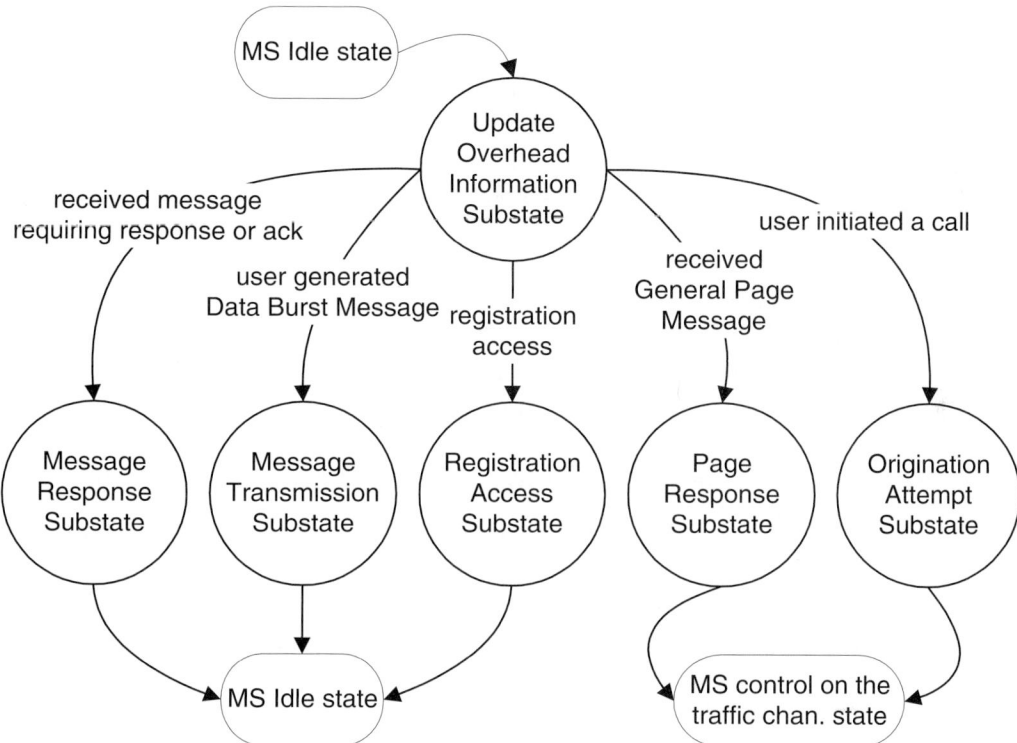

Figure 6.6 Mobile station system access state (not all states or transitions shown).

is received, the mobile station sets a 12-second timer while waiting to be assigned a traffic channel. If resources are available, the base station that detects the mobile station origination attempt sends the *Extended Channel Assignment Message* (ECAM) on the F-PCH or F-CCCH. This message contains, among other information, the radio configuration to be used on the forward and reverse traffic channels; the traffic channel frame offset; the pilot offsets corresponding to the CDMA channels on which a traffic channel has been granted; the Walsh code index of the traffic channels being assigned; and forward-link power-control parameters. On receipt of the message, the mobile station enters the mobile station control on the traffic channel state. The mobile station aborts the access attempt and reenters the idle state if the 12-second timer expires prior to receipt of the ECAM.

Page Response Substate. In this substate the mobile station sends a *Page Response Message* (PRM) or an RCNM on the access channel in response to a GPM from the base station. On completion of message transmission, the procedure is similar to that used in the origination attempt substate. The mobile station enters the mobile station control on the traffic channel state once it is assigned a traffic channel.

Message Response Substate. In this substate the mobile station sends a message on the access channel that is in response to a message received from the base station other than a page. Typical scenarios that require the mobile station to enter this substate include receipt of the *Authentication Challenge Message* (AUCM); see also Section 6.7.1.3. The message response depends on the message the mobile station received. The mobile station reenters the idle state when it receives base station acknowledgment of the response message.

Registration Access Substate. In this substate the mobile station sends a *Registration Message* (RGM) on the access channel (see also Section 6.6). The mobile station reenters the idle state when it receives base station acknowledgment of the RGM.

Mobile Station Message Transmission Substate. In this substate the mobile station sends a *Data Burst Message* (DBM) on the access channel, for example, for SMS message delivery to the base station. The mobile station reenters the idle state when it receives base station acknowledgment of the DBM.

When sending a signaling message on the R-ACH or R-EACH while in the system access state, the mobile station uses the random access procedures discussed in Chapter 5. While in this state, the mobile station continues its pilot search and may perform handoff procedures aimed at guaranteeing that the mobile station is always communicating with the best possible base station. Discussion of handoffs while in the system access state is deferred to Chapter 7.

6.2.4 Mobile Station Control of the Traffic Channel State

In this state the mobile station and base station communicate using the forward and reverse traffic channels that have been granted at call setup. As shown in Figure 6.7, it comprises three substates.

Traffic Channel Initialization Substate. In this substate the base station sends null frames at eighth rate, and the mobile station verifies that it can reliably demodulate the forward traffic channel before starting transmission on the reverse traffic channel. This to ensure that the forward-link power-control channel is of sufficient quality to ensure that reverse-link power control can be effective. The power-control subchannel gain is slaved to that of the traffic channel it is multiplexed onto, so that its quality is deemed sufficient if that of the traffic channel is sufficient. A straightforward method to estimate traffic channel quality is to measure its frame error rate. If the mobile station receives at least two consecutive good frames, the forward link is acquired, and the mobile can start transmission on the reverse traffic channel. Reverse pilot channel only is enabled at the time reverse-link transmission starts, while the data component of the Reverse Fundamental Channel (R-FCH) is gated off. The mobile station aborts the call and reenters the idle state if no two consecutive good frames are received within a 1-second period. Meanwhile the base station

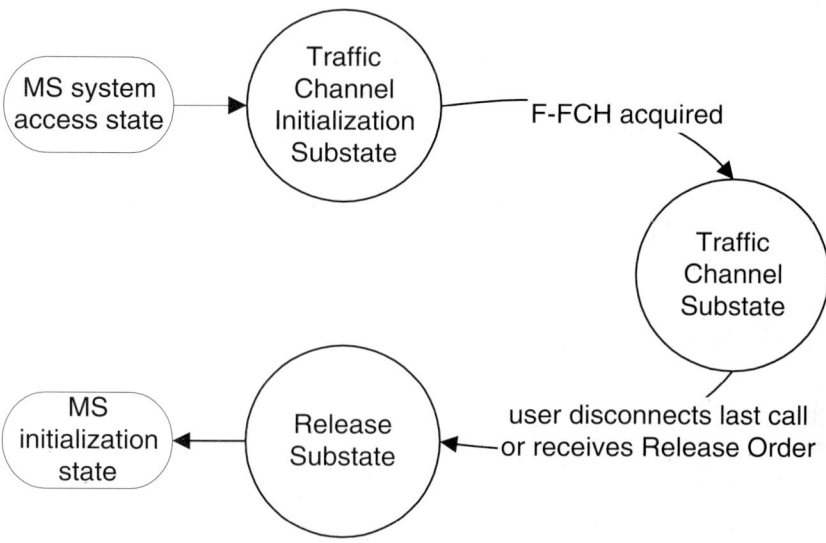

Figure 6.7 Mobile station control on the traffic channel state.

attempts to acquire the reverse pilot channel. Once acquired, the base station sends the *Base Station* (BS) *Acknowledgment Order Message* (ORDM)[11] on the Forward Fundamental Channel (F-FCH), indication to the mobile station that it can now start transmitting null data frames in addition to the reverse pilot. From a Layer 3 perspective, the mobile station and base station instantiate a *call instance* and enter the traffic channel substate once the traffic channel initialization is complete.

Traffic Channel Substate. In this substate the mobile station and base station transmit signaling and data traffic on the reverse and forward dedicated channels. The mobile station and base station perform handoff processing as described in Chapter 7. Forward- and reverse-link power control is enabled using the procedures described in Chapter 8. The mobile station and base station perform supervision procedures by monitoring the quality of the traffic channel. While in this state, other calls, each corresponding to a new service, may be connected in a mode of operation called *concurrent services* (see Section 6.5). When the call is disconnected (the last call in the case of concurrent services), either by the base station sending the *Release* ORDM or by user intervention, the mobile station and base station enter the release substate.

---

11.   The ORDM is a generic message used in a variety of signaling procedures. ORDMs are distinguished by means of their ORDER code. In the following, when referring to an order we use the function corresponding to the order code, in italics, followed by the acronym ORDM.

Release Substate. In this substate the call and all associated physical channels are disconnected. The release process requires handshaking between mobile station and base station to avoid dangling resources. If the release is initiated by the base station by sending the *Release* ORDM, the mobile station responds with a *Release* ORDM, shuts off its transmitter, and enters the idle state. If a user directs the mobile station to release the call, the mobile station sends the *Release* ORDM to the base station and sets a 2-second timer waiting for a confirmation from the base station. While in this state, the mobile station sends null frames at eighth rate on the reverse traffic channel while still performing power-control and handoff processing. On receipt of the *Release* ORDM or expiration of the release timer, the mobile station shuts off its transmitter and enters the initialization state. As already mentioned, provisions were introduced in IS-2000B that allow the mobile station to enter the idle state directly, without first entering the initialization state. That minimizes both the time between successive calls and the probability of missing a page just after call release.

While in the traffic channel state, the mobile station and base station perform special functions, such as service negotiation and channel supervision, described in the next sections.

6.2.4.1   Service Negotiation

In the traffic channel state the mobile station and base station communicate through the exchange of traffic channel frames. The mobile station and base station use a common set of attributes for building and interpreting such frames. This set of attributes is called the service configuration. The service configuration attributes consist of both negotiable and nonnegotiable parameters.

The set of negotiable service configuration parameters consists of the multiplex option, radio configuration, and data rates (all discussed in previous chapters), and the service option connections. A service option connection represents a service that is *connected*, or in simpler terms, in use. A service option connection is associated with a service option, a traffic channel type (e.g., primary or secondary), and a service option connection reference. Since several services may be in use simultaneously, the connection reference provides a means for uniquely identifying the service option connection.

The nonnegotiable service configuration parameters are so called because their setting is solely under control of the base station. They consist of the reverse pilot gating rate, the power-control parameters, and the logical-to-physical channels mapping.

Service negotiation is that set of procedures used to negotiate the service configuration. Its need arises from the fact that base station and mobile station capabilities may differ, for example, because support of some configurations and services is optional. Several messages may be used for service negotiation. Service negotiation may be initiated by either the mobile station or the base station. Typically, the mobile station chooses the default service configuration associated with a service option at call origination and may negotiate different attributes once on the traffic channel.

The base station can do the same when it pages the mobile station. Once either entity receives a request for a new service configuration, it may grant the request, reject it, or respond with yet another service configuration request. The process ends when the mobile station and base station mutually agree on the service configuration to use or either party rejects the request.

Note that as soon as the traffic channel is established but no service is yet connected, for example, because service negotiation has not been completed, the mobile station and base station transmit (beside signaling messages) null data using eighth-rate frames to minimize resource consumption. The rule above applies to the F/R-FCH only, as the Forward Packet Data Channel (F-PDCH) and Forward/Reverse Dedicated Control Channel (F/R-DCCH) never carry null data.

The attributes of the agreed-upon configuration are stored in the service configuration record, which finds an important application with packet data services. Such services are characterized by bursty traffic and by relatively long periods of user inactivity. During these periods, traffic channels may be released to avoid wasting resources. Once user activity starts again, it is desirable to reconnect the service as quickly as possible. In these scenarios the mobile station can ask the base station, or vice versa, to reconnect a previously established connection using the same service configuration record, thus avoiding the latency associated with service negotiation procedures.

6.2.4.2   *Channel Supervision*

While in the traffic channel state, both mobile and base station perform supervision of the forward and reverse traffic channels, respectively, to detect loss of connection due to radio link failure.[12] A radio link failure may occur, for example, when the mobile station moves outside the coverage area of the access network. Forward-link supervision procedures differ depending on whether the F-FCH/DCH or F-PDCH is allocated.

Forward-link supervision when the F-FCH/DCCH is allocated. On the mobile station side, traffic channel supervision primarily aims at avoiding erratic reverse-link power control that may lead to jamming the base station receiver.[13] Quality of the forward-link power-control subchannel cannot be reliably assessed by estimating its received energy. However, it can be inferred from that of the associated forward traffic channel, whose quality is easily assessed by monitoring occurrence of frames received in error. The standard, therefore, specifies that the mobile station must disable its transmitter if it receives 12 consecutive traffic channel bad frames. Thereafter,

---

12.   IS-2000 channel supervision procedures are mandated only for the mobile station. For the base station, supervision is implementation dependent.

13.   Whenever the power-control command is received in error, the mobile station will change its transmit power in a direction opposite to that required to counteract the change of reverse-link path loss. In the extreme case of complete fade of the forward primary traffic channel, the mobile station's transmit power will wander in random-walk fashion.

the mobile station reenables its transmitter only when receiving two consecutive good frames. In addition to disabling its transmitter, the mobile station must decide when to release the connection. That is achieved by means of the *fade timer*. The timer is enabled when the mobile station first enables its transmitter upon entering the traffic channel state. The fade timer is reset for 5 seconds whenever the mobile station receives two consecutive good frames on the traffic channel. If the timer expires, the mobile station declares a loss of the traffic channel and releases the connection.

Forward-link supervision when the F-FCH/DCH is not allocated. When the mobile station is assigned to the F-PDCH (see also Chapter 9), the F-FCH and F-DCCH are optional. If neither the F-FCH or F-DCCH are assigned, the F-PDCH by itself cannot be used for supervision because it is a shared channel. The mobile station uses instead the Forward Common Power Control Channel (F-CPCCH) sub-channel (allocated per mobile station) quality for supervision purpose. Note that the F-CPCCH must be always assigned if neither the F-FCH or F-DCCH are assigned.

The F-CPCCH is monitored, and if for 480 ms the channel conditions are considered bad, the reverse-link transmitter is turned off. When the reverse-link transmitter is turned off, the mobile station still monitors the allocated F-CPCCH subchannel. If within a 5-second interval, after turning off the transmitter, the channel quality improves and it is considered good for 80 ms, the transmitter is turned on. If not, the call is released.

The specific metric to decide on F-CPCCH quality is, however, implementation-specific. A reasonable implementation would be that the mobile station estimates the F-CPCCH subchannel average received bit energy over a 20-ms interval. If the average energy is below a set threshold, it declares bad channel conditions for the entire 20-ms interval. If the average energy is above the threshold, the channel conditions are considered good. Now, 24 consecutive bad intervals would cause the transmitter to be turned off and, thereafter, four consecutive good intervals would cause the transmitter to be reactivated. The supervision algorithm then becomes equivalent to that depicted in Figure 6.8 for the F-FCH, except that transitions between transmitter enabled/disabled states are made more stringent to compensate for the lower reliability of quality measurements obtained by means of energy estimate relative to those obtained by means of frame erasure count.

Reverse-link supervision. On the base station side, supervision is implementation-dependent and is typically performed by monitoring the quality of the reverse traffic channel after frame selection. The base station may also use a fade timer. When reverse traffic channel frames are consecutively received in error for a period equal to the base station's fade timer, the base station declares loss of traffic channel and releases the connection. During this period, the base station maintains the forward traffic channel transmitter enabled since the considerations about random power

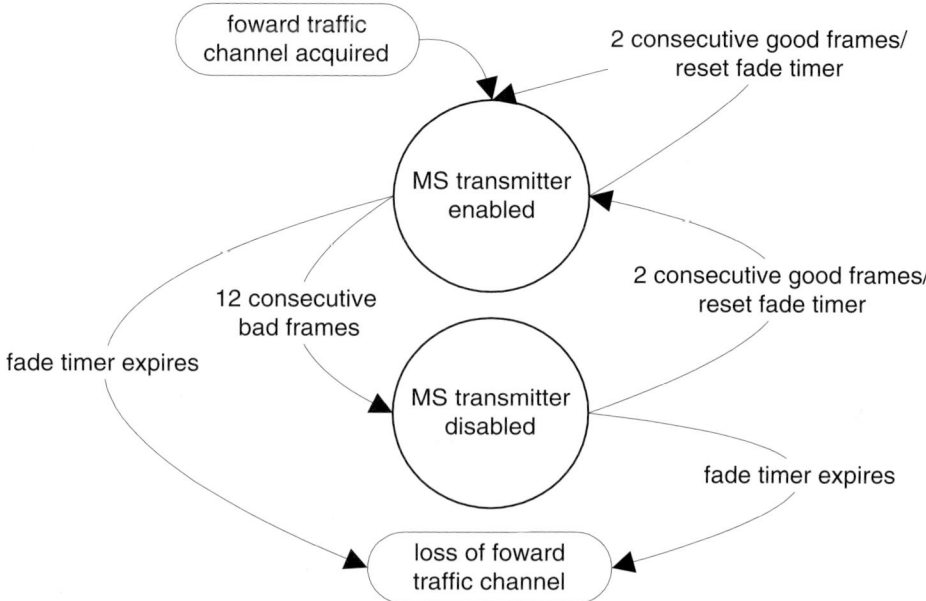

Figure 6.8   Forward traffic channel supervision when the F-FCH is allocated.

control and jamming do not apply to the forward traffic channel, which has much smaller power-control range than that of the reverse traffic channel. Note also that since the reverse-link failure may be a consequence of the mobile station disabling its transmitter because of poor forward-link quality, the base station's fade timer should be set to a value larger than that of the mobile station.

6.2.5 *Call Control Processing*

Call control processing takes place once the mobile station and the base station are in the traffic channel state. Call control is performed on a per-call basis, where the term *call* stands generically for any connected service, either voice or data. Each call instance is assigned a unique identifier, or connection reference (CON_REF). Processing of a call involves exchange of messages to control connection and disconnection of the call as well as messages needed for specialized call treatments. The actions taken by the mobile station and base station on transmission or receipt of any such message is specified in the Layer 3 protocol by means of a state machine with the substates and transition triggers depicted in Figure 6.9.

   Waiting for Order Substate. If the call being instantiated is a mobile station–terminated voice call, the call-control instance enters the waiting for order substate. On entering this substate, the mobile station sets a 5-second timer and waits to receive

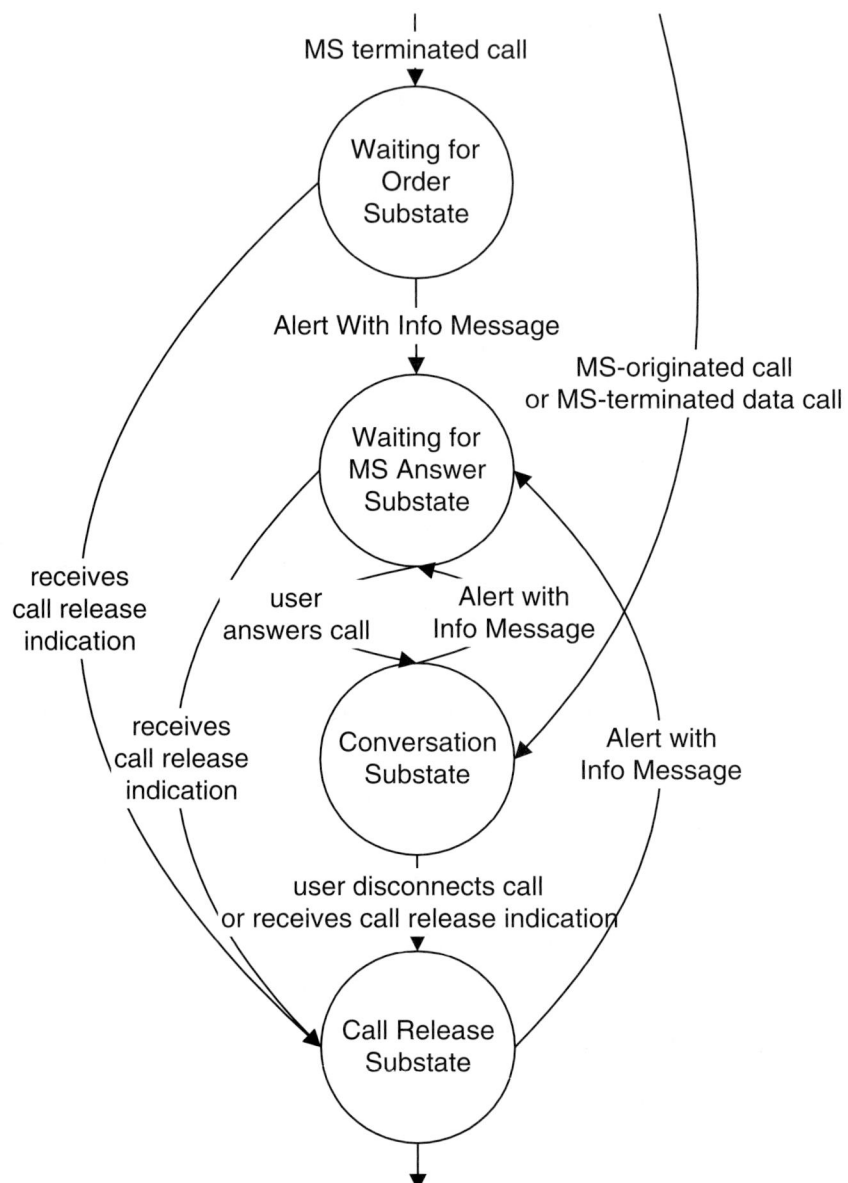

Figure 6.9 Call control.

the *Alert with Information Message* (AWIM). If the timer expires before the message is received, the call-control instance enters the call release substate; otherwise the mobile station alerts the user of the incoming call and enters the waiting for mobile station answer substate. The AWIM is also used to carry the calling party directory number.

Waiting for Mobile Station Answer Substate. In this substate the mobile station waits for the user to answer the call. The mobile station sets a 65-second timer upon entering the substate. If the user answers the call, the mobile station sends a *Connect* ORDM, and the call-control instance enters the conversation substate. Note that the user may also invoke specialized treatment, such as call-forwarding to a preprogrammed number or to voice mail. If the 65-second timer expires before user intervention, the call-control instance enters the call release substate. Typically, the network supports a much shorter answer timer so that in case of network time-out, the call is released by the base station on receipt of the network command even before expiration of the Layer 3 timer.

Conversation Substate. In this substate mobile station and base station are in a call. In case of a mobile-originated call or mobile-terminated data call, this substate is entered into directly on instantiation of the call-control instance, since user alerting is not applicable in these cases. While in this substate the call-control instance processes messages to connect the service to which the call is assigned (service negotiation). Other messages processed while in this substate include those to support supplementary services that may be invoked by the user, such as three-way calling, or those used to carry encoded DTMF signals triggered when the user presses the keypad for remote control of, for example, a voice mail system.

Call Release Substate. In this substate the call instance waits to be released. If this is the only call presently instantiated, the mobile station enters the release substate of the mobile station control on the traffic channel state, and physical resources are also released. If other services are concurrently connected, only the physical resources associated with the call being released and that are not shared by other calls are released.

6.3 Call Processing Examples: Voice Calls

Call-flow diagrams are compact and useful representations of call processing and in this section we present a few related to basic voice calls. These should be regarded as examples only. Failures and/or alternative scenarios are not described, as a simple call flow cannot capture all aspects of the rather complex Layer 3 protocol state machine.

6.3.1 Mobile Station–Originated Voice Call

In the example illustrated in Figure 6.10, the mobile station originates a voice call. The base station assigns the mobile station onto the traffic channel. After service negotiation is completed, two-way conversation can take place. Specialized call treatments are also shown.

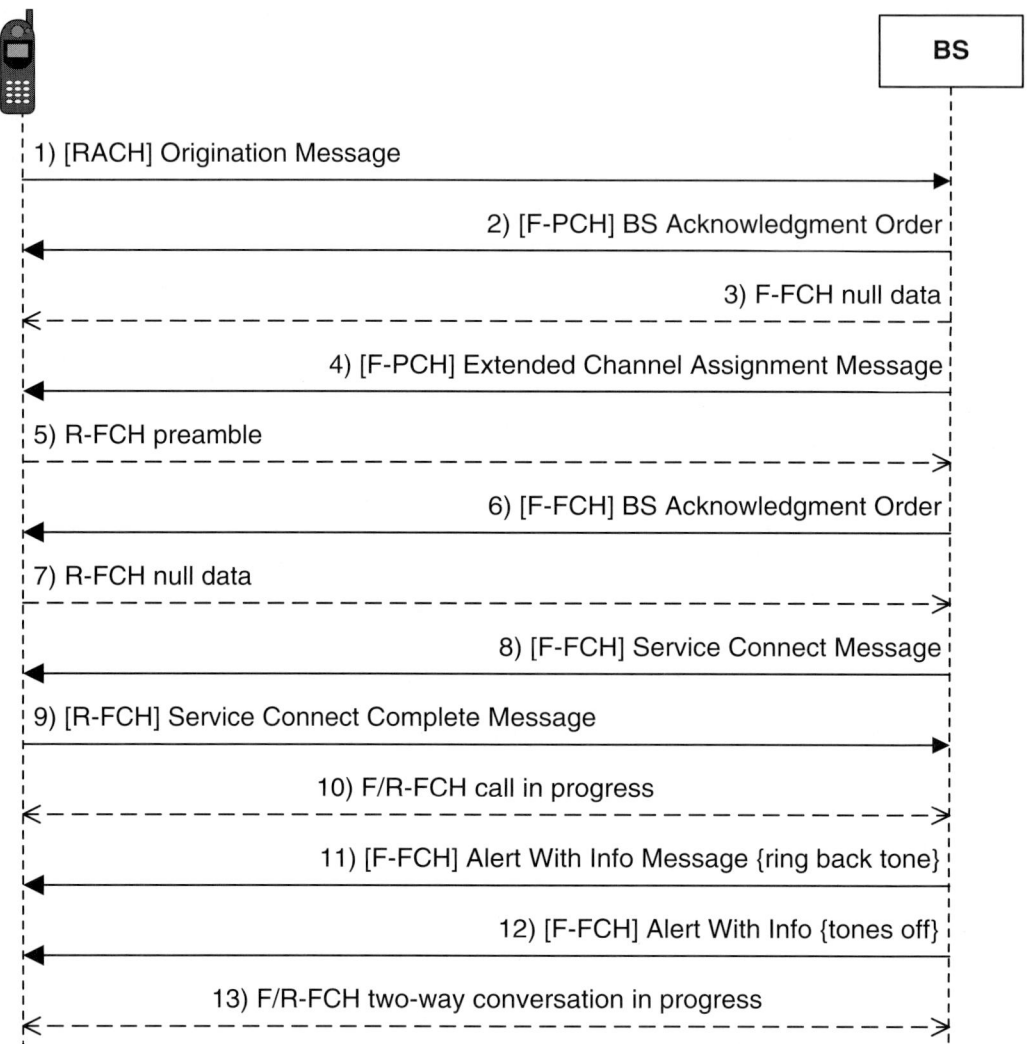

Figure 6.10 Mobile station-originated call.

1. The mobile station in idle state is directed by the user to originate a call. The mobile station enters the system access state and sends the ORM on the R-ACH. The message contains the dialed digits.

2. The base station receives the ORM and sends the *Base Station Acknowledgment* ORDM on the F-PCH. On receipt of the order, the mobile station stops the access procedure and starts a 12-second timer.

3. The base station runs its admission control algorithm. If resources are available, the base station initializes the required resources and sets up the F-FCH. At this point the F-FCH carries null data at the lowest possible data rate (eighth-rate frames).

4. The base station sends the ECAM on the F-PCH, directing the mobile station to a traffic channel. The base station waits to detect mobile station transmission on the reverse link. In the meantime, on receipt of the message, the mobile station clears the 12-second timer and attempts to acquire the F-FCH. From this point on, signaling messages are exchanged only on the dedicated channels.

5. Before any data is transmitted on the reverse link, the mobile station must receive at least two consecutive good frames to ensure that the forward link (with its power-control subchannel) is of sufficient quality to guarantee reliable reverse-link power control. If the condition is not met within 1 sec, the call fails and the mobile station reenters the idle state. Once the condition is met, the mobile station starts transmitting the reverse pilot (the data part of the R-FCH is gated off).

6. The base station sends the mobile station the *Base Station Acknowledgment* ORDM on the F-FCH when it is able to detect the R-FCH. The detection procedure is left implementation-dependent, but typically involves receiving at least one good frame. Once the R-FCH is detected, the base station effectively initiates reverse-link inner-loop power control.

7. The mobile station, on receipt of the order, stops transmitting the preamble and starts transmitting null frames at eighth rate on the R-FCH.

8. The base station initiates service negotiation procedure by sending the *Service Connect Message* (SCM).

9. The mobile station accepts the proposed service configuration and sends the *Service Connect Complete Message* (SCCM).

10. At this point the call is, from the point of view of the base station, connected. However, two-way conversation may not happen, as the called party has yet to answer the call.

11. When directed by the network, the base station sends the AWIM that instructs the mobile station to generate locally the ring-back tone. The message contains the tone characteristics. Often, this message is not sent, as the network opts to play the tone in-band. In such case, the tone is inserted into the audio path transparently to the base station.

12. Once the called party has answered the call, and when directed by the network, the base station sends the AWIM that instructs the mobile station to halt the ring-back tone.

13. Two-way conversation is now in progress.

6.3.2 Mobile Station–Terminated Voice Call

In the example illustrated in Figure 6.11, the mobile station receives a voice call. The base station assigns the mobile station onto the traffic channel. After service negotiation is completed, two-way conversation can take place. Specialized call treatments are also shown.

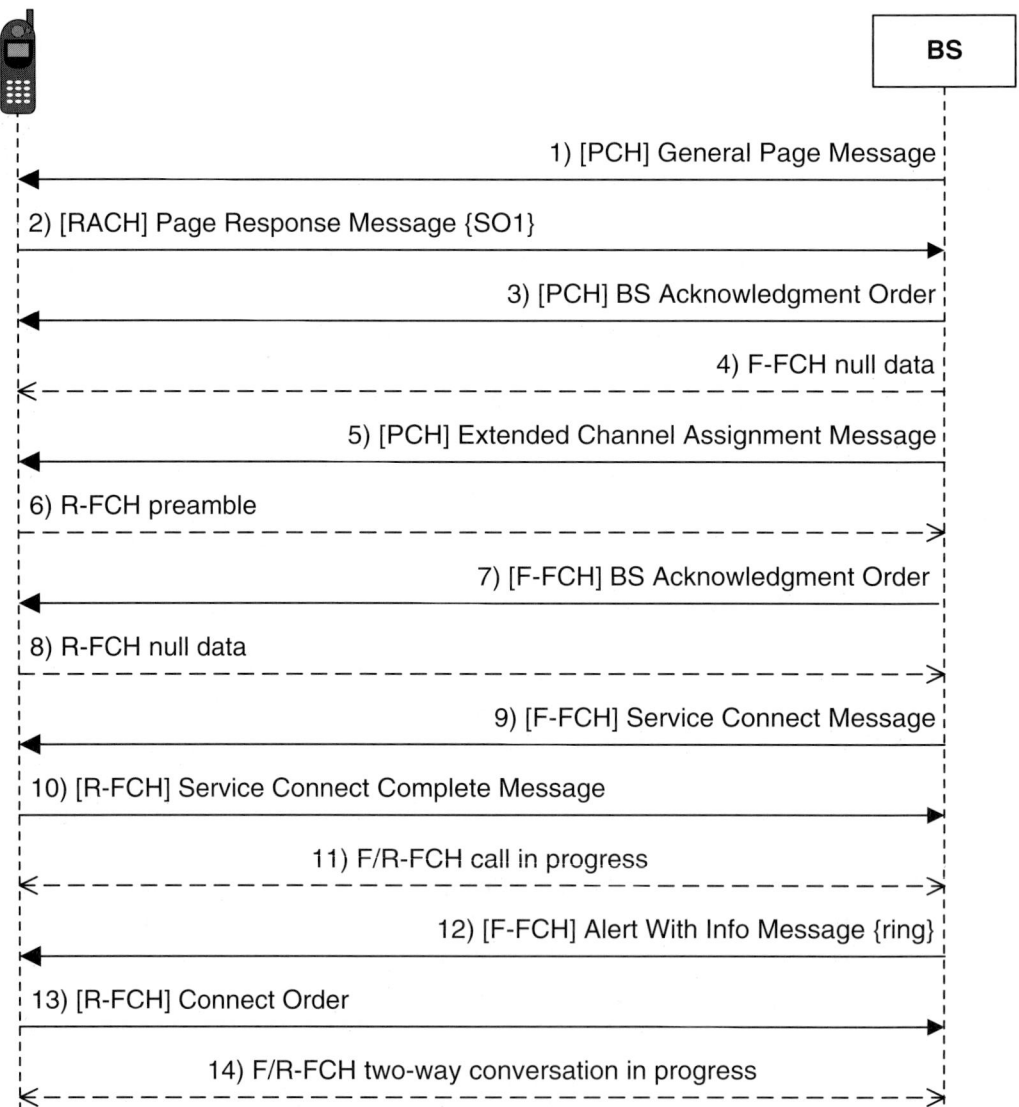

Figure 6.11 Mobile station-terminated call.

1. The mobile station in idle state wakes up when it detects that its assigned F-QPCH paging indicators are set to ON, and proceeds with demodulating the assigned F-PCH slot.

2. The mobile station receives a Page Message whose address matches its own. It enters the system access state and sends the PRM on the R-ACH

3. The base station receives the PRM and sends the *Base Station Acknowledgment* ORDM on the PCH. On receipt of the order, the mobile station stops the access procedure and starts a 12-second timer (T42m).

4. The base station runs its admission-control algorithm. If resources are available, the base station initializes the required resources and sets up the F-FCH. At this point the F-FCH carries null data at the lowest possible data rate (eighth-rate frames).

5. The base station sends the ECAM on the F-PCH, directing the mobile station to a traffic channel. The base station waits to detect mobile station transmission on the reverse link. In the meantime, on receipt of the message, the mobile station clears the T42m timer and attempts to acquire the F-FCH. From this point on, signaling messages are exchanged only on the dedicated channels.

6. Before any data is transmitted on the reverse link, the mobile station must receive at least two consecutive good frames to ensure that the forward link (with its power-control subchannel) is of sufficient quality to guarantee reliable reverse-link power control. If the condition is not met within 1 second (T_{51m}), the call fails and the mobile station reenters the idle state. Once the condition is met, the mobile station starts transmitting the reverse pilot (the data part of the R-FCH is gated off).

7. The base station sends the mobile station the *Base Station Acknowledgment* ORDM on the F-FCH when it is able to detect the R-FCH. The detection procedure is left implementation-dependent, but typically involves receiving at least one good frame.

8. The mobile station, on receipt of the order, stops transmitting the preamble and starts transmitting null frames at eighth rate on the R-FCH.

9. The base station initiates the service negotiation procedure by sending the SCM.

10. The mobile station accepts the proposed service configuration and sends the SCCM.

11. At this point the call is, from the point of view of the base station, connected. However, two-way conversation may not happen yet, as the mobile station has yet to answer the call.

12. The base station sends the AWIM that instructs the mobile station to generate locally the alert tone. The message may also contain the calling party directory number that was conveyed to the base station by the network.

13. Once the mobile station's user has answered the call, the mobile station sends the *Connect* ORDM.

14. Two-way conversation is now in progress.

6.3.3 Mobile Station–Initiated Call Release

In the example illustrated in Figure 6.12, a voice call is in progress on the F/R-FCH when the mobile station initiates call release.

1. The mobile station in the traffic channel state is directed by the user to end the call.
2. The mobile station enters the release substate, sets a 2-second timer, and sends the *Release* ORDM.
3. While waiting for base station confirmation, the mobile station sends null frames at eighth rate and discards voice frames. Power control, handoff processing, and supervision are still enabled.
4. Within 0.8 second after receipt of the *Release* ORDM, the base station replies with the *Release* ORDM. The *Release* ORDM is typically sent in quick repeats. Resources are not immediately released, however, since during the handshaking procedures, the mobile station still needs to be power controlled.
5. Upon receipt of the *Release* ORDM or expiration of the timer, the mobile station shuts off its transmitter and enters the idle state.
6. The base station starts transmitting null frames at eighth rate and the power-control subchannel for at least 0.3 sec. During this period, the base station may transmit all downs power-control command to minimize reverse-link capacity consumption.
7. The base station stops F-FCH transmission and releases all associated resources.

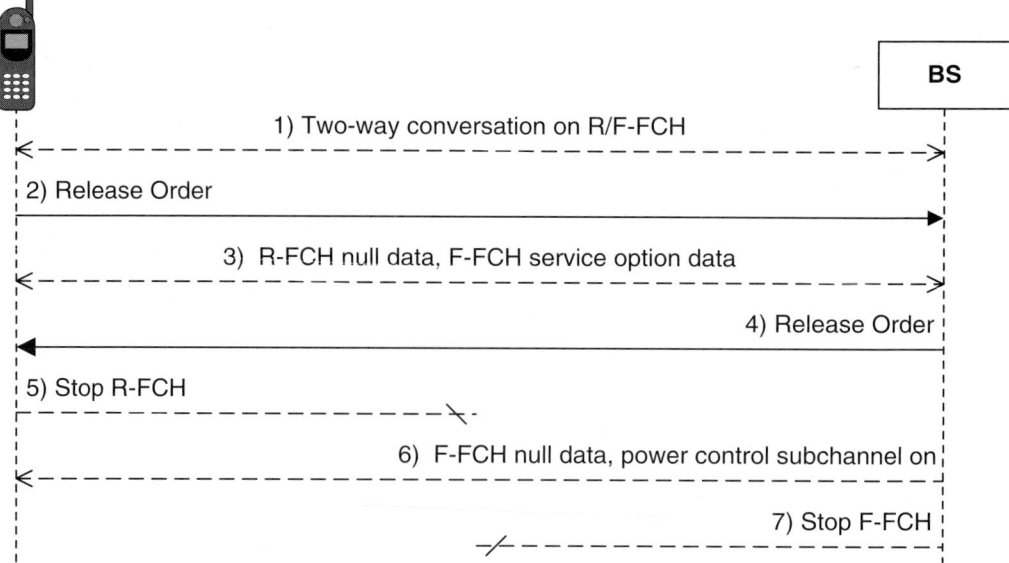

Figure 6.12 Mobile station-initiated call release.

6.4 Packet Data Service Protocol

From a Layer 3 perspective, a data call is realized by means of the traffic channel setup and service negotiation procedures described in generic terms (for both voice and data services) in the previous sections. However, in the case of packet data calls using service option 33 (SO33), specialized procedures are in place to manage air interface resources, quality of service, and end-to-end connectivity with the Packet Data Serving Node (PDSN). These procedures are specified in the IS-707 standard [5] rather than in IS-2000 and are collectively known as the air interface packet data services protocol. The most relevant of such procedures are summarized in the following sections.

6.4.1 Packet Data Dormancy Support

A dedicated traffic channel is not always needed during a packet data session, as the end-user data traffic is bursty. A typical example is a Web-browsing session, in which *reading periods* with little or no user activity follow the download of each Web page. During a reading period, no data traffic is generated, and the mobile station and base station exchange idle Radio Link Protocol (RLP) frames. After a short period, equal to up to round-trip time between the mobile and the base stations, idle RLP frames can be suppressed unless the physical channel is FCH. However, the R-PICH must remain active as long as there is a traffic channel connection between the mobile and the base stations. The reading period greatly varies in duration, as it depends on many unpredictable factors, such as end-user behavior, but it can be, as described in Chapter 3, characterized with exponential distribution. Hence, after a period of prolonged user inactivity, it is reasonable to anticipate that the user has entered a long reading period. Then, the packet data call may be disconnected without user intervention to conserve radio resources. The associated physical channels are released, and the mobile station enters the Layer 3 idle state once the call is disconnected. From a packet data service protocol standpoint, however, the packet data service is still active, while the associated call is said to be *dormant*. The call is reconnected, and its associated physical channels reestablished once user activity generates new data traffic. To prevent frequent oscillation between the connected and dormant states, the mobile station could maintain a packet data *dormant timer*. In this case, once the packet data call goes dormant, the mobile delays a request for packet data service until the timer expires irrespective of new data being available for transmission. The dormant timer is controllable by the base station through the *Service Option Control Message* (SOCM) and DORM_TIME field. The timer can be set in the range of 0.1 to 25.5 sec.

At service setup, the SCM includes a service configuration synchronization identifier, SYNC_ID. When the packet data service at the mobile station goes dormant, the entire service configuration is stored and associated with SYNC_ID. Note that in general, one service configuration may be associated with several services, or service reference identifiers (SR_ID). Multiple packet data services with distinct SR_IDs may be simultaneously active. To activate a particular service, it is not necessary that the base station and the mobile station go through a full service-negotiation procedure. Instead, the packet data service protocol allows mobile station and base station to use SYNC_ID when requesting transition out of dormancy. At the network side, the service configuration and SYNC_ID are stored at the Packet Data Control Function (PCF). The base station, unaware of dormancy, simply releases traffic channels. An example of the expedited call-setup procedure is illustrated in Figure 6.13.

1. The mobile station can request service reactivation from dormancy with the RCNM. The service configuration is indicated with SYNC_ID.
2. The base station acknowledges the reception of the message.
3. The base station sends ECAM and instructs the mobile station to restore the service configuration with SYNC_ID reported in the RCNM. The SR_ID_RESTORE field indicates the particular service option that needs to be restored. SR_ID_RESTORE = 111 instructs all service options associated with the SYNC_ID to be activated.
4. Traffic channel setup follows.
5. After the traffic channels are set up, service negotiation is not necessary. Data exchange follows immediately.

The packet data connection is terminated at the PDSN, as explained in Chapter 2. Throughout the dormant state, a logical connection between mobile station and PDSN is maintained, although no data is being exchanged. Since two base stations may be connected to different PDSNs, a mobile station with a dormant packet data call that moves into the service area of a base station connected to a different PDSN needs to establish a new Point-to-Point (PPP) connection. The base station advertises the PCF it is connected with, by means of the PACKET_ZONE_ID field of the *Extended* (or *ANSI-41*) *System Parameters Message* sent on the

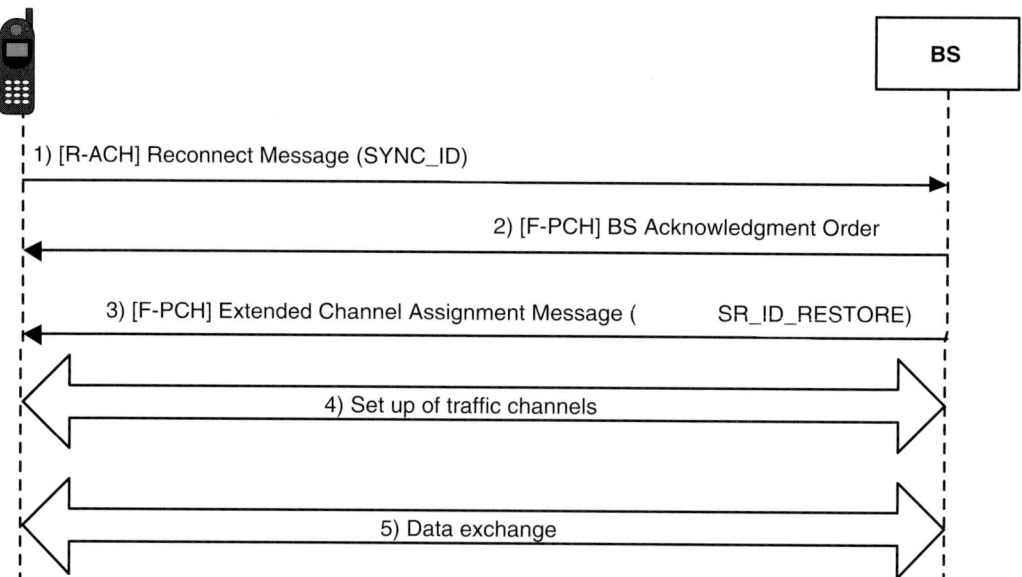

Figure 6.13 Fast call setup using SYNC_ID.

F-PCH (or F-BCCH). When the mobile station detects a change in the PACKET_ZONE_ID, the mobile station needs to register with the base station. The change in the PACKET_ZONE_ID indicates a change of the serving PCF, but not necessarily a change of the serving PDSN (see Chapter 2). If the target PCF selects a PDSN different from the one on which a PPP negotiation was last performed, the mobile station must tear down the PPP connection with the source PDSN and reestablish the PPP connection with the target PDSN.

6.4.2 High-Speed Operation

High-speed operation uses the Reverse Supplemental Channel (R-SCH), the Forward Supplemental Channel (F-SCH), and/or the F-PDCH to increase transmission bandwidth available to a connected SO33 service. In the sections that follow, we examine the Layer 3 and packet data protocol mechanisms for R/F-SCH and F-PDCH operation. Note that here we focus on the protocol aspects only, as aspects of resource management for packet data operation, such as scheduling and power control, are detailed in Chapter 9.

6.4.2.1 Reverse Supplemental Channel

This section describes protocol procedures to set up, reconfigure, and release the F-SCH. References are made throughout this section to the message parameters listed in Table 6.1.

Table 6.1 R-SCH Assignment Parameters*

Field	Message	Description
PREFERRED_RATE	SCRM	Mobile station requested R-SCH data rate.
DURATION	SCRM	Mobile station requested R-SCH assignment duration.
SIZE_OF_REQUEST_BLOB	SCRM	Indicates the aggregate size of the request parameters. If set to 0, it indicates that the mobile station releases the R-SCH prior to expiration of the assigned period.
NUM_REV_SCH	ESCAM	Number of R-SCH assigned by this message. If set to 0, it indicates that the current R-SCH assignment is terminated.
REV_SCH_RATE	ESCAM	Granted R-SCH data rate.
REV_SCH_START_TIME	ESCAM	R-SCH transmission start time.
REV_SCH_DURATION	ESCAM	R-SCH assignment duration. It can be finite or infinite. If finite, it varies from 20 ms to 5.12 sec.
REV_SCH_DTX_DURATION	ESCAM	Maximum period of time the mobile station is allowed to stop R-SCH transmission and then resume it.

*Not all R-SCH assignment-related fields are listed.

The mobile station may request reverse-link high-speed operation on the R-SCH by sending the *Supplemental Channel Request Message* (SCRM). Request transmission is triggered, for example, when the amount of data queued for transmission exceeds an implementation-dependent threshold. The request includes the preferred data rate and assignment duration, whose selection is implementation-dependent. The mobile station uses the SCRM also to report a change in the desired data rate, for example, when it can no longer sustain the currently assigned data rate because the transmitter has reached the maximum power level.

The base station controls reverse-link high-speed operation by sending the *Extended Supplemental Channel Assignment Message* (ESCAM).[14] Up to two R-SCH can be allocated concurrently and their bandwidth bundled together to support the same service instance if very high data rates are required, but this option is rarely used in practice. The ESCAM contains the R-SCH configuration record and the assignment information. The configuration includes the Walsh function identifier and the assigned data rate (REV_SCH_RATE). The assignment information includes the transmission start time (REV_SCH_START_TIME) and the assignment duration. The assignment duration ranges from 20 ms to 5.12 sec, but the base station also has the option to grant the R-SCH indefinitely. If the assignment duration is finite, the R-SCH is released implicitly at the assignment expiration time; otherwise the R-SCH must be explicitly released. Even before the assigned period has elapsed, R-SCH release can be initiated by the base station by sending the ESCAM with NUM_REV_SCH set to 0, or by the mobile station by sending the SCRM with the SIZE_OF_REQUEST_BLOB set to 0.

The relevance of configuration versus assignment information is that once the R-SCH is configured, it can be assigned and de-assigned repeatedly without the need of conveying configuration information at each assignment. This option allows using shorter signaling messages that carry only assignment information. The *Reverse Supplemental Channel Assignment Mini-Message* (RSCAMM), for example, contains assignment information only and can be carried over a 5-ms frame, therefore reducing assignment latency over the ESCAM that may span more than one frame.

If R-SCH discontinuous transmission is allowed by the base station within the assigned transmission period, the mobile station may autonomously stop R-SCH transmission when the transmit data queue is empty and resume it when data become available again for transmission. The base station controls the maximum duration the R-SCH can be gated off by means of the REV_SCH_DTX_DURATION field of the ESCAM. If discontinuous transmission is not allowed or the maximum allowed DTX period has elapsed, and the mobile station no longer has any data to send, the mobile station must explicitly release the R-SCH by sending the SCRM. By controlling the maximum DTX duration, the base station can then trade off resource consumption (e.g., base station hardware resources must be reserved at the base station even when the R-SCH is gated off) with signaling load and latency.

---

14.   Messages other than the ESCAM may also be used for this purpose.

An exemplary procedure of R-SCH setup and release is depicted in Figure 6.14. In this example, the mobile station has a SO33 packet data call connected. The mobile station autonomously initiates data transmission on the R-FCH when data is available for transmission. When the amount of data queued for transmission at the mobile station becomes large, the mobile station requests permission to switch to high-speed operation on the R-SCH. The request is granted, and scheduled transmission of user data takes place on the R-SCH. The mobile station releases the R-SCH when no data is available for transmission.

1. Transmission of RLP frames carrying SO33 data is initiated autonomously by the base station and mobile station on the R/F-FCH.
2. When the transmit data queue size at the mobile station exceeds an implementation-dependent threshold, the mobile station requests high-speed operation on the R-SCH by sending the SCRM. The message carries the preferred data rate—for example, 153.6 kbps (PREFERRED_RATE = 4), and assignment duration—for example, infinite duration (DURATION = 511).

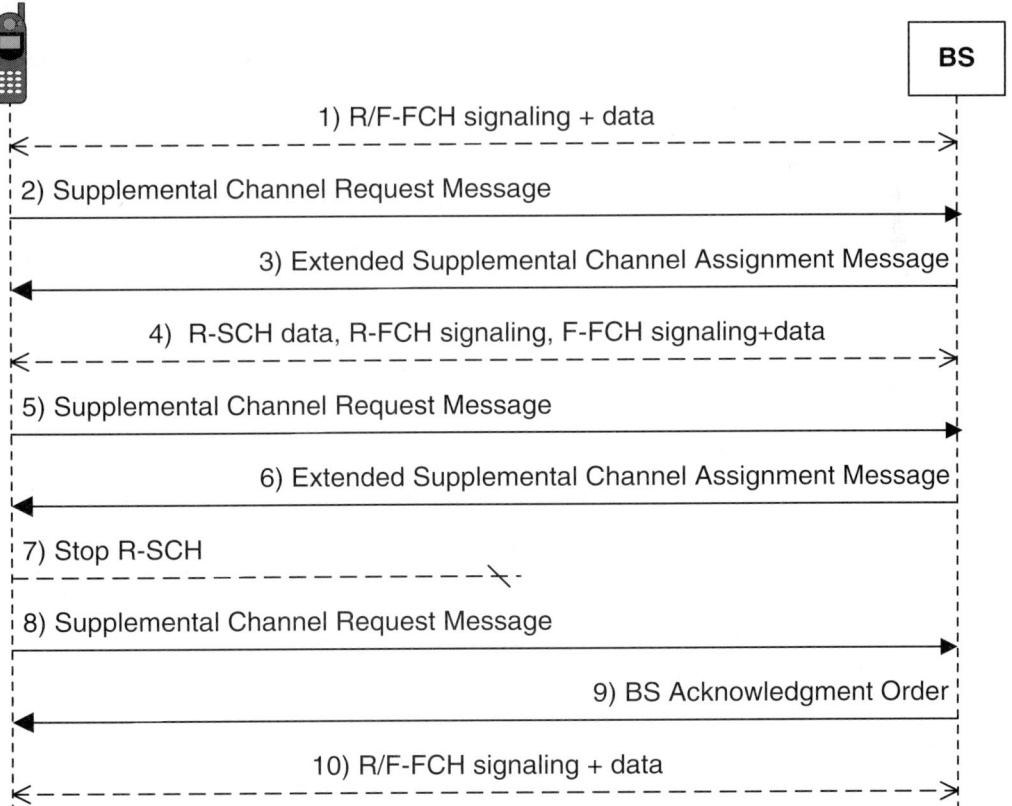

Figure 6.14 R-SCH setup and release.

3. The base station grants the request. It sends the ESCAM assigning the R-SCH for a finite period of time—for example, 5.12 second (REV_SCH_DURATION = 14). The data rate may be set to a value smaller than that requested by the mobile station—for example, 76.8 kbps (REV_SCH_RATE = 3) station if limited resources are available. The base station opts not to allow the mobile station to operate the R-SCH in discontinuous transmission for more than, say, 200 ms (REV_SCH_DTX_DURATION = 10).

4. Scheduled transmission of RLP frames takes place on the R-SCH.

5. As the amount of data queued for transmission is still above the threshold when the granted R-SCH period is about to expire, the mobile station renews its request by sending a SCRM with same parameters as those used in the previous request.

6. The base station grants the request and sends the ESCAM with same parameters as those used in the previous assignment.

7. The mobile station gates off the R-SCH when there are no data available for transmission.

8. As no data are available for transmission after 200 ms since the last time the R-SCH was gated off, the mobile station sends the SCRM indicating that the R-SCH is released (SIZE_REQ_BLOB = 0).

9. The base station acknowledges the message and releases all resources associated with the R-SCH.

10. Autonomous transmission of RLP frames continues on the R/F-FCH.

6.4.2.2   Forward Supplemental Channel

This section describes protocol procedures to set up, reconfigure, and release the F-SCH. References are made throughout this section to the message parameters listed in Table 6.2.

Table 6.2   F-SCH Assignment Parameters*

Field	Message	Description
FOR_SCH_DURATION	ESCAM	F-SCH assignment duration.
FOR_SCH_START_TIME	ESCAM	F-SCH processing start time.
FOR_SCH_RATE	ESCAM	F-SCH data rate.
SCCL_INDEX	ESCAM	Index of the record in the F-SCH code list corresponding to this assignment.

*Not all F-SCH assignment-related fields are listed

The base station may invoke forward-link high-speed operation on the F-SCH when, for example, the amount of data queued for transmission exceeds an implementation-dependent threshold. The base station controls forward-link high-speed operation by sending the

ESCAM.[15] The ESCAM contains the F-SCH configuration record and the assignment information, just as for the R-SCH, but with slightly different parameters. The configuration includes the assigned data rate (FOR_SCH_RATE) and the Walsh function identifier (FOR_SCH_CC_INDEX). As the F-SCH active set size may be different than that of the F-FCH, the configuration information also lists the PILOT_PN of the base stations in soft handoff carrying the F-SCH. The assignment information includes the transmission start time (FOR_SCH_START_TIME) and the assignment duration.

As for the R-SCH, a configured F-SCH can also be assigned and de-assigned using a 5-ms mini-messages carrying only assignment information, the *Forward Supplemental Channel Assignment Mini-Message* (FSCAMM). Using the FSAMM instead of ESCAM allows for much faster and efficient scheduling of the F-SCH.

An exemplary procedure of F-SCH setup and release is depicted in Figure 6.15. In this example, the mobile station has an SO33 packet data call connected. Data exchange takes place on the F/R-FCH. When the amount of data queued for transmission at the base station (or PCF) becomes large, the base station switches to high-speed operation on the F-SCH. The base station releases the F-SCH when no data is available for transmission.

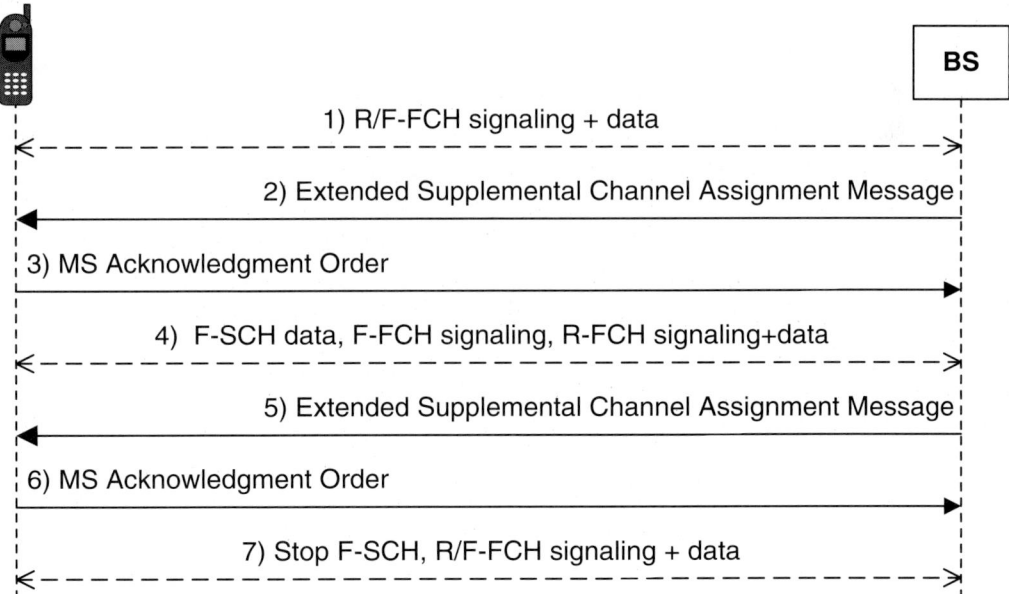

Figure 6.15 F-SCH setup and release.

---

15.   Messages other than the ESCAM may also be used for this purpose.

1. Transmission of RLP frames carrying SO33 data is initiated autonomously by the base station and mobile station on the R/F-FCH.

2. When the transmit data queue size at the base station exceeds an implementation-dependent threshold, the base station commands high-speed operation on the F-SCH by sending the ESCAM. The F-SCH is assigned for a finite period of time—for example, 1.92 second (FOR_SCH_DURATION = 12). The data rate is set to, for example, 153.6 kbps (REV_SCH_RATE = 4).

3. The mobile station acknowledges receipt of the ESCAM.

4. Scheduled transmission of RLP frames takes place on the F-SCH.

5. As the amount of data queued for transmission is still above the threshold when the assigned F-SCH period is about to expire, the base station renews the assignment by sending the ESCAM with same duration and data rate as those used in the previous request.

6. The mobile station acknowledges receipt of the ESCAM.

7. As no or little data is available for transmission at the time of expiration of the assigned period, the base station stops transmitting the F-SCH. The mobile station stops receiving the F-SCH. Transmission of RLP frames continues at a lower data rate on the F/R-FCH.

6.4.2.3   *Forward Packet Data Channel*

The call setup for F-PDCH without a forward dedicated channel is similar to call setup with F-FCH. The same messages are involved, and the sequence of events is the same. However, because dedicated channels are not involved, the channel acquisition procedure is different. Figure 6.16 shows a basic call setup for F-PDCH operation without a dedicated forward channel.

The base station broadcast over the F-PCH the *Extended System Parameters Message* (ESPM) that indicates whether F-PDCH is supported and supplies the WALSH_TABLE_ID and the Walsh codes for Forward Packet Data Control Channel (F-PDCCH).[16] Alternately, the same information can be obtained over F-BCCH through the *MC-RR Parameters Message*. The mobile station then registers with the system over the R-ACH or R-EACH using RGM. In this message the mobile station indicates whether or not F-PDCH is supported. Provided it is supported, the F-PDCH can be set up. The setup procedure is as follows.

1. The mobile station sends a PRM responding to a page or simply sends an ORM if it originates the call. Both of these messages are sent over either R-ACH or R-EACH and indicate whether radio configuration with F-PDCH is supported, and if F-PDCH is sup-

---

16.   Up to two F-PDCCHs are allowed.

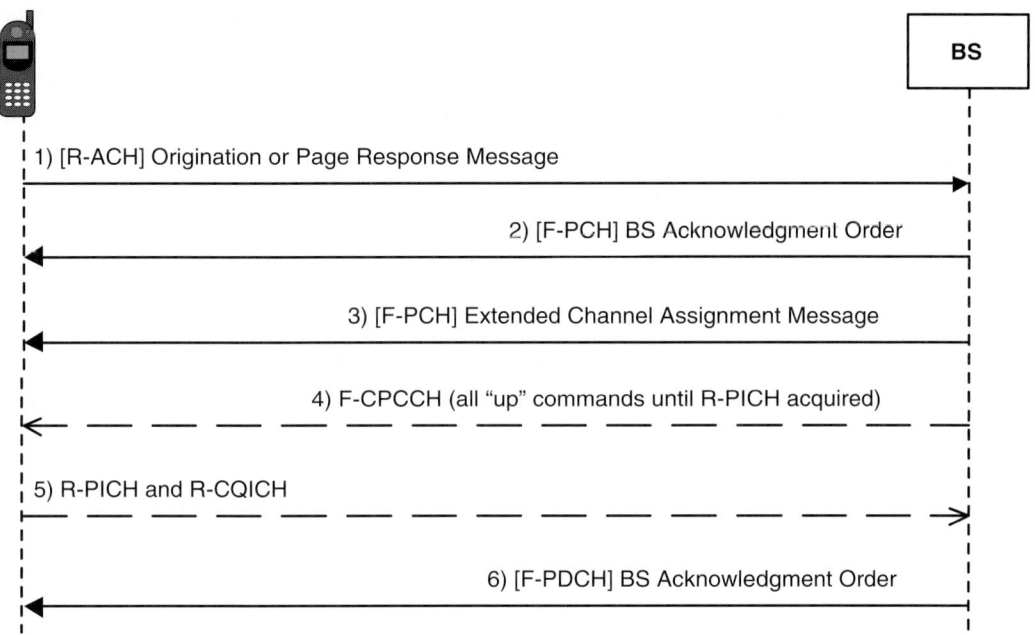

Figure 6.16 F-PDCH setup.

ported, how many parallel ARQ channels are allowed and what the value for ACK_DELAY is.

2. *The base station then responds with the Base Station Acknowledgment ORDM, which* acknowledges the reception of the PRM or ORM.

3. The base station sends the ECAM over the F-PCH or F-CCCH. The ECAM indicates the F-PDCH mode of operation, for example, F-PDCH+F-CPCCH+R-FCH, mobile station MAC_ID per pilot, REV_CQICH_COVER or Walsh code for the Reverse Channel Quality Indicator Channel (R-CQICH) per pilot, full or differential C/I feedback with associated parameters, such as frame offset, number of slots for switching, Walsh codes for F-PDCCH and F-CPCCH, and so on.

4. After sending the ECAM the base station start sending all up power control commands over F-CPCCH subchannel associated with the mobile station.

5. Upon receiving ECAM, the mobile station starts monitoring F-CPCCH. The mobile starts transmitting on R-PICH and R-CQICH upon acquisition of the F-CPCCH.

6. When the base station acquires R-PICH, it sends the *Base Station Acknowledgment ORDM* over F-PDCH and from now one the mobile station is on the traffic channel. The service negotiation proceeds over the traffic channels.

The release of F-PDCH is quite simple, and as shown in Figure 6.17, it is not different from the release procedure for dedicated channels. For example, the base station can send an Extended Release Message, with EXT_CH_IND indicating release of F-PDCH and associated channels. The mobile station responds with an Extended Response Release Message confirming the release. Alternately, all physical resources can be released using a *Release* ORDM. In this case, the mobile station and the base station exchange the *Release* ORDM.

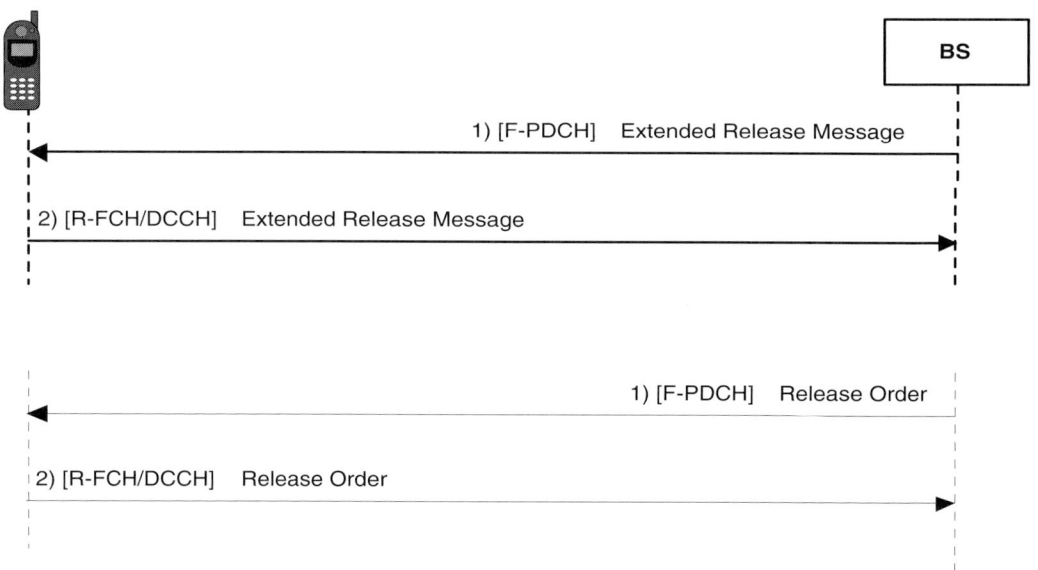

Figure 6.17 F-PDCH release.

6.4.3 Quality of Service

Packet data service provided by SO 33 allows for a set of quality of service (QoS) parameters. For each parameter, a set of values is defined. These parameters are negotiated during service negotiation, and they are included in the QOS_BLOB. The QOS_BLOB may be part of the service configuration information record when the SO field is set to 33. The QOS_BLOB does not have to be included in the service configuration information record. If the QOS_BLOB is not included, best-effort packet data service is assumed. The QoS for SO33 can be provided in two modes: nonassured mode and assured mode.

The nonassured mode is the default QoS mode. There is only one parameter that may be included, and that is user's priority, NON_ASSURED_PRI_ADJ. This parameter can take 14 different values. It is 4-bit field, and if omitted, the default or lowest priority is assumed.

In the assured mode the user's priority is defined for the forward and reverse links separately. In Table 6.3, we list the assured mode parameters and the set of values they can take. In addition to the user's priority, the assured mode defines the data rate, data loss, and delay parameter. Forward and reverse links are denoted with 'F' and 'R', respectively. The assured mode QOS_BLOB also includes required and acceptable values, denoted with REQ and ACC, respectively.

Table 6.3   Assured Mode QOS_BLOB

Parameter	Field	Set of values
Priority	F_ASSURED_PRI_ADJ	14 different priorities
	R_ASSURED_PRI_ADJ	
Data rate	F_REQ_DATARATE	8, 32, 64, 96, 144, 288, and 384 kbps
	F_ACC_DATARATE	
	R_REQ_DATARATE	
	R_ACC_DATARATE	
Data loss	F_REQ_DATALOSS	10^{-8}, 10^{-6}, 10^{-4}, 10^{-3}, 10^{-2}, $2\cdot10^{-2}$, $5\cdot10^{-2}$, and 10^{-1}
	F_ACC_DATALOSS	
	R_REQ_DATALOSS	
	R_ACC_DATALOSS	
Delay	F_REQ_DELAY	40, 120, 360, 1000, 2000, 3000, and 5000 ms
	F_ACC_DELAY	
	R_REQ_DELAY	40, 120, 360, 1000, 2000, 3000, and 5000 ms
	R_ACC_DELAY	

After the parameters have been negotiated, the base station must provide the desired level of QoS to the mobile station. Of course, the provided QoS level is subject to feasibility due to the existing radio conditions. Let us now examine in more detail what each specified parameter means.

The priority parameter simply provides differentiated services. For example, higher priority traffic has precedence over traffic with lower priority. This is true both for the assured and nonassured mode. The assured mode has precedence over nonassured mode.

The data rate parameter explicitly defines the data rate that the mobile station and the base station are committed to. For example, the negotiated data rate may be 64 kbps. The negotiated rate refers to the received data octets excluding physical, multiplex sublayer, and RLP overhead. To satisfy the data rate requirement in the given example, the mobile station and the base station may set up SCH at 76.8 kbps.

The data loss rate refers to the ratio of erroneously received octets to total number of transmitted octets as measured above RLP. The low loss rates typically cannot be achieved at the physical layer, and RLP retransmissions are necessary. As we discussed in Chapter 5, the retransmissions reduce the residual data loss rate at the expense of delay.

The delay is another parameter specified by the QOS_BLOB. It represents the maximum acceptable air interface delay. If the data cannot be delivered within the specified time interval, it should be discarded.

6.4.4 Data Bursts Transmission on Common Channels

Depending on availability of dedicated traffic channel resource or on the QoS associated with a dormant packet data call, the base station may choose to exchange SO33 data traffic on the common channels. This option is also called *short burst teleservice*, as user data is sent encapsulated in a DBM with data burst type set to Short Data Burst. The message is sent in assured mode on the F-PCH or F-CCCH. The message header contains the SR_ID that allows the base station to route the message payload to the corresponding service instance.

This service allows transmission of only one burst at time, as a burst must be acknowledged before a new one can be sent. Moreover, the data rate is limited to that offered by the F-PCH or F-CCCH. Hence, this service is well suited for applications that are delay–tolerant and that require relatively low bandwidth.

6.5 Concurrent Services

As wireless terminals are used for an increasing variety of services and applications, the capability to support multiple services simultaneously has become increasingly desirable. Such capability, first introduced in Revision A of IS-2000, is commonly called *concurrent services* (CCS) support. The CCS support includes all the Layer 3 functionality necessary to establish, maintain, and release multiple, independent, and simultaneous services. The introduction of CCS motivates the split between the state machine governing resource management, mobility management, and other lower level functionalities of the Layer 3 protocol and the state machine governing call control that we described in Section 6.2.5. A simplified view of the interaction between Layer 3 and call control is depicted in Figure 6.18, where it is assumed that three simultaneous services, voice, packet, and circuit-switched data are connected.

Layer 3 is responsible for instantiation and termination of calls. Each call is assigned a unique identifier, the CON_REF, by the base station. The identifier is attached to all call control-related signaling messages so that Layer 3 can route them to the intended call instance. Both mobile station and base station can initiate a new call while calls are already in progress. Once the call is instantiated, it must go through service negotiation before the service can be connected, just like in the case when CCS is not supported. Calls can be released individually, or they can be released all at once together with the associated physical resources.

With CCS, the logical-to-physical channel mapping, or radio bearer profile, is more elaborate than in the case of a single call. Various choices of radio bearer profiles exist, consisting of

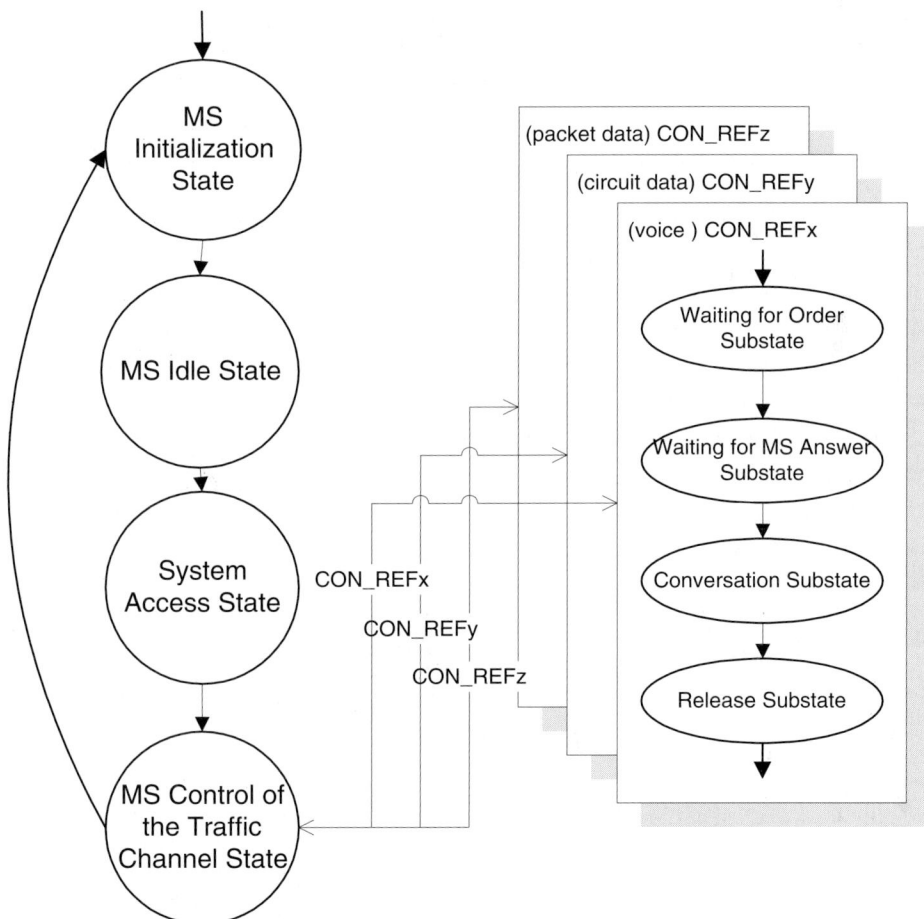

Figure 6.18 Layer 3 and call-control state machines relationship with CCS.

different combinations of F/R-FCH and/or F/R-DCCH, and/or F/R-SCH, and/or F-PDCH used to transport signaling and user data. Consider the case of simultaneous packet data and voice calls, for example. Obviously, the data service requires establishment of a traffic channel to carry user data in addition to the F/R-FCH used to carry voice, as the real-time voice service demands the full bandwidth provided by the F/R-FCH. On the other hand, signaling traffic for both voice and data calls can be multiplexed onto the F/R-FCH, but then voice quality would suffer due to the high frequency of voice frames blanked by the signaling messages necessary operate the F/R-SCH in bursts. A better approach is to also set up the F/R-DCCH, which would carry signaling traffic for both the voice and packet data call. If the F/R-DCCH pair is used, then it can carry packet data, thus making the F/R-SCH needed only for high-speed bursts.

The CCS procedures differ depending on whether the service requested is voice or data, whether the request is base station- or mobile station-originated, and whether the request is to set up or release a call for such service. The CCS procedures make use of the following messages. The *Enhanced Origination Message* (EOM) is used by the mobile station to request addition of a new call. The *Call Assignment Message* (CLAM) is used by the base station to assign a new call, either in response to an EOM or in case of a mobile station-terminated call. The SCM is used to connect the new service. Note that the SCM may also be used in lieu of the CLAM if the call is assigned as part of the service option connection establishment. If the new call requires change in the radio bearer profile and setup of additional physical channels, the base station may use the *Universal Handoff Direction Message* (UHDM) in lieu of both the CLAM and the SCM. In this case, the UHDM carries call-assignment information, the service configuration record for the new call, and the configuration of the new physical channel(s) to be set up. The *Service Request Message* (SRQM) is used by the mobile station to initiate the release of a call, either voice or data. In the response to the SRQM, the base station may send the SCM or the UHDM. If a change in the radio bearer profile and release of physical channels is not required, then the SCM is used; otherwise the UHDM is used. In case all calls are to be released at once, including associated physical channels, the *Release* ORDM is used just as discussed in Section 6.3.3. In the following sections we describe call flows related only to the most common CCS procedures.

6.5.1 Mobile Station–Terminated Voice Call While Data Call is in Progress

In the example illustrated in Figure 6.19, the mobile station is in a packet data call when it receives a voice call. The base station decides to reconfigure the radio bearer profile by establishing the F/R-DCCH at the same time the new call is connected.

1. An SO33 packet data call is in progress using the F-FSCH for high-speed data bursts and the F/R-DCCH for both signaling and user data. When an incoming voice call is terminated at the mobile station, the base station sends the UHDM. The UHDM con-

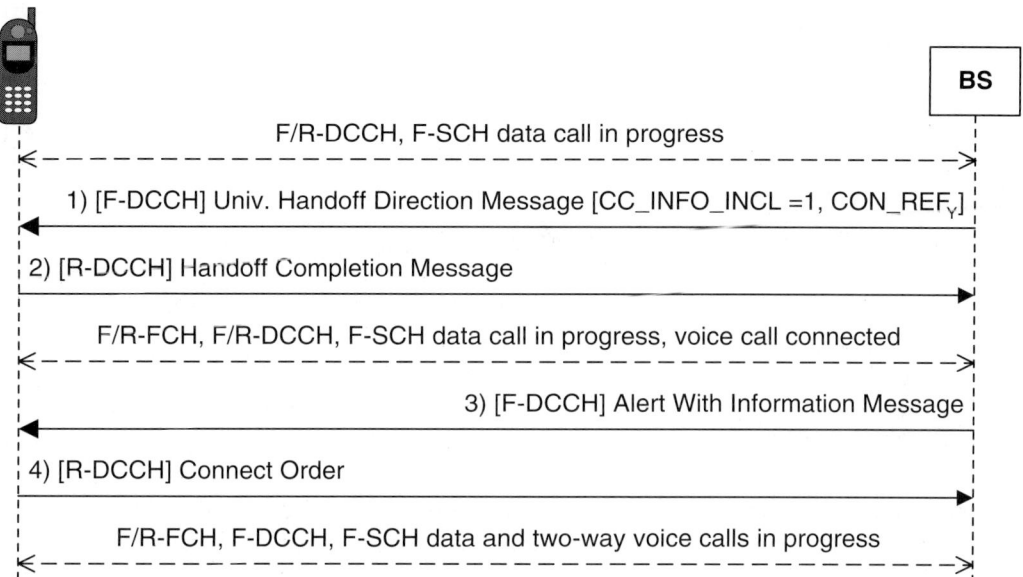

Figure 6.19 Mobile station-terminated voice call while data call is in progress.

tains the CC_INFO_INCL set to 1 to indicate that call assignment information is included. The UHDM also contains the service configuration record for the new voice call. The UHDM carries the CON_REF corresponding to the voice call, which is set to the same value as that used in the service configuration record. The UHDM also directs the mobile station to set up the F/R-FCH.

2. At the action time of the UHDM, the mobile station sends the *Handoff Completion Message* (HOCM). At this time, voice data is carried by the F/R-FCH, user packet data is carried by the F/R-DCCH and F-SCH, and signaling for both voice and packet data calls is carried by the F/R-DCCH. Note that although the voice service is connected from the signaling layer point of view, two-way voice path is not yet established, as the user has not been alerted yet.

3. The base station sends the AWIM. On receipt of the message, the mobile station alerts the user of the incoming call.

4. Once the user answers the call, the mobile station sends the *Connect* ORDM. Two-way voice and packet data calls are now in progress.

6.5.2 Mobile Station–Originated Data Call While Voice Call Is in Progress

In the example illustrated in Figure 6.20, the mobile station is in a voice call when data become ready to be sent, and the mobile station sends a request to set up a packet data call. It is assumed that at the time the data call is originated, the radio bearer profile includes both F/R-FCH and F/R-DCCH and, therefore, no new physical channel is required to connect the new call. The F-SCH is established after packet data call setup only when needed to send a high-speed data burst.

1. A voice call is in progress using the F/R-FCH for user data and the F/R-DCCH for signaling. When data becomes ready for transmission, the mobile station sends the EOM requesting the base station to set up an SO33 packet data call. The EOM contains the TAG identifier for this transaction. In the EOM, the DRS bit is set to 1 to indicate that data is ready to be sent.
2. In response to the EOM, the base station sends the CLAM. The CLAM contains the same TAG used by the mobile station in the EOM. The ACCEPT_IND field is set to 1 to indicate that the call request is granted. The CLAM contains the unique CON_REF identifier for this call instance.

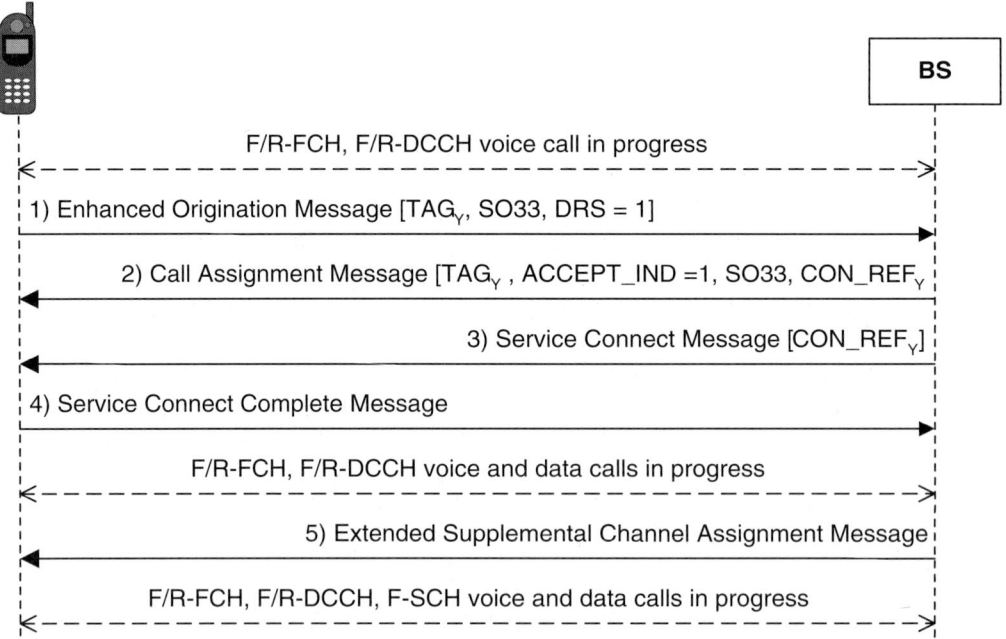

Figure 6.20 Mobile station–originated data call while voice call is in progress.

3. The base station sends the SCM that contains the service configuration record for the CON_REF corresponding to the packet data call.

4. The mobile station sends the SCCM in response to the SCM. Both voice and data calls are now in progress. The F/R-FCH is used solely to transport voice data, while the F/R-DCCH transports user packet data and signaling for both the voice and the packet data calls.

5. The base station decides to send a data burst at high speed using the F-SCH, for example, because the amount of data received from the packet core network exceeds the bandwidth available on the F-DCCH. The base station sends the ESCAM to set up the F-SCH. At the action time of the ESCAM, user packet data is carried by the F/R-DCCH and F-SCH, voice data is carried by the F/R-FCH, and signaling for both voice and packet data calls is carried by the F/R-DCCH.

6.5.3 Mobile Station Voice Call Release While Voice and Data Calls Are in Progress

In the example illustrated in Figure 6.21, the mobile station initiates release of the voice call while a voice call and a packet data call are in progress.

1. A voice call and a data call are in progress. The voice data is carried by the F/R-FCH, user packet data is carried by the F/R-DCCH and F-SCH, and signaling for both voice and packet data calls is carried by the F/R-DCCH. The mobile station initiated release of the voice call by sending the SRQM. The service option connection record corre-

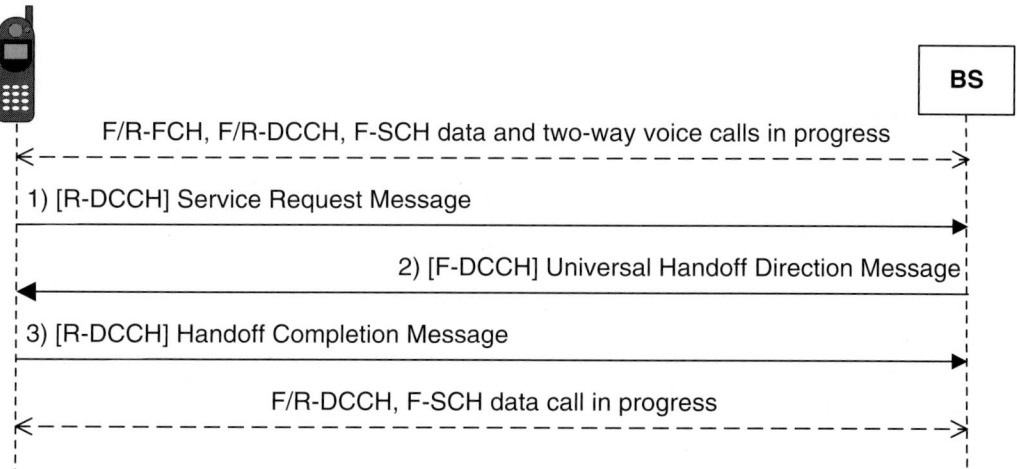

Figure 6.21 Mobile station voice call release while voice and data calls are in progress.

sponding to the voice call is omitted from the service configuration record included in this message.

2. On receipt of the SRQM, the base station sends the UHDM with a service configuration. The service configuration record does not include the service option connection record corresponding to the voice call. The UHDM also directs the mobile station to release the F/R-FCH.

3. At the action time of the UHDM, the mobile station releases the voice call, releases the F/R-FCH, and sends the HCM. At this time the packet data call continues on the F/R-DCCH, used to carry both signaling and user data, and the F-SCH, used to carry high-speed data bursts.

6.6 Registration

Registration is the process by which the mobile station notifies the base station of its identity, location, status, slot cycle, and capabilities. Mobile station identity and location is needed so that the access network can efficiently page the mobile station using *multicell paging* rather than *flood paging*. Efficient multicell paging means that the page message is sent only from that (small) set of base stations that are likely to be serving the mobile station, whereas with flood paging, all base stations in the network send the same page message, thereby wasting paging channel capacity. Slot cycle information is needed because when the mobile station operates in slotted mode, the base station needs to determine which slot the mobile station is monitoring. The mobile station class mark and protocol revision number represent mobile station capabilities and are also supplied to the base station with the RGM. There are 11 forms of registration, each corresponding to a different triggering condition and/or procedure. Of the 11, we focus on the *autonomous registration* procedures that are enabled by the mobile station roaming status and can be optionally enabled by the access network. They are called autonomous because once enabled, they are triggered by events other than explicit end-user or access network intervention.

Regardless of the type, autonomous registration procedures involve the same signaling messages exchange over the air interface, as explained in the following with reference to Figure 6.22.

1. When registration is triggered from the idle state, the mobile station sends the RGM to the base station on the access channel. The message includes the REG_TYPE field determining the type of registration (power-down, power-up, and so on). The message also includes information about mobile station current configuration (e.g., SLOT_CYCLE_INDEX), protocol revision, capability (e.g., QPCH and transmit diversity support), and security (e.g., encryption and message integrity).

2. On receipt of the RGM, the base station sends the *Base Station Acknowledgment ORDM*. On receipt of the order, the mobile station returns to the idle state. The base station forwards the registration message to the MSC that initiates procedures according to its type.

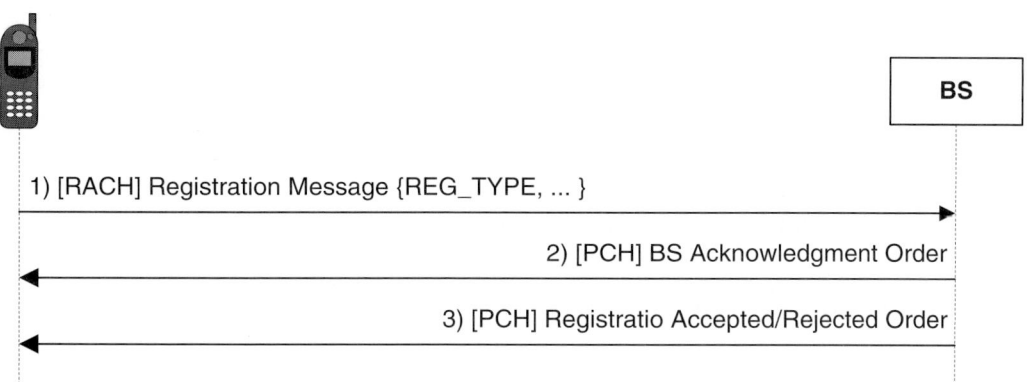

Figure 6.22 Autonomous registration procedure example.

3. If the registration is accepted, the base station sends the *Registration Accepted* ORDM to the mobile station. Otherwise, the base station sends the *Registration Rejected* ORDM. The registration may be rejected, for example, in the case of authentication failure or if the base station decides that the mobile station, possibly a roamer, cannot receive normal service on this system. In case of rejection, the mobile station transitions to the initialization state and selects another system.

Although in this section we focus on radio interface only, it is important to note that registration messages are propagated throughout the core network to the home location register (HLR) that is ultimately the repository of mobile station's status and location information. We address the impact of registration procedures on the core network side in Chapter 11.

6.6.1 Power-Up Registration

Power-up registration is performed when the mobile station is turned on, thus triggering RGM transmission. At the network side, receipt of the RGM has the effect of toggling the mobile station status to active and enables call delivery as well as specialized call treatment. To prevent multiple registrations when the mobile station is power-cycled repeatedly in a short period of time, RGM transmission is inhibited for a period of 20 second after entering the idle state.

6.6.2 Power-Down Registration

Power-down registration is performed when the mobile station is switched off by the user. If this registration type is performed, the mobile station does not actually power down until completion of the procedure, even though that may go unnoticed to the end user, who may see, for example, the terminal's screen turning blank immediately after power down. Power-down registration, if enabled, is performed only if the mobile station previously registered in the serving system.

On the network side, power-down registration has the effect of toggling the mobile station status to inactive and enables specialized call treatment. Consider, for example, a land subscriber attempting to call a mobile station that has deregistered from the system by means of power-down registration. The core network will not even attempt call delivery, but will immediately apply a reorder tone in the audio path of the land user or play an announcement indicating that the cellular subscriber has turned off his or her terminal. That has a positive effect on the radio access network in that it avoids the waste of resources involved with paging inactive terminals.

6.6.3 Timer-Based Registration

Timer-based registration causes the mobile station to register at regular intervals even if its location or status has not changed from the time of the previous registration. This form of registration is complementary to power-down registration in that it allows the network to automatically deregister a mobile station that did not successfully perform power-down registration. The network may maintain an inactivity timer that is reset each time mobile station activity, such as an origination attempt or a registration attempt, is detected. A mobile station is deregistered when the timer expires. Call delivery is then disabled, and the same call treatment is applied to calling parties as in the case of a mobile station deregistered because of power-down registration. The network inactivity timer should be set to a value two or three times larger than the registration period to avoid deregistering a subscriber that only momentarily was in an area out of coverage and missed its chance to register.

The registration period is in units of 80-ms paging channel slots. The mobile station maintains a slot counter. Registration is triggered when the counter reaches a maximum value (REG_COUNT_MAX). The maximum counter value is obtained from the 7-bit REG_PRD field of the *System Parameters Message* (SPM) as REG_COUNT_MAX = $\lfloor 2^{REG\_PRD/4} \rfloor$.

REG_PRD can be set to zero or from 29 to 85 inclusive, the latter range corresponding to a registration period from 12.16 second to 55 hours. Setting REG_PRD to zero disables timer-base registration. A typical setting of REG_PRD is equal to 66, corresponding to a registration period of ~2 hours. A smaller registration period would cause unnecessarily large signaling traffic, consuming valuable resources on the access and paging channels as well as on the IS-41 network.

Registration periods of different mobile stations are randomly staggered, since mobile stations initialize their counters to a random value uniformly distributed between 0 and REG_COUNT_MAX – 1 (or else mobile stations will flood the base station with simultaneous registration attempts at the end of the registration period). The counter is reset with every successful registration of any type.

6.6.4 Zone-Based Registration

Zone-based registration causes the mobile station to register when it moves into a new zone, that is, a zone that has not been recently visited. The REG_ZONE field of the SPM identifies a base station assignment. A zone is uniquely identified by the REG_ZONE and the network it belongs to, or SID/NID pair. A mobile station maintains a list of zones it recently visited and registered in, and an age timer for each zone in the list. The timer is enabled for each zone except the last one the mobile station registered in. The zone is removed from the visited zone list when its timer expires. Timer duration is determined by the ZONE_TIMER field of the SPM broadcast by the base station on which the mobile station registered most recently. The size of the visited zone list is bounded by the TOTAL_ZONES parameter of the SPM. If a new zone is to be added to the list when the list is already full, then the zone corresponding to the timer that has aged the most is removed from the list to make room for the new zone. Given the zone configuration and maintenance rules above, zone-based registration procedures simply require an idle-state mobile station to register if the serving base station belongs to a zone not included in the visited zone list.

From a network administration and maintenance perspective, zone-based registration has the desirable property of being simple to configure, as it requires a one-time mapping of registration/paging areas to base stations (other schemes do not have this property, as we shall see shortly). However, it also suffers one major drawback: *registration hot spots*. The signaling load caused by registration traffic on the access and paging channels is concentrated on the cells along the registration zone boundary.

A second (potential) drawback is registration ping-pong effect, because of which a mobile station moving along the edge of a zone boundary may frequently cross back and forth two or more zones. The ping-pong effect can be mitigated by letting the mobile station register in multiple zones (TOTAL_ZONES set to 2 or more). The degree to which the ping-pong effect is mitigated depends on the setting of the ZONE_TIMER: The larger the timer setting, the less frequent the registration attempts for a mobile station located at the zone boundary. But this solution comes with a price, since a mobile station registered in multiple zones will have to be paged on all zones, thereby increasing the paging traffic and detracting from the advantage of registration.[17] Use of the ZONE_TIMER requires the network to maintain a paging area list—the mirror image of the mobile station registration zone list—for each registered subscriber. The paging area list is updated with time as the age timer associated to an area expires, mimicking the procedure implemented at the mobile station. An alternative, and simpler, method is for the base station to maintain only the last registered zone in the subscriber paging area list. If the time since the last registration is less than the zone timer, then the paging area comprises the cells in the registration zone as well as the border cells of the adjacent zones; else, the paging area and the registration zone are one and the same. This method also increases the paging load, as it increases the average paging area size.

An exemplary sequence of events related to zone-based registration is depicted in Figure 6.23. The example shows that although most of the registration attempts are triggered at the zone

---

17.   The ZONE_TIMER should be set to a small value relative to the mean subscriber velocity-to-cell radius ratio. The ratio determines the registration rate (see Section 6.6.6). As a rule of the thumb, the average number of zones in the mobile station's zone list is approximately 1 plus the product of registration rate times the timer duration.

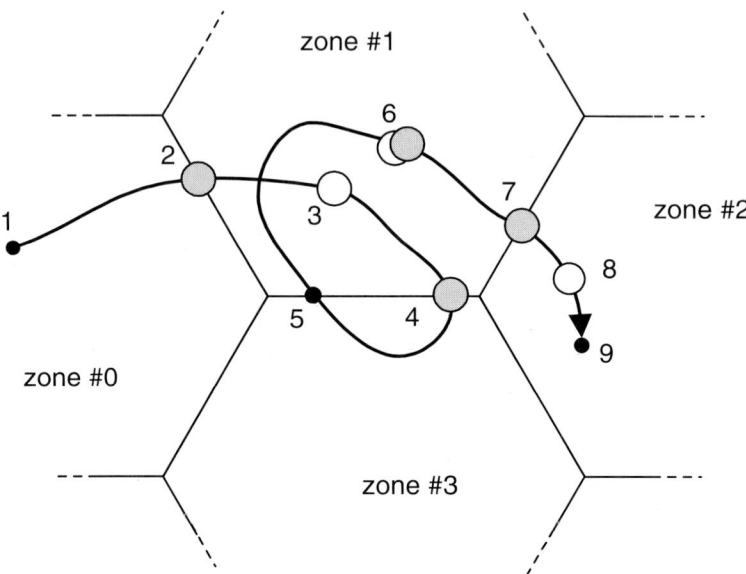

Figure 6.23 Exemplary zone-based registration procedures, from power up (step 1) to call delivery (step 9). Shaded dots represent locations where registration is triggered. Blank dots represent zone timer expiration.

boundaries, some attempts may also be triggered within a zone due to the effect of the zone timer expiration. The sequence of events is described in the following:

1. The mobile station powers up and registers with zone #0.
2. The mobile station enters zone #1 and registers. The zone list now includes zones #0 and #1. Zone #0 timer is started.
3. Zone #0 timer expires and zone 0 is delisted. The zone list includes zone #1 only.
4. The mobile station crosses the boundary with zone #3 and registers. The zone list includes zones #1 and #3. Zone #1 timer is started.
5. The mobile station reenters zone #1. However, the mobile station does not reregister, since zone #1 is already in the zone list.
6. The zone timer for zone #1 expires, and zone #1 is delisted. However, since the mobile station is currently in zone #1, it reregisters with zone #1. The zone list includes zone #1 and #3. The zone timer for zone #3 is started.
7. The mobile station enters zone #2 and registers. The zone list includes zones #1, #2, and #3. Zone #1 timer is started.
8. Zone #3 timer expires, and zone #3 is delisted. The zone list includes zones #1 and #2.
9. The mobile station receives a call and is paged on all base stations in zones #1 and #2.

6.6.5 Distance-Based Registration

Distance-based registration causes a mobile station to register when the distance between the current serving base station and the base station it last registered exceeds a threshold. The distance is computed by means of the base station latitude and longitude information broadcasted on the SPM of the F-PCH (BASE_LAT and BASE_LONG, in seconds/4). When it registers, the mobile station stores base station latitude and longitude information (BASE_LAT_REG and BASE_LONG_REG, in degrees/14400) and the distance threshold (REG_DIST_REG). After each idle handoff to a new base station, it reads the base station location information and computes the distance as

$$DISTANCE = \left\lfloor \frac{\sqrt{(\Delta lat)^2 - (\Delta long)^2}}{16} \right\rfloor \qquad (6.3)$$

where

$$\Delta lat = BASE\_LAT - BASE\_LAT\_REG$$

$$\Delta long = \left(BASE\_LONG - BASE\_LONG\_REG\right)\cos\left(\frac{\pi}{180}\frac{BASE\_LAT\_REG}{14400}\right) \qquad (6.4)$$

If DISTANCE exceeds the threshold REG_DIST_REG, the mobile station initiates registration procedures. Note that assuming a flat-earth model, the computed distance is in units of $\pi/(180 \times 3,600) \times R$ [Km], where R is the earth radius.

When distance-based registration is used, the access network pages the mobile station on all and only the base stations within registration distance from the base station successfully registered by the last access. Therefore, there may be as many paging areas as there are base stations, requiring the access network to maintain a large database. Despite the relatively complex configuration requirements imposed on the access network, distance-based registration is particularly attractive because it spreads the registration load uniformly across all base stations in the network. This property distinguishes it from zone-based registration.

6.6.6 Excess Signaling Due to Subscriber Location Uncertainty

Because of user mobility, call delivery requires paging the mobile station from multiple base stations. The set of such base stations is called the paging area. The paging area typically corresponds to a registration area. The mapping between base stations and paging/registration areas depends on the registration scheme in use, distance-based or zone-based. In this section we are interested in estimating the paging channel load caused by registration and paging messages, and determining the optimum registration area size that minimizes such load.

The user mobility model is based on the uniform traffic model. This model assumes that the subscriber density, ρ, is uniform over the entire network area and that the direction along which any given mobile station is moving is uniformly distributed between 0 and 2π. The model predicts that the expected rate at which mobile stations with mean velocity v cross any curve (in either direction) of length L is given by [6].

$$\text{crossing rate} = \rho \frac{\overline{v}\, L}{\pi} \tag{6.5}$$

We can use this model for an idealized cellular network in which base station transceivers (BTS) comprise three cells each and are laid out on a hexagonal grid (see also Chapter 10). We then consider a registration area as that consisting of the first k concentric rings of cells centered on the reference cell. This network model is suited to study both distance-based and zone-based registration. In the case of distance-based registration, we consider distances that cause a registration to take place when a mobile station travels from the center cell and crosses the k-th ring boundary, at which time we "redraw" the rings around the new reference cell for determining the next registration.

It can be seen that the number of cells in a zone consisting of k rings is equal to $3 + 9k(k - 1)$. Accounting for the fact that a cell at the zone boundary contributes with either 2 or 3 edges to the area perimeter, L, it can be seen that the perimeter is related to the number of rings as in $L = 6[3 + 2(k - 2)]r$. The cell area is $\sqrt{3}r^2/2$. If N is the average number of (idle) subscribers per cell, the subscriber density is $2N/(\sqrt{3}r^2)$. From Eq. (6.5) we obtain the registration rate per cell per unit of time as

$$\lambda_{reg} = N \frac{\overline{v}}{\pi r} \frac{2\sqrt{3}\left[3 + 2(k-2)\right]}{1 + 3k(k-1)} \tag{6.6}$$

If v is expressed in Km/h and r in Km, then λ_{reg} is the registration rate per cell per hour. We want to assess the signaling load offered to the paging channel and caused by registration attempts. We define the excess load due to registration, λ_{reg}^*, as the combined signaling rate of messages required per registration attempt. Each successful registration attempt requires transmission of a Layer 2 acknowledgment (carried by the *Base Station Acknowledgment* ORDM) and a *Registration Accepted* ORDM. Then, we simply have $\lambda_{reg}^* = 2\lambda_{reg}$. Note that with zone-based registration, the registration load is concentrated on the perimeter cells of each registration area. For distance-based registration, the registration load is instead uniformly spread over all cells in the network.

The paging area is equal to the registration area. Indicate with λ_{call} the rate of arrivals of *call-termination* attempts per subscriber per hour. Each successful attempt requires transmission of one GPM in each cell in the paging area, one *Base Station Acknowledgment* ORDM, and one

ECAM (only in the cell where the PRM is received). Then, the signaling message rate per cell per hour due to call-termination attempts is

$$\lambda_{page} = N\lambda_{call}\left[3+9k\left(k-1\right)\right]+2N\lambda_{call} \tag{6.7}$$

Ideally, if we knew the exact location of the subscriber, the page message would be transmitted from one cell only. When the paging area comprises multiple base stations, all pages but one are to be considered excess paging due to subscriber location uncertainty. The excess paging rate per cell per hour is then

$$\lambda_{page}^{*} = N\lambda_{call}\left[2+9k\left(k-1\right)\right] \tag{6.8}$$

Let us pause for some observations. The registration rate decreases nearly as $1/k$ and is proportional to $\bar{v}/r$ so that, especially when the subscribers are highly mobile and/or the cell size is small, one would want to increase the number of rings to decrease the registration load. However, increasing the number of rings soon becomes counterproductive, as the excess paging grows nearly as k^2. We also note that while the excess paging is proportional to both subscriber density and call-termination rate, the registration load depends only on subscriber density. Then, we may dynamically adjust the registration area size, for example, by using smaller values of k during the busy hour. Figure 6.24 shows, by numerical examples, the trade-off involved in selecting the registration area size.

From the point of view of the IS-41 core network, the excess paging is null, since only one message is sent to the base station over the A-interface for every call-delivery attempt, regardless of the paging area size. Then, the IS-41 objective is solely to maintain the registration load below a maximum tolerable level. From the point of view of the radio access network, however, we are interested in minimizing the combined excess paging and registration load. To that goal, we compute the relative signaling increase, $\Delta_{increase}$, as a function of the registration area size in rings. The relative increase is defined as the sum of the excess registration and paging load normalized by the minimum ideal load (i.e., when subscriber location is known a priori). We obtain

$$\Delta_{increase} = \frac{\lambda_{reg}^{*}+\lambda_{page}^{*}}{3N\lambda_{call}} = \underbrace{\frac{1}{\lambda_{call}}\frac{\bar{v}}{\pi r}\frac{4\left[3+2\left(k-2\right)\right]}{\sqrt{3}\left[1+3k\left(k-1\right)\right]}}_{\text{increase due to registrations}}+\underbrace{\frac{2+9k\left(k-1\right)}{3}}_{\text{increase due to excess paging}} \tag{6.9}$$

Numerical examples of the relative signaling increase are plotted in Figure 6.25, where it is apparent that for values of $\bar{v}/r$ in the range of interest, the optimum registration area size is equal to 2 rings, or 7(21) BTSs (cells).

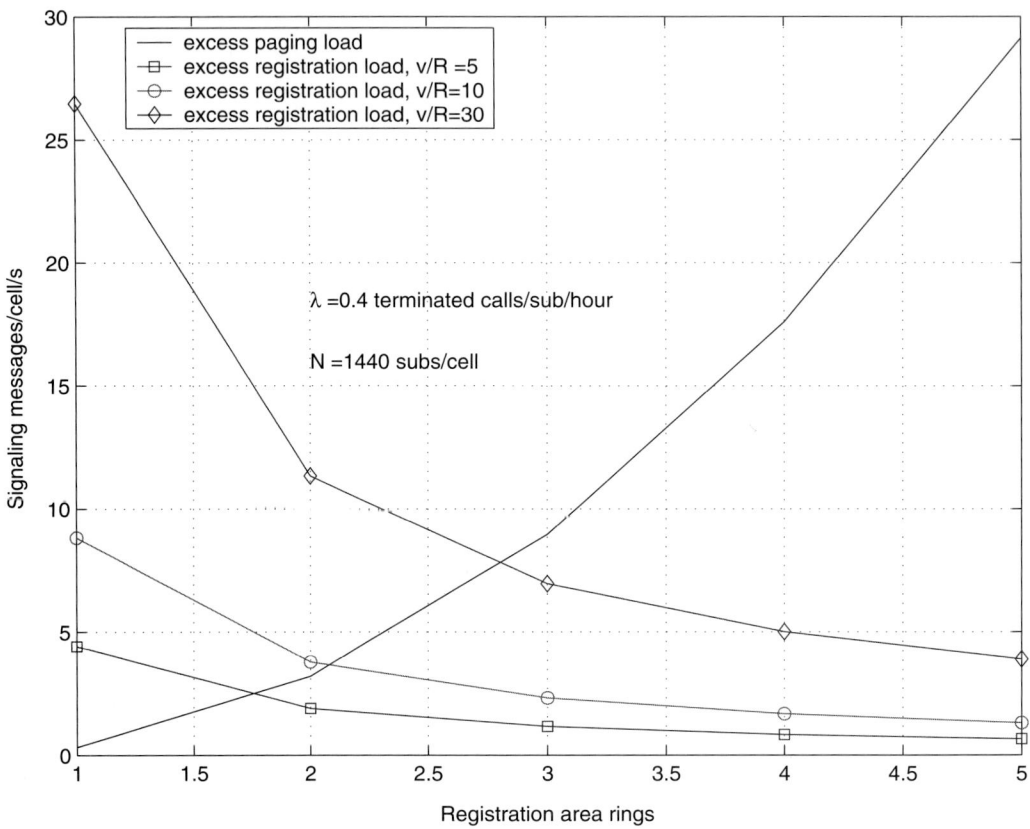

Figure 6.24 Registration and paging signaling load.

A word of caution is needed when applying the results obtained above to a real network. The uniform traffic model we used is adapted from fluid dynamics models in which the density

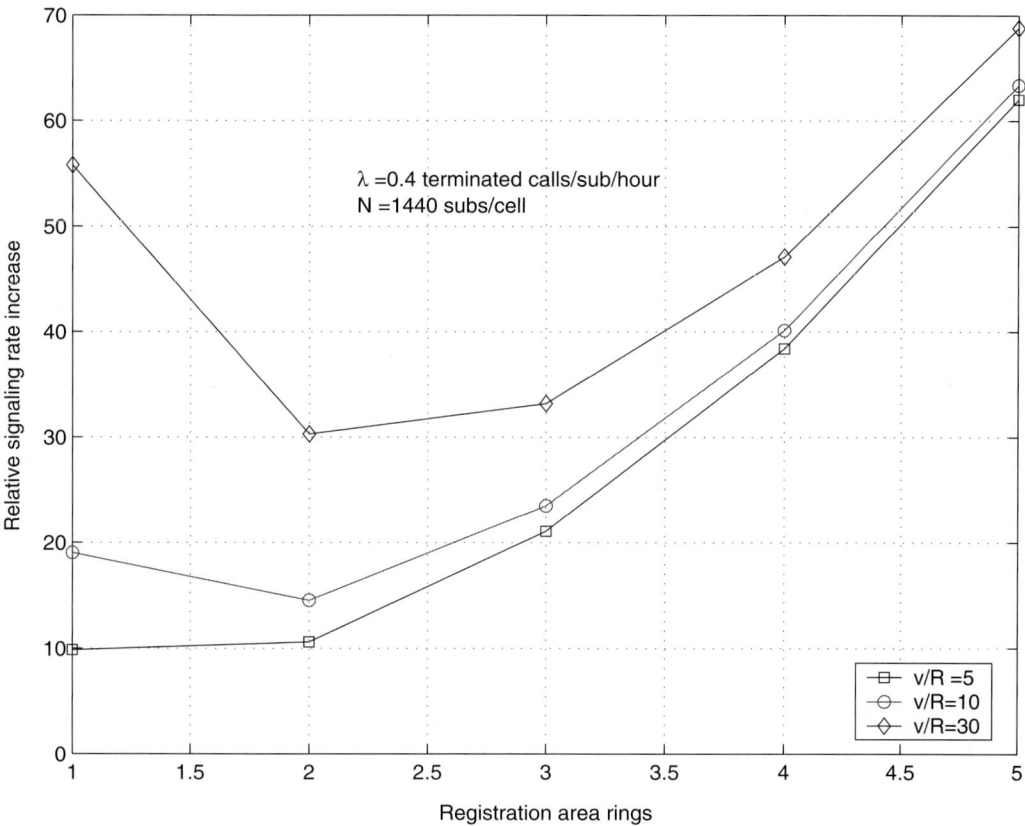

Figure 6.25 Relative signaling increase due to subscriber location uncertainty.

and velocity are roughly inversely proportional. That is not always a good model for a cellular network. Along a freeway for example, both subscriber density and velocity can be high relative to other areas of the network. The hexagonal cell layout model also has some impact on results, as it tends to underestimate the perimeter size (and, therefore, the registration rate) relative to the area size.

6.7 Security

Security is of paramount importance to both the network operator and the subscribers. The operator is primarily interested in precluding unauthorized access to his or her network from cloned or fraudulent mobile stations to avoid loss of revenue. Subscribers may be interested in a secure means of transferring confidential information, such as PINs, or may want to keep their conversation private. Security features have evolved significantly from the early introduction of cellular systems (when even the most basic security feature was optional) and to this date continue to evolve due to two forces: the evolution of services and application, with increasingly stringent security requirements; and the need to outpace the security threats that are growing in number and sophistication.

Security features can be grouped into three categories: those aimed at preventing unauthorized subscriber access (by means of authentication), eavesdropping (by means of encryption), and altering or tampering with user information (by means of message integrity). Below is the list of the cdma2000 security features. Selected feature are detailed in the following sections:

Authentication is the process by which information is exchanged between a mobile station and a base station for the purpose of confirming the identity of the mobile station, and it can be used to detect cloned mobile stations. A successful outcome of the authentication process occurs only when it can be demonstrated that the mobile station and base station possess identical sets of shared secret data.

Signaling message encryption is used to encipher signaling messages transmitted on the forward and reverse traffic channel only, so that no one else but the intended party can access the content of such messages (such as *Burst DTMF Messages* which may carry users' PINs).

Voice privacy is used to encipher the traffic channel frames and prevent eavesdropping, for example, of voice conversation.

Extended encryption for signaling messages and for user data is the enhanced security mechanism used to encipher both user data and signaling messages. Unlike the earlier signaling message encryption, it also can be used to encipher signaling message sent on the common channels (*f/r-csch*).

Authentication and Key Agreement (AKA), also called third-generation authentication, is an enhanced method to authenticate both subscriber and base station.

Message integrity is used to prevent a legitimate message transmitted in clear from being modified while on its way to the intended recipient, or an illegitimate message from being introduced into the channel.

In the following sections we discuss only selected security features: authentication, voice privacy, and AKA.

6.7.1 Authentication

The base station may choose to enable authentication to prevent fraudulent access of cloned mobile stations that use the identity of legitimate subscribers. Subscriber identity can be stolen, for example, when transmitted in clear on over-the-air messages that may be monitored by an illegitimate receiver. Two authentication procedures can be used: the global challenge and the unique challenge. The global and unique challenges allow only the base station to authenticate the mobile station and are called second-generation authentication procedure. While they differ in their triggering mechanisms and in the signaling messages they use, they are conceptually identical, as they use the same method to validate the mobile station's signature. This method is described next, followed by the description of authentication procedures.

6.7.1.1   Secret Shared Data and Authentication Signatures

Once authentication is triggered (when and how is detailed in the following sections), both mobile station and base station compute an authorization signature using a common cryptographic algorithm. The algorithm uses two inputs (see Figure 6.26). The first input is a set of non-secret parameters that may be transmitted in clear on over-the-air messages. The second input is a secret data known only by the legitimate subscriber and the network. The signature generated by the cryptographic algorithm will differ when different inputs are used. A fraudulent mobile station, even when using the same non-secret parameters and same signature-generation algorithm, will not be able to generate the same signature as that of the legitimate mobile station without access to the secret key. Therefore, when comparison of the mobile station and base station signatures results in a match, the base station has assurance that the mobile station possesses the valid secret key, which ultimately is proof of identity.

The secret data input to the authentication signature-generation algorithm is the 64 most significant bits (SSD_A) of a 128-bit quantity called *shared secret data* (SSD) that is stored in semi-permanent memory in the mobile station. The SSD is shared between mobile station and authentication center (AC), and may also be shared with the serving system visitor location reg-

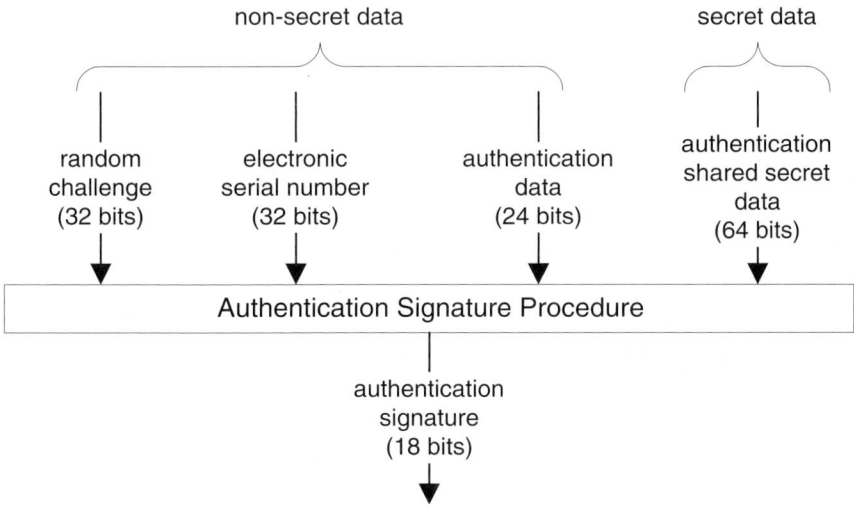

Figure 6.26 Authentication signature generation.

ister (VLR). No illegitimate party can access the SSD. since it is never transmitted over the air. The 64 least significant bits (SSD_B) are used for message encryption.

The signature generation algorithm is described in a restricted distribution document [7]. As with most cryptographic algorithms, the signature is generated by a *one-way function* to protect confidentiality of the secret key. A one-way function is one that can be easily computed and whose inverse is computationally infeasible to find [8]. This property is key to the authentication process. Not only is the illegitimate subscriber unable to generate the correct signature without access to the secret data, but also he or she is unable to generate the secret data by eavesdropping the signature.

6.7.1.2   Global Challenge

The authentication procedure dubbed global challenge[18] is applied indiscriminately to all mobile stations that attempt to access the network (hence the term *global*) for registration, mobile-originated or mobile-terminated calls, or data burst transmission. Global challenge is described hereafter with reference to the exemplary signaling flow in Figure 6.27 for call origination.

1. If global challenge is enabled, as indicated by the information in overhead messages broadcast by the base station, the mobile station generates the authorization signature. The signature generation algorithm (see Figure 6.26) uses the following inputs. The random challenge is set equal to the 32-bit RAND, a random number broadcast by the base station on the paging channel. The authentication data[19] is set to the binary-coded version of the last six dialed digits. The secret data is set equal to SSD_A.

2. The mobile station transmits the ORM with the authentication fields populated as fol-

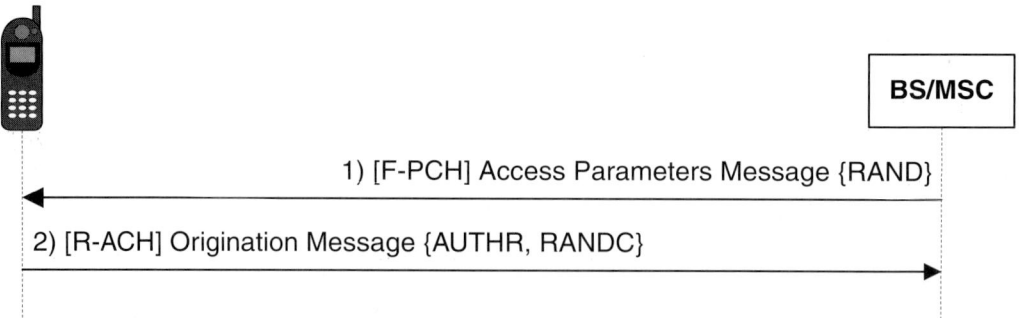

1) [F-PCH] Access Parameters Message {RAND}

2) [R-ACH] Origination Message {AUTHR, RANDC}

Figure 6.27 Subscriber authentication at call origination using global challenge.

---

18.   The global challenge authentication is specified in the LAC protocol because it can be viewed as related to access. We chose, however, to discuss it together with the other security features in the context of the Layer 3 protocol.

19.   In case of mobile-terminated call or registration, the authentication data is equal to the mobile station's IMSI_S1.

lows. The authorization response (AUTHR) corresponds to the authentication signature. The RANDC is set equal to the 8 most significant bits of the RAND value last received by the mobile station.

On receipt of the message, the base station compares the RANDC with the active RAND on the paging channel. In case of mismatch, the message may be discarded. If RAND and RANDC match, the base station forwards the authentication data to the VLR that computes the authentication signature using the same input parameters that the legitimate subscriber would have used. If the locally generated signature and the AUTHR match, the mobile station is successfully authenticated, and Layer 3 further processes the message. In case of mismatch, the base station may deny access to the mobile station.

But what prevents a fraudulent mobile station from stealing the AUTHR recently used (and transmitted unprotected) by a legitimate subscriber together with its identity? Nothing prevents that, but that is not a problem if the AUTHR is made unusable after its first use. This is accomplished by periodically changing the RAND. Furthermore, since for mobile-originated calls the dialed digits are used as input to the signature generation, a stolen AUTHR is unusable for call origination to a different directory number even during the short period of time when the RAND remains unchanged.

6.7.1.3   Unique Challenge

The unique challenge is initiated by the base station and is directed to a specific mobile station only. This authentication method can be used in addition to or in lieu of the global challenge. The unique challenge can take place either on the common channels (e.g., paging and access channels) or on the dedicated traffic channel, the latter being a more common scenario. Unique challenge is described hereafter with reference to the exemplary signaling flow in Figure 6.28.

1. If the base station is directed by the serving system's VLR to initiate unique challenge, it sends the mobile station the AUCM carrying a randomly generated 24-bit quantity, the RANDU.

2. On receipt of the AUCM, the mobile station generates the authorization signature. The

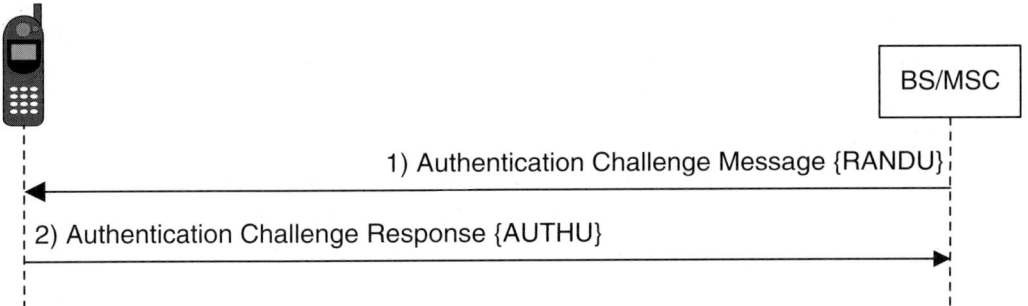

Figure 6.28 Subscriber authentication on the traffic channel using unique challenge.

signature-generation algorithm (see Figure 6.26) uses the following inputs. The 24 most significant bits of the random challenge are set equal to the RANDU, and the remaining 8 least significant bits are set equal to the mobile station's IMSI_S2. The authentication data is set equal to IMSI_S1. The secret data is set equal to SSD_A. Then, the mobile station sends the *Authentication Challenge Response Message* (AUCRM), whose AUTHU field is populated with the authorization signature generated as above.

On receipt of the message, the serving system computes the authentication signature using the same input parameters that the legitimate subscriber would have used. If the locally generated signature and the AUTHU match, the mobile station is successfully authenticated and Layer 3 further proceeds with the call. In case of mismatch, the base station may drop the call in progress or initiate the process of updating the SSD.

6.7.1.4   SSD Update and the Role of the A-Key

Maintaining secrecy of the SSD is of paramount importance, as the authentication algorithm relies on the fact that only legitimate parties know the SSD. Two such parties are, of course, the mobile station and the AC. However, if the AC were to authenticate all registration and call origination events taking place throughout the systems it serves, it may be inundated by authentication requests and become a bottleneck.

An alternative approach is to share the burden of performing authentication with the VLRs. When authentication signature computation and matching operation is conducted at the VLR, the signaling and processing load is distributed among many entities and is confined at the edge of the IS-41 network instead of being concentrated in its core. The drawback is that the secret data must now be shared with many other entities, increasing the chances of a security breach. To solve this problem, the A-key is introduced.

The A-key, or authentication key, is the real secret key. The A-key is stored in permanent memory in the mobile station and in the AC only. It is never shared with other entities, not even with the serving systems. The A-key is not accessible to the mobile station user and is typically programmed into the phone once and for all using nonmanual methods. If the AC decides to share authentication responsibility with the serving system, it generates the SSD using a cryptographic algorithm similar to that used for signature generation, but with different inputs (see Figure 6.29).

The SSD is then shared with the serving system VLR. But how can it be shared with the mobile station without being sent over the air? That is the role of the SSD update procedure that is explained next, with reference to the exemplary signaling flow in Figure 6.30.

1. The AC initiates SSD update, typically when it detects a mobile station that has moved into the service area of a new system. In such a case the AC updates the SSD and then passes it to the new serving system's VLR (core network procedures are detailed in

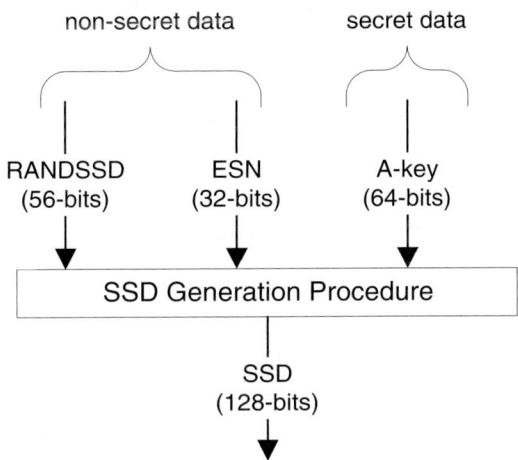

Figure 6.29 SSD generation.

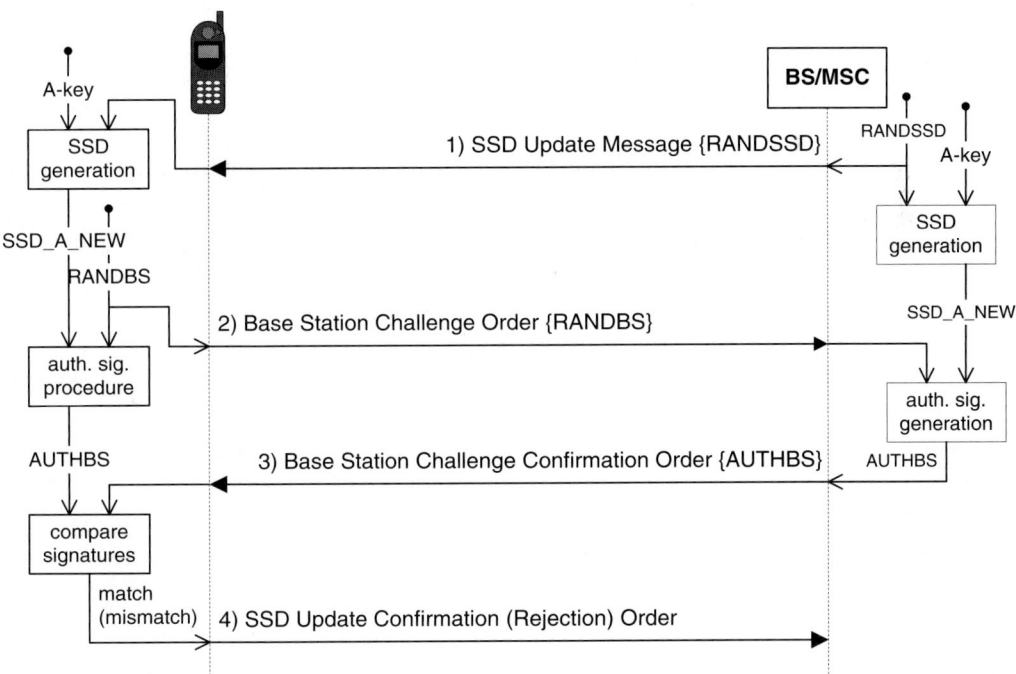

Figure 6.30 SSD update procedure.

Chapter 11). For what concerns the over-the-air procedure, SSD updates begin with the base station sending the *SSD Update Message* (SSDUM) to the mobile station. The message contains a 56-bit random quantity, RANDSSD. Upon message transmission, the base station generates a new SSD, the SSD_A_NEW, using the SSD-generation algorithm with the inputs described in Figure 6.29.

2. On receipt of the message, the mobile station also generates the new SSD using the SSD-generation algorithm with the inputs described in Figure 6.29. Then, the mobile station sends the *Base Station Challenge* ORDM to the base station. The ORDM contains yet another random number, the RANDBS. Upon message transmission, the mobile station generates a new authentication signature (AUTHBS) using the signature-generation algorithm (see Figure 6.26) with the following inputs: the random challenge bits, set to RANDBS; the ESN; the authentication data, set to the IMSI_S1; the secret data, set to SSD_A_NEW.

3. On receipt of the *Base Station Challenge* ORDM, the base station also generates a new authentication signature (AUTHBS) using the signature-generation algorithm with the same inputs used by the mobile station. Then, the base station sends the *Base Station Challenge Confirmation* ORDM to the mobile station. The message contains the newly generated authentication signature, AUTHBS.

4. On receipt of the *Base Station Challenge Confirmation* ORDM, the mobile station compares the authentication signature generated by the base station with the one locally generated. If the result is a match, the mobile station updates SSD_A to the new value and sends the *SSD Update Confirmation* ORDM to the base station. If the result is a mismatch, the mobile station discards the SSD_A_NEW and sends the *SSD Update Rejection* ORDM to the base station. Similarly, the base station updates the SSD only if the *SSD Update Confirmation* ORDM is received.

Note that the SSD is successful if and only if the mobile station possesses the same A-key stored at the AC. Moreover, the SSD is never transmitted over the air.[20] One may ask why steps 2 and 3 of the procedure above—that is, the base station challenge—are necessary. Base station challenge is done to verify that the base station has the correct current SSD and is therefore a legitimate base station, thus preventing false base stations from forcing SSD update.

6.7.1.5   *Authentication and Key Agreement (AKA)*

AKA allows the mobile station and the base station to perform mutual authentication. AKA, or third-generation authentication, is summarized in the following.

The root key, called the K-key, is stored at the User Identity Module (UIM) and the AC. The AC precomputes the set of authentication vectors (AVs) based on the K-key. When the mobile station registers with the base station, the base station sends an AV to the mobile station.

---

20.   However, the SSD is carried by the ANSI-41 network unprotected.

The mobile station authenticates the base station based on the AV and sends back the response, which is used by the base station to authenticate the mobile station.

After the authentication is completed, the base station and the mobile station set up an Integrity Key (IK) and UIM Authentication Key (UAK). The IK is used to support message integrity (see Chapter 5), and it is stored in the mobile equipment (ME) part of the mobile station. Recall from Chapter 2 that the mobile station consists of ME and UIM. The UAK is stored in the UIM, and it is never passed to the ME. It is used during the setup of the IK. The UAK is introduced to support authentication with the Removable UIM (R-UIM). It generates UIM Message Authentication Code (UMAC), which is transmitted with a message and used by the base station to authenticate the mobile station. Now, if the R-UIM is removed, the UMAC cannot be computed and validated by the base station. Note that without a UAK, a rogue mobile station could be programmed to send valid messages after the R-UIM is removed.

6.7.2 Voice Privacy

Without security measures in place, it is conceivable that the user conversation may be eavesdropped on by an unintended receiver equipped with the user long code used for spreading. The long code is a public code in that its mask is generated using either non-secret data such as the mobile station ESN or other data sent in clear on over-the-air messages. Therefore, using a private long code for signal spreading guarantees voice privacy, since no unintended receiver would be able to despread the signal. A cryptographic algorithm using the mobile station A-key generates the private long-code mask.

When the private long-code mask is used, there is no information sent in clear that may be used to generate the long code. Either the base station or the mobile station may initiate the procedure to transition to the private long code at any time during the call. The mobile station may also inform the base station right at call setup of its intent to use voice privacy. The procedure for such a case is explained in the following with reference to the exemplary call flow in Figure 6.31.

1. The mobile station originates a call and requests the base station to enable voice privacy by setting the privacy mode (PM) field of the *Origination* (or *Page Response*) *Message* to 1.

2. The base station and mobile station proceed with the normal traffic channel establishment procedures. The traffic channel is initially set up using the public long-code mask.

3. If the base station decides to accept the mobile station request for privacy, it sends the *Long Code Transition Request* ORDM. Although the standard allows the base station to send such order at any time during the call, the order is sent prior to the connection of the service option, that is, before two-way conversation is enabled. Note that base station and mobile station must synchronously switch to the private long code, or else all traffic channel frames will be received in error and the call may drop. To allow for synchronous switching, the order includes an explicit action time (the time at which the transition to the private long code should happen).

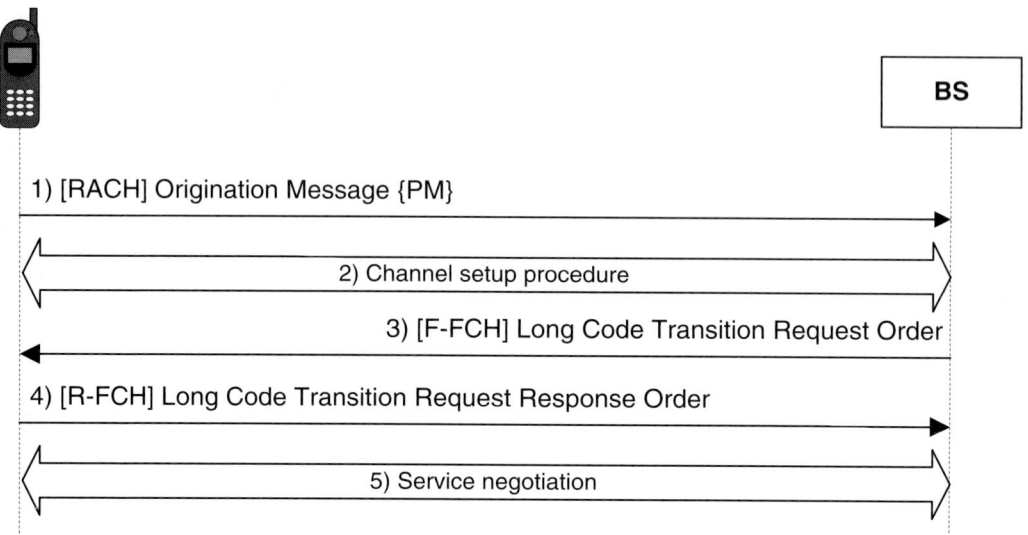

Figure 6.31 Voice privacy procedure following mobile station request at origination.

4. On receipt of the *Long Code Transition Request* ORDM, the mobile station replies with the *Long Code Transition Response* ORDM and waits for the action time. The base station, on receipt of the *Long Code Transition Response* ORDM, waits for the previously specified action time to switch to the private long code (synchronously with the mobile station).

5. The base station and mobile station perform service negotiation and connect the service. Private two-conversation takes place.

6.8 CDMA Tiered Services

CDMA tiered services provide the user custom services and special features based on the mobile station location. Tiered services also provide private network support. Tiered services are offered in user zones. Each user zone, identified by a user zone ID (UZID), is associated with a set of features/services that are made available to the customer. There are two types of user zones: broadcast user zones and mobile-specific user zones.

Broadcast user zones are identified to the mobile station using the *User Zone Identification Message* transmitted on the PCH, and they allow for permanent and temporary subscription.

Mobile-specific user zones are not explicitly identified to the mobile station using overhead signaling. The mobile station uses other base station parameters, such as BASE_ID, BASE_LAT, or BASE_LONG, and compares such parameters with an internally stored list of user zone parameters to identify the user zone.

One CDMA tiered service that is likely to be requested by operators is zone-based billing. With zone-based billing, the mobile station in idle state that enters a new "active" zone, as identified by the UZID of the *User Zone Identification Message*, performs user zone registration. The base responds with a *Feature Notification Message* that carries the User Zone Update Information record. At this point the mobile station display informs the user of the billing rates in that zone. Similar procedure can be performed if the mobile station is already active on the traffic channel.

If the base station supports CDMA tiered services, it shall support the following procedures. The base station shall broadcast the *User Zone Identification Message* and *Private Neighbor List Message* to help the MS identify the presence of user zones.

To validate or update the user zone requested by a mobile station, the base station sends the mobile station the *Flash with Information Message* or the *Feature Notification Message* if the mobile station is operating on a dedicated channel or common channel, respectively. Such messages contain the User Zone Update Information record.

The base station maintains the set of accessible user zones that corresponds to the current active set. When a change in an active set corresponds to a change in the user zones that are accessible, the base station notifies the mobile station by sending the *Flash with Information Message* with the User Zone Update Information record containing the new UZID.

REFERENCES

[1] "TIA/EIA/IS-2000.5 Upper Layer Signaling Standard for cdma2000 Spread Spectrum Systems, Release C," *Telecommunications Industry Association, 2002.*

[2] "TIA/EIA/IS-2000.4 Signaling Link Access Control (LAC) Standard for cdma2000 Spread Spectrum Systems, Release C," *Telecommunications Industry Association, 2002.*

[3] ITU-T Recommendation E.212, Identification Plan for Land Mobile Stations, 1988.

[4] Sarkar, Sandip, and B. Butler, "Phone Stand-by Time and the Quick Paging Channel," in the proceedings of *PIMRC*, Osaka, Japan, September 1999.

[5] "TIA/EIA-IS-707-B Data Service Options for Spread Spectrum Systems," *Telecommunications Industry Association, 2004.*

[6] Thomas, R., H. Gilbert, and G. Mazziotto, "Influence of the Moving of Mobile Stations on the Performance of a Radio Cellular Network," in the proceedings of *Third Nordic Seminar on Digital Land Mobile Radio Communications*, Copenhagen, Denmark, September 1988.

[7] *Common Cryptographic Algorithms*, Rev. C, 1997. An EAR-controlled document subject to restricted distribution.

[8] Newman, David, J. Omura, and R. Pickholtz, "Public Key Management for Network Security," *IEEE Network Magazine*, 1(2), April 1987.

Handoffs

Handoffs are those procedures designed to maintain continuity of the connection between the access network and mobile station while the latter moves from one cell to another. Although several handoff types exist, the key one that has contributed much to the success of CDMA is the *soft handoff*. Soft handoff allows the mobile station (MS) to commence communication with a new base station without interrupting communication with the current serving one. This *make-before-drop* technique differs drastically from those employed by earlier technologies, in which a connection with the target base station can be established only after releasing the connection with the serving one. Soft handoff allows for universal frequency reuse, which increases capacity; provides spatial macro-diversity gain, which increases cell range; and, by allowing smooth transition from one cell to the next, improves quality of service.

In this chapter we study the handoff procedures specified by the Layer 3 protocol and various handoff-related algorithms. We begin with describing the pilot search procedures that enable handoffs. Then, we study the types of handoffs taking place when the mobile station is in the idle, access, and traffic channel states. Finally, we discuss interfrequency handoff procedures, which are needed because cdma2000 systems operate on multiple carriers and sometimes require a mobile station's handoff to a different carrier. Throughout this chapter we focus on handoff techniques applicable to the F/R-FCH only, as those are well suited to explain the general principles. Discussion of specialized techniques applicable to the Forward/Reverse Supplemental Channel (F/R-SCH) and Forward Packet Data Channel (F-PDCH) is deferred to Chapter 9 in the context of packet data services.

7.1 Handoff Principles

In a cellular network, a mobile station that moves away from the serving base station will eventually find itself in a location where the serving base station is no longer the one with the best possible signal quality. At such location the mobile station should be directed to switch from the current serving base station to the one that can provide the best quality of service. If the serving and target base stations operate on different frequency assignments, the switching procedure entails releasing the connection with the serving base station and then establishing a new one with the target base station. This *drop-before-make* technique is called *hard handoff*. Hard handoff is both undesirable, as it momentarily disrupts transmission continuity, and impractical if the serving and target base stations operate on the same frequency assignment. In the latter case, which is the norm in CDMA networks with universal frequency reuse, a handoff technique is needed that allows the mobile station to initiate communications with a new base station without interrupting communications with the one it is currently being served by. This make-before-drop technique is called *soft handoff*. Soft handoff is made possible by the spread spectrum nature of CDMA. Signals transmitted from multiple base stations can be demodulated in parallel by the Rake receiver fingers, as the despreading process greatly reduces mutual interference. Furthermore, the demodulators' outputs can be combined because the base stations involved in soft handoff transmit identical traffic channels data symbols in a synchronous fashion. Note that this process is akin to that used by the Rake receiver for multipath combining. In the reverse direction, the traffic channel is demodulated by the base station transceivers (BTS) involved in soft handoff and the decoded frames are sent to the base station controller (BSC). The BSC selects the one with the highest quality metric. This process is called selection combining, and allows for a weaker form of spatial diversity than that enjoyed by the mobile station on the forward link.

7.1.1 Types of Handoff

There are many handoff types in cdma2000. A useful categorization of such types, which the organization of this chapter is a reflection of, is in terms of IS-2000 Layer 3 protocol state.

Idle handoff. While in idle state, the mobile station may perform an *idle handoff* when it moves from the coverage area of one base station to that of another base station. Idle handoff allows the mobile station to monitor the Forward Paging Channel (F-PCH) or Forward Broadcast Control Channel (F-BCCH) and Forward Common Control Channel (F-CCCH) of the base station providing the best quality of service in a given location. Idle handoff is carried out independently by the mobile station and does not require any explicit signaling between mobile station and base station. However, the base station supports idle handoff procedures by broadcasting the set of base stations that are feasible handoff targets. Idle handoff can occur between base stations supporting either the same or different frequency assignments.

Access handoff. While transitioning to or in the system access state, the mobile station may perform one of three types of *access handoffs* to increase the probability of successful access. Access handoffs are motivated by the fact that channel conditions can drastically change during the access procedure, which can last for up to a few seconds. Then, access handoffs allow the mobile station to monitor the best F-PCH or F-CCCH at all times. Access handoffs are initiated by the mobile station but are governed by the access network that dictates which conditions must be satisfied for access handoff to occur and specifies the set of base stations that are allowed for handoff.

Traffic channel handoff. While in the control of the traffic channel state, the access network may direct the mobile station into soft handoff with multiple traffic channels of different base stations. All forward traffic channels associated with base stations in soft handoff carry identical modulation symbols, possibly with the exception of the power control subchannels. That allows the mobile station to provide soft combining of the associated traffic channels, which provides for spatial diversity gain. The access network may also direct the mobile station to initiate a hard handoff to a traffic channel on a different frequency assignment. Handoffs on the traffic channel are initiated by the access network, but the process is typically triggered by mobile station measurements; that is, they are mobile-assisted handoffs. Traffic channel handoffs require explicit signaling procedures.

7.1.2 Pilot Sequence Offsets

The forward pilot channel, or pilot for short, plays many fundamental roles in handoffs. The first role is that it allows identification of the forward CDMA channel with which it is associated. That is, the pilot identifies the set of code channels transmitted from a particular base station to the mobile station on a frequency assignment. Clearly, CDMA channel identification is a prerequisite for base station-to-mobile station interoperability procedures during handoffs.

As seen in Chapter 4, only one pilot pseudonoise (PN) sequence generator polynomial is specified by the physical layer protocol. Every base station is assigned identical pilot PN sequence, which may appear at odds with the need of using the pilot as a base station identifier. However, recall that in cdma2000, base stations' transmission timing is synchronized to the same time reference. Therefore, by staggering their pilot sequence-transmission time, base stations can be distinguished by the relative delay at which the associated pilots are received. The delay, in units of chip duration, between a pilot sequence and that of a reference pilot sequence (the zero-offset sequence) is called pilot PN sequence offset, or pilot offset for short. Pilot offsets are defined in multiples of 64 chips and are assigned an index called PILOT_PN. For example, PILOT_PN = 12 means that transmission of the associated pilot is delayed by $12 \times 64 = 768$ chips relative to the reference pilot. There are at most 512 possible offsets, because the pilot sequence period is 32,768 chips. In practice, for reasons that will become apparent shortly, a smaller set of pilot offsets is employed. Specifically, the access network selects only pilot offsets in increments that are multiples of 64 chips. The multiplier is a system configuration parameter called PILOT_INC. The pilot offset index is then used as base station identifier. However, since there are only a finite number of offsets, it may be necessary to assign the same pilot offset to multiple base stations in the network. That is called *pilot offset reuse*. The implications of pilot offset reuse are addressed in Section 7.6.2.

7.1.3 Pilot Sets

Pilot search and handoff procedures are defined in terms of four pilot categories, or sets. The *active set* consists of those pilots associated with the traffic channels assigned to the mobile station in the traffic channel state or, when in the idle state, to that (single) pilot associated with the F-PCH or BCCH being monitored. Active-set pilots are those that have the best quality of all detectable pilots and are of sufficient quality to allow reliable demodulation of the associated control and traffic channel. The *candidate set*, applicable to the traffic channel state only, consists of those pilots that are not yet included in the active set but whose strength is high enough to make them good candidates for promotion to the active set. The *neighbor set*, applicable to both the idle and the traffic channel states, consists of those pilots that are likely candidates for handoff because the corresponding base stations are located in proximity (i.e., they are neighbors) of the active-set pilot(s). All other pilots with permissible pilot offset are included in the *remaining set*.

7.1.4 Pilot Search Fundamentals

Handoff procedures rely on the mobile station's capability to search for, detect, and measure the strength of pilots that are possible handoff targets. Note that the pilot is designed to facilitate such operation. It is an unmodulated signal transmitted at relatively high power that can be reliably detected employing a simple receiver. It is also transmitted at constant power so that its received energy is a measure of forward-link quality that can be used to evaluate handoffs. Such capabilities are typically bundled in a single functional element of the mobile station called the *pilot searcher*. The pilot searcher consists of a pilot detector and a search controller. The detector performs static tests to decide whether or not a pilot signal is present at a given offset from the reference time; the controller selects the hypothesis to be tested. The modes of operation of the pilot searcher depend on the state the mobile station is in, for example, idle or traffic channel state. However, some basic operations apply to all states. Those are examined in this section.

The detector is implemented in the form of a filter matched to the chip waveform followed by a sampling device and by the hypothesis-testing device. The hypothesis-testing device (see Figure 7.1) uses a despreader, an accumulator, a squaring device, and a threshold detector. The despreader is implemented in the form of a complex multiplier and uses a locally generated replica of the in-phase, PN_I, and quadrature, PN_Q, spreading sequences used at the transmitter side. The in-phase and quadrature components of the despreader are independently accumulated over a number of chips to smooth out the noise, and then fed into the squaring devices. The number of

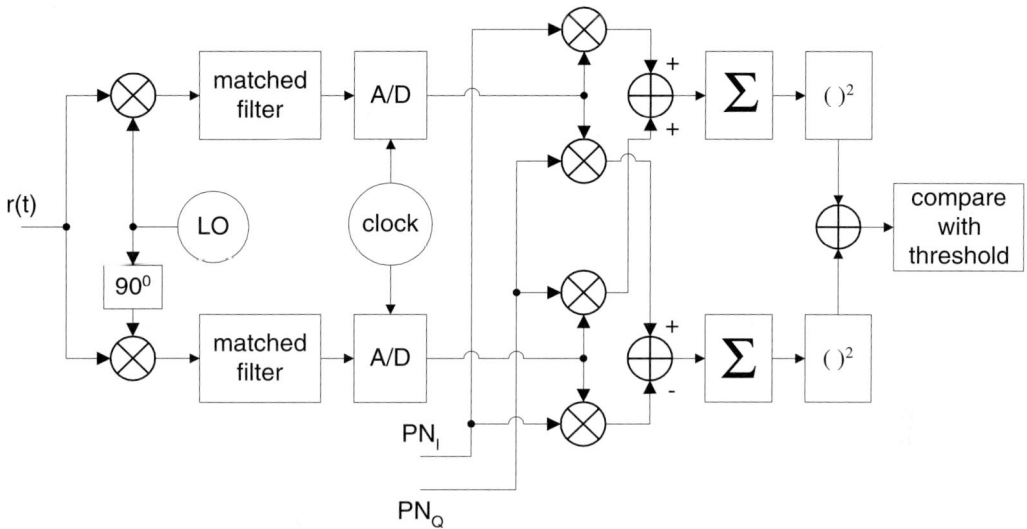

Figure 7.1 Pilot-detection block diagram.

chips that are accumulated prior to squaring is called the *coherent integration period*. The outputs of the squaring devices are added together to yield an estimate of the received pilot energy.
The estimate is then fed into the threshold device to test the hypothesis of a pilot signal being
present. Alternately, energy estimates obtained over successive integration periods may be
summed together and the result fed into the threshold device. This method is called *post-detection combining*, and the number of energy estimates being combined is called *noncoherent integration length*. The coherent and noncoherent integration length and the detection threshold
setting are design parameters that affects searcher performance.

   If the local oscillator is perfectly tuned to the carrier frequency, the signal-to-noise ratio
(SNR) of the decision variable increases with increasing length of the coherent integration
period. However, in case of frequency mismatch, the residual frequency error causes the signal
to rotate during the integration period and decreases the effective SNR. In fact, if the coherent
integration period is equal to the inverse of the frequency error or a multiple thereof, the useful
signal at the input of the squaring device cancels out. The detection-threshold setting is also a
design parameter, affecting the trade-off between the probability of falsely detecting a pilot and
the probability of misdetection. Various methods exist to cope with residual frequency error,
including post-detection combining. With this method, the decision statistic is formed by summing the signal energy estimated over subsequent integration periods.

   The hypothesis-testing device performs only static tests, and the search controller selects
the pilot offset hypothesis to be tested. Assume, for example, that a neighbor pilot is being
searched. The search controller defines a set of hypotheses corresponding to that pilot. The set of
hypotheses is centered on the nominal neighbor pilot offset relative to that of the reference pilot.

The hypotheses in the set are spaced apart by a fraction of a chip (typically, a half-chip), thus forming a *search window* centered on the nominal offset. The search window compensates for the timing uncertainty due to the propagation-delay differential between the reference pilot and the one being searched. The search window must also be wide enough to capture all multipath components. The searcher then "sweeps" the search window by shifting the local pilot sequence used for despreading in half-chip increments, while the detector tests the hypothesis of the pilot being presented at each increment. This process may be done serially or in parallel. In a serial search, the test is done each time using a different set of signal samples. When a new hypothesis is tested, the previous set of samples is discarded and a new set is collected. In a parallel search, on the other hand, the same set of signal samples is used to test multiple hypotheses. In practice, current signal-processing technology allows for efficient parallel searchers. The search window size depends on the set the pilot belongs to and is determined by the upper layers. General principles for window size selection are discussed in Section 7.6. The searcher must also schedule visits to the pilots according to the corresponding set priority. The scheduling algorithm depends on the mobile station state and is discussed in the following sections.

7.2 Initial Pilot Acquisition

When in the pilot acquisition substate of the Layer 3 protocol initialization state (see Chapter 6), for instance, just after power up, the mobile station has yet to acquire any time reference and therefore must scan the entire pilot sequence space. The initial time reference is chosen at random, and the search windows are positioned contiguously to scan the entire pilot sequence space. The search window width W is an implementation-dependent parameter. In this state the frequency error caused by inaccurate tuning of the local oscillator is unknown and cannot be compensated for. Hence, the coherent integration time is relatively small, and noncoherent accumulation is used more heavily. Once the first strong energy peak is detected (not necessarily the strongest of the entire pilot sequence space so that the acquisition process can be expedited), the controller tentatively assigns a demodulating element to that peak so that the frequency-tracking loop can be enabled. Once the frequency error is estimated and compensated for, the searcher validates its tentative decision and searches for other peaks within the search window. In the example of Figure 7.2, two peaks detected within the search window have energy higher than the threshold.

Once the hypothesis is validated, the controller reports the event to the upper layers that initiate Forward Synchronization Channel (F-SYNCH) acquisition. As seen in Chapter 6, in this state the mobile station demodulates the F-SYNCH and receives system information, including the pilot sequence offset index, PILOT_PN. The pilot offset, in turn, is used to determine frame timing necessary to decode messages carried by the overhead channels. Now that it has fully acquired system timing (at both chip and frame levels), the mobile station enters the idle state and demodulates the F-PCH (or F-BCCH) to receive system configuration. The pilot associated with this F-PCH becomes the active-set pilot.

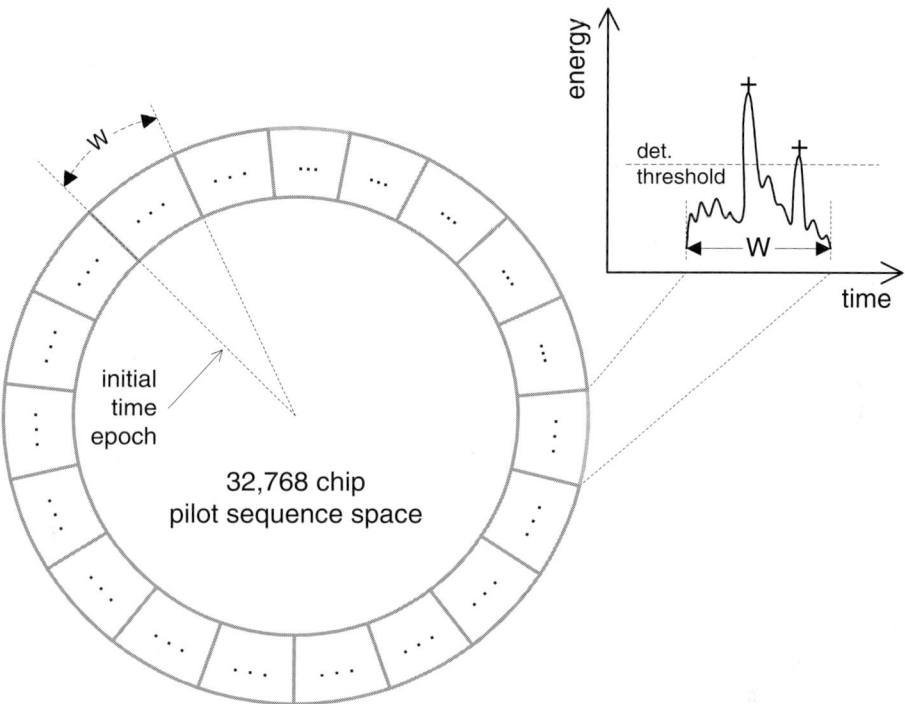

Figure 7.2 Pilot search space in acquisition state.

7.3 Idle Handoff

As discussed in Chapter 6, the mobile station in idle state must periodically monitor the F-PCH (or F-BCCH) to receive updated configuration information and page messages. Since only one F-PCH can be monitored at one time, when the mobile station moves across the cell boundary, a mechanism must be in place to switch from the F-PCH of one base station to that of another. This process is called *idle handoff*. An idle handoff is triggered whenever the pilot associated with the F-PCH being monitored becomes weak and the mobile station detects a better quality pilot corresponding to a neighboring base station. The decision as to how much better the neighboring pilot must be to be worthy of an idle handoff is implementation-dependent. Idle handoff procedures are best described by distinguishing two phases: the pilot search and the idle handoff phase.

7.3.1 Pilot Search in Idle State

There are two pieces of information that the mobile station in idle state needs to operate its pilot searcher: a list of pilot offsets to be searched and the corresponding search windows. That information, summarized in Table 7.1, is carried by overhead messages like the *Extended System Parameter Message* (ESPM) and the *Extended Neighbor List Message* (ENLM).

Table 7.1 Pilot Search Parameters*

Field	Message	Description
SRCH_WIN_A	ESPM	Search window to be used for the active- and/or candidate-set pilot.
SRCH_WIN_N	ESPM	Search window to be used for the neighbor-set pilot.
SRCH_WIN_R	ESPM	Search window to be used for the remaining-set pilot.
PILOT_INC	ENLM	Pilot PN sequence offset increment in units of 64 chips that the mobile station must use to search for pilots in the remaining set.
NGHBR_PN	ENLM	Pilot PN sequence offset for this neighbor.

*Not all handoff-related fields are listed.

The ESPM information elements include the search window size corresponding to the various pilot sets. The ENLM information elements include the neighbor list, that is, the list of pilot offset indexes (NGHBR_PN) associated with the neighboring pilots. The base station controller maintains a database that contains the information used to populate the neighbor list, the pilot database. The pilot database is provisioned and maintained by the network operator using network planning tools and/or field measurements. The mobile station still needs to identify the pilots in the remaining set. Those pilots are not listed one by one in the ENLM, as there are a large number of them and that would result in an ENLM of excessive length. Rather, the base station specifies the PILOT_INC with the implicit rule that all pilots in the network can only be assigned an offset that is a multiple of PILOT_INC. Any other pilot sequence offset is not admissible; that is, it cannot be used by any base station in the network. With the information contained in the ESPM and ENLM, the searcher can now define the search space. Consider the example in Figure 7.3. In this example the active-set pilot has PILOT_PN 12, the neighbor set includes PILOT_PN 4 and 8, and PILOT_INC is set to 4. Then, the remaining-set pilots have offset index $n \times$ PILOT_INC, with $n = 4, 5 \dots 127$. The labels W_A, W_N, and W_R represent the active pilot, neighbor, and remaining-set pilot search windows' widths, respectively. W_A, W_N, and W_R are obtained by means of a lookup table operation using the SEARCH_WIN_A, SEARCH_WIN_N, and SEARCH_WIN_R parameters of the ESPM. Figure 7.3 highlights that the mobile station, once in idle state, no longer needs to search the entire pilot space as it does during initial pilot acquisition. Rather, it has to sweep through relatively narrow search windows centered on a well-defined subset of pilot offsets.

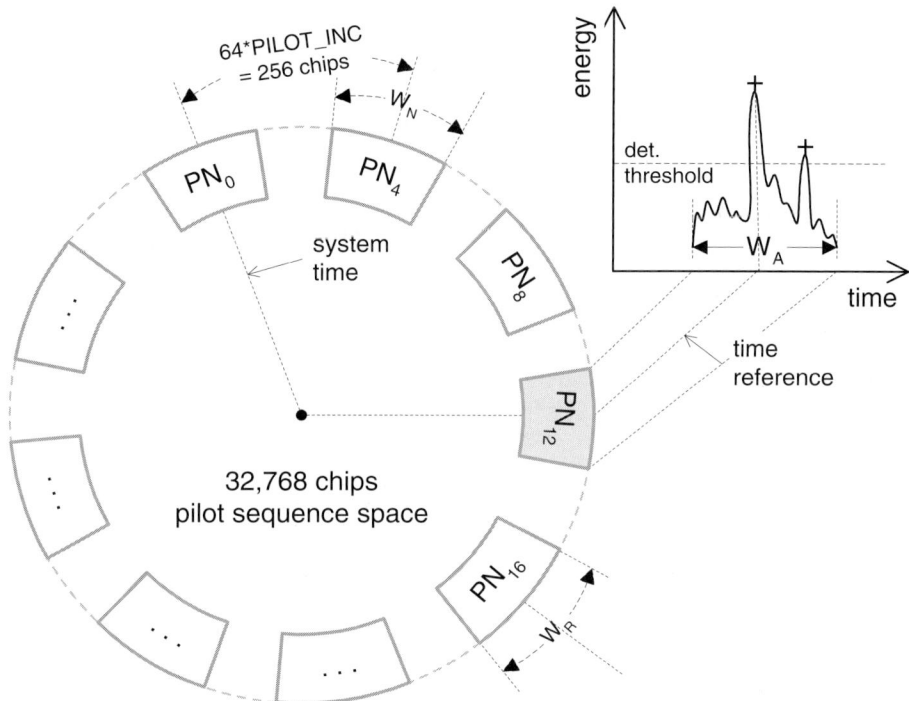

Figure 7.3 Pilot search space in idle state.

Pilot search in idle state does not happen continuously because the mobile station operates in slotted mode to conserve battery power (see Chapter 6). When sleeping in between its assigned F-PCH slots, the searcher is disabled. The pilot searcher is re-enabled just prior to the assigned paging slot, when the mobile stations wakes up and prepares to demodulate the Forward Quick Paging Channel (F-QPCH). During this period, the mobile station attempts to reacquire the active-set pilot and the pilots that were detected during the previous slot cycle (either belonging to the neighbor or remaining set). Then, throughout the reminder of the wake-up period, the searcher keeps monitoring the detected cells as well as searching for new neighbor and remaining-set pilots. The frequency and the priority with which a pilot is searched depend on an implementation-dependent scheduling algorithm. It is clear, however, that such frequency is negatively affected by a large neighbor-set size or a large search window size. Each time a pilot is detected, the energy of all its peaks within the search window is summed together, yielding the estimated pilot energy, which is used for idle handoff evaluation.

7.3.2 Idle Handoff Evaluation

When the searcher has detected the presence of a neighbor-set pilot whose strength is sufficiently stronger than the active-set pilot, the mobile station initiates an idle handoff. The process is governed by hysteresis to avoid a ping-pong effect between two neighboring base stations. That would be the case when there is no dominant server and pilots received from different base stations, although nearly the same on average, fade independently causing one base station's pilot to become only momentarily stronger than all others. IS-2000 does not mandate any particular idle handoff decision rule, which is left implementation dependent. A suitable rule consists in comparing the neighbor pilot strength against a monotonically increasing and convex function of the active-set pilot strength. One such function [1] is as in Eq. (7.1).

$$Pilot\,\frac{E_{c,N}}{I_0} \geq \beta\left(Pilot\,\frac{E_{c,A}}{I_0}\right)^{\alpha} \tag{7.1}$$

In Eq. (7.1) the pilot strength, or quality, is expressed in terms of received pilot-chip energy, E_c, to total received power spectral–density ratio, I_0. The method to estimate such ratio is implementation-dependent and nontrivial. In most common mobile stations' implementations, the estimator relies on the receiver's automatic gain control (AGC) unit to keep the I_0 constant and centered within the receiver's analog-to-digital converter (ADC) dynamic range, and then performs filtering of the output of the squaring devices in Figure 7.1.

Note that setting $\alpha > 1$ in Eq. (7.1) is such that conditions for idle handoff become more stringent for increasing quality of the active-set pilot. That is reasonable, as handing off to a strong neighboring pilot may not be advantageous if the current serving pilot is already strong enough. The idle handoff hysteresis is controlled by both α and β parameters. The parameters should be set to a value that corresponds to 2~3 dB hystereses when the active-set pilot strength is the one typically experienced by a mobile station in proximity of the handoff boundary. For example, if the nominal active-set pilot strength at the cell boundary is –8 dB, a suitable parameters' setting is $\alpha = 1.5$ and $\beta = 10^{(7/10)} = 7$ dB. As a result, the neighbor-set pilot strength must be at least -5 dB (or 3 dB larger than the active-set pilot) to cause an idle handoff. The importance of proper design of the idle handoff decision rule cannot be overstated. If the rule is too stringent, the mobile station may be camping on a weak pilot and increase the probability of missing a page. If the rule is too relaxed, idle handoff will be triggered too frequently and cause excessive battery drain that decreases standby time, as explained below.

Once idle handoff is triggered, the mobile station must operate in nonslotted mode[1] while monitoring the new F-PCH or F-BCCH and decode at least one valid configuration message. If the current configuration parameters are deemed not current (see Chapter 6), the mobile station

---

1.   The procedure is different if the F-QPCH with configuration change indicators is supported by both mobile station and base station. Use of configuration change indicators, however, is not discussed in this book.

updates its configuration parameters to match those of the new base station. The list of neighbor-set pilots is also updated, and the search for new pilots to which the mobile station can handoff continues thereafter.

7.4 Access Handoffs

At the cell edge, with fast-changing channel propagation conditions and significant interference from adjacent cells, system access procedures have a much greater probability of failure. System access failures are particularly worrisome during call setup, since they lower overall network QoS and user satisfaction. There are three main stages during the call setup procedure that contribute to the system access failure rate.

Stage 1. The mobile station sends access probes until it receives the *Base Station Acknowledgment Order Message* (ORDM).

Stage 2. The mobile station waits for the *Extended Channel Assignment Message* (ECAM) while it continuously monitors the F-PCH or F-CCCH.

Stage 3. The mobile station is assigned the traffic channel, and there is further latency until it can initiate a soft handoff procedure.

As idle handoff is prohibited while in any of these stages, system access failures in a cell area where multiple pilots of similar strength are received may be caused by the following events.

F-PCH or F-CCCH lost before receipt of Base Station Acknowledgment ORDM. The mobile station is not monitoring the F-PCH or F-BCCH corresponding to the strongest pilot. Then, in case of mobile-terminated calls, the mobile station enters the page response substate but cannot reliably demodulate the F-PCH or F-CCCH and receive the *Base Station Acknowledgment* ORDM.

F-PCH or F-CCCH lost before receipt of ECAM. The active-set pilot fades while a neighbor-set pilot becomes stronger. While the mobile station is waiting for the ECAM, the F-PCH or F-CCCH corresponding to the weakening active-set pilot is lost.

F-FCH lost upon entering the traffic channel state. The active-set pilot fades when the mobile station is assigned the traffic channel while a neighbor-set pilot becomes stronger. The mobile station cannot receive any good frame on the traffic channel corresponding to the weakening active-set pilot.

In Table 7.2 we summarize the call setup failure rate estimated from field measurements. Such measurements were taken in a network characterized by an urban environment with harsh channel conditions at the cell edge. The failure rate is broken down in its main components according to the three stages defined above.

Table 7.2 Typical Call Setup Failure Rate When Access Handoffs Are Not Supported

Call setup failure reason	Call setup failure rate
Stage 1 – F-PCH loss	1.4%
Stage 2 – F-PCH loss	1.9%
Stage 3 – F-FCH not acquired	1.9%
Other	0.4%
TOTAL	5.6%

It is clear that if handoff was not allowed during the system access state and if the call setup time was large due to processing delay at the base station, the probability of access failure may be significant. To cope with the problems above, the standard supports four types of access handoff.

Access Entry Handoff. The mobile station may be permitted to perform this handoff, after a page or message reception and before beginning an access attempt, to a base station in the access entry handoff list. The main benefit in supporting this feature is that the mobile station has additional time to search for the strongest pilot after waking up and then send access probes to the base station corresponding to the strongest pilot.

Access Probe Handoff. The mobile station may be permitted to perform this handoff to certain base stations between probes of an access attempt before the mobile station receives a *Base Station Acknowledgment* ORDM. The mobile station is permitted to perform an access probe handoff on loss of the F-PCH after the acknowledgment timer for the previous probe has expired. Support of this feature can reduce the call setup failure rate, especially when there is a long delay between the mobile station sending the first access probe and receiving the *Base Station Acknowledgment* ORDM. Such delay may be caused by harsh channel conditions on the forward or reverse links.

Access Handoff. The mobile station may be permitted to perform this handoff to certain base stations after a successful access attempt. Support of this feature can reduce the call setup failure rate due to the long delay between receipt of the *Base Station Acknowledgment* ORDM and the *ECAM*, and to harsh channel conditions on the forward link.

Channel Assignment into Soft Handoff (CASH). The mobile station may be directed to multiple traffic channels in soft handoff at the time of traffic channel assignment itself to facilitate forward-link acquisition. Thus, the access network includes multiple pilots in the active set of the mobile station. This feature is mobile station-assisted, as pilot measurements of certain neighbor pilots (with strength above a certain threshold) are included in the *Origination Message* or *Page Response Message* (PRM). The access network then determines the active set based on these pilot measurements.

Let us see by means of an example how the access entry handoff, access handoff, and channel assignment in soft handoff procedures work. The relevant signaling messages' parameters are listed in Table 7.3.

Table 7.3 Access Handoff Parameters*

Field	Message	Description
ACCESS_HO	ESPM	Access handoff permitted indicator. Handoff permitted if set to 1.
ACCESS_PROBE_HO	ESPM	Access probe handoff permitted indicator. Handoff permitted if set to 1.
MAX_NUM_PROBE_HO	ESPM	Maximum number of times the mobile station is permitted to perform an access probe handoff in a given access attempt.
ACCESS_ENTRY_HO	ESPM	If set to 1, access entry handoff to the corresponding base station is permitted when entering the system access state. The i-th occurrence of this field corresponds to the i-th pilot in the ENLM.
ACCESS_HO_ALLOWED	ESPM	If set to 1, access handoff and access probe handoff to the corresponding base station is permitted when entering the system access state. The i-th occurrence of this field corresponds to the i-th pilot in the ENLM.
T_ADD	SPM	Minimum pilot strength for inclusion in the access handoff list.
ACTIVE_PILOT_STRENGTH	ORM	Pilot strength of the active-set pilot in the system access state.
NUM_ADD_PILOTS	ORM	Number of neighbor pilots forming the access handoff list.
PILOT_PN_PHASE	ORM	PN phase of a pilot in the access handoff list.
PILOT_STRENGTH	ORM	Strength of a pilot in the access handoff list

*Not all access handoff–related fields are listed

In this example, the ACCESS_ENTRY_HO and the ACCESS_HO fields sent by both base stations 1 and 2 indicate that access entry handoff and access handoff are permitted. The PILOT_PN of the pilots corresponding to base stations 1 and 2 are included in each other's ENLM and are PILOT_PN 12 and 48, respectively. The ACCESS_HO_ALLOWED fields indicate that the corresponding pilot can be included in the ACCESS_HO_LIST built by the mobile station on entering the system access state. Before time t1, the mobile station is in idle state and the pilot corresponding to base station 1 is the active-set pilot, although its strength is less than that of the neighbor PILOT_PN 48 corresponding to base station 2 (e.g., because of idle handoff hystereses; see Section 7.3). The sequence of events is described below with reference to Figure 7.4 and Figure 7.5.

1. At time t1, the access network pages the mobile station. The paging area includes both base stations 1 and 2 as well other neighboring base stations. The mobile station receives the *General Page Message* (GPM) from base station 1.

2. As PILOT_PN 48 is much stronger than PILOT_PN 12, the mobile station initiates access entry handoff and demodulates the F-PCH of base station 2 to update its configuration.

3. At time t3, the mobile station sends the PRM to base station 2. Since both PILOT_PN 12 and 48 are above T_ADD, the PRM includes the pilot strength of both pilots.

4. The access network receives the PRM on the RACH of base station 2 and determines the ACCESS_HO_LIST used by the mobile station by means of the PILOT_PN_PHASE fields. At time t4, the base station sends the *Base Station*

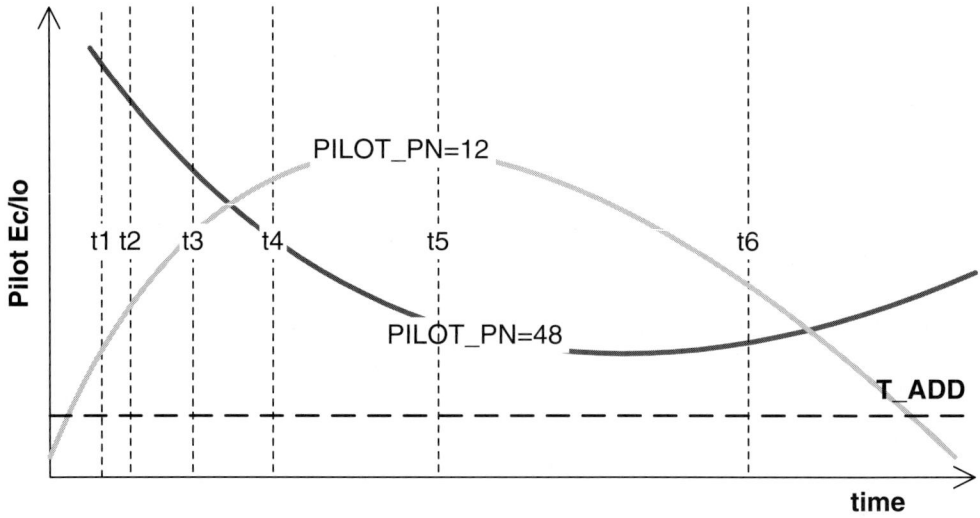

Figure 7.4 Example of access entry handoff, followed by access handoff and CASH.

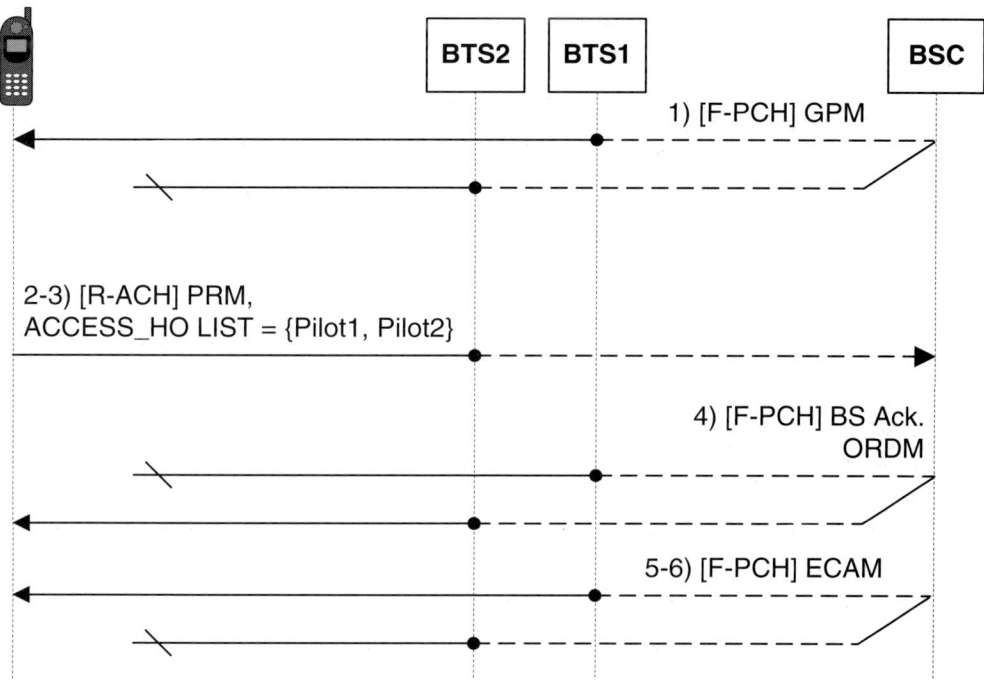

Figure 7.5 Mobile station-terminated call with access entry handoff, access handoff, and CASH.

Acknowledgment ORDM. The message is sent on the F-PCH of *all* base stations in the access handoff list in case the mobile station may be attempting an access probe or access handoff to a base station other than that on which the PRM was received. The mobile station receives the *Base Station Acknowledgment* ORDM on the F-PCH of base station 2.

5. As PILOT_PN 12 has now become much stronger than PILOT_PN 48, the mobile station initiates an access handoff to base station 1 and demodulates its F-PCH while waiting for the ECAM.

6. In the meantime, the access network has set up traffic channels on both base stations 1 and 2. No other base station in the access handoff list is selected because their corresponding pilot strength is not sufficiently high relative to that of PILOT_PN 12 and 48 to justify additional resource consumption. At time t6 the base station sends the ECAM on the F-PCH of all base stations in the access handoff list, directing the mobile station to demodulate the traffic channels setup on base station 1 and 2. The mobile station receives the ECAM on the F-PCH of base station 1 and enters the traffic channel state in soft handoff with base stations 1 and 2.

We can assess the performance improvement achievable with access handoffs. In Table 7.4 we summarize the call setup failure rate estimated from field measurements obtained from a network in an urban environment when different types of access handoff combinations are supported. Compared to that of an access network not supporting access handoffs (see Table 7.2), performance is greatly improved.

Table 7.4   Exemplary Call Setup Failure Rate When Access Handoffs Are Supported

Access handoff types supported	Call setup failure rate
None	5.5%
Access entry handoff	3.3%
Access probe handoff	2.9%
CASH	2.4%
Access handoff	2.3%
Access handoff and CASH	1.6%

7.5 Soft Handoff

When in soft handoff, the mobile station simultaneously demodulates the traffic channels transmitted by multiple base stations. By definition, those pilots belong to the active set. All active-set base stations transmit identical traffic channels data symbols, which allows the mobile station to combine[2] the demodulated signals prior to frame decoding. Since same data symbols must be sent simultaneously on the traffic channels corresponding to the active-set pilots, the BSC multicasts the data frame to all base stations involved in handoff. In the reverse direction, the traffic channel is demodulated by each of the active-set base stations, and the decoded frames are sent to the BSC along with a quality metric. A simple quality metric may be the result of the frame cyclic redundant code (CRC) checksum, that is, pass or fail. The BSC receives multiple copies of the reverse traffic channel frame and selects the one with the highest quality metric. This process is called *selection combining* and allows for a weaker form of spatial diversity than that provided by combining the demodulator data symbols as in the forward-link case.

In the case of a BTS comprising multiple cells, the RF signals from the cells' receive antennas can be fed into the same demodulating element. Therefore, the BTS can combine the demodulated signals prior to frame decoding, just as the mobile station does on the forward link. The decoded frame is then sent to the BSC for the selection process. In this case, the handoff is

---

2.   The Rake receiver utilizes the received pilot strength to perform maximum ratio combining of the soft symbols.

called *softer handoff*. Softer handoff is more effective than soft handoff in terms of frame-detection performance.[3] However, it does not provide the same level of macro-diversity gain as that offered by soft handoff, since with softer handoff the shadow-fading experienced by the cells of the same BTS is highly correlated. In general we say that the mobile station is in n-way soft, m-way softer handoff when the mobile station has $n + m$ pilots in the active set and n of them correspond to cells belonging to different BTSs. Consider Figure 7.6, for example, where a dashed line represents the coverage area of a cell. Within the intersection of $\alpha1$, $\beta2$, and $\gamma2$ coverage areas, the mobile station is in one-way soft, two-way softer handoff. That is, one cell of BTS1 and two cells of BTS2 correspond to active-set pilots.

Soft handoff procedures are best described from the viewpoint of the mobile station. We can then distinguish three types of functionalities that enable soft handoffs: pilot search, pilot measurement reporting, and pilot set maintenance.

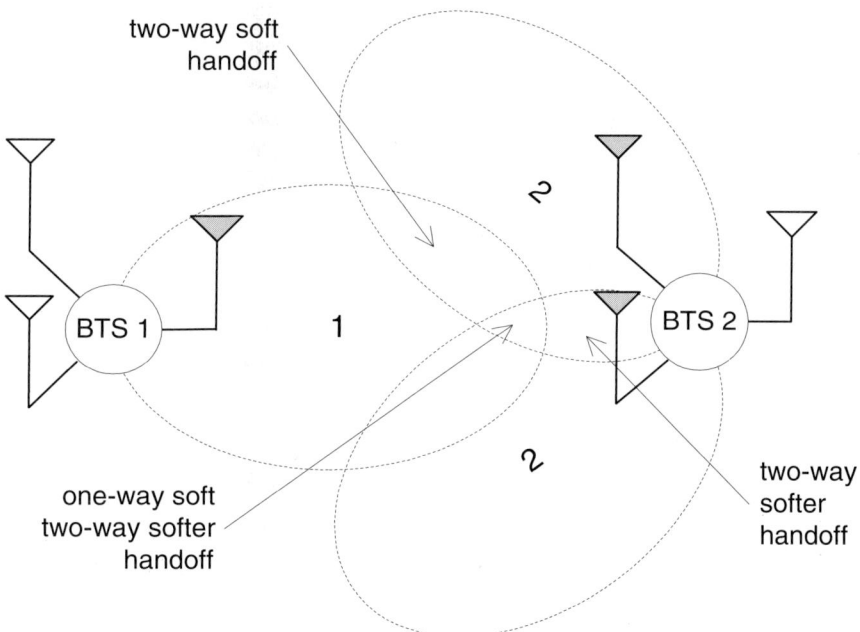

Figure 7.6 Soft handoff combinations.

---

3.   Another advantage of softer handoff is it allows coherent combining of the forward power-control sub-channels, which has a beneficial impact on power control (see Chapter 8).

7.5.1 *Pilot Search in the Traffic Channel State*

When in the traffic channel state, the mobile station continuously monitors pilots that have been detected and search for new peaks as well as new pilots. Just as it does while in idle state, a mobile station in the traffic channel state does not scan the entire pilot sequence space. Rather, the searcher sweeps only the search windows centered on the pilot offsets of the active, candidate, neighbor, and remaining-sets' pilot (refer again to the example of Figure 7.3). Reducing the search space is fundamental because soft handoff is subject to a critical constraint: handoff latency. Therefore, pilot-detection latency must be kept to a minimum to guarantee prompt handoff execution. Unfortunately, when the neighbor-set size increases and the search windows are wide, the search space may still be too large to be scanned frequently enough. In such cases, the searcher scheduler must prioritize the pilot offsets to be searched.

The scheduling algorithm is implementation-dependent, but we can envision a simple implementation in which active and candidate-set pilots have priority over neighbor and remaining-set pilots. For example, the searcher may schedule visits to the pilots in the active and candidate sets in a round-robin fashion. At the end of each cycle, it serves one pilot in the neighbor set, again using round-robin discipline. Similarly, once all neighbor-set pilots are cycled through, a pilot in the remaining set is searched. Say, for example, the active set is {A1, A2}, the candidate set is {C1}, the neighbor set is {N1, N2}, and the remaining set is {R1,R2, ...}. The scheduled visits would then follow a pattern like A1, A2, C1, N1, A1, A2, C1, N2, A1, A2, C1, R1 ..., and so on.

It is interesting to estimate how often a pilot is searched when using the scheduling algorithm above. Clearly, the duration of the detector's coherent integration period and the search window size affect the achievable search cycle period. Their relationship is, however, difficult to quantify in practice, as the searcher implementation affects the cycle period the most. Now simply assume that with a nominal search window size, the searcher is able to sweep search pilot offsets at a rate R in units of pilot offset search per seconds. If N_A, N_C, and N_N represent the size of the active, candidate, and neighbor sets respectively, the corresponding search cycle times can be estimated using the formula in Eq. (7.2), where it is assumed for simplicity that all sets have identical search window size.

$$T_A = T_C = \frac{N_A + N_C + 1}{R}$$

$$T_N = \frac{(N_A + N_C + 1)(N_N + 1)}{R}$$

$$T_R = \frac{(N_A + N_C + 1)N_N\left(\dfrac{512}{\text{PILOT\_INC}} - N_A - N_C - N_N\right)}{R} \tag{7.2}$$

T_A, T_N, and T_R represent the search cycle times of the active, neighbor, and remaining-set pilots respectively. Numerical examples using Eq. (7.2) are summarized in Figure 7.7. It can be seen that in a typical scenario with three active and candidate set pilots and 15 neighbor-set pilots, the search rate must be at least 128 searches per second to achieve a neighbor-set pilot search-cycle time below 0.5 second. Now consider that the cycle time is roughly equivalent to the maximum detection delay. A delay of 0.5 second may not seem excessive. However, in the case of a mobile station moving at high speed, the interference caused by a fast-rising pilot may increase quite rapidly and cause a call drop before handoff can be completed. The active-set pilot cycle time represents another constraint on the minimum search rate. The search results are used to assign, de-assign, or swap demodulating elements of the Rake receiver to the paths associated with active-set pilots. As the life of a path can be as short as a few tens of milliseconds, the active pilot search cycle cannot be longer than ~20 to 40 ms; otherwise demodulation resources will be wasted and valuable other path energy will not be collected.

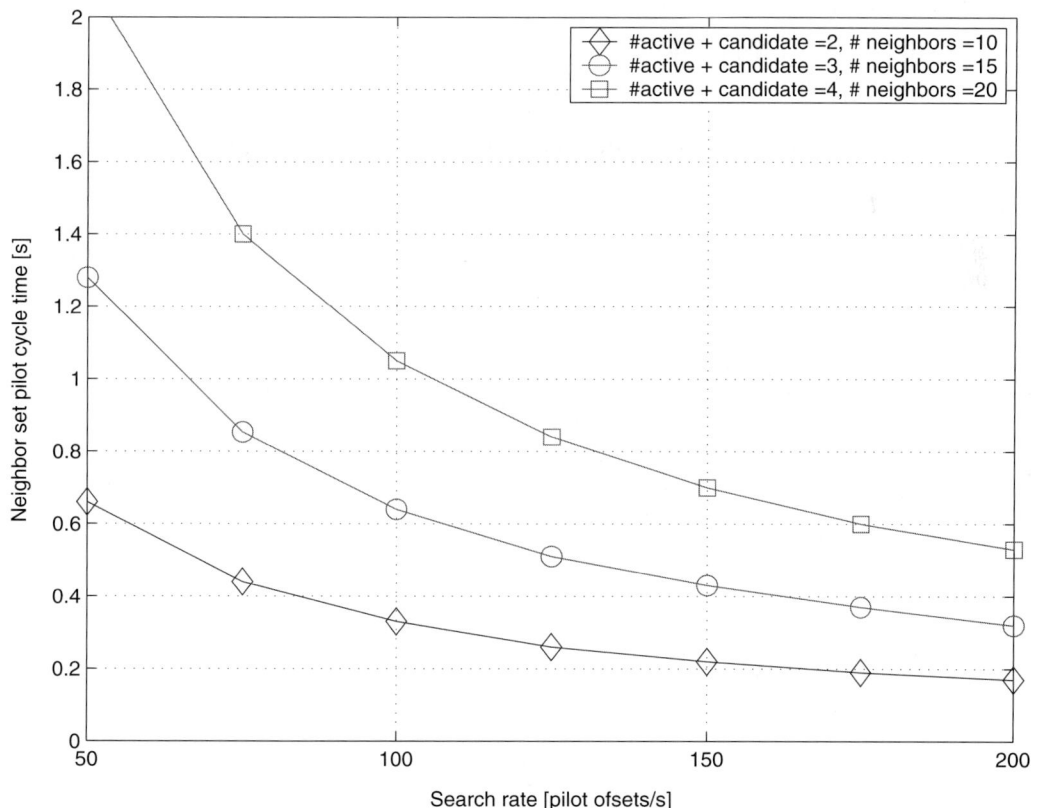

Figure 7.7 Neighbor-set pilot search-cycle time.

7.5.2 Pilot Measurement Reporting

Once the searcher detects one or more paths corresponding to a pilot offset, the received chip energy, E_c, to total power spectral density, I_o, ratio is summed over all paths to obtain an estimate of the total Pilot E_c/I_o. The IS-2000 Layer 3 standard refers to this ratio as the pilot strength. The pilot strength estimate is fed to a low-pass filter to reduce excess variability and then compared to a threshold. Depending on the set the pilot belongs to and its strength, the mobile station may trigger transmission of the *Pilot Strength Measurement Message* (PSMM). This process is key to soft handoff because PSMM reporting is the trigger used by the base station either to promote a pilot to the active set (*soft handoff add*) or demote an active-set pilot to either the candidate or neighbor set (*soft handoff drop*). In this section we often refer to the handoff parameters listed in Table 7.5 that are carried by the *System Parameters Message* (SPM), the *Extended System Parameters Message* (ESPM), and *In Traffic System Parameters Message* (ITSPM).

Table 7.5   Pilot Measurement Reporting Parameters

Field	Message*	Description
T_ADD	SPM/ITSPM	Pilot strength threshold to trigger transition from the neighbor to the candidate set.
T_DROP	SPM/ITSPM	Pilot strength threshold to trigger the handoff drop timer.
T_COMP	SPM/ITSPM	Active versus candidate set comparison threshold to trigger transmission of PSMM.
T_TDROP	SPM/ITSPM	Handoff drop timer.
SOFT_SLOPE	ESPM/ITSPM	Slope in the inequality criterion to add/drop an active-set pilot.
ADD_INTERCEPT	ESPM/ITSPM	Intercept in the inequality criterion to add an active-set pilot.
DROP_INTERCEPT	ITSPM	Intercept in the inequality criterion to drop an active-set pilot.

*Other messages may also include this field

Consider first a pilot in the candidate or neighbor set. If the strength of such pilot increases above a threshold, the mobile station sends the PSMM, including the pilot offset and strength. On receipt of the PSMM, the BSC may decide (depending on resource availability and other factors) to assign the mobile station an additional traffic channel on the base station corresponding to the reported pilot and initiate a soft handoff. The reporting threshold accounts for the pilot strength as well that of the active-set pilots. A pilot should be reported (allowing the base station to initiate soft handoff and promote it to the active set) only if it can contribute appreciably to the forward-link quality, that is, only if its strength relative to that of the pilots already in the active set is relatively large. Then, a PSMM is sent only if the strength of one or more neighbor or candidate-set pilots satisfies the following inequality criterion:

$$\text{Pilot}\ \frac{E_c}{I_o} > \max\left\{\beta_{\text{add}}\left(\sum_{i=A}\text{Pilot}\ \frac{E_{c,i}}{I_o}\right)^\alpha ; \gamma_{\text{add}}\right\} \tag{7.3}$$

The right-hand side of Eq. (7.3) consists of the combination of a dynamic and a static threshold. The exponential behavior of the dynamic threshold prevents weak pilots from triggering the PSMM when the active-set pilots are already strong. The dynamic threshold has two parameters, the slope α and the intercept β. The static threshold is based on the parameter γ. The parameters are set by the base station, which has full control of the handoff triggering conditions. Their setting is conveyed to the mobile station via the ESPM or the ITSPM. Specifically, the mobile station sets α = SOFT_SLOPE/8, β_{add} = 10 ^ (ADD_INTERCEPT/20), and γ_{add} = 10 ^ (T_ADD/20). Note that Eq. (7.3) when computed in decibels becomes a linear function of the sum of the active-set pilots' strengths in decibels, as depicted in Figure 7.8. The slope parameter is typically set to a value larger than one. Therefore, if the combined active-set pilots' strength is fairly large, condition for measurement reporting is more stringent than in the case of weak active-set pilots. The absolute threshold γ_{add} is used to avoid reporting of unsuitable weak pilots regardless of the active-set pilots' strength.

Another condition that, if satisfied, triggers PSMM transmission is when the strength of a candidate-set pilot exceeds both the dynamic threshold and the strength of an active-set pilot by an amount equal to T_COMP/2 in decibels. This reporting mechanism allows, for example, the BSC to swap a relatively weak active-set pilot with a better candidate-set pilot. T_COMP is typically set to 4 to 6, equivalent to 2 to 3 dB candidate-to-active-set pilot comparison threshold.

When an active-set pilot becomes weak and does not contribute significantly to the total active-set strength, it is desirable to drop it from the active set and release the corresponding traffic channel to avoid wasting forward-link capacity. The PSMM reporting mechanisms for weak active-set pilots is governed by a dynamic threshold and a timer. The dynamic threshold for dropping a pilot is similar to that used to promote a pilot to the active set but uses different parameters. Specifically, the mobile station starts the handoff drop timer (T_TDROP) for the i-th active-set pilot if its strength satisfies the inequality in Eq. (7.4), where the active-set pilots are ordered in ascending strength.

$$\text{Pilot}\ \frac{E_{c,i}}{I_o} < \max\left\{\beta_{\text{drop}}\left(\sum_{j>i}\text{Pilot}\ \frac{E_{c,j}}{I_o}\right)^\alpha ; \gamma_{\text{drop}}\right\} \tag{7.4}$$

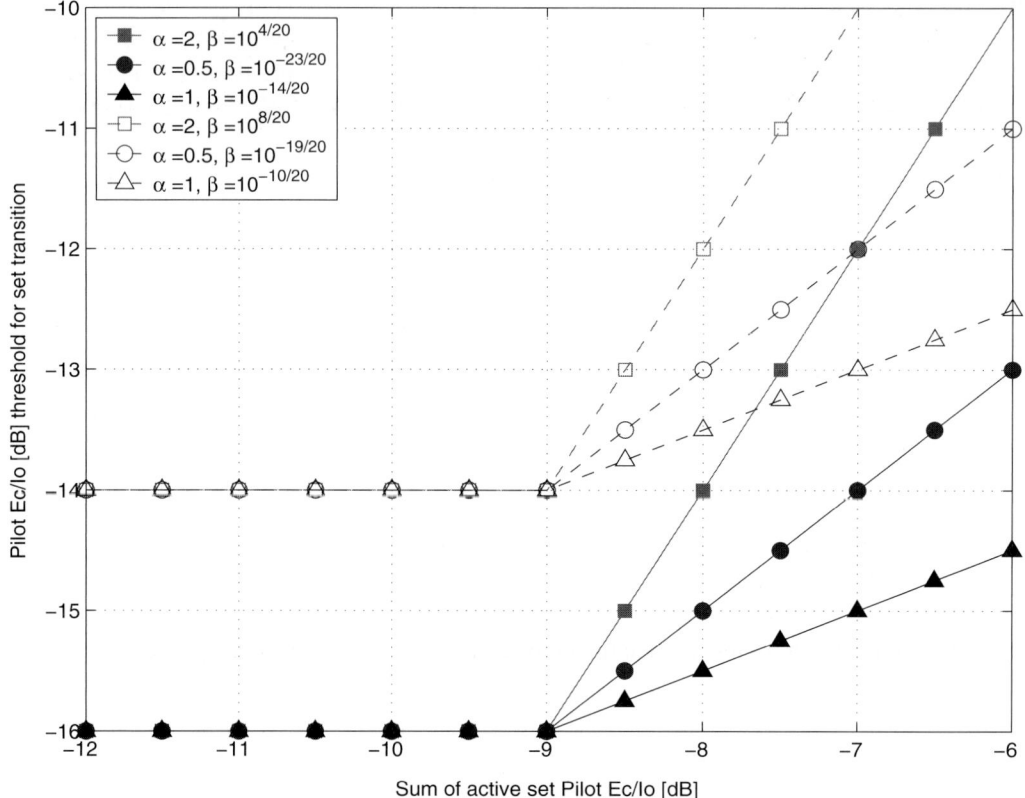

Figure 7.8 Dynamic handoff thresholds (T_ADD/2 = −14 dB, T_DROP/2 = −16 dB).

If the condition above persists until timer expiration, the mobile station sends the base station a PSMM requesting to drop the pilot. The setting of the drop-threshold parameters is also included in the ESPM or the ITSM. Specifically, it is $\beta_{drop} = 10$ ^ (DROP_INTERCEPT/20) and $\gamma_{drop} = 10$ ^ (T_DROP/20). Figure 7.8 depicts the dynamic drop threshold for different parameters' settings.

Proper setting of the dynamic threshold parameters requires a number of field measurements and a great deal tuning. The dynamic thresholds settings greatly affect forward-link capacity because they determine the average active-set size, also called handoff reduction factor, h. The handoff reduction factor, in turns, affects the total transmit power per call. The optimum handoff thresholds are the ones that minimize the total transmit power per call while meeting the QoS requirement. Now, the transmit power per traffic channel decreases with h because of diversity gain. However, the diversity gain may not compensate for the additional power required to sustain multiple traffic channels for the same call. It turns out that the total transmit power is a convex function of the handoff reduction factor. For relatively small values of h, the total transmit power tends to decrease when h is increased, as the incremental diversity gain is large. However, as the incremental handoff diversity gain diminishes with an increasing number of links, the total transmit power tends to increase with h when h exceeds the optimum operating point. The optimum value of h, and the handoff thresholds that would achieve it, greatly depend on the mobile station speed, terrain morphology, and cell layout. In many commercial networks, the handoff reduction is in the range ~1.5 to 2 after tuning of the handoff thresholds within the range values shown in Figure 7.8.

7.5.3 Pilot-Set Maintenance

Pilot-set maintenance is the process for which pilots are moved from one set to the other. Pilot-set maintenance is executed in response to Layer 3 signaling messages, pilot strength measurements, and timers.

The mobile station supports a maximum active-set size of six pilots. When the mobile station is first assigned a traffic channel, the mobile station initializes the active set to contain all pilots associated with the assigned traffic channels in the ECAM. Thereafter, the active set can change when the mobile station receives an *Extended* or *Universal Handoff Direction Message* (EHDM or UHDM), typically sent by the base station in response to a PSMM. All pilots listed in the EHDM replace the pilots in the current active set.

The candidate set is empty when the mobile station is first assigned a traffic channel. Once the strength of a neighbor or remaining-set pilots exceeds T_ADD, the pilot is promoted to the candidate set. Note that the PSMM is not necessarily triggered at this time, as the strength may not be sufficiently large to exceed the dynamic threshold. If an EHDM is received that does not list a pilot currently in the active set, and the handoff drop timer associated with such pilot has not expired, then the pilot is demoted to the candidate set. If the handoff drop timer associated with a candidate pilot expires, the pilot is deleted from the candidate set.

The mobile station supports a neighbor-set size of at least 40 pilots. When the mobile station is first assigned a traffic channel, the mobile station initializes the neighbor set to include all pilots specified in the most recent ENLM, that is, the neighbor set that was last used in idle state. The neighbor set is updated when one of the following events occur: a neighbor pilot with strength exceeding T_ADD is promoted to the candidate set; the handoff timer of a candidate pilot expires and that pilot is moved to the neighbor set; an active-set pilot whose handoff timer has expired and is not listed in the EHDM is moved to the neighbor set; on receipt of the *Extended Neighbor List Update Message* (ENLUM). The ENLUM is typically sent by the base station after handoff completion. The neighbor-set maintenance procedure following receipt of the ENLUM is controlled by an age counter. Each neighbor pilot has its own age counter. The counter is incremented each time the ENLUM is received. Pilots whose age exceeds a maximum value (the NGHBR_MAX_AGE) and that are not listed in the ENLUM are deleted from the neighbor set, while pilots that are newly listed in the ENLUM are promoted from the remaining to the neighbor set. The age counter is initialized to 0 whenever the pilot moves from the active or candidate set to the neighbor set. However, the age counter of pilots that are moved from the remaining to the neighbor set is initialized to NGHBR_MAX_AGE. As it appears from the procedures above, the age counter is used to control the "stickiness" of neighbor pilots. Pilots that are repeatedly excluded from the ENLUM are deemed not useful and demoted to the remaining set. Typically, NGHBR_MAX_AGE is set to 1 or 2. Finally, if the maximum neighbor set is exceeded, the mobile station deletes pilots that have aged the most, starting from the weakest ones. Figure 7.9 depicts the pilot sets and the events causing transition from one set to another.

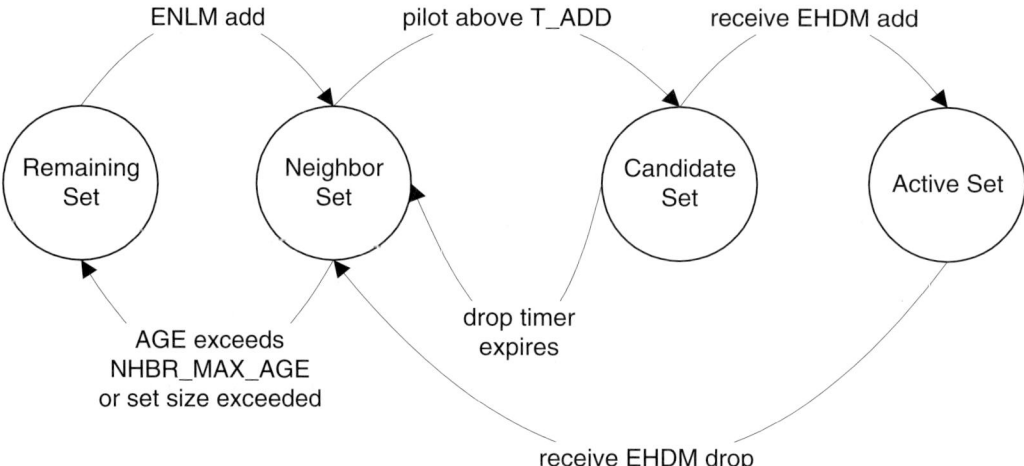

Figure 7.9 Pilot-set transitions (not all transitions shown).

7.5.4 Handoff Procedures

Pilot search, pilot-set maintenance, and pilot-measurement reporting are the responsibility of the mobile station. The base station becomes involved in soft handoff only on receipt of a PSMM. The PSMM may trigger a handoff to add one or more pilots to the active set (soft handoff add) or to drop one or more pilots from the active set (soft handoff drop). The fields of the PSMM are summarized in Table 7.6.

Table 7.6 PSMM Parameters

Field	Message	Description
REF_PN	PSMM	The pilot sequence offset index of the pilot used by the mobile station to derive its time reference.
PILOT_PN_PHASE	PSMM	The phase of the pilot being reported relative to the zero offset pilot, in chips.
PILOT_STRENGTH	PSMM	Pilot E_c/I_o, with 0.5 dB resolution.
KEEP	PSMM	Set to 0 if the handoff drop timer has expired; else, set to 1.

When the mobile station sends the PSMM, it includes the pilot Ec/Io (PILOT_STRENGTH), a flag indicating if the handoff drop timer has expired (KEEP), and the pilot phase (PILOT_PN_PHASE) is computed as

$$PILOT\_PN\_PHASE = \left(\Delta\phi + 64 \times PILOT\_PN \right) \bmod 2^{15} \tag{7.5}$$

The offset $\Delta\phi$ in Eq. (7.5) is obtained by comparing the phase of the earliest arriving usable multipath component of the candidate pilot to that of earliest arriving usable component of the reference pilot (with pilot offset REF_PN).

The base station evaluates whether or not handoff should be granted when a PSMM is received that includes one or more strong candidate pilots with KEEP = 1. The decision may be based solely on availability of resources, as the PSMM reporting is controlled by the handoff dynamic threshold whose parameters are chosen by the base station in the first place. If handoff is granted, the BSC initializes the F/R-FCH resources on the base station associated with the candidate pilot and starts transmitting the F-FCH. It then sends the EHDM (or GHDM/UHDM) directing the mobile station in soft handoff with the new base station and waits for a *Handoff Completion Message* (HCM) confirming successful handoff completion. Meanwhile, the base station associated with the candidate pilot attempts to acquire the R-FCH. R-FCH acquisition[4] can be expedited if the search window of the base station demodulator searcher is centered exactly on the expected R-FCH arrival time, so that a relatively narrow search width can be used. The algorithm to estimate the R-FCH arrival time is discussed in Section 7.5.4.2. Once the HCM is received, the BSC may update the neighbor set by sending the ENLUM. The updated list should include all pilots that are likely to become candidate pilots because they are associated with base stations in proximity of the current mobile station location. The list, however, cannot be too large, or else the search-cycle time

---

4. The base station receiver uses the reverse pilot to detect the presence of the R-FCH and acquire chip-level timing. The process is similar to that used by the mobile station for forward-link pilot search.

increases and the probability of misdetection or delayed handoff increases. An algorithm for neighbor-list updating is discussed in Section 7.5.4.3.

If the PSMM includes a weak pilot whose handoff timer has expired (KEEP = 0), the BSC initiates handoff drop procedures. It first sends the EHDM, which includes all pilots except those included in the PSMM with KEEP = 0. Upon receipt of the EHDM, the pilots whose handoff timer has expired and that are not listed in the EHDM are deleted from the active set. The mobile station sends the HCM to signal handoff completion. Only on receipt of the HCM can the BSC release the resources associated with the pilots deleted from the active set. Note that the same EHDM can be used to add or drop multiple pilots, or to add and simultaneously drop one or more pilots. After the HCM is received, the BSC evaluates the need to update the neighbor list by sending the ENLUM.

7.5.4.1   Soft Handoff Add and Soft Handoff Drop Example

We now describe pilot-set maintenance and handoff signaling procedures by means of an example. The pilot-set maintenance procedures are described with reference to Figure 7.10; the signaling procedures are described with reference to Figure 7.11. In this example, a soft handoff add is followed by a soft handoff drop.

1. At time t1, the active set and candidate set include one pilot each. The active and candidate pilot offset index are PILOT_PN 12 and 48, respectively. PILOT_PN 48 is promoted to the candidate set when its strength exceeds T_ADD.

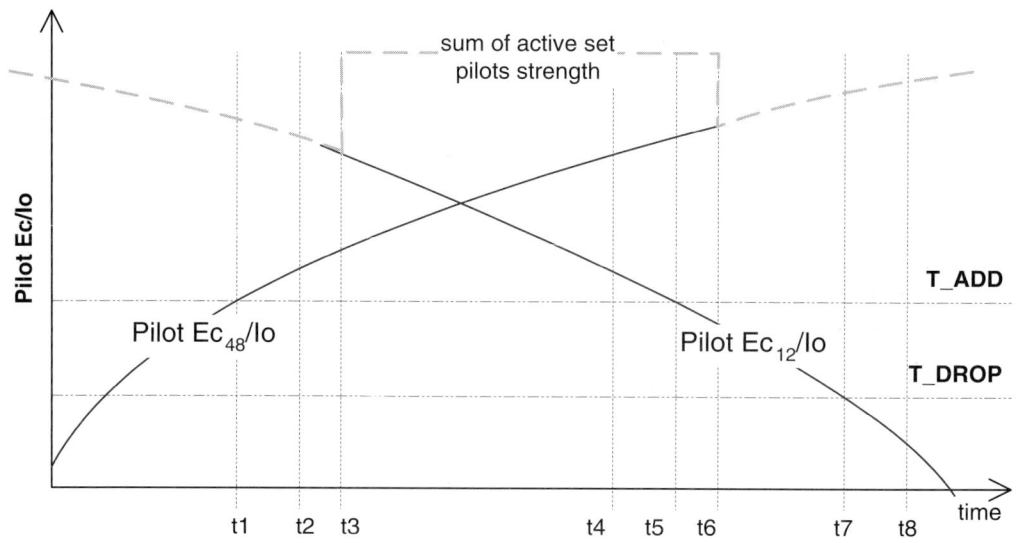

Figure 7.10 Example of soft handoff add and drop procedure.

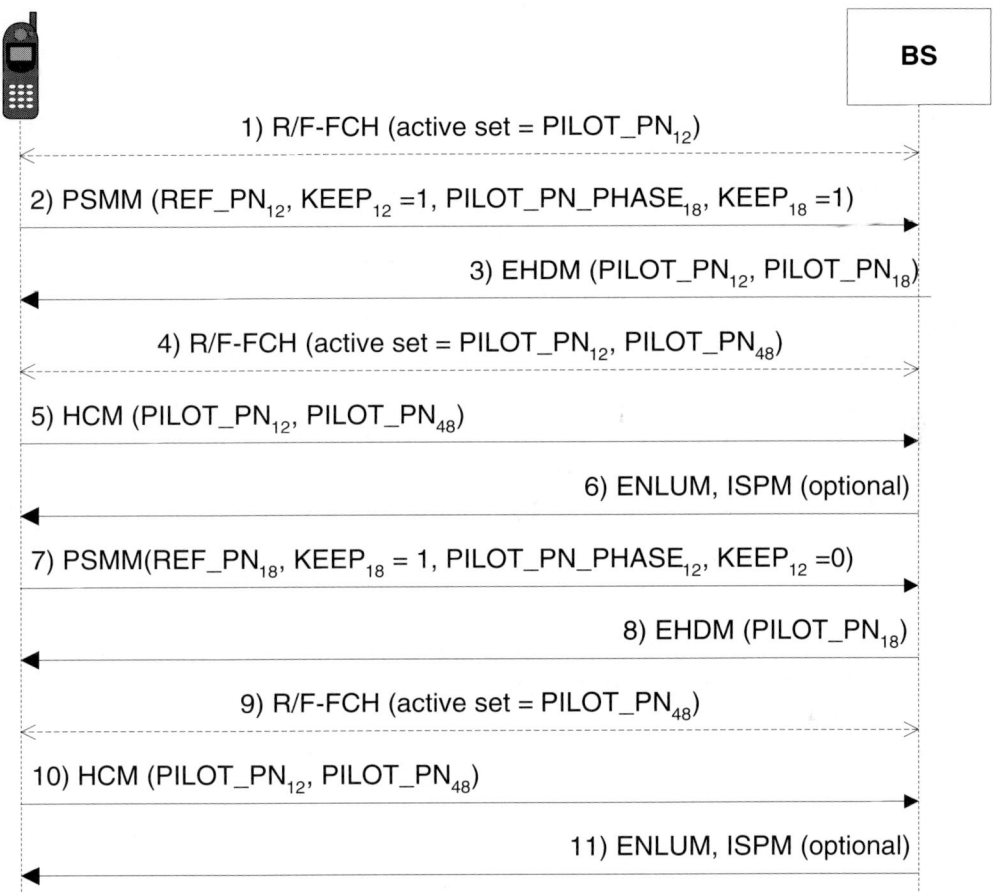

Figure 7.11 Example of soft handoff call processing.

2. At time t2, PILOT_PN 48 strength satisfies the inequality in Eq. (7.3), and the mobile station sends the PSMM (step 2 in Figure 7.11). The PSMM has REF_PN = 12, and the corresponding KEEP flag is set to 1. The PILOT_PN_PHASE corresponds to that of the candidate pilot, and its KEEP flag is set to 1. The BSC uses the PILOT_PN_PHASE to determine the candidate pilot offset index. The candidate pilot offset index is used to determine the identity of the associated base station. Resources are available on such base station so that handoff can be granted. The F/R-FCH are initialized, and transmission on the F-FCH starts. The BSC sends the EHDM containing both PILOT_PN 12 and 48 (step 3 in Figure 7.11).

3. At time t3, the mobile station receives the EHDM and promotes PILOT_PN 48 to the active set. At that time, it starts soft-combining the symbols carried by the F-FCH associated with the active-set pilots. Note the sudden improvement in total received pilot

Ec/Io. The mobile station sends the HCM containing both PILOT_PN 12 and 48. On receipt of the HCM, the BSC may send the ENLUM and/or the ISPM (steps 5 and 6 in Figure 7.11).

4. At time t4, PILOT_PN 12 strength drops below the threshold specified in Eq. (7.4). The mobile station starts the handoff drop timer.

5. At time t5, the PILOT_PN 12 handoff drop timer expires. The mobile station sends the PSMM (step 7 in Figure 7.11). Assuming that the mobile station has moved closer to PILOT_PN 48 base station, the earliest arriving usable multipath component belongs to PILOT_PN 48, which has become the new reference pilot. Hence, the PSMM now has REF_PN = 48. The PILOT_PN_PHASE corresponds to the pilot with PILOT_PN 12, and its KEEP flag is set to 0.

6. At time t6, the base station sends the EHDM that includes PILOT_PN 48 only (step 8 in Figure 7.11). Meanwhile, however, F/R-FCH processing on the base station associated with PILOT_PN 12 continues. Once the mobile station receives the EHDM, PILOT_PN 12 moves to the candidate set because it is not listed in the EHDM and its strength is above T_DROP. The mobile station sends the HCM containing PILOT_PN 48 only. F-FCH transmission from the base station associated with PILOT_PN 12 halts, and associated resources are released. The BSC may update the neighbor list and/or the systems parameters by sending the ENLUM/ISPM (steps 10 and 11 in Figure 7.11).

7. At time t7, PILOT_PN 12 strength drops below T_DROP, and the handoff drop timer is again started.

8. The handoff drop timer expires, and PILOT_PN 12 moves to the neighbor set.

7.5.4.2 Reverse Traffic Channel Acquisition in Soft Handoff

During soft handoff add, when the base station associated with the candidate pilot attempts to acquire the R-FCH, the base station receiver searches for the reverse-link pilot within the acquisition search window. If the mobile station distance to the candidate pilot's base station is unknown, the acquisition search window size would need to be at least as large as the maximum two-way propagation delay. That can be quite large (e.g., ~123 chips if the cell radius is 15 Km) and significantly slow down R-FCH acquisition. Acquisition can be expedited if the search window is centered exactly on the expected R-FCH arrival time so that a narrower search width accounting for only multipath can be used.

The expected arrival time is offset relative to the forward-link transmission timing by an amount equal to the round-trip delay (RTD). The RTD has two components. The first component is the two-way propagation delay, $\_$, which depends of the distance between mobile station and base station. The second component is the base station forward and reverse distribution delay, D, which can be accurately measured. The distribution delay depends on the base station hardware implementation and may be significant, for example, if surface acoustic wave (SAW) filters are used. The RTD at the base station receiver associated with the candidate pilot can be estimated using the following procedure.

When it receives the PSMM, the BSC computes the offset $\Delta\phi$ from the PILOT_PN_PHASE and the PILOT_PN fields. The offset $\Delta\phi$ represents the estimated difference between the one-way propagation delay of the candidate and that of the reference pilots. The two-way propagation delay difference is then $2 \cdot \Delta\phi$. Once added to the RTD of the reference pilot, it yields the estimated RTD of the candidate pilot. That is,

$$
\begin{aligned}
\text{RTD}_{\text{cand}} &= D + \tau_{\text{cand}} = D + \tau_{\text{ref}} + 2 \cdot \Delta\phi \\
&= \text{RTD}_{\text{ref}} + 2(\text{PILOT\_PN\_PHASE} - 64 \cdot \text{PILOT\_PN})
\end{aligned}
\tag{7.6}
$$

The reference pilot RTD, as well as that of any other pilot in the active set, is measured at the base station receiver by comparing the R-FCH timing to that of the forward pilot. RTD measurements are then reported to the BSC via signaling on the Abis interface, either periodically or whenever a significant RTD change relative to the previously reported RTD is measured at the receiver. Using Eq. (7.6), the RTD can be accurately estimated to within a few chips. The R-FCH acquisition window size can be set to the maximum expected multipath spread plus a margin to account for inaccurate window centering. If Eq. (7.6) was not used, the acquisition search width would have to be at least as large as the maximum two-way propagation delay.

7.5.4.3 Neighbor-List Update Algorithm

As discussed in Section 7.3.1, a large number of neighbor pilots will cause the mobile station to increase the search period and increase the risk of delaying handoff triggering. It is therefore desirable to limit the neighbor set to those pilots that are likely handoff targets. At the same time, the cost that comes with excluding a pilot from the neighbor set can be high if in fact the mobile station happens to move within its coverage area. In such a case, the pilot would be in the remaining set and given low search priority. It will likely go undetected and not be added to the active set, therefore causing significant interference and possibly a call drop. Since with every handoff event new information about the mobile station position and surrounding pilots is gathered by means of the PSMM, the BSC can update the members of the neighbor set by excluding those pilots that are no longer likely handoff targets and including new ones that are potential handoff targets. As we have seen, the base station sends the ENLUM to update the neighbor list. The algorithm used by the base station for neighbor list updating is implementation-dependent. Hereafter, we propose some guidelines.

Every active-set pilot is associated with a neighbor list. The association is maintained by the BSC in a database, the same that is used to determine the pilots to be included in the ENLM sent on the F-PCH/F-BCCH and used by the mobile station in idle mode. When the active set is updated following a handoff add or drop, the BSC forms the union of the neighbor lists corresponding to the active-set pilots. Each pilot in the union set gets a number of "votes" equal to the number of its neighboring active-set pilots. The pilots in the union set are then sorted based on the number of votes they have received, and the BSC can select the top few pilots, say, up to 15.

That is a "majority voting" rule. This rule is simple and has the property that all pilots belonging to the intersection of the active-set pilots' neighbor lists will be ranked at the top. That is desirable because those pilots are most likely candidates for handoff. The drawback of this rule is that it may discard a neighbor of a dominant active-set pilot, which is therefore a likely handoff target but happens to be voted for by relatively few active-set members.

To limit such drawback, the majority voting rule can be enhanced by giving preferred voting power to active-set pilots that are relatively strong.[5] According to this new rule, the vote assigned by an active-set pilot is weighted proportionally to its strength. We may call this rule the "rigged election" rule. For example, if the active set contains pilots PN_i, PN_j, and PN_k, each with received Ec/Io equal to P_i, P_j, and P_k, a neighbor PN_n in the neighbor list of PN_j and PN_k would be given a total vote equal to $V_n = f(P_j) + f(P_k)$, where f is some monotonically increasing function of P. The choice of f is left to the designer, but a simple one is $f(P) = \max\{P - T\_DROP; 0\}$, where T_DROP is the handoff drop threshold defined in Section 7.5.1. This metric seems reasonable, as a pilot whose strength is below T_DROP (but whose handoff drop timer has not yet expired and, therefore, is still in the active set) should be given no voting power. Finally, the sorted list can be truncated to exclude pilots that have totaled less than a minimum combined vote, V_{MIN}.

The rigged election rule maintains the desirable property of the simpler majority voting rule for which pilots belonging to the intersection of the active-set pilots' neighbor lists will be ranked at the top. Moreover, the parameter V_{MIN} can be tuned to eliminate the clutter in the neighbor set generated by weak pilots. The resulting updated neighbor list is typically neither too large (to slow down pilot search) nor too small (to miss worthy neighbors).

7.6 Pilot Planning

There are two aspects of network planning that involve pilot-related configuration parameters: selection of the pilot search window sizes and assignment of pilot sequence offsets to the base stations. Those are related topics and are discussed in the following sections.

7.6.1 Pilot Search Window Selection

As discussed in Section 7.1.4, the pilot searcher continuously scans the search windows centered on well-defined offsets in search of energy peaks. In this section we explore the tradeoffs involved with selection of the pilot search windows.

The search window size must be large enough to accommodate the multipath spread and allow the mobile station to detect the presence of paths caused by distant reflectors that are received with significant delay relative to the earliest arriving multipath component. Consider Figure 7.12, for example. The waveform propagates at the speed of light, so one chip duration

---

5.   This method was first proposed by Walid Hamdy, of QUALCOMM, Inc.

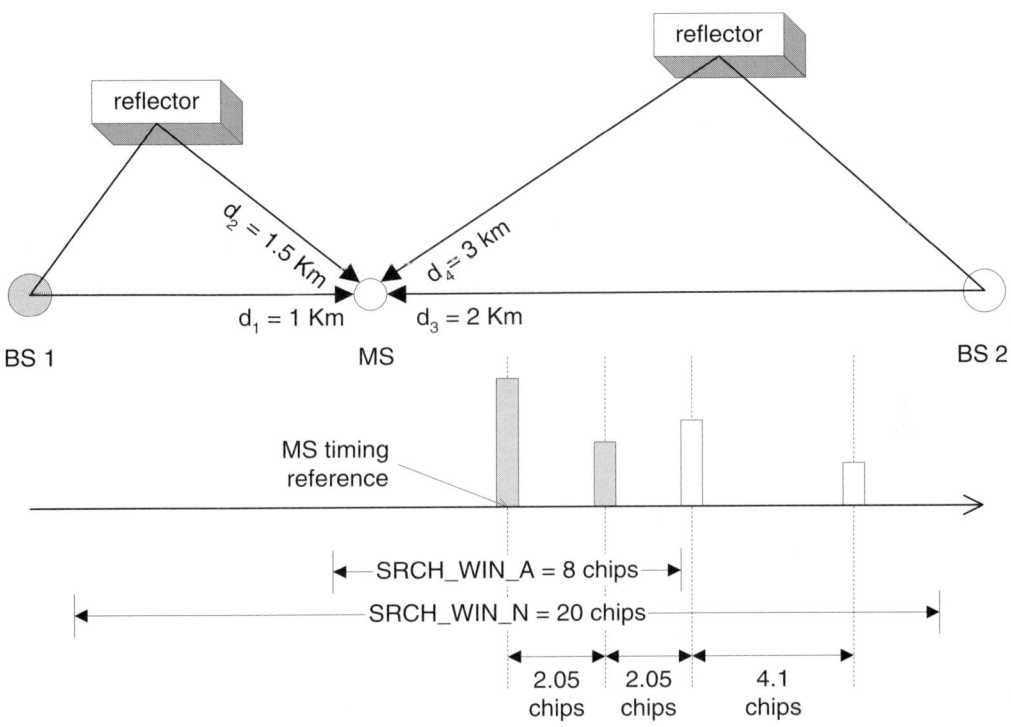

Figure 7.12 Pilot search example.

(0.8138 μs) is equivalent to a propagation distance of ~ 244.14 m. The mobile station is located at a distance $d_1 = 1$ Km in line of sight from the serving base station, BS1. The line-of-sight multipath component is the earliest arriving and is, therefore, used by the mobile station as a timing reference. A reflector is located far enough away that the reflected path must travel a distance $d_2 = 1.5$ Km. The reflected component will be seen by the mobile station arriving with $(d_2 - d_1)/244.14 = 2.05$ chips delay. An active-set pilot search window size equal to 8 chips centered on the earliest arriving multipath component will suffice in this scenario. In general, the larger the cell size, the wider the search window should be. Other factors to be considered are the presence of large reflectors and their distance relative to the cell size. Now, let's assume that a neighboring base station, BS2, corresponding to a pilot in the neighbor set has a reflector located within its range. It is desirable to detect the components of the neighboring pilot well before the mobile station approaches the idle handoff boundary. Referring again to the example, the line-of-sight component and the reflected component will be seen by the mobile stations with a delay $(d_3 - d_1)/244.14 = 4.1$ chips and $(d_4 - d_1)/244.14 = 8.2$ chips, respectively. The neighbor-set search window size is set to 20 chips, and the mobile station may, therefore, detect its presence provided it is received with sufficient strength. Note that the neighbor-set search window size is typically set to a value larger than the active-set one to account for the greater distance of non-serving base stations from the serving ones.[6]

7.6.1.1   Neighbor Search Window Offset

In a previous example (see Figure 7.12) we discussed how the neighbor search window is centered on the first multipath component of the reference pilot in the active set. In that example the search window size had to be set to 20 chips even though the delay spread is only 4.1 chips. That is clearly wasteful. To make more efficient use of the mobile station searcher resources, the ENLUM signaling message can be optionally used to set the search window size and window center position for each neighbor pilot *individually*. In the ENLUM, the neighbor-specific search window size is called SRCH_WIN_NGHBR. The window center is offset relative to the mobile station reference time by an amount indicated by the SRCH_OFFSET_NGHBR parameter. The offset can be either ±0.5, ±1, or ±1.5 times the neighbor search window size. The mobile station receiving the ENLUM uses such information in lieu of the default SRCH_WIN_N value obtained from the overhead channel messages. To optimally set a neighbor pilot's search window parameters, we must estimate the mobile station location and its distance to the neighbor base station. When the mobile station is in handoff with more than one base station, we may be able to do just that.

Assume the mobile station to be in soft handoff with N base stations and that its distance to said base stations is known. The mobile station coordinates in the Cartesian plane, $(x.y)$, can be estimated by solving the system of quadratic equations:

$$\begin{cases} \left(x - X_1\right)^2 + \left(y - Y_1\right)^2 = d_1 \\ \left(x - X_2\right)^2 + \left(y - Y_2\right)^2 = d_2 \\ \\ \left(x - X_N\right)^2 + \left(y - Y_N\right)^2 = d_N \end{cases} \tag{7.7}$$

In Eq. (7.7) the i-th base stations coordinates are represented by (X_i, Y_i) and the i-th base station to mobile station distance is d_i. The base station coordinates are known to the BSC. The BSC can also estimate the distances d_i from the PILOT_PN_PHASE measurements in the PSMM or from the RTD measurements taken by the active-set base stations, or both, as explained in Section 7.5.4. Note that Eq. (7.7) has a unique solution if $N \geq 3$ and two feasible solutions if $N = 2$. If the mobile station is not in handoff, $N = 1$, Eq. (7.7) has an infinite number of solutions corresponding to all points on the circle of radius d_1 centered on (X_1, Y_1). Once (x, y) is estimated, the distance from any neighbor base station with known coordinates (X_{N+1}, Y_{N+1}) and the mobile station can be obtained as

---

6.   That is generally true provided that the base stations have the same transmit-power rating.

$$d_{N+1} = \sqrt{(x - X_{N+1})^2 + (y - Y_{N+1})^2} \qquad (7.8)$$

The difference d_{N+1} minus d_1 is proportional to the differential propagation delay between the neighbor base station and the reference pilot's base station. The differential propagation delay is exactly the desired search window offset relative to the reference pilot's search window center. We describe the process above by means of the simple example depicted in Figure 7.13. The mobile station is in three-way soft handoff. The active-set base stations' coordinates, rela-

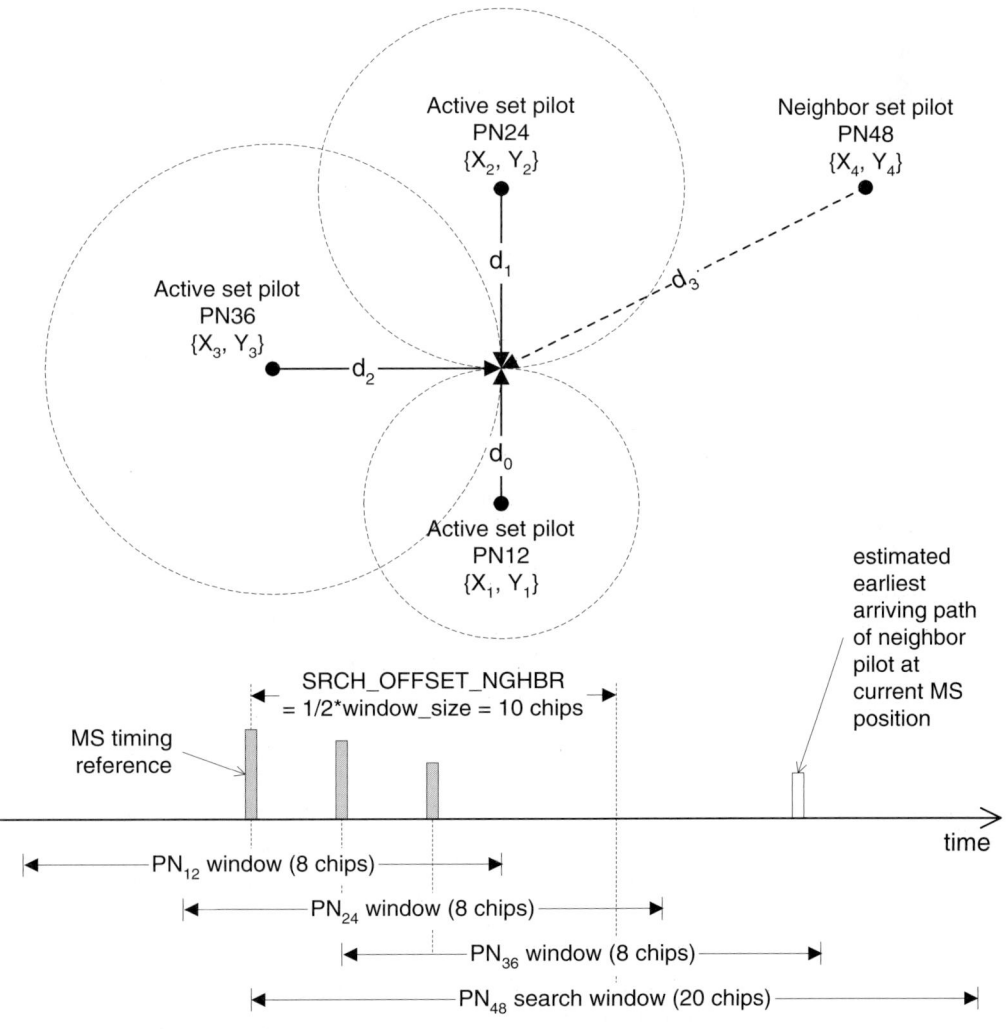

Figure 7.13 Neighbor search window offset setting example.

tive to the reference pilot's base station, are $(X_1, Y_1) = (0, 0)$, $(X_2, Y_2) = (0, 4.2)$, and $(X_3, Y_3) = (-3, 1.8)$, in units of Km. The mobile station distances to the active-set base stations are $d_1 = 1.8$ Km, $d_2 = 2.4$ Km, and $d_3 = 3$ Km. Solving Eq. (7.7), the mobile station's coordinates can be estimated as (0, 1.8). Now, consider a neighboring base station of coordinates $(X_4, Y_4) = (4.8, 4.2)$. Its distance to the mobile station is obtained from Eq. (7.8) as ~5.35 Km. As 244.14 m corresponds to a propagation delay of 1 chip, the earliest multipath component of the neighbor pilot will be delayed by (5.35-1.8)/0.244 ~ 14.5 chips relative to the mobile station time reference. Figure 7.13 shows the delay profile of the earliest arriving multipath components of the active-set and neighbor-set pilots. The delay profile shows that suitable search window parameters for the neighbor pilot are SRCH_WIN_N = 20 chips, SRCH_OFFSET_NGHBR = 1/2 × SRCH_WIN_NGHBR = 10 chips. SRCH_WIN_A = 8 chips will suffice for all active-set pilots. Note that if the neighbor offset was unspecified, the neighbor search window size would have to be 40 chips, or twice as large.

7.6.2 Pilot Offset Assignment

In Section 7.1.2 we discussed how base stations can be distinguished from each other by staggering transmission of their pilot sequences. However, as the pilot sequence repeats itself periodically, there are only a finite number of offsets that can be assigned, which leads to pilot offset reuse. Moreover, when two base stations are assigned different pilot offsets, the differential propagation delay between such base stations and the mobile station may be large enough to cancel the offset, and their pilots will be received at the mobile station with identical code phase. If pilot offsets are not properly assigned to neighboring base stations, the mobile station may detect a pilot signal and mistakenly associate it with the wrong base station. That is called pilot *aliasing*. Pilot aliasing severely disrupts soft handoff and likely leads to call drop. The scope of this section is to identify the conditions that may lead to pilot aliasing and give guidelines on how to avoid them.

As hinted above, pilot aliasing may be caused by either a base station that is assigned the same pilot offset as that of a base station in the active set (called the co-PN offset pilot) or by a base station that is assigned a pilot offset that is sufficiently close to that of a base station in the active set (called the adjacent PN offset pilot) (see also [2]). Conditions for co-PN and adjacent PN offset aliasing are better explained with reference to Figure 7.14. Assume that the active-set reference pilot is assigned PILOT_PN$_X$. Such pilot is received at the mobile station with propagation delay τ_0 in chips. Further consider a co-PN offset pilot that is not in the active set and is received at the mobile station with delay τ_1. If the delay differential, $\Delta\tau = \tau_1 - \tau_0$, is smaller than half the active-set search window, that is, $\Delta\tau < W_A/2$, the co-PN offset pilot signal may be confused as multipath from the base station in the active set. The mobile station Rake receiver will then combine the legitimate path with noise, giving rise to a high level of interference. Now, consider an adjacent PN offset base station with pilot offset index PILOT_PN$_Y$, and whose pilot is received at the mobile station with delay τ_2. The delay differential is $\Delta\tau = \tau_2 - \tau_0$. The pilot PN sequence phase difference at the transmitter is (PILOT_PN$_X$ – PILOT_PN$_Y$)·64 = $\Delta\phi$ in

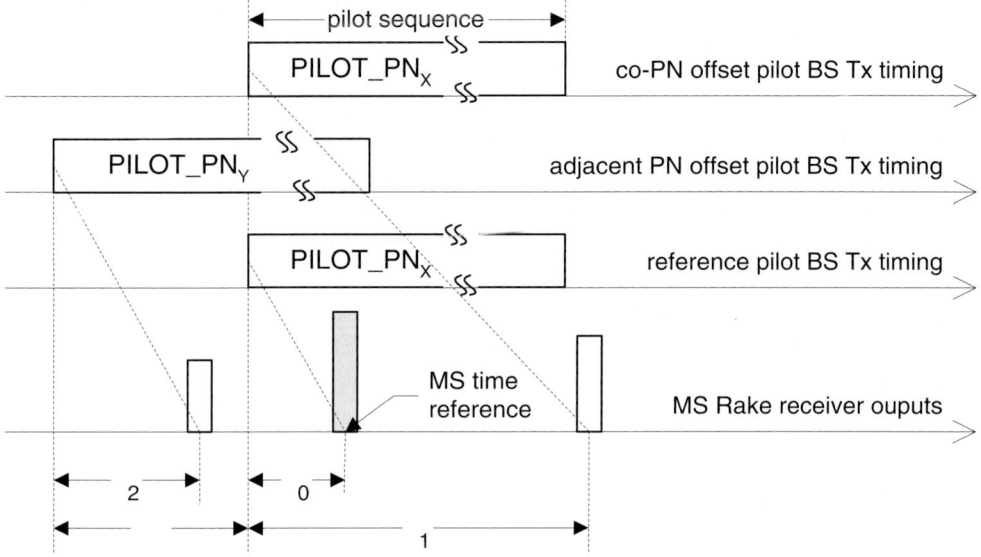

Figure 7.14 Pilot aliasing example.

chips. However, if $\Delta\phi < \Delta\tau + W_A/2$, the adjacent PN offset pilot is received at the mobile station within the active-set search window and may be confused with multipath.

It is also insightful to express the aliasing conditions above in terms of cell radius and cell-to-cell distance rather than differential delays. To that purpose, think of a mobile station located at the cell edge and on the straight line that goes from the serving base station to the co-PN or adjacent-PN offset base station. If the cell radius is R and the distance between base stations is D, in units of 244.14 m, the differential delay becomes equal to $\Delta\tau = (D - R) - R = D - 2R$, in units of one chip delay. Therefore, sufficient condition to avoid aliasing is

$$\frac{W_A}{2} < D - 2R < \Delta\phi - \frac{W_A}{2} \tag{7.9}$$

The left-hand side of Eq. (7.9) suggests that co-PN offset should be assigned to base stations located as far apart as possible, while the right-hand side suggests that adjacent-PN offsets should be assigned to base stations located as near as possible. As an example, assume that the search window size is 10 chips, the cell radius is 1 Km (or ~ 4.1 chip delay), and the minimum phase offset is $\Delta\phi = 256$ chips. According to Eq. (7.9), the *minimum* co-PN offset reuse distance is ~3.22 Km, and the *maximum* adjacent PN offset reuse distance is ~63.2 Km. Clearly, the adjacent PN offset reuse distance is not a concern in this case, while the co-PN offset reuse distance must be seriously considered when assigning pilot offsets to the base stations.

We now address one last question. How many pilot offsets are at our disposal for offset planning? Note that Eq. (7.9) represents a sufficient but not necessary condition to avoid aliasing. In fact, even if from a timing standpoint aliasing may occur, the adjacent PN offset pilot must be received with sufficient strength to be detected and confused with multipath. Consider as we did before a mobile station located at the cell edge and on the straight line that goes from the serving base station to the adjacent PN offset base station. Assume that the propagation loss is proportional to the α-th power of distance. Further, assume that the ratio of the adjacent PN offset to reference PN offset base stations' transmit-power level is ΔP. The adjacent PN offset pilot received SNR is equal to

$$\gamma = \Delta P \left(\frac{D - R}{R} \right)^{-\alpha} \tag{7.10}$$

Distances are again expressed in units of 244.14 m (1 chip delay). Now, let γ_{min} represent the pilot-detection threshold. Combining the right-hand side of Eq. (7.9) with Eq. (7.10) and then solving for the pilot offset, we obtain

$$\Delta\phi \ge \left(R + \frac{W_A}{2} \right) \left(\Delta P^{1/\alpha} \gamma_{min}^{-1/\alpha} - 1 \right) = \Delta\phi_{min} \tag{7.11}$$

$\Delta\phi_{min}$ represents the minimum offset guaranteeing that the adjacent PN offset pilot's received SNR is lower than the detection threshold and therefore cannot cause pilot confusion. The number of available pilot offsets is then bounded by $512/(\Delta\phi_{min}/64)$. The worst case is represented by very large cells with flat terrain (small propagation loss exponent). If, for example, $\gamma_{min} = -24$ dB, $W_A = 28$ chips, $R = 50$ Km, $\Delta P = 3$ dB, and $\alpha = 3.5$, the number of available phase offsets is 30. Even in this worst-case scenario, the network planner has a sufficient number of offsets for "color-coding" the network base stations.

7.7 Hard Handoff

cdma2000 supports operation on different carriers and frequency bands; thus, it is sometimes necessary to handoff from one frequency assignment to another. This is called inter-frequency handoff, or most simply *hard handoff*.[7] Unlike soft handoff, hard handoff does not allow preservation of transmission continuity, since the mobile station must discontinue operation on the traffic channel on the source frequency assignment in order to tune its RF circuitry to the target

---

7. Strictly speaking, hard handoff and interfrequency handoff are not synonymous, as the former is a general term for any form of handoff that results in loss of traffic channel continuity, such as change or frame offset or complete change of active-set pilots. In practice, however, hard handoffs are always of the interfrequency type.

frequency assignment. Only after it is tuned on the new carrier and the new traffic channel is acquired can the mobile station resume operation. Hard handoff is used in many operational scenarios, but the most typical are the following three.

Intersystem handoff [1]. Consider the scenario in which the mobile station moves across the boundary between base stations operating on a disjointed set of frequency assignments. Soft handoff is not possible, and interfrequency handoff must be performed. This scenario is most typical of neighboring base stations belonging to different systems operating on different carriers or even different frequency bands. In this case the source and target base stations are controlled by a different BSC and mobile switching center (MSC) pair. The interfrequency hard handoff then becomes an intersystem hard handoff, which requires specialized procedures on the A (BSC to MSC) and E (MSC to MSC) interfaces (see Chapter 11).

Frequency-rich to frequency-poor transition. Consider a network where base stations located in an area with high traffic, such as a downtown area, are provisioned with multiple carrier frequencies (the frequency-rich area), while cells located in the outer rings where traffic is lower are provisioned with a smaller number of carriers (the frequency-poor area). The number of carriers decreases in a "wedding cake" fashion as base stations are located further away from the inner ring. In such a scenario, mobile stations that are in a call in the inner ring and move outward have to be handed off to a different carrier if the carrier frequency currently in use is not available at the base stations in the outer ring. The source and target base stations may belong either to the same or different BSCs, that is, the interfrequency handoff may be of the intrabase station or interbase station type. In the latter case, the BSCs may interface to the same (intrasystem) or different MSC (intersystem).

Load balancing. Another use of interfrequency handoff is for load-balancing purposes. Load balancing may be used when one carrier of a multicarrier base station becomes significantly more loaded than another carrier of the same base station. In this case, some load may be shed from the loaded carrier and handed off to the unloaded carrier via an interfrequency handoff. As this procedure requires centralized resource management, this type of interfrequency handoff is typically an intrabase station handoff (same BSC).

Whichever scenarios apply, the challenges with hard handoff are the determination of the triggering instant and selection of the target base stations. The triggering instant is important, since the connection on the source carrier must be suspended before it can resume on the target carrier; doing so too early or too late may either lead to a ping-pong effect (the mobile station is handed off back and forth between the two carriers) or cause a call drop. Methods to control hard handoff fall into two categories: *measurement-directed hard handoff*, or MDHHO [5], and *mobile-assisted hard handoff*, or MAHHO [4]. The two types are described in some detail in the next sections.

7.7.1 Measurement-Directed Hard Handoff

With MDHHO, the BSC tracks the position of the mobile station and channel quality on the source carrier by means of base station and mobile station measurements to determine whether hard handoff is both necessary and likely to be accomplished successfully.

Certain cells are marked as *boundary cells*. An active set in which all cells are boundary cells is denoted as a *boundary active set*. When the mobile station active set is a boundary active set, the BSC initiates periodic measurements to determine if hard handoff should be triggered. The first set of measurements is to determine mobile station position more precisely than can be inferred from its current active set. To that purpose, the BSC receives RTD measurements from each base station in the active set. Recall that mobile station reverse traffic channel transmission is slaved to the time of arrival of the first usable multipath component on the forward direction. Therefore, the base station can estimate the RTD by comparing reverse traffic channel timing with that of its internal time reference (the one used for forward-link timing). The BSC may also request the mobile station to periodically report pilot strengths by sending the *Periodic Pilot Strength Measurement Message* (PPSMM). If the RTD of the reference pilot exceeds a configurable threshold and/or the active set combined pilot strength falls below a configurable threshold, the BSC triggers hard handoff to the new carrier. The target cell(s) are selected from a data base maintained at the BSC. Although many algorithms are possible, the general framework for target selection is the following. All boundary cells are associated with an RTD threshold and a target cell. The RTD and target base cell are determined in such a way that the handoff is triggered in a location well within the coverage area of the target cell. If the boundary active set consists of multiple cells, the association that prevails in determining the triggering instant and the target cell is the one corresponding to the reference pilot's cell. The source and target BSCs may support hard handoff-into-soft handoff in which the mobile station is assigned multiple traffic channels on the target frequency. Once the mobile station tunes to the target frequency, it starts to demodulate all assigned traffic channels as it would do while in soft handoff. If hard handoff-into-soft handoff is supported, the association may be between a boundary cell and multiple target cells. In this case the target selection is done by means of a majority voting rule. Such rule entails selecting, for example, the two cells that have collected more votes. Each vote may be weighted by the strength of the corresponding boundary cell.

Another practical MDHHO technique makes use of *pilot beacons* instead of border cells. A pilot beacon is a cell that does not carry any user traffic and transmits only the pilot channel (and possibly the overhead channels). When the mobile station reports the PILOT_PN corresponding to the beacon cell in the PSMM, the BSC triggers handoff to a target cell. Also in this case the association between pilot beacons and target cells is maintained at the base station's pilot data base. The pilot beacon is typically collocated with the target cell.

MDHHO is relatively simple but suffers from performance limitations, and it poses restrictions to cell planning. With MDHHO, target selection is basically blind, as the BSC can only make an educated guess about the suitability of the selected target cells. The RTD estimate itself is of limited reliability because it is biased by the base station distribution delay, it is affected by the presence of multipath, and it is typically computed with a resolution of one-eighth of a chip (~30 m). These factors combined make the RTD estimate accurate within one chip at best. MDHHO, then, is suited for boundary cells whose radius is significantly larger than the RTD measurement accuracy, say, ~3 Km or larger. More importantly, the coverage area of a boundary cell cannot overlap that of many cells on the target carrier, or else target selection is mere guesswork. A more reliable handoff method is the MAHHO, which is examined in the next section.

7.7.2 Mobile-Assisted Hard Handoff

MAHHO was designed to overcome the limitation of MDHHO discussed in the previous section. With MAHHO, the mobile station performs pilot strength and timing measurements on a candidate frequency and reports such measurements to the base station. Armed with knowledge of pilots' strength on the candidate frequency, the base station can more reliably select the target CDMA channels and trigger handoff at the optimum time. Moreover, timing measurements can be used to center the reverse traffic channel's search window of the target base station's receiver and expedite acquisition (see also Section 7.5.4.2). The catch here is that a mobile station equipped with only one receiver (as those commercially available today) cannot take measurements on the candidate frequency and demodulate the forward traffic channel on the serving frequency at the same time. It must first suspend traffic channel demodulation to tune to the candidate frequency, perform pilot measurements on the candidate frequency, and tune back to the serving frequency before it can resume normal operation. If this process takes more than a few frames, voice quality degrades and operation of the base station becomes more complex, as some of its functions, mainly power control and supervision, are greatly impacted.

When MAHHO was first introduced in an earlier version of the standard, it did not gain much popularity exactly because of such performance concerns and implementation complexities. MAHHO has been improved with cdma2000, which relies on a more efficient and sophisticated method to perform interfrequency measurements. In cdma2000 a mobile station that is directed to perform searches on a candidate frequency, f_C, performs the following steps. The mobile station first tunes to f_C and collects a sequence of consecutive chip samples. It then tunes back to the serving frequency, f_S, where it resumes full-duplex transmission. Note that the mobile station performs postprocessing of the chip samples to detect candidate pilots and measure their strengths only when it has resumed operation on f_S. The processing takes place in the background and is called *offline pilot detection*. With offline pilot detection, the time spent performing the search-only visit to the candidate frequency is much reduced relative to the time required with earlier implementations of MAHHO. The search time is only that spent to tune the synthesizer (from f_S to f_C and back to f_S), to allow the AGC tracking loops to settle (on f_C and then back on f_S), and to sample the received signal on f_C. The overall time spent on the candidate frequency can be kept well below 5 ms by a mobile station implemented with existing technology, and it is expected to decrease further as technology improves. Since this period is much smaller than a traffic channel frame period, forward error correction may be able to recover the frames affected by the visit to the candidate frequency. Moreover, the mobile station may perform special power-control procedures to decrease the probability of frame erasure even further [5]. For example, the mobile station may boost the transmitted power just before and just after the candidate frequency search. The consequence is that the search procedure affects only the mobile station's Layer 1 functions. Upper layers are oblivious to the search process, and no coordination between mobile station and base station is required.

In summary, MAHHO with offline pilot detection allows the BSC to select the optimum hard handoff triggering instant and target base station because it can rely on exact knowledge of the channel conditions on the candidate frequency seen by the mobile station. It is also simple to operate, since it does not require extensive tuning and does not pose constraints to network planning.

7.7.2.1 MAHHO Procedures

In this section we describe the signaling procedure used to trigger candidate frequency searches and to direct the mobile station to perform hard handoff. MAHHO procedures can be divided in two stages: candidate frequency (CF) search and hard handoff.

Candidate frequency search is triggered once the mobile station is in proximity of a likely handoff boundary. A mobile station is in proximity of the handoff boundary if, for example, its active set is a boundary active set (see also Section 7.7.1). At the onset of such condition, the base station sends the mobile station the *Candidate Frequency Search Request Message* (CFS-RQM). Although the CFSRQM can be used to start either a single or periodic CF search, its very first transmission serves two other purposes: to convey the mobile station the search configuration parameters and to interrogate the mobile station about its search capability. The configuration parameters carry information for CF search information, CF search measurement and reporting control, and hard handoff control.

The configuration parameters related to CF search include the band class and CDMA frequency assignment corresponding to the candidate frequency, the search type (single search or periodic), and a list of pilots to be searched on the candidate frequency (the search list) with corresponding search window size and priority. The search list contains all pilots that may be feasible hard-handoff targets. Note that each boundary cell is associated with its own list of hard-handoff targets. The association may be simply determined based on geographical information and stored in a database at the BSC. The search list is then computed by the BSC by merging the lists of hard-handoff targets associated with each boundary cell in the active set. The merging algorithm is implementation-dependant and may be similar to that used for neighbor-list update in soft handoff (see Section 7.5.4.3).

The configuration parameters related to CF search measurement and reporting control determine the conditions that must be met for a CF search to be initiated and for a pilot on the CF to be reported. As each visit requires momentarily interrupting two-way communication with the serving base station, thereby degrading QoS, it is desirable to perform such visits if and only if the channel quality on the serving frequency is poor. The channel quality is measured in terms of the combined active-set received signal energy and/or active-set received SNR. If any of the two measures is above the corresponding threshold, SF_TOTAL_EC_THRESH and

SF_TOTAL_EC_IO_THRESH, the mobile station is not to perform the CF search. After each CF search, the mobile station reports to the BSC the identity of all strong pilots that have been detected. The parameter CF_T_ADD represents the minimum pilot Ec/Io for a pilot to be included in the measurement report message.

The configuration parameters related to hard handoff control are used by the mobile station when directed to perform hard handoff to assess the quality of the target frequency and declare hard handoff success or failure. This information is in support of an optional procedure called *return if fail* that allows early detection of hard handoff failure with fast return to the previously serving frequency. More details on this procedure follow shortly.

On receipt of the CFSRQM, the mobile station estimates the time it will have to interrupt operation on the serving frequency to perform a CF search. Such time depends both on mobile station capabilities and on the number of pilots that are in the search list. If the mobile station does not support offline pilot detection and the search list is large, the mobile station may perform one CF search in multiple visits. In such a case it computes the maximum interruption time on the forward and reverse link corresponding to a single visit (MAX_OFF_TIME_FWD/MAX_OFF_TIME_REV) and total interruption time for a complete CF search (TOTAL_OFF_TIME_FWD/TOTAL_OFF_TIME_REV). A mobile station supporting offline pilot detection can limit the maximum interruption time to a few (two to four) power-control groups and thus avoid complex procedures to coordinate base station and mobile station during the CF search. In the following we assume that the mobile station supports offline pilot detection.

Once the mobile station is configured with CF search parameters and the base station is notified of the mobile station MAHHO capabilities, the base station can initiate CF search by sending the *Candidate Frequency Search Control Message* (CFSCM). The CFSCM is used to start or stop a periodic or single search using the configuration parameters specified in the last CFSRQM. At the end of each search, the mobile station sends the *Candidate Frequency Search Report Message* (CFSRPM). The CFSRPM contains a list of pilots detected on the CF whose strength exceeds CF_T_ADD. If periodic CF search is enabled, the mobile station maintains a periodic search timer. The timer is disabled each time the search conditions are not met. Once search conditions are met again, the periodic search timer is reset and re-enabled.

Based on the information contained in the CFSRPM, the BSC may decide that a hard handoff is both necessary (serving frequency channel quality is poor) and likely to succeed (one or more suitable target cells are found on the candidate frequency). It then selects the hard-handoff target cell(s) and sends the mobile station the GHDM or UHDM. The hard handoff is sometimes called hard handoff-into-soft handoff if multiple pilots are included in the new active set. In such a case the mobile station attempts to acquire, demodulate, and combine all such pilots on the target frequency as it would normally do in soft handoff.

Channel conditions at the time of handoff triggering may change by the time the mobile station attempts to acquire the pilot(s) on the target frequency, resulting in a handoff failure. That is more likely when the mobile station speed and the handoff signaling latency are high. To prevent the call from dropping in such a situation, the base station can optionally allow the mobile station to return to the serving system to continue the call. The mobile station can then continue searching the CF to retry a hard handoff again if possible. This process if called return if fail. It is resource-intensive, as the base station must reserve the resources allocated on the serving frequency, at least momentarily, in anticipation of a handoff failure. To minimize resource consumption, the mobile station must be able to detect hard handoff failure in the shortest time possible. The MIN_TOT_PILOT_EC_IO serves that purpose. If the combined active-set pilot E_c/I_o on the target frequency is found to be below such threshold, the mobile station does not attempt to demodulate the traffic channel but returns immediately to the source frequency. Otherwise, it attempts to demodulate the traffic channel. If at least one good frame is not received within TF_WAIT_TIME, the mobile station returns to the source frequency.

The procedures above are summarized in the following example, with reference to the call-flow diagram in Figure 7.15. MAHHO parameters are summarized in Table 7.7.

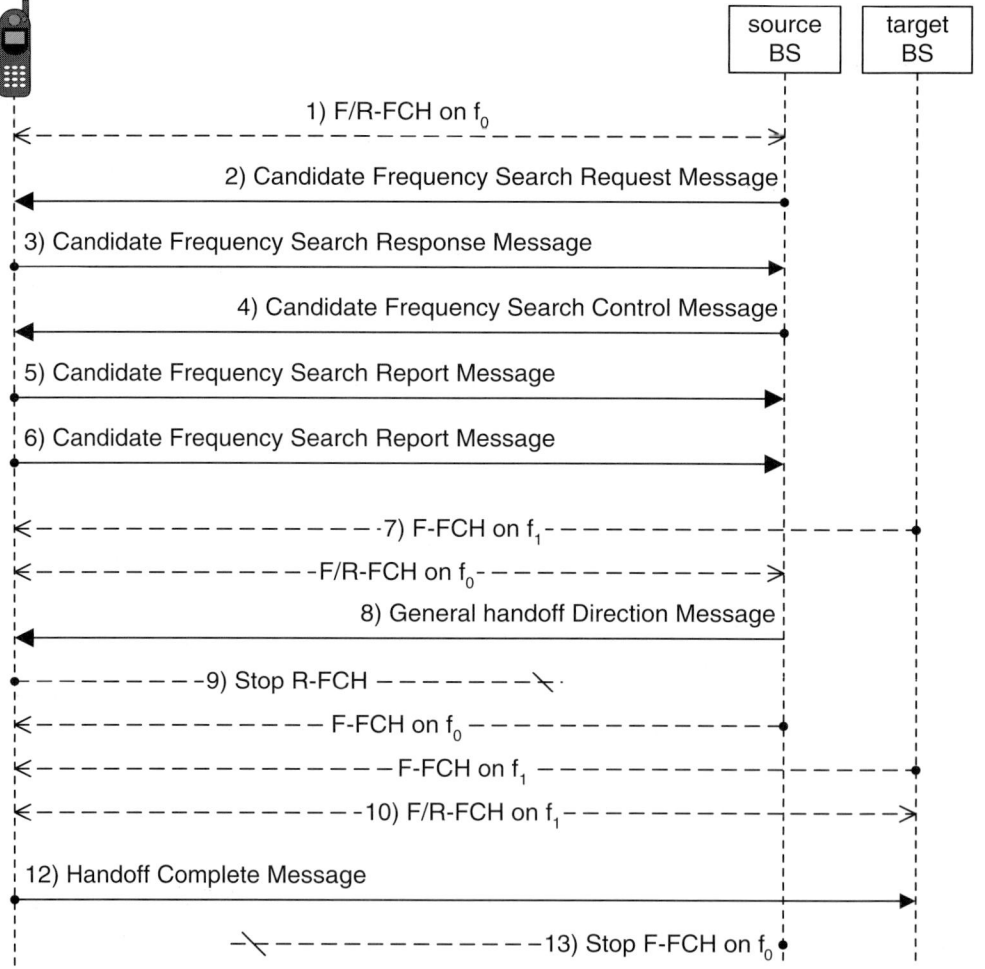

Figure 7.15 Periodic CF search followed by successful hard handoff.

Table 7.7   MAHHO Parameters*

Field	Message	Description
SEARCH_TYPE	CFSRQM	Indication to either stop the current search or start a single or periodic search.
SF_TOTAL_EC_THRESH	CFSRQM	Threshold controlling search procedures; it is the combined chip energy of all active-set pilots on the serving frequency.
SF_TOTAL_EC_IO_THRESH	CFSRQM	Threshold controlling search procedures; it is the combined Ec/Io of all active-set pilots on the serving frequency.
CF_T_ADD	CFSRQM	CF pilot-detection threshold.
TF_WAIT_TIME	CFSRQM	Maximum time spent on the target frequency waiting for one good frame after hard handoff.
MIN_TOTAL_PILOT_EC_IO	CFSRQM	Minimum combined Ec/Io of all active-set pilots; used to determine handoff success.
MAX_OFF_TIME_FWD/ MAX_OFF_TIME_REV	CFSRSM	The maximum time the mobile station will suspend operation of the serving frequency to perform a CF search or part thereof.
TOTAL_OFF_TIME_FWD/ TOTAL_OFF_TIME_REV	CFSRSM	The total time the mobile station will suspend operation of the serving frequency to perform a CF search.

*Not all MAHHO related fields are listed.

In this example a mobile station capable of offline pilot detection is in a voice call on frequency f_0 when it is directed by the base station to perform periodic CF searches on f_1. When conditions are met, the base station triggers hard handoff to f_1, which is successfully completed. It is assumed that the hard handoff is an interbase hard handoff. The signaling that takes place on core network interfaces is not shown because it is out of the scope of this chapter (intersystem hard handoff procedures are explained in Chapter 11).

1. The mobile station is on a voice call on f_0.
2. The serving base station detects conditions for CF search are met, for example, the current active set is a boundary active set. It sends the CFSRQM to the mobile station containing the candidate frequency, f_1, the pilot search list, and the CF search configuration parameters.
3. On receipt of the CFSRQM, the mobile station estimates duration of the maximum (MAX_OFF_TIME_FWD/MAX_OFF_TIME_REV) and total TOTAL_OFF_TIME_

FWD/TOTAL_OFF_TIME_REV) interruption periods on f_0 and sends that information to the base station via the CFSRSM.

4. The base station sends the CFSCNM with SEARCH_TYPE set to 11, directing the mobile station to perform searches on f_1.

5. The mobile station at the action time of the CFSCNM performs the first search and sends the CFSRPM containing the strength of the detected pilots whose strength exceeds CF_T_ADD and the received power on the serving frequency (SF_TOTAL_RX_POWER). As channel quality on the serving frequency is relatively good, and channel quality on the candidate frequency (candidate pilots' combined E_c/I_o) is relatively poor, the base station does not trigger hard handoff yet.

6. The mobile station performs periodic searches with the search interval specified by the base station in the CFRQM and reports measurements in the CFSRPM. Once conditions for hard handoff are met, the source base station initiates a procedure on the core network interfaces (not shown in Figure 7.15).

7. The target base station is informed by the core network that it is the designated target of a hard handoff. The information includes all elements to set up the air interface resources necessary to support the call. The base station sets up radio resources on f_1 accordingly and starts transmission of the F-FCH while it attempts to acquire the R-FCH. In the meantime, the serving (from now on called the source) base station and the mobile station continue operation on f_0.

8. When directed by the core network (responsible for coordinating source and base station procedures), the source base station sends the GHDM with RETURN_IF_FAIL set to 1, indicating that the mobile station is to return to the source frequency upon detection of hard handoff failure.

9. The mobile station, at the action time of the GHDM, stops transmission and reception on f_0, and tunes to f_1, attempting to acquire the F-FCH. The source base station continues transmitting the F-FCH on f_0 and starts a handoff timer. The timer duration is implementation-dependent but should be set no less than the minimum time the mobile station will take to revert to f_0 (i.e., TF_WAIT_TIME). The source base station attempts to acquire the R-FCH in order to detect the mobile station in case it reverts to f_0 following a hard handoff failure. The target base station continues transmitting the F-FCH on f_1.

10. The mobile station can detect at least one good F-FCH frame on f_1 before expiration of its TF_WAIT_TIME timer (hard handoff successful!) and start transmitting the R-FCH. The target base station acquires the R-FCH. Two-way voice call is now in progress on f_1 between the mobile station and the target base station.

11. The mobile station sends the HCM to the target base station, completing hard handoff procedures. The target base station may optionally update the system parameter used by the mobile station for this connection or update the neighbor list (not shown in Figure 7.15).

12. The source base station, either because the handoff timer has expired or because it has received indication from the core network that the handoff has been completed, whichever happens first, stops transmitting the F-FCH and releases all resources associated with the voice call.

REFERENCES

[1] Kim, Dongwoo, and Kyunam Kim, "Improving Idle Handoff in CDMA Mobile Systems," *IEEE Communications Letters*, 2(11), November 1998.

[2] Chang, Chu Rui, J. Wan, and M. Yee, "PN Offset Planning Strategy for Non-uniform CDMA Networks," *In Proc. Of IEEE VTC*, June 1997.

[3] Holcman, Alejandro, and Ed Tiedemann, "CDMA Intersystem Operations," *In Proc. of IEEE VTC*, June 8, 1994.

[4] Weaver, Lindsay, "Method and Apparatus for Measurement Directed Hard Handoff in a CDMA System," *U.S. Patent number 5,917,811*, June 29, 1999.

[5] Sarkar, Sandip, and Ed Tiedemann, "Interfrequency Hard Handoff in cdma2000," *In. Proc. of MWCN*, Recife, Brazil, August 2001.

CHAPTER 8

Power Control

Power control is crucial to the operation of CDMA systems. As the channel path loss varies due to mobile station movement, the mobile station's and base station's traffic channel transmit power must be changed accordingly to ensure that the received signal strength is equal to the receiver's sensitivity, that is, the level required for reliable signal demodulation. If the transmit power is not increased in response to an increase of path loss, the received signal power will fall below receiver sensitivity, thus degrading quality of service (QoS). Conversely, if the transmit power is not decreased in response to a decrease of path loss, the received signal power will exceed the receiver sensitivity, thus creating excessive interference that limits system capacity. In the latter case, another drawback applicable to the reverse link is that the excessive current is drained from the mobile station battery, thus decreasing talk-time. Then, accurate power control is necessary to guarantee that the received signal strength is no less and no more than the receiver sensitivity. We distinguish between closed- and open-loop power control. With closed-loop power control, the transmitter relies on the receiver's feedback to regulate the transmit power. With open-loop power control, the transmitter autonomously detects path loss variations and adjusts the transmit power level accordingly. The adjustments caused by open- and closed-loop power control are compounded.

In this chapter we examine the fundamental power-control techniques. Although many such techniques are of general applicability, the examples we provide are applicable to the Reverse and Forward Fundamental Channels (R/F-FCH) in the context of voice services. Specialized techniques applicable to the Reverse and Forward Supplemental Channels (R/F-SCH), although conceptually equivalent, differ in their implementation details and, for ease of exposition, are treated separately in Chapter 9 in the context of packet data services.

8.1 Power-Control Fundamentals

Before we explore in detail the power-control algorithms, it is useful to set the framework. A combination of open- and closed-loop power-control techniques are used on both forward and reverse links. In either channel direction, both mobile station and base station are responsible for some aspects of the power-control operation. The simplified functional diagrams in Figure 8.1 and Figure 8.2 depict the power-control functions implemented at the mobile station and base station, respectively. In the sections that follow, we refer to these diagrams often.

The mobile station performs reverse-link-loop power control by controlling the transmitter chain voltage gain prior to feeding the analog transmit waveform into the power amplifier. The combination of two components determines the control voltage. The first one is the open-loop component, which is a function of the estimated reverse-path loss. Reverse-path loss estimation is based on the received power as measured by the automatic gain control (AGC) device. The second component is the closed-loop component, obtained by processing the power-control commands received from the base station via a feedback channel. The feedback channel, or Forward Power Control Subchannel (F-PCSCH), is time-multiplexed with the F-FCH. The mobile station decodes the power-control commands and adjusts the reverse-link gain accordingly. Note that when the F-FCH is not assigned—for example, when the mobile station operates only on the Forward Packet Data Channel (F-PDCH)—the Forward Common Power Control Channel (F-CPCCH) is used in lieu of the F-PCSCH.

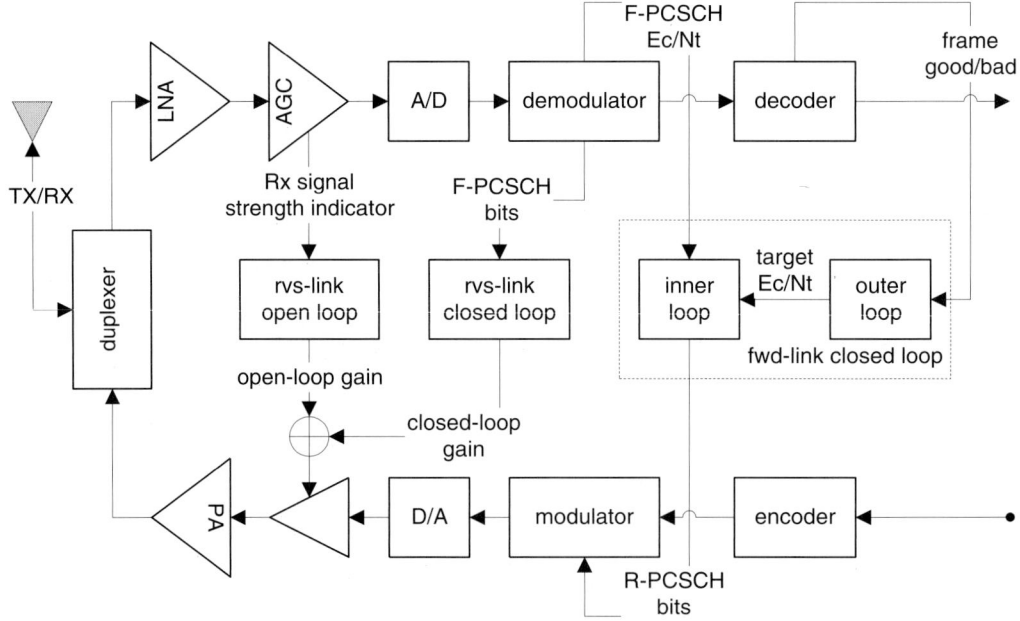

Figure 8.1 High-level mobile station power-control functional diagram.

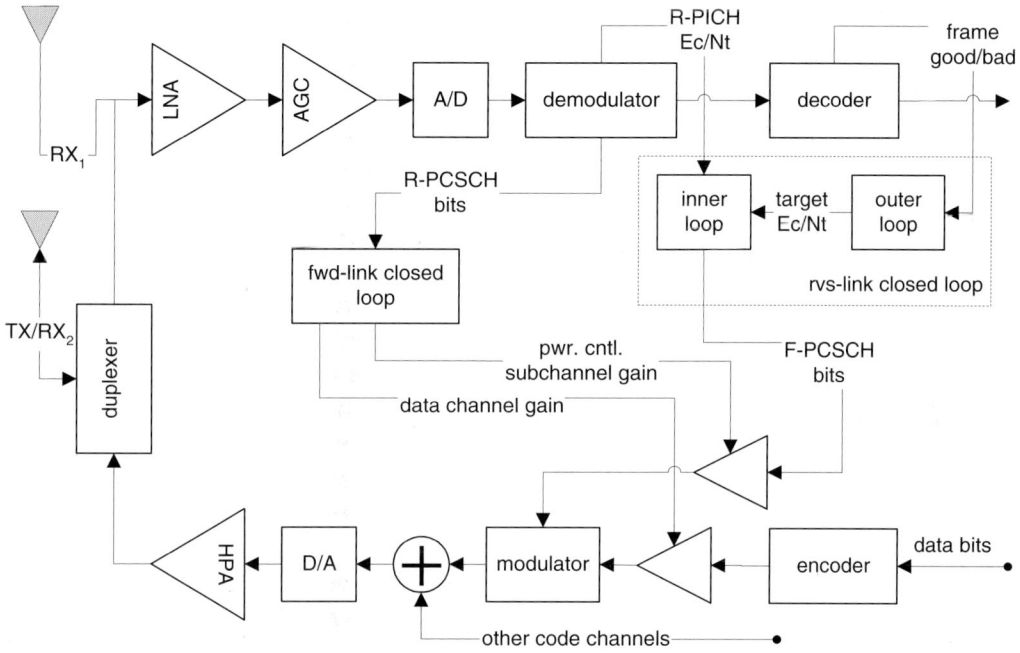

Figure 8.2 High-level base station power-control functional diagram.

The mobile station also performs forward-link closed power control. It estimates the quality of the received signal and sends the base station power-control commands via a feedback channel. Forward channel quality is estimated by combining signal to noise ratio (SNR) measurements and frame error rate (FER) estimates. The feedback channel, or Reverse Power Control Subchannel (R-PCSCH), is time-multiplexed with the Reverse Pilot Channel (R-PICH).

At the base station, power-control functions are symmetric to those performed by the mobile station, although implemented differently. The code channel gain, set by the closed-loop function, is obtained by processing the power-control commands that are sent by the mobile station on the R-PCSCH.

The base station also performs closed-loop power control. It estimates the quality of the received signal and sends the mobile station power-control commands via the F-PCSCH. Reverse channel quality is estimated by combining SNR measurements and FER estimates. The F-PCSCH is multiplexed with the traffic channel, and its transmit power follows that of the F-FCH with a configurable offset. The base station combines all forward-link code channels prior to converting the resulting signal into an analog waveform, which is then fed into the high-power amplifier.

From the high-level description above, we notice that reverse- and forward-link power control are coupled. For reverse-link power control to work, the F-PCSCH must be received reliably, which implies correct functioning of the forward-link closed-loop power control. Analogously, forward-link power control relies on reliable reception, at the base station, of the R-PCSCH whose transmit power is regulated by the reverse-link closed-loop power-control function.

Power-control functions and corresponding algorithms greatly vary depending on the channel they are applied to. Our objective in the sections that follow is to highlight the concepts of power control and the fundamental enabling mechanisms. Therefore, we limit the scope of the discussion to the power control of the R-FCH and F-FCH.

8.2 Reverse-Link Power Control

Reverse-link power control adjusts the mobile station transmit power to achieve the target SNR at the base station receiver. The target SNR is the one corresponding to the desired QoS or FER. Capacity is consumed unnecessarily if the transmit power is larger than the minimum required. Conversely, if the transmit power is smaller than the minimum required, QoS is degraded and the call may drop. Path loss variations can be categorized as one of two types. Of the first type are those long-term variations caused by changes in mobile-to-base station distance and shadowing. Of the second type are those short-term variations caused by fast fading. This differentiation motivates the use of two distinct types of reverse-link power-control mechanisms: the *open-loop* and the *closed-loop* power control. The two control loops operate concurrently, and their effect is compounded to determine the mobile station transmit power adjustments. The open-loop component is the reciprocal part of the long-term channel path loss, while the closed component is the adjustment necessary to account for fast fading and open-loop inaccuracy.

As depicted in Figure 8.1 and Figure 8.2, the reverse-link open-loop power control is implemented at the mobile station, while the closed-loop functionality is implemented at the base station.

8.2.1 Open Loop

With open-loop power control, the mobile station sets its nominal transmit power to a level inversely proportional to the total received power. The principle is that the larger the received signal strength, the smaller the transmission loss between mobile station and base station is expected to be, and vice versa. Despite the simplicity of this principle, determining the optimum proportionality constant between received and transmit power requires some ingenious thinking. In [1], the rule for setting such a constant was first devised for IS-95 systems. The rule is optimum in the sense of maximizing the minimum SNR at the base station receiver or, equivalently, minimizing the average mobile station transmit power. We derive the optimum open-loop power-control rule for IS-2000 systems following an approach similar to that used in [1]. Denote with R_c the chip rate, L the forward-link transmission loss between the serving base station and the mobile station, P_{Rx}^{MS} the mobile station total received power, and P_{Tx}^{BS} the base station total transmit power. We see that P_{Rx}^{MS} can be written as

$$P_{Rx}^{MS} = P_{Tx}^{BS} \cdot L + I_{oc} R_c = \left(1 + I_{oc} / \hat{I}_{or}\right) P_{Tx}^{BS} \cdot L \tag{8.1}$$

where I_{oc} represents the other cell interference plus thermal-noise power spectral density, and $\hat{I}_{or}$ is the received energy from the serving base station. Thus,

$$\hat{I}_{or} = P_{Tx}^{BS}/R_c \cdot L \tag{8.2}$$

Then, the transmission loss can be expressed as

$$L = \frac{P_{Rx}^{MS}}{\left(1 + I_{oc}/\hat{I}_{or}\right)P_{Tx}^{BS}} \tag{8.3}$$

Now consider R-FCH transmission. The R-FCH received SNR depends on the transmission loss and the mobile station transmit power, as in

$$SNR = \frac{P_{Tx}^{MS} \cdot L}{P_{Rx}^{BS} - P_{Tx}^{MS} \cdot L} \quad \frac{P_{Tx}^{MS} \cdot L}{P_{Rx}^{BS}} \tag{8.4}$$

Substituting Eq. (8.3) into Eq. (8.4), we obtain

$$SNR = \frac{P_{Tx}^{MS}}{P_{Rx}^{BS}} \cdot \frac{P_{Rx}^{MS}}{\left(1 + I_{oc}/\hat{I}_{or}\right)P_{Tx}^{BS}} \tag{8.5}$$

If the left-hand side of Eq. (8.5) represents the desired minimum received SNR, then the required mobile station transmit power that achieves such SNR is equal to

$$P_{Tx}^{MS} = \underbrace{\frac{1}{P_{Rx}^{MS}}}_{} \cdot \underbrace{\left(1 + I_{oc}/\hat{I}_{or}\right)}_{\substack{\text{interference} \\ \text{correction}}} \cdot SNR_{min} \cdot \underbrace{P_{Rx}^{BS} \cdot P_{Tx}^{BS}}_{\text{offset power}} \tag{8.6}$$

We distinguish three terms in Eq. (8.6): the total received power, the *interference correction*, and the *offset power*. The first one tells us that the mobile station transmit power should be inversely proportional to the total received power. However, the total received power includes both that received from the serving cell and the interfering cells and therefore is a biased estimate of the forward-link transmission loss. The interference correction is used to remove such bias. The product of the first two factors is equal to the inverse of the power received from the serving cell. Finally, the offset power represents the optimum setting of the proportionality constant that we mentioned above. Let us now see a practical implementation of the open-loop power-control function specified by Eq. (8.6).

The first term is the inverse of the total received power, which is measured by the mobile station by means of its AGC circuitry (see Figure 8.1).

The second term is the interference correction. This term can be roughly estimated from the Forward Pilot Channel (F-PICH) received SNR, or received chip energy, E_c, to total power spectral density, I_0, ratio, which we denote as F-PICH(E_c/I_o). All other things equal, the lower the F-PICH SNR, the larger the other cell interference is expected to be and therefore the larger the interference correction. However, the F-PICH SNR is a biased estimator of the other cell interference because it also depends on the serving cell transmit power. We see that the interference correction can be expressed as

$$1 + I_{oc}/\hat{I}_{or} = \frac{I_0}{\hat{I}_{or}} = \frac{I_0}{I_{or} \cdot L} = \underbrace{\text{F-PICH } E_c/I_{or}}_{\substack{\text{interference correction} \\ \text{threshold}}} \cdot \frac{1}{\text{F-PICH}\left(E_c/I_0\right)} \tag{8.7}$$

where F-PICH E_c/I_{or} denotes the F-PICH transmit chip energy to the total base station transmit energy, I_{or}, ratio. Such ratio can be measured periodically by the base station, and the most current measurement can be signaled over the Forward Paging Channel (F-PCH) or Broadcast Control Channel (F-BCCH) so that it can be used by all mobile stations to compute the interference correction using Eq. (8.7). The F-PICH E_c/I_{or} ratio, in the context of open-loop power control, is called the *interference correction threshold*. As we see shortly, it can be seen as the threshold level at which the interference correction begins to be applied. The second term on the right-hand side of Eq. (8.7) is the F-PICH received SNR, which can be measured directly by the mobile station.

We are left with the third term, the offset power. This term, just like the interference correction, cannot be computed autonomously by the mobile station, as it depends on the base station transmit and received power. One solution is for the mobile station to use a default value for the offset power. The default value can then be adjusted by means of an offset power-correction parameter signaled by the base station on the F-PCH or F-BCCH. The base station should set the offset power correction equal to the difference between the current measured offset power and the default one, known as a priori, used by the mobile station.

After these considerations, the algorithm specified in the IS-2000 standard should be less cryptic. The algorithm in its generic form is as follows:

$$
\begin{aligned}
\text{mean R-PICH Tx power [dBm]} = \quad &- \text{ mean Rx power [dBm]} \\
&+ \text{ default offset power} \\
&+ \text{ interference correction} \\
&+ \text{ offset power correction} \\
&+ \text{ other corrections}
\end{aligned}
\tag{8.8}
$$

Note that the channel being power-controlled as per Eq. (8.8) is the R-PICH. The transmit power of all reverse-link channels that are code-multiplexed with the R-PICH is referenced to that of the R-PICH, with a fixed offset. The offsets depend on the channel type and configuration, for example, data rate. The offset's default values are specified in the physical layer protocol and can also be changed by the base station via Layer 3 signaling. The right-hand side of Eq. (8.8) takes specialized forms depending on the reverse channel being considered and the current call-processing state. In the case of the Reverse Enhanced Access Channel (R-EACH), the relevant parameters used to compute the mean R-PICH transmit power are carried by the *Extended Access Parameters Message* (EAPM) that is sent on the F-BCCH. Such parameters are summarized in Table 8.1.

Table 8.1   R-EACH Open-Loop Power-Control Parameters

Field	Message	Description
Offset power	N/A	Set to –81.5 when using band class 0 and –84.5 when using band class 1.
EACH_NOM_PWR	EAPM	Correction applied to the offset power
EACH_INIT_PWR	EAPM	Additional correction applied to the offset power
IC_THRES	EAPM	Interference correction threshold
IC_MAX	EAPM	Maximum interference correction that can be applied
RLGAIN_ADJ	ECAM	Traffic channel transmit power adjustment relative to the access channel.

Consider, for example, the initial mobile station transmission on the R-EACH, followed by transmission on the R-FCH. Before sending the first access probe on the R-EACH, the mobile station reads the F-BCCH to get the current setting of IC_THRES, IC_MAX, EACH_NOM_PWR, and EACH_INIT_PWR. The interference correction term is computed based on the $\text{F-PICH}(E_c/I_o)$ measurement and the interference correction threshold. The offset power correction is set equal to the sum of EACH_NOM_PWR and EACH_INIT_PWR. The R-PICH transmit power, in decibels relative to 1 mW, is then set equal to the sum of all correction factors minus the total received power. The transmit power is increased above the open-loop estimate by a fixed amount with every access probe transmission (see also Chapter 5). In summary, the transmit power is adjusted according to Eq. (8.9):[1]

$$
\begin{aligned}
\text{mean R-PICH Tx power [dBm]} = \ & - \text{mean Rx power [dBm]} \\
& -81.5 \\
& + \min\left\{\max\left\{\text{IC\_THRES} - \text{Pilot}\frac{E_c}{I_o}\text{ [dB] ; } 0\right\}, \text{IC\_MAX}\right\} \\
& + \text{EACH\_NOM\_PWR} + \text{EACH\_INIT\_PWR} \\
& + \text{sum of all access probes corrections}
\end{aligned}
\tag{8.9}
$$

1.   During transmission of the preamble portion of the access probe, the transmit power is further increased by 6 dB.

The offset power is set to –81.5, since we assume operation in Band Class 0 (cellular band). Once the access attempt is terminated successfully and the mobile station starts transmission of the R-FCH, the open-loop transmit power is set equal to Eq. (8.10). The interference correction term is updated with the most current F-PICH SNR measurement. The access correction term is equal to the offset power correction plus the sum of all the access-probe corrections that occurred while transmitting on the R-EACH. The additional correction factor (RLGAIN_ADJ) accounts for the different received SNR requirement of the R-FCH relative to the R-EACH.

$$
\begin{aligned}
\text{mean R-PICH Tx power [dBm]} = \quad & - \text{mean Rx power [dBm]} \\
& -81.5 \\
& + \min\left\{ \max\left\{ \text{IC\_THRES} - \text{Pilot}\frac{E_c}{I_0}\ [\text{dB}], 0 \right\}, 7 \right\} \\
& + \text{access corrections} \\
& + \text{RLGAIN\_ADJ}
\end{aligned}
\tag{8.10}
$$

Once the F-FCH is acquired, the mobile station demodulates the power-control subchannel and adjusts the R-PICH in response to the power-control commands. That is the closed-loop power control, which we discuss in the next section. At this stage the mobile station no longer updates the interference correction term.

We now discuss how the base station can aid the mobile station in adjusting the open-loop power-control parameters. As mentioned above, the base station should set the IC_THRES field of the EAPM to the current value[2] of F-PICH E_c/I_{or}. Setting of the EACH_NOM_PWR field depends on the current value of base station received and transmit power. Assume that the required R-PICH SNR is equal to −19 dB, and the base station current transmit and received power levels are +43 and −106 dBm respectively.[3] Then, the EACH_NOM_PWR field should be set to a value equal to $−19 + 43 − 106 + 81.5 = 0$. That is, under such scenario the default offset power used by the mobile station is optimum and no further correction is necessary. Now assume that due, for example, to decreasing cell loading, base station transmit and received power levels decrease to +37 dBm and −108.5 dBm, respectively. The EACH_NOM_PWR field should now be set to $−19 + 37 − 108.5 + 81.5 = −9$ dB.

8.2.2 Closed Loop

Even with an ideal implementation, open-loop power control alone cannot guarantee that the R-FCH received power at the base station is equal to receiver sensitivity. That is because the open-loop relies on forward- and reverse-link path loss symmetry. But such symmetry does not

---

2.   The IC_THRES is typically set equal to the low-pass filtered version of F-PICH E_c/I_{or} to avoid too frequent updates of the overhead channels' messages.

3.   Assuming a base station transmit power rating equal to 20W and a receiver noise figure equal to 4 dB, it can be seen that such scenario corresponds to a heavily loaded cell.

exist in practice. Although forward and reverse path loss due to distance and shadowing are highly correlated, Rayleigh fading strongly depends on carrier frequency, which is different in the two directions. Other factors that limit the effectiveness of the open loop are the inaccuracies in setting the open-loop correction factors and in calibration of the mobile station's power amplifier. Thus, a closed-loop control must be devised that varies the mobile station transmit power based on received SNR measurements made at the base station. Such measurements enable the base station to continuously send the mobile station commands to either increase or decrease its transmit power depending on whether the received SNR was lower or higher than desired.

With closed-loop power control, the mobile station receives power-control commands on the F-PCSCH and increases or decreases the transmit power accordingly. The R-PICH transmit power in the traffic channel state, when closed loop is active, is given by

$$
\begin{aligned}
\text{mean R-PICH Tx power [dBm]} = \quad & - \text{mean Rx power [dBm]} \\
& + \text{open loop corrections} \\
& + \text{sum of closed loop power control corrections}
\end{aligned}
\tag{8.11}
$$

As mentioned earlier, the R-FCH is transmitted with a fixed power offset relative to the R-PICH. If the received power-control bit is equal to 1, the transmit power is decreased by an amount equal to the power-control step size; if the received power-control bit is equal to 0, the transmit power is increased by an amount equal to the power-control step size. The default power-control step size is equal to 1 dB. However, the base station can direct the mobile station to use a step size equal to 0.5 dB or 0.25 dB,[4] as specified by the PWR_CNTL_STEP field of the *Power Control Message.*

At the base station, closed-loop power control consists of an inner and an outer loop. The base station sets the power-control bits to track the received R-PICH SNR (inner loop's function) and allocate each active mobile station just enough SNR to close the reverse link (outer loop's function).

8.2.2.1 Inner Loop

The inner loop power-control function is performed at the radio base station receiver in two steps. In a typical implementation, at first it estimates the received R-PICH SNR by accumulating the R-PICH energy over 1,152 consecutive chips, or 1.25 ms periods. Such periods are time-aligned with the frame boundary and are called *power-control groups*. Then, it compares the estimated SNR against that allocated to the mobile station. If the received SNR is higher than the target, the base station sets the power-control bit to 1 (*down* command). If the received SNR is lower than the target, or if the receiver is out of lock, the power-control bit is set to 0 (*up* com-

---

4.   A smaller step size may be used when the base station knows that the user terminal is fixed, as in the case of wireless local loop applications. In such scenario, path loss variations are very slow and of limited magnitude and can be tracked by a slower closed-loop. A slower loop has the advantage of smaller variance.

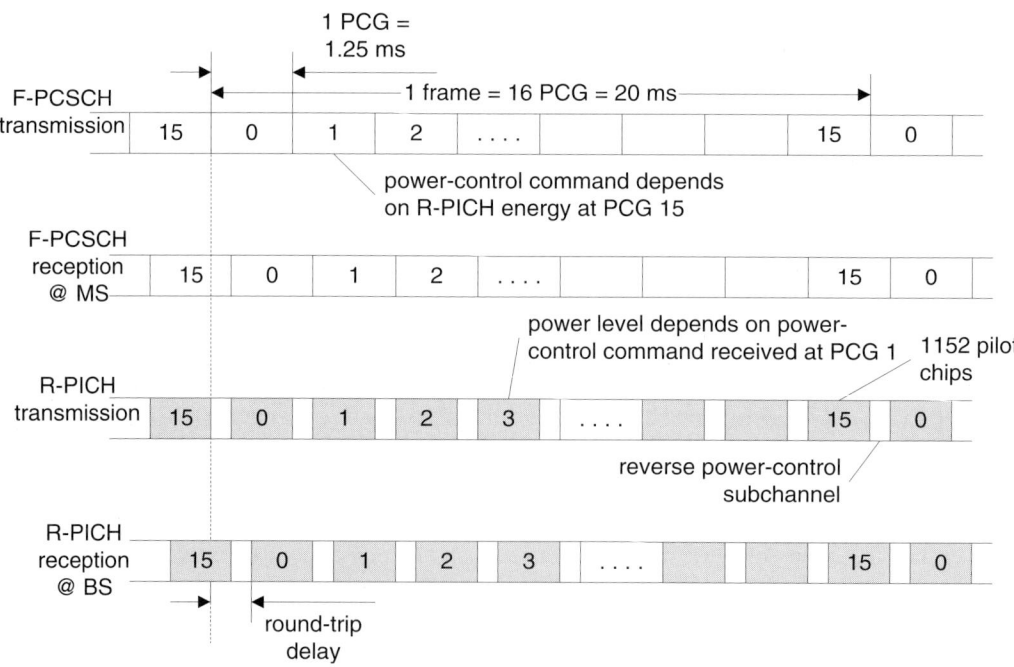# Chapter 8 • Power Control

mand). Since there are 16 power-control groups in a 20-ms frame, the power-control command rate is equal to 800 bps. Figure 8.3 depicts reverse-link power-control timing from both mobile and base station perspective.

We conclude this section with an example of how open-loop and closed inner-loop power control jointly work according to Eq. (8.11). Figure 8.4 depicts a scenario in which a sudden improvement in channel condition is followed by a sudden degradation. The speed at which the mobile station tracks variations in received signal strength is equal to the inverse of the AGC filter time constant, assumed equal to 10 ms. The power-control step size is equal to 1 dB. It can be seen that in steady state the sum of the closed power-control corrections is equal to 0 and that the base station transmits alternately up and down commands.

In reality, reverse-link power control is less ideal than Figure 8.4 shows, since we must account for the impact that fast fading has on the joint operation of the open and closed inner loop (see Section 8.2.4) and closed-loop imperfections. Closed-loop imperfections are mainly the effect of power-control commands received in error and of the SNR estimate such commands are based on. Note that the power-control commands are transmitted uncoded to minimize the turnaround time. When power-control commands are received in error, power control suffers, as the change in transmit power is exactly opposite in direction to that required to counteract the path loss variation. When the SNR estimate is incorrect, the wrong power-control command may

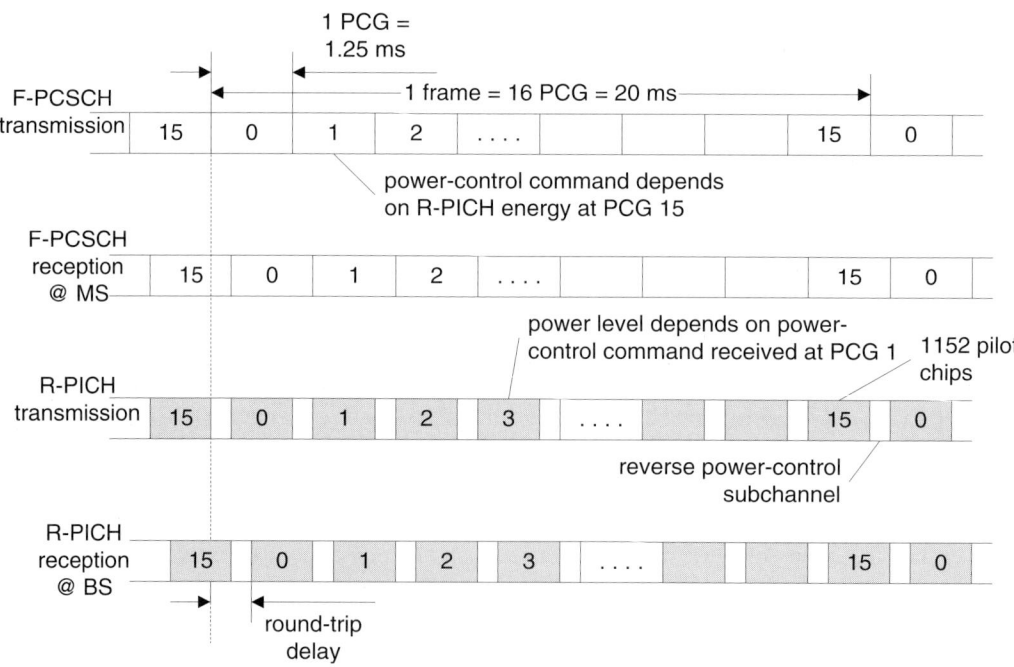

Figure 8.3 Reverse-link power-control timing.

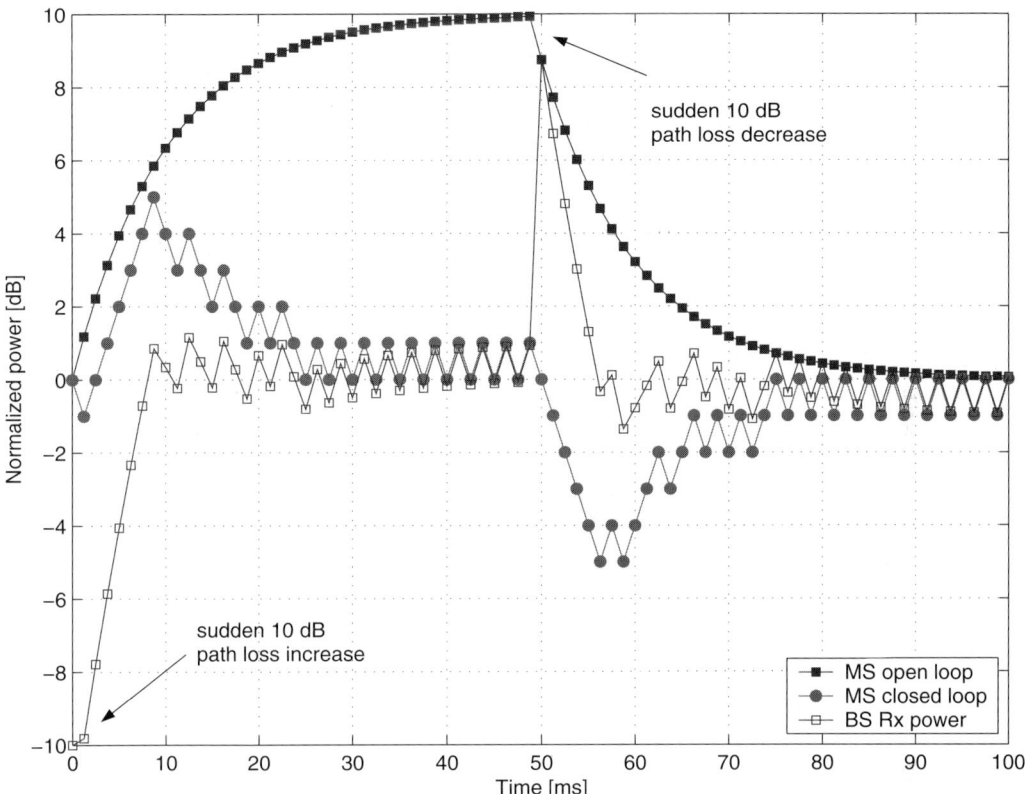

Figure 8.4   Relative signal levels for sudden path loss changes.

be sent to the mobile station. These errors, and the finite power-control rate, are such that the received R-PICH SNR is only on average equal to the target but exhibits fluctuations that are more pronounced when the mobile station moves at high speed.

8.2.2.2   *Outer Loop*

The outer-loop power-control function tracks QoS and adjusts the R-PICH target SNR used by the inner loop accordingly. Allocating too little SNR would degrade QoS. Allocating too much SNR would degrade cell capacity. An obvious measure of QoS is the R-FCH FER. The outer loop must estimate the actual FER and compare it against the desired one. If the estimated FER is higher than the target, then the outer loop allocates a larger SNR to the mobile station and commands the inner loop to raise its SNR threshold. Otherwise, the outer loop commands the inner loop to lower its SNR threshold. The outer loop is implemented at the base station controller since, when in soft handoff, the FER must be estimated after selecting the best quality frame among those received from the links in soft handoff. A typical implementation of the outer-loop algorithm is described next.

Ideally, the outer loop should use the estimate of the instantaneous FER to adjust the allocated SNR. Since the instantaneous FER cannot be known, a pragmatic approach is to assume that the FER is low if the current frame is good and, therefore, decrease the allocated SNR. Conversely, if the current frame is bad, we can assume the FER to be high and, therefore, increase the allocated SNR. Furthermore, we must decide upon the rate of change of the allocated SNR due to changes in channel conditions and mobile station speed. The rate of increase of allocated SNR, R_{UP}, should be such that the loop responds quickly to degrading channel conditions to avoid call drop. The rate of decrease of allocated SNR, R_{DOWN}, should be such that the loop responds quickly to improving channel conditions to avoid waste of capacity. Once we have decided on the rate of change, how much we increase or decrease the allocated SNR, Δ_{UP} and Δ_{DOWN}, respectively, in response to the received frame quality depends on the rate of bad and good frames. Finally, outer-loop operation is subject to delays, τ, for example, due to frame transmission to the base station controller and, therefore, its response must be slowed to match those delays and avoid an increase of loop variance. After these considerations, we can appreciate the rationale for the proposed algorithm. If the allocated SNR is to increase with slew rate R_{UP} and considering that after receipt of a bad frame the allocated SNR is not increased for a period of at least the τ frames loop delay, we have

$$R_{UP} \cdot \tau = \Delta_{UP} - \Delta_{DOWN}\left(\tau - 1\right) \tag{8.12}$$

Consider now the average number of steps per loop period. Clearly, the loop period must be equal to the inverse of the target FER. Assume, for example, that the target FER is 1%. On average, we expect to receive 1 bad frame followed by 99 good frames in each loop period. Then, the loop period is equal to 100 frames, and the step up must be 99 times larger than the step down if want the allocated SNR to return to the same level at the beginning of each loop period. That is, step up and step down are such that

$$\Delta_{UP} = \left(\frac{1}{\mathrm{FER}} - 1\right) \cdot \Delta_{DOWN} \tag{8.13}$$

From Eq. (8.12) and (8.13), we obtain

$$\Delta_{DOWN} = \frac{R_{UP} \cdot \tau}{1/\mathrm{FER} - \tau} \tag{8.14}$$

In summary, the outer loop operates as follows. If a bad frame is received and at least $\tau - 1$ frames have been received since the last bad frame, the allocated R-PICH SNR is increased by an amount Δ_{UP} dB; otherwise the allocated SNR is decreased by an amount Δ_{DOWN} dB. The allocated SNR is sent by the base station controller to the radio base station receiver performing inner-loop power control. Furthermore, the allocated SNR is constrained to be within an upper and a lower limit. The upper limit is needed to avoid allocating excessive SNR to a mobile station that can no longer sustain its reverse link because, for example, it is moving outside of the base station coverage area. The lower limit is needed to avoid allocating too little SNR to a mobile station experiencing such favorable channel conditions that are likely to be only temporary, such as when the mobile station stops and then suddenly resumes its motion at accelerating speed. The outer loop parameters and their significance are summarized in Figure 8.5. Typical parameters' settings are FER = 1%, R_{UP} = 0.055 dB/frame, and τ = 4 frames (including the radio base station to base station controller transmission delay). Based on these settings we obtain Δ_{UP} = 0.22 dB, and Δ_{DOWN} = $0.23 \cdot 10^{-1}$ dB.

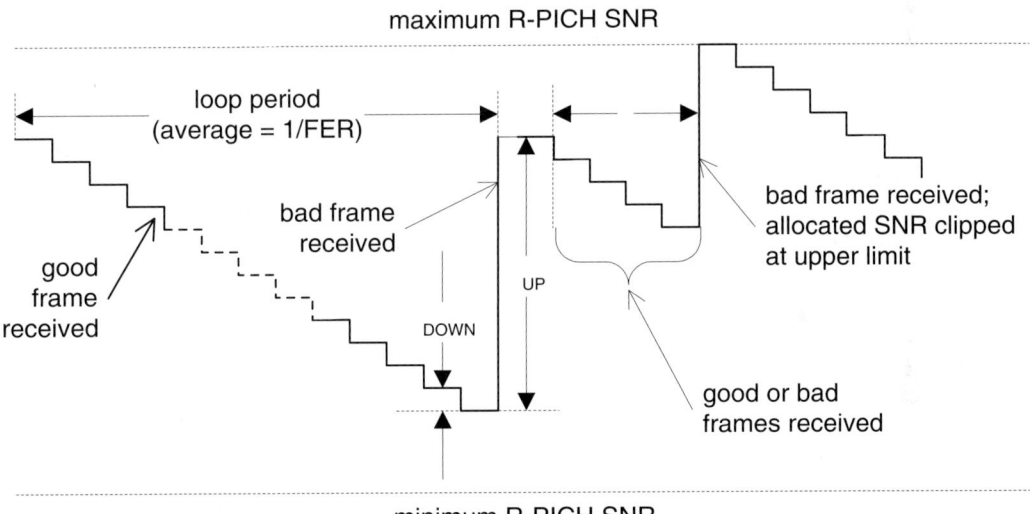

Figure 8.5   Reverse-link closed-loop power control: exemplary outer loop trajectory.

8.2.3 Reverse-Link Power Control in Handoff

When in soft handoff, the R-PICH received by the active-set base stations differ in strength because of different transmission loss between mobile station and each base station involved in the handoff. Then, while the R-PICH is received at some base stations with high SNR and such base stations send a down power-control command, some other base stations may receive the R-PICH with insufficient strength and decide on an up command. The implication is that the mobile station cannot perform coherent combing of the power-control bits carried by the F-FCH as it does for the data bits. Instead, it must demodulate them individually and then decide on the correction to be applied to the R-PICH transmit power on the basis of all received power-control commands.

The rule used by the mobile station is the so-called *OR-of-the-downs*, for which the mobile station increases the transmit power if and only if all received power-control commands are up commands. The rule is devised so that at least one base station receives the R-PICH at the allocated SNR, while all other base stations receive it at the lower level. That is, the mobile station is effectively power controlled by the base station with the best reverse link. The rule maximizes reverse-link capacity while guaranteeing that the mobile station can close the reverse link at all times with at least one base station. Note that the standard mandates that only "reliable" power-control commands are combined in the OR-of-the-downs fashion. In fact, power control in soft handoff may fail altogether in the presence of unreliable commands, which may happen when the path loss difference of the links in handoff is relatively large or because of poor forward-link power-control performance.[5] Assume, for example, a two-way handoff scenario in which the quality of one of the links in handoff is so poor that the power-control subchannel bit-error rate is 50%. This is of course an extreme and unrealistic case but serves well as an example. The best serving base station has instead a nominal power-control subchannel error rate equal to p. Further assume that the reverse-link path loss increases as the mobile station moves further away from the base station. In this case both base stations will send up commands, but as a result of combining valid with erroneous bits, the mobile station will decrease its transmit power instead of increasing it! Let us see how that happens. The probability P_U that the mobile station increases its transmit power is equal to the probability that both bits received from the base station are decoded as up commands, that is, $P_U = 0.5(1-p) \le 0.5$. Unless one of the power-control subchannels is error free, the mobile station transmit power control is downward biased, and slowly but surely the reverse-link connectivity will be lost with both base stations. In general, when the multiple soft links are combined the i-th power-control subchannels' bit-error rate, p_i must be such that $\prod_i (1-p_i) \le 0.5$. To ensure that this constraint is satisfied, only reliable bits are fed into the bit decision logic, as the physical layer protocol requires. Determination of bit reliability is implementation-dependent. An approach is to use the estimated pilot and traffic channel SNRs to infer the quality of the power-control bit. Another approach is to use an erasure threshold against which the soft decoded bits are compared. If the magnitude of the soft value of the decoded bit is less than the threshold, the bit is erased; otherwise, it is fed into the bit combiner.

Before concluding this section, we need to mention the peculiarity of power control in softer handoff. Recall from Chapter 7 that softer handoff is the situation in which two or more cells of the same base station transceiver (BTS) correspond to pilots in the active set. The signals from the cells' receive antennas can be fed into the same demodulating element and soft-combined. Then, inner-loop power control is performed based on the combined received SNR, and the power-control bit is set to the same value on all F-PCSCHs of the links in softer handoff. The mobile station coherently combines the power-control subchannels corresponding to the links in softer handoff to obtain a single power-control command. The resulting command is then used

---

5.   See, for example, Section 8.3.3 for discussion of forward-link imbalance when in soft handoff.

in the OR-of-the-downs rule with power-control commands received from the remaining links in the active set. The base station informs the mobile station whether a link is in soft of softer handoff by setting the corresponding power-control combining indicator (PWR_COMB_IND) to either 0 or 1, respectively, in the *Universal Handoff Direction Message* (UHDM).

8.2.4 Open-Loop Time Constant Selection

Before concluding the discussion on reverse-link power control, there are important remarks to be made on the interaction between open and closed loops and its impact on mobile station receiver design. The speed at which the mobile station tracks variations in received signal strength depends on the time constant of the AGC filter (see Figure 8.1). Proper setting of the time constant depends on the correlation between forward and reverse links. If the correlation is high, then a fast open loop can be used. Conversely, a slow open loop is preferred if the correlation is weak. In practice, the AGC filter time constant must be selected so that fast channel variations due to Rayleigh fading are filtered out to avoid upsetting the closed loop.

Assume, for example, that the forward channel and reverse channel transmission loss change by the same amount, say, 10 dB, but in opposite directions. If the open loop were to react instantaneously to the forward transmission loss change, then the transmit power would be adjusted in the opposite direction and the open-loop error would be equal to 20 dB, which may not be compensated for fast enough by the closed loop. However, if the mobile station had not reacted instantaneously to such change, then the closed loop would have had to compensate only for an open-loop error equal to 10 dB. Due to the above considerations, the AGC time constant is typically set between 10 and 20 ms. An example of open-loop power transition in response to 10-dB path loss change is depicted in Figure 8.6 for two different values of the AGC filter time constant. The physical layer protocol specifies a mask for the open-loop transient characteristic, whose limits are also depicted in the same figure.

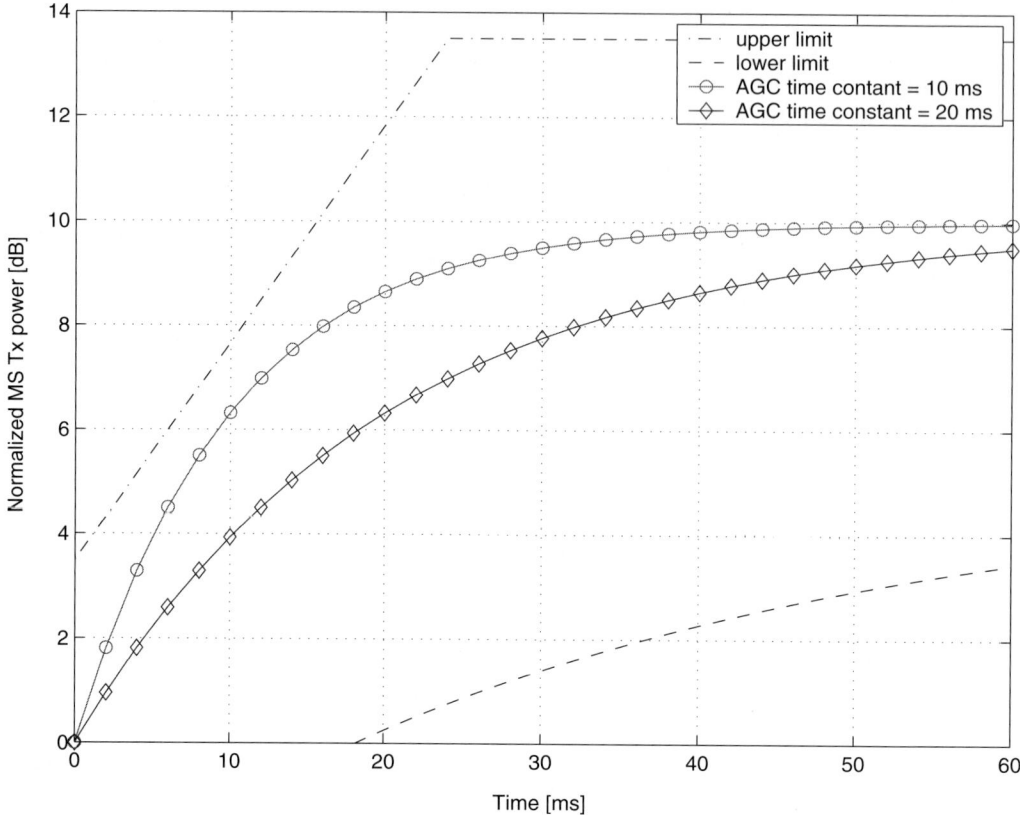

Figure 8.6 Reverse-link open-loop power transitions characteristics.

8.3 Forward-Link Power Control

Forward-link power control is similar in nature to its reverse-link counterpart, although the implementation details are quite different. Equivalently to the reverse link, forward-link power control adjusts the base station code channel transmit power to achieve the target SNR at the mobile station receiver. The target SNR is the one corresponding to the desired QoS. Capacity would be consumed unnecessarily if the transmit power were larger than required. On the other hand, if the transmit power were smaller than required, connection quality would be degraded and the call may drop. Just like the reverse link, forward-link power control consists of an open loop and a closed loop. Forward-link open-loop functionality is implemented at the base station, while closed-loop functionality is implemented at the mobile station.

8.3.1 Open Loop

When the forward traffic channel is initialized, for instance, at call setup, the base station must select its initial transmit power level and initialize the code channel gain accordingly. We call such operation *forward-link open-loop power control*. Note that no reference is made to forward-link open-loop power control in the physical layer specification, since it is an optional feature whose design is implementation-dependent. The F-FCH initial transmit power should be high enough to allow the mobile station to acquire and reliably detect the F-FCH and should be low enough to avoid unnecessary interference to other mobile stations. The first consideration is typically given more weight by system designers, since a lower than needed initial F-FCH transmit power is likely to cause connection establishment failure. It is not before the mobile station has detected two good frames on the forward link that it starts transmission of the R-PICH and power-control subchannel, allowing forward-link closed-loop power control to be activated. Since forward-link path loss is not known a priori, the simplest solution is to initialize the F-FCH transmit power to the maximum allowable level to maximize call setup reliability. Such blind initialization, however, is wasteful of forward-link capacity and increases the peak-to-average of the base station total transmit power. In this section we describe an algorithm to determine the initial F-FCH transmit power following the method proposed in [3]. We consider the simplest case of a single path channel and a single base station in the active set at the time of channel assignment. Results can be extended to the case of a multipath channel and channel assignment in soft handoff (channel assignment in soft handoff is discussed in Chapter 7).

The received F-PICH chip energy to total power spectrum interference density ratio is

$$\text{F-PICH}\frac{E_c}{I_o} = \frac{\text{F-PICH}\,E_c/I_{or}}{I_{oc}/\hat{I}_{or} + 1} \tag{8.15}$$

Similarly, for the F-FCH, we can obtain

$$\text{F-FCH}\frac{E_b}{N_t} = \frac{\text{F-FCH}\,E_c/I_{or}}{I_{oc}/\hat{I}_{or}} \cdot \frac{R_c}{R_b} \tag{8.16}$$

where E_b/N_t is the received bit energy to effective received noise power spectral density ratio.[6] From Eq. (8.15) and (8.16) we obtain

$$\frac{\text{F-FCH}\,E_c}{I_{or}} = \frac{\text{F-FCH}\,(E_b/N_t)}{\text{F-PICH}\,(E_c/I_o)}\frac{R_b}{R_c}\left(\frac{\text{F-PICH}\,E_c}{I_{or}} - \text{F-PICH}\frac{E_c}{I_o}\right) \tag{8.17}$$

---

6.   Note that in the case of multipath, Eq. (8.15) may be used as an upper bound of the F-FCH E_b/N_t.

Eq. (8.17) is the result we were looking for, since the F-FCH transmit power is expressed in terms of quantities known to the base station: The received $\text{F-PICH}(E_c/I_o)$ is reported by the mobile station in either the *Origination Message* (ORM) or *Page Response Message* (PRM); the transmit $\text{F-PICH}\,E_c/I_{or}$ can be measured by the base station; and the target F-FCH received E_b/N_t is a design parameter. Then, the base station can initialize the code channel transmit power as in Eq. (8.17) and achieve the desired forward-link quality.

8.3.2 Closed Loop

Closed loop is used to continuously adapt the forward-link code channel transmit power to varying channel conditions. Unlike the reverse link in which open- and closed-loop power control jointly determine the required transmit power, the open-loop adjustment of the F-FCH transmit power is done once and only once at call setup. Thereafter, the F-FCH transmit power can be adjusted only by means of the closed loop. The closed loop is activated once the mobile station receives two consecutive good frames at call setup and starts reverse-link transmission. At that time the base station receives power-control commands on the R-PCSCH and increases or decreases the F-FCH transmit power accordingly.

At the mobile station, closed-loop power control consists of an inner and an outer loop, which are described in the following sections. Table 8.1 summarizes the forward power-control parameters specified in IS-2000.

Table 8.2 F-FCH Closed-Loop Power Control Parameters

Field	Message	Description
FPC_SUBCHAN_GAIN	ECAM	F-PCSCH transmit power relative to the F-FCH transmit power.
FPC_FCH_INIT_SETPT	ECAM	Initial F-FCH outer loop E_b/N_t setpoint value.
FPC_FCH_FER	ECAM	F-FCH target frame error rate.
FPC_FCH_MIN_SETPT	ECAM	Minimum F-FCH outer loop E_b/N_t setpoint value.
FPC_FCH_MAX_SETPT	ECAM	Maximum F-FCH outer loop E_b/N_t setpoint value.

The forward-link power-control mode discussed in this chapter, mode 000, is the basic mode used when only the F-FCH is allocated to the call. That is the case, for example, for a voice call. When the F-SCH is also allocated, as in the case of data calls, other specialized power-control modes can be used. Those are discussed in Chapter 9 in the context of packet data services.

8.3.2.1   Inner Loop

The mobile station performs inner-loop power control by comparing the estimated E_b/N_t of the received F-FCH against the target value set by the outer loop. If the estimated value is less than the target, the mobile station sets the power-control bit of the R-PCSCH to 0; otherwise, it sets it to 1. This operation is performed for every power-control group interval, corresponding to a power-control rate of 800 bps.

The main challenge for the mobile station is to reliably estimate the F-FCH received E_b/N_t. Recall that the F-FCH is operated at variable data rate, and the mobile station does not know a priori the transmission rate used by the base station in any given frame. The frame rate can be determined only after frame decoding by selecting that rate for which the corresponding cyclic redundancy code (CRC) checks. Without knowledge of the frame rate, bit energy estimation is not possible. Furthermore, as we discuss shortly, the base station sets the transmit power depending on, among other factors, the frame rate. To overcome this problem, the mobile station relies on the forward-link power-control subchannel. This subchannel, consisting of power-control bits multiplexed within (by puncturing) the data bearing F-FCH, is sent at constant rate, and the bit energy is constant regardless of the data frame rate. As the gain of the power-control subchannel relative to the F-FCH is known, the mobile station can then infer the F-FCH received bit energy from that of the power-control subchannel. That is the reason for the base station to advertise the relative gain of the power-control subchannel at the time of traffic channel assignment. The information is carried by the FPC_SUBCHAN_GAIN field of the *Extended Channel Assignment Message* (ECAM). The mobile station may use the received F-PICH energy in addition to that of the power-control subchannel to estimate the F-FCH E_b/N_t, as the F-PICH is transmitted continuously at high power, thus allowing for better SNR estimation. The E_b/N_t estimation algorithm is not specified by the physical layer protocol and is instead implementation-dependent.

At the base station, if the received power-control bit is equal to 1, the F-FCH transmit power is decreased by an amount equal to the power-control step size; if the received power-control bit is equal to 0, the F-FCH transmit power is increased by an amount equal to the power-control step size. The base station may also decide that the received power-control bit is unreliable because, for example, the R-PICH energy received in the corresponding power-control group is lower than a given threshold. In such a case, the base station disregards the power-control bit. Unlike reverse-link power control, the IS-2000 standard does not mandate the power-control step size to be used by the base station (and the base station may decide not to perform power control at all). Typically, the step size is set to 1 dB. A smaller step would limit the ability of the closed loop to track fast path loss variations, while a larger step would increase the loop variance at the expense of capacity. The base station can adjust the F-FCH transmit power only at the power-control group boundaries. This constraint is necessary because the mobile station decides on the power-control command on the basis of the received F-FCH SNR estimated over a power-control group period. During this period, the transmit power must remain constant to make the command meaningful. Once the transmit power change is decided based on the received power-control bit, the base station adjusts the F-FCH transmit power depending on the frame rate. Specifically, the transmit power is scaled relative to that used for full-rate frames so that the energy-per-modulation symbol is the same regardless of the data rate. Furthermore, the base station may choose to reduce the transmit power of all non-full-rate frames by an additional amount because such frames, in case of voice transmission, do not need to be received with the same quality as full-rate frames. This method, called *subrate adjustment*, effectively increases the average frame error rate of the non-full-rate frames with little impact on voice quality, but also increases forward-link capacity, as the average F-FCH transmit power is reduced.

The F-FCH transmit power is constrained within a range. An upper limit is imposed to preserve forward-link capacity. Since forward capacity is limited by the finite amount of power at the base station's disposal, we want to avoid allocating excessive transmit power to a single mobile station that may be located in unfavorable channel conditions so that the entire cell capacity is not compromised for the benefit of a single user.[7] Settings of the upper and lower transmit power limits is implementation-dependent. The upper limit for the F-FCH is typically set around 5% to 10% of the total base station transmit power, as such level provides adequate signal quality to a mobile station located near the cell edge.

8.3.2.2   Outer Loop

As does the reverse link, the forward-link outer loop continuously adapts the inner loop E_b/N_t setpoint to achieve the target F-FCH FER. The target FER is chosen by the base station and signaled to the mobile station by means of the FPC_FCH_FER field of the *ECAM*. The same message also carries the setpoint minimum and maximum values, FPC_FCH_MIN_SETPT and FPC_FCH_MAX_SETPT, respectively. Although the outer-loop algorithm used by the mobile station is implementation-dependent, a typical implementation is equivalent to that explained in Section 8.2.2.2 for the reverse link, which we do not repeat here. A difference exists, however, that is worth discussing.

Consider the case when the forward-link path loss exceeds momentarily the maximum level that can be compensated by inner-loop power control. In this condition the traffic channel transmit power allocated to the mobile station is the maximum permissible by the inner-loop power-control range. As the frame error rate increases, the target setpoint also increases and the mobile station continuously sends up commands which, however, are not honored by the base station. The target setpoint keeps increasing to the point that it reaches the upper limit of the outer-loop range, that is, FPC_FCH_MAX_SETPT. In this situation, the setpoint may reach a level that is several decibels higher than the one actually required to achieve the target FER. When channel conditions improve and the FER decreases to a level below the target FER, the setpoint decreases, albeit slowly. Recall that the rate of setpoint increase is much larger than the rate of decrease. Then, during the transient period, the mobile station still requests more power than it needs. As a result, the F-FCH transmit power remains at its maximum limit throughout

---

7.   Another reason for the upper limit is to reduce the excess transmit power caused by power control when it counteracts deep fades, as explained in Section 8.3.5.

the transient period. If this situation occurs periodically, with a small period relative to the outer-loop downward slew rate, the traffic channel transmit power remains at its upper limit at all times and forward-link capacity is degraded. This phenomenon is called *wind-up*. Wind-up is particularly deleterious in the event of forward-link congestion. In such a situation the outer loop of a large number of mobile stations in the cell simultaneously experience wind-up. The wind-up then acts as a positive feedback, exacerbating forward-link congestion. A remedy to this scenario is to use provisions against wind-up. The basic premise is that a mobile station that asks for F-FCH power but does not receive it should not ask for more. The mobile station can detect wind-up by monitoring the rate at which it sends up commands. In steady state when the inner loop is working, the ratio of up commands to down commands is approximately equal to one. The mobile station can detect wind-up by monitoring the short-term average of that ratio. When wind-up is detected, the mobile station can either freeze the outer loop or significantly decrease its rate of increase. Only when the inner loop settles again can the outer loop be restarted in normal mode. The wind-up detection threshold must be carefully tuned to avoid false alarms. A false alarm would inhibit the outer loop from reacting to channel degradations that require a higher received SNR. Note that wind-up effect is most typical in the forward link, not the reverse link, because of its much smaller inner-loop range (approximately 17 dB and 70 dB for the forward and reverse inner loops, respectively).

8.3.3 Forward-Link Power-Control in Soft Handoff

While in soft handoff, the reverse-link power-control subchannel is received and demodulated independently by all the base stations in the handoff active set. Power-control commands are transmitted uncoded and therefore can be received in error. Then, while one base station receives an up command and increases the F-FCH transmit power, another base station may demodulate the power-control bit as a down command and decrease the F-FCH transmit power. The result is a drift of the F-FCH transmit power of one base station relative to the other. Such drift is akin to a random walk process.

The difference in F-FCH transmit power has serious consequences on the functioning of the mobile station receiver. Recall that when in handoff, the mobile station Rake receiver performs soft-combining of the traffic channel symbols received from all base stations in the active set. Soft-combining is implemented in the form of maximum ratio combining, whereby the Rake receiver's fingers are assigned a weight proportional to the corresponding pilot channel's received energy. Due to differences in transmit power, the mobile station may weight more heavily the traffic channel symbols received from one base station because its corresponding pilot is the strongest of all in the active set, while in fact its traffic channel is the weakest of all. Then, soft-combining is suboptimum and receiver performance is degraded.

A simple method to reduce the transmit power drift is the one in which the base station controller periodically re-syncs the base stations in the active set to a common transmit power level. This method is, however, impractical because of the excessive amount of signaling that is required and because the sudden transmit power adjustments would upset closed-loop operation.

In [4], a method is proposed to reduce the transmit power imbalance in soft handoff that performs well and has a simple implementation. We refer to this algorithm as the *rubber-band algorithm* for reasons that becomes apparent later.

The algorithm's block diagram is depicted in Figure 8.7. With each adjustment period, k, or power-control group, the F-FCH transmit power of the i-th base station in the active set, P_i, is adjusted relative to that of the previous period by an amount that depends on two components. The first component is the inner-loop power correction, which is equal to $\pm\delta$ depending on the polarity of the received power-control bit, or equal to 0 if the power-control bit is deemed unreliable. The second component is the drift reduction component, which is proportional to the difference between the transmit power and a nominal, or reference, transmit power level, P_{REF}. In mathematical terms the algorithm is as in Eq. (8.18).

$$P_i(k) = P_i(k-1) + (1-\gamma)\left[P_{i,\,REF} - P_i(k-1)\right] + u_i(k) \tag{8.18}$$

The term $u_i(k)$ represents the inner-loop power correction at the k-th interval. Note that the drift reduction component *pulls* back the transmit power toward a reference power level as if the transmit power was anchored to the reference level by means of a rubber band. The amount of the pull depends on the band *stiffness*, $1 - \gamma$. At the same time, the inner-loop component (controlled by the mobile station's power-up/down commands) *pushes* the transmit power toward the desired level. If the push is stronger than the pull, as it ought to be, the net effect is that the transmit power in steady state wonders around the desired level. Moreover, even though power-control commands are received in error, such level will be the same for all base stations in the active set. The last statement is better explained resorting to some math. Let us consider the simplest case of two base stations in soft handoff. The reference power is assumed equal for both. Then, at the time of the k-th adjustment, the instantaneous transmit power difference between the two base stations is given by Eq. (8.19).

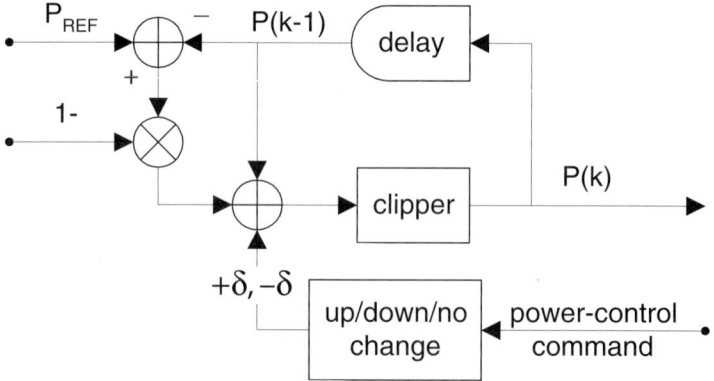

Figure 8.7 Forward-link closed-loop power control with adjustment loop.

$$P_1(k) - P_2(k) = \begin{cases} \gamma \cdot \left[P_1(k-1) - P_2(k-1) \right] & \text{if } u_1(k) = u_2(k) \\ 2\delta + \gamma \cdot \left[P_1(k-1) - P_2(k-1) \right] & \text{if } u_1(k) = 0, u_2(k) = 1 \\ -2\delta + \gamma \cdot \left[P_1(k-1) - P_2(k-1) \right] & \text{if } u_1(k) = 1, u_2(k) = 0 \end{cases} \quad (8.19)$$

Eq. (8.19) tells us that the transmit power difference approaches zero in the long term. Power-control errors may either decrease or increase such difference and therefore average out each other, while the difference decreases by an amount depending on the rubber-band *elasticity*, γ, whenever the received power-control commands are the same. As seen in Eq. (8.20), for example, after n consecutive periods in which the received power-control commands are identical, the transmit power difference is reduced by a factor γ^n relative to the initial difference.

$$\begin{aligned} P_1(k) - P_2(k) &= \gamma \cdot \left[P_1(k-1) - P_2(k-1) \right] \\ &= \ldots \\ &= \gamma^n \cdot \left[P_1(k-n) - P_2(k-n) \right] \end{aligned} \quad (8.20)$$
$$\text{if } u_1(i) = u_2(i), \forall i = n+1, \ldots, k$$

Typically, the rubber-band elasticity, γ, is set equal[8] to 0.94~0.98. A stiffer band guarantees faster drift reduction but also decreases the effective rate of power control. The reference power is set equal to the middle of the power-control range.

How effective is this algorithm? In [4], simulation results indicate that the average transmit power difference between the base stations in the active set is up to 1.7 dB and 6.6 dB with and without the rubber-band algorithm, respectively. When averaging overall possible channel conditions and active set states, the forward-link capacity gain obtained with the rubber-band algorithm can be up to 17%.

8.3.4 Power-Control Subchannel Gain Control

As seen in Section 8.2.2, the F-PCSCH carries up or down commands that are used by the mobile station to adjust the R-PICH transmit power. The forward power control is time-multiplexed onto the F-FCH. The power-control bits are uncoded and, therefore, have a different received SNR requirement than that of the coded traffic channel bits. Therefore, the gain of the F-PCSCH needs to be set differently from that of the forward traffic channel it is multiplexed onto. That is called F-PCSCH gain control.

The F-FCH symbol error rate determines the rate of both power-control commands and data frames received in error. Since the FER depends on the traffic channel transmit power, it is

---

8.   For example, if $\gamma = 0.96$ and if no power control errors occur in a frame, the difference is reduced by γ^{16}, or one-half, with every frame.

reasonable to relate the gain of the F-PCSCH to that of the traffic channel. The F-PCSCH is then referenced to the forward traffic channel; as the traffic channel transmit power changes by a certain amount in decibels due to inner-loop power control, so does the F-PCSCH in the same direction and by the same relative amount.

When in soft handoff, the mobile station soft-combines the F-PCSCHs for all pilots in the active set that are in softer handoff, but cannot combine those corresponding to pilots in soft handoff because they may carry different power-control bits. As discussed in Section 8.2.3, the mobile station performs the OR-of-the-downs rule. It is because of this rule that reliability of the F-PCSCH decreases with increasing size of the handoff active set. That is in sharp contrast to the traffic channel whose reliability increases in soft handoff due to soft-combining. An obvious remedy is to increase the F-PCSCH to F-FCH power offset with increasing handoff active-set size, h. However, in soft handoff the relation between each individual F-PCSCH symbol error rate and the traffic channel FER is difficult to quantify. The implication is that when in soft handoff, it is not advantageous to set the F-PCSCH gain proportional to that of the traffic channel. In this case, a better approach is to set it to a constant value that is proportional to that of the corresponding F-PICH.

We then resort to a heuristic algorithm that accounts for all the considerations above. We indicate with k_w and k_0 the proportionality constants of the F-PCSCH with regard to the F-FCH and F-PICH, respectively. Then, the F-PCSCH transmit power relative to the pilot channel becomes

$$10 \cdot \text{Log}\left(\frac{\text{F-PCSCH } E_c}{\text{F-PICH } E_c}\right) = \max\left\{10 \cdot \text{Log}\left(\frac{\text{F-FCH } E_c}{\text{F-PICH } E_c}\right) + k_w + 10 \cdot \text{Log}(h) ; k_o\right\} \qquad (8.21)$$

A sensible setting of the parameters above is $k_w = 0$ dB and $k_0 = -15$ dB. Note that if the F-PCSCH gain is changed, the base station must inform the mobile station, as the gain affects the F-FCH E_b/N_t estimation used for inner-loop power control. The UHDM carries the FPC_SUBCHAN_GAIN for that purpose.

8.3.5 Excess Transmit Power and Power-Control Gain

Fast power control attempts to invert channel fades and make the mobile station received signal energy nearly constant. In doing so, power control decreases the average received SNR required for the desired quality of service, or FER. However, power control also has a drawback. When a deep fade occurs, channel inversion requires the base station to increase substantially the F-FCH transmit power. Conversely, when the channel gain increases sharply, channel inversion requires the base station to decrease substantially the F-FCH transmit power. As we demonstrate in this section, the net effect of these transmit power adjustments is to increase the average transmit power relative to the unfaded channel case to achieve a given SNR. We call such average transmit power increase the *excess transmit power* due to power control. Then, the net *power-control gain* is equal to the SNR improvement less the excess transmit power due to power control. In what follows, we use the Rayleigh fading channel model to estimate the excess transmit power.

Consider an *L*-path Rayleigh fading channel with *l*-th path received energy denoted with α_l^2. The instantaneous E_b/N_t at the Rake receiver output after maximum ratio combining can be written as

$$\frac{E_b}{N_t} = \frac{\overline{E_b} \sum_{l=1}^{L} \alpha_l^2}{N_t} \tag{8.22}$$

where we assumed that the N_t is constant, the average received path energy, $\overline{\alpha_l^2}$, is normalized such that $\sum_{l=1}^{L} \overline{\alpha_l^2} = 1$, and the average received bit energy is equal to $\overline{E_b}$.

Let γ_L be the instantaneous received energy, $\gamma_L = \sum_{l=1}^{L} \alpha_l^2$. The excess transmit power required for channel inversion is defined as

$$\Gamma = E\left\{ \frac{1}{\gamma_L} \right\} \tag{8.23}$$

A practical example that leads to a closed-form expression of the excess transmit power is that of an equal-strength *L*-path channel. For such channel, the sum of the paths' energies is a chi-square distributed random variable with 2*L* degrees of freedom with probability density function given by

$$f(\gamma_L) = \frac{L^L}{(L-1)!} \gamma_L^{L-1} e^{-L\gamma_L} \tag{8.24}$$

In this case, the excess transmit power due to channel inversion is

$$\Gamma = \int_0^{\infty} \frac{1}{\gamma_L} f(\gamma_L) \, d\gamma_L = \begin{cases} \infty & \text{if } L=1 \\ L/L-1 & \text{if } L \geq 2 \end{cases} \tag{8.25}$$

Eq. (8.25) suggests that in the single path case the excess transmit power is infinite,[9] while in the two-path channel case the excess transmit power is 3 dB (i.e., the average transmit power

---

9. In practice, the average transmit power is always finite because the peak power allocated to a single user is bounded. Moreover, if the fading is fast, the channel cannot be inverted because of the limited transmit power-control step size (1 dB) and rate (800 bps).

is doubled relative to the unfaded channel case). If we were to compute the cumulative distribution function using Eq.(8.24), we would also find that the relative transmit power increase exceeds ~13 dB and ~8 dB for single- and two-path channels respectively 5% of the time. When such deep fades occur, it is more efficient to rely on interleaving and coding to limit the probability of frame erasure rather than to boost the transmit power. Thus, by limiting the F-FCH peak power, the excess transmit power is significantly reduced at the expense of a moderate increase in FER. As mentioned in Section 8.3.2.1, the F-FCH peak power is typically set between 5% and 10% of the maximum base station transmit power. In practice, the excess transmit power, or average transmit power increase, on F-FCH could be as high as 4 dB for Rayleigh fading channels with a single dominant path and a single pilot in the active set.

The net power-control gain depends on many factors, but it is typically (even after excess transmit power) larger for low mobility—for example, pedestrian users—and for Rayleigh channels that have one dominant path.

REFERENCES

[1] Soliman, Samir, Charles Wheatley, and Roberto Padovani, "CDMA Reverse Link Open Loop Power Control," *In Proc. of Globecom'92,* December 1992.

[2] Viterbi, Andrew, *CDMA Principles of Spread Spectrum Communications.* Reading, MA: Addison Wesley, 1995.

[3] Vanghi, Vieri, and Aleksandar Damnjanovic, "cdma2000 Forward Link Open Loop Power Control," *In Proc. of CDMA International Conference,* Seoul, Korea, 2000.

[4] Hamabe, Kojiro, "Adjustment Loop Transmit Power Control During Soft Handover in CDMA Cellular Systems," *In Proc. of VTC,* Spring 2000.

CHAPTER 9

Packet Data Operation

Packet data service enables high-speed always-on connection between the mobile station and the Internet. To efficiently provide the desired quality of service (QoS), cdma2000 traffic channels' characteristics and procedures for their operation are tailored to data traffic requirements. In this chapter we begin with a generic introduction to the concepts that play an important role in the design of a system supporting packet data services. As packet data services can be provided simultaneously with voice services on the same carrier, emphasis is given to the concepts of resource management and scheduling. Such concepts are then applied in the discussion of practical aspects related to the design and operation of the Forward/Reverse Supplemental Channel (F/R-SCH) and Forward Packet Data Channel (F-PDCH).

9.1 Fundamentals

Data and voice traffic have vastly different characteristics and QoS requirements in terms of frame error rate (FER), delay, and data rate. The design objectives are to achieve the target QoS for both services while maximizing efficiency of radio resource utilization. With these objectives in mind, we begin the chapter by revisiting the main issues concerning packet data operation.

9.1.1 User Experience

Knowing the factors affecting the user's experience and understanding how to enhance them are vital to implementing and deploying a successful packet data network. At a high level, the most relevant aspects of the user experience are latency, throughput, and always-on connectivity.

The latency experienced by a user must be minimized. A service that minimizes latency or masks it by conducting network operations in the background creates the most positive experience for users. Minimizing the time it takes for the service to deliver a response to a user's request for information (for example, Web page download) is critical for providing an acceptable packet data service. The overall delay is a result of setting up radio traffic channels and call processing, initializing link layer protocols such as Radio Link Protocol (RLP) and Point-to-Point Protocol (PPP), throughput limitations, network distribution delays, content-rendering delay on the terminal (for complex graphics and content), and server response times and Internet traffic delay.

Maximizing throughput between the user and the network is also essential. Interactive applications with frequent exchanges between the user and other locations on the Internet have noticeably higher latency when throughput is not adequate. Dependence on throughput becomes apparent when applications exchange larger amounts of data. Throughput determines the QoS (for example, high bandwidth vs. low bandwidth video) for conversational and streaming types of applications. Throughput at the physical layer depends on a number of factors, such as efficient power control and data rate adaptation. These factors are described in detail in the sections that follow.

The user must be given the perception of being always connected to the network. Once a mobile station has established a data session with the network, dormancy provides an always-on connection without requiring a dedicated traffic channel. Instead, the system maintains the PPP state, using traffic connections only when data between the mobile station and the network is actively exchanged. The bursty nature and delay tolerance of packet data traffic allow the cdma2000 packet data network to provide many users with an always-on connection to the Internet while using fewer resources than needed for circuit-switched traffic.

9.1.2 Prioritization of Voice and Data Services

Typically, high-speed data traffic has bursty characteristics resulting in a bursty utilization of the radio interface capacity. This impacts the amount of capacity available to voice calls, possibly degrading their quality if the system is not engineered correctly. Hence, a key design issue is assuring that a base station carrying voice and data calls simultaneously on the same carrier frequency negligibly impacts the QoS of both, particularly voice.

Mobile users increasingly expect wire-line voice quality and reliability from their mobile phones. As a result, maintaining the QoS of voice calls is of paramount importance. High-speed data calls should be supported without a noticeable degradation to the quality of voice calls. Hence, maintaining voice QoS at a higher priority than that of packet data is expected.

In addition to minimizing the impact to the voice service QoS, operators may also require that the voice grade of service (GoS), which refers to the probability of call blocking, is not affected by the support of high-speed data services. If all data applications are delay-tolerant (for example, Web browsing, file downloads), the desired voice GoS may be possible to achieve with little impact to the required data QoS. However, it is more difficult to achieve if the data applications are delay-intolerant (for example, video streaming).

Another degree of freedom is the choice of prioritizing new voice calls over a handoff of a data call, or vice versa. In general, serving an active data call is more critical than serving a potential new voice call. From the above, a typical order of priority would be voice call hand-offs, voice call originations, data call handoff, and data call originations. A key step is to set the resource allocation priority of all voice calls higher than that of any data call. For example, if after all voice users are allocated their requested share of transmit power no power is left for data calls, transmission of data calls can be discontinued for a short duration of time without significantly compromising user experience. In contrast, if data calls were assigned power before voice calls, the probability increases that one or more voice calls will not have sufficient resources allocated, possibly resulting in voice quality degradation or call drop.

9.1.3 Voice and Data Multiplexing Efficiency

It is apparent that there is interplay among a number of variables, including voice and data services QoS/GoS. If designed incorrectly, the overall system performance will be degraded when both types of services are supported simultaneously. At the same time, however, the stringent voice requirements leave a margin of unused air-link capacity. Moreover, the voice activity of the active calls also results in additional unused air-link capacity. Both forms of unused capacity may be used in principle to support packet data services without directly impacting the voice calls if packet data is always treated as a lower priority service. This concept is turned into practical examples in the sections that follow.

9.2 Resource Management for Mixed Voice and Data Traffic

Voice traffic has very stringent QoS/GoS requirements. The probability of blocking a voice call should not be more than 2%. In addition to real-time transport, the FER should be around 1%. Unlike voice, most data traffic is to some extent delay-tolerant. Therefore, when voice and data share common radio resources, priority may be given to voice users and only residual resources be allocated to data users.[1] In cdma2000 systems, the main radio resources that need to be managed for efficient and effective support of mixed voice and data traffic are Walsh codes and power. The Walsh code space is a limited resource on the forward direction because it is shared by numerous forward-link channels comprising the CDMA channel. Power can be seen as a resource for both the forward and the reverse links, although how it is managed greatly differs between the two. On the forward link, the base station transmit power is limited by the base station transceiver (BTS) power amplifier capability and must be shared by all active users. On the reverse link, the received power from multiple users creates mutual interference. When the aggregate interference from all such users exceeds a certain level, no additional service can be provided by that base station without degrading the QoS provided to the existing users. Then, reverse-link power management requires measuring the total base station received power and

---

1.   In general, resource allocation is based on the service QoS requirements.

estimating the additional interference, or received power, that can be tolerated. Such additional power is thus a residual resource that is shared by voice and data users. In the following sections we describe the forward and reverse links radio resource management principles for mixed voice and data traffic support.

9.2.1 Base Station Transmit Power Management

The rating of the BTS power amplifier determines the maximum base station transmit power. At a given instant, the total transmit power relative to the maximum determines the base station loading, and its complement determines the residual forward-link capacity. Residual capacity can be allocated to new calls or redistributed to existing calls. In case of mixed voice and data traffic, we further distinguish between the base station transmit power component consumed by overhead channels, channels carrying data for real-time services, and those carrying data for best-effort or delay-tolerant services.

A minimum relative transmit power must be reserved for use by the overhead channels—for example, Forward Pilot Channel (F-PICH), Forward Paging Channel (F-PCH), Forward Broadcast Control Channel (F-BCCH), Forward Common Control Channel (F-CCCH), and Forward Quick Paging Channel (F-QPCH)—to ensure that any service can be provided by that base station. The power consumed by the overhead channels is nearly constant if we neglect the variability introduced by those (low-power) channels that are operated with discontinuous transmission, such as the F-QPCH or F-CCCH.

Dedicated channels carrying real-time data, such as the Forward Fundamental Channel (F-FCH), Forward Dedicated Control Channel (F-DCCH), and F-SCH, are allocated the residual power available after the needs of the overhead channels are met. The transmit power consumed by the F-FCH, F-DCCH, and F-SCH varies with time. Its variability has a fast and a slow component. The fast component is due to forward-link power control and voice activity. The slow component is due to links being set up or released because of call origination or release events, handoffs, and F-SCH data rate changes. As explained in Section 9.4.1.2, F-SCH data rate control is relatively slow because of the latency associated with Layer 3 signaling.

Finally, the F-PDCH that carries nonreal-time or best-effort data is allocated the residual power available after the needs of the all other channels are met. The transmit power consumed by the F-PDCH varies due to the effect of data rate adaptation. Unlike the F-SCH, the F-PDCH allows very fast resource allocation (every 1.25 ms) by means of MAC layer signaling handled by Packet Data Channel Control Function (PDCHCF), implemented at the BTS.

In light of the above, base station power management is distributed between BTS and base station controller (BSC). The base station measures the transmit power allocated to voice and data calls, and allocates residual power to the F-PDCH, while the BSC performs call admission and congestion control.[2] Hence, resource management at the BSC has priority over that at the BTS, although the latter is more efficient than the former, as explained below.

---

2.   Transmit power is not the only resources considered by call admission and congestion control.

9.2.2 Residual Forward-Link Capacity Estimation

The base station computes the residual capacity, C_{TX}, as the difference between the base station's nominal maximum transmit power and the transmit power consumed by all channels other than the F-PDCH and Forward Packet Data Control Channel (F-PDCCH). The computation can be at baseband (using the modulator outputs) by summing together the square of the digital gains of all channels in use on the forward link. Implementation considerations aside, the computation may be done every power-control group,[3] or 1.25 ms. The residual capacity is filtered to remove the effects of power control and voice activity. The filter response should not be too slow to respond to time-varying load. The BTS sends the BSC an indication of the current residual capacity, either periodically or on request or both.

9.2.3 Forward Link Call Admission Control

When a request to establish or reconfigure a traffic channel is received, the BSC estimates the transmit power required on average to sustain such channel, $\hat{P}$. A method to estimate $\hat{P}$ is presented in Section 9.4.1. The BSC grants the request if the residual capacity is sufficiently larger than the required power.

$$C_{TX} - \hat{P} \geq C_{TX,\min} \tag{9.1}$$

The threshold $C_{TX,min}$ is used to prevent depletion of the available transmit power and to accommodate transmit power fluctuations. The threshold may depend on the QoS profile of the user or the QoS requirements of the traffic. The threshold may also depend on whether the request for channel establishment is for call origination or soft handoff. The threshold in case of call origination is larger than in case of soft handoff, as from a user perspective a blocked call is better than a dropped call (the likely result of denying handoff). Unlike all other traffic channels, F-PDCH admission control is not based on computation of residual capacity because the F-PDCH is allocated residual power by the BTS and its average required transmit power P is undefined. F-PDCH admission control simply ensures that the number of calls "sharing" the F-PDCH does not become excessively large.

9.2.4 Forward-Link Congestion Control

Despite call admission control, the short-term average of the transmit power may exceed the maximum nominal value (residual capacity becomes a negative number), an event that we call congestion. Congestion may happen when some active mobile stations move in locations with unfavorable channel conditions and require more transmit power. At the onset of congestion, the BSC can take the some actions to restore normal operation—for example, selected

---

3.   The base station transmit power is constant over the duration of one power-control group.

mobile stations may be handed off to an alternate frequency with available residual capacity; the F-SCH of selected mobile station may be reconfigured with a lower data rate; or selected traffic channels may be released entirely.

9.2.5 F-PDCH/F-PDCCH Power Allocation

The BTS uses the residual capacity, C_{TX}, to decide how much power is available to the F-PDCH. The residual capacity is unfiltered, as in principle it can be estimated every 1.25 ms, which is also the shortest F-PDCH scheduling period. The residual capacity is used to decide upon the maximum data rate that can be granted to the F-PDCH at the next assignment interval. Allocation is fast enough to take advantage of the variability of the other traffic channels' transmit power.

Figure 9.1 and Figure 9.2 illustrate typical base station power consumption with mixed voice and data traffic when the latter is carried by the F-SCH[4] (scheduled at the BSC) and F-PDCH (scheduled at the BTS), respectively. As the figures indicate, the base station transmit power is fully utilized when the F-PDCH is used, but that is not the case when the F-SCH is used.

The F-PDCCH power is set in such a way to insure sufficient correct decoding probability by the targeted mobile station. Note that the base station gets an indication whether the mobile station has decoded F-PDCCH or not through the Reverse Acknowledgment Channel (R-ACKCH). If the R-ACKCH is positive or negative, the F-PDCCH has been decoded. However, if the R-ACKCH indicates null—that is, no signal—it means that R-PDCCH has not been decoded.

---

4.   A portion of data traffic is also transferred over F-FCH.

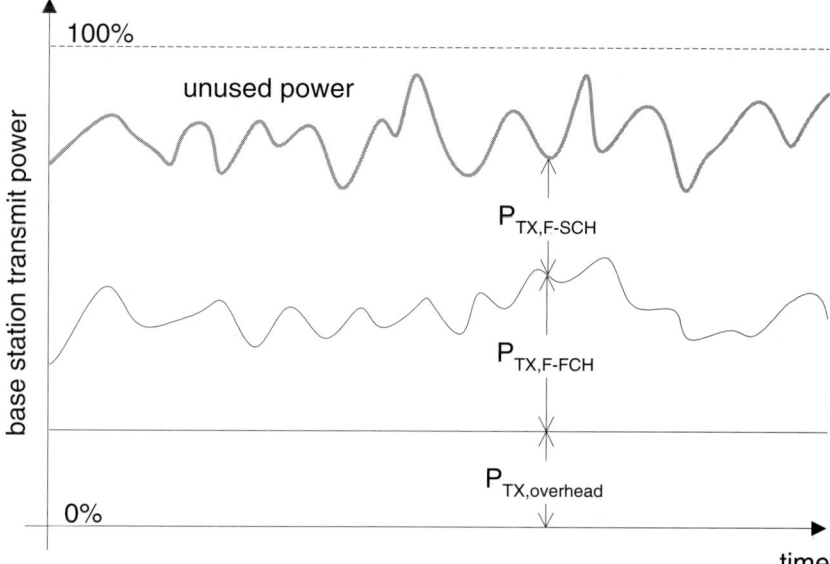

Figure 9.1   Mixing voice and data with F-SCH.

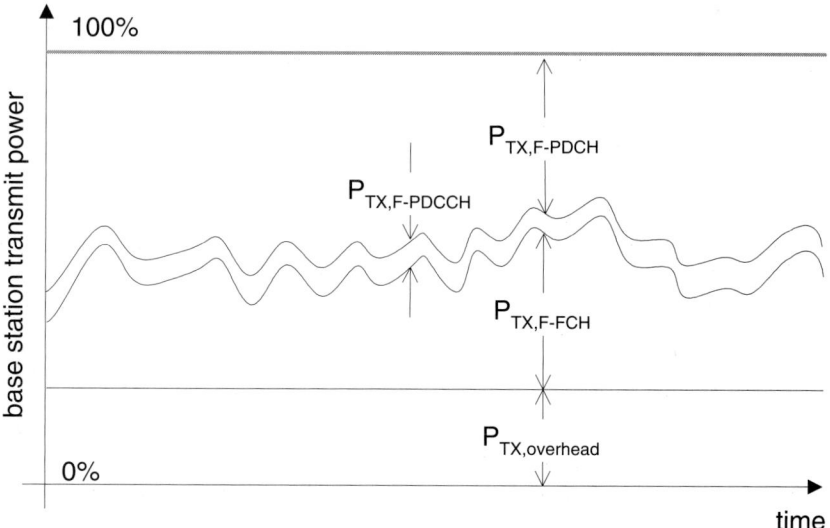

Figure 9.2   Mixing voice and data with F-PDCH.

9.2.6 Forward-Link Walsh Code Management

In the case of mixed voice and services, Walsh code assignment policy minimizes the Walsh code space fragmentation and maximizes the peak data rate supported on F-SCH. A single voice user occupies one Walsh code of length 64 or length 128 for Radio Configuration (RC) 3 and 4, respectively. A high-rate F-SCH, however, requires a large, contiguous Walsh space. For example, a full quadrant of contiguous Walsh code space is needed for the highest rate F-SCH (153.6 kbps), corresponding to a spreading factor equal to 4. Therefore, a good Walsh code allocation policy for voice users is one that avoids fragmentation of the Walsh code space, thus maximizing the availability of shorter Walsh codes that can be assigned to high rate F-SCH. The allocation policy is illustrated in Figure 9.3, where the branches terminating with a shaded dot correspond to Walsh codes in use, those terminating in an empty dot correspond to an available Wash code, and those that do not terminate in any dot represent Walsh codes that are blocked. Blocked Walsh codes are those for which one or more descendants are in use and those for which their parent is in use. Blocked Walsh codes cannot be assigned in order to preserve orthogonality of the codes in use. In this example, the Walsh code usage reflects the following scenario:[5]

- One data user is assigned a length-4 Walsh code enabling the F-SCH at 153.6 kbps plus one length-64 Walsh code enabling the F-FCH at 9.6 kbps (RC3). The F-FCH is used to carry signaling and user data.

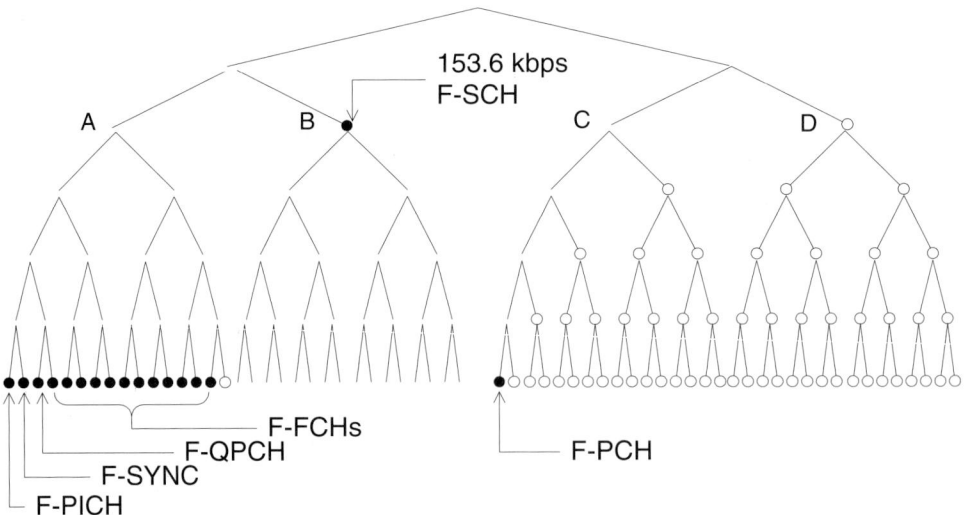

Figure 9.3 Example of Walsh code tree usage with mixed voice and data services.

5.   Note that the Walsh code index does not directly indicate the node used in the Walsh code tree in the figure. The figure uses the position as determined by the bit-reversed index.

- Each of the 11 active voice users is assigned a length-64 Walsh code enabling the F-FCH at 9.6 kbps (RC3).
- Four length-64 Walsh codes are assigned to the overhead channels, namely the F-PICH, F-SYNCH, F-QPCH, and F-PCH.

In this scenario, the next length-64 Walsh code to be allocated to the F-FCH (either for a voice call or to carry signaling of a data call) should be the remaining available one in the sub-tree corresponding to node A. Such allocation increases the used space without increasing the number of blocked codes. Additional voice calls should be assigned to fill out the subtree labeled C.

In case of a data call at rate 19.2, 38.4, or 76.8 kbps, the F-SCH should be allocated at the leftmost available Walsh code of length 32, 16, or 8, respectively in the subtree corresponding to node C. As a result of such allocation, there is no shorter Walsh code that becomes blocked. A data call at 153.6 kbps can be assigned the only available length-4 Walsh code corresponding to the subtree labeled D.

The F-PDCH can also benefit from the same Walsh code allocation policy. As we see in later sections, unlike the F-SCH, the F-PDCH does not require contiguous Walsh codes. It is only required that the Walsh code length of at least 32, or preferably 16, is available.

An additional aspect that must be considered is the impact on Walsh code utilization if additional F-QPCH and F-PCH channels are used. The second F-QPCH, if used, occupies the Walsh code channel immediately to the right of the first F-QPCH, so the second F-QPCH has no effect on high-rate channels. Having additional F-PCHs, however, results in a significant impact to the high-rate channels, especially for RC3. Adding two extra F-PCHs, bringing the total number to three, blocks the use of any F-SCH completely, as each takes the leftmost node of the subtrees labeled B and D.

As voice calls arrive and depart, the Walsh code tree will be inevitably fragmented. When the need arises for an F-SCH at 153.6 kbps, for example, but only 15 adjacent Walsh codes are available, the base station resource manager should request the reassignment of the voice call blocking the length-4 Walsh code to another length-64 Walsh code if available. The opposite scenario is also possible. If a voice call cannot be served because length-64 Walsh codes are either in use or blocked, then an active F-SCH may reassigned to a longer Walsh code (lower data rate) to make one or more length-64 Walsh codes is available to the voice call.

Some Walsh codes may also be reserved for handoff requests of voice calls. If the number of Walsh code of length 64 in use exceeds a certain threshold, the admission controller may deny resources to a call origination, as it is desirable to minimize the probability to deny Walsh code requests for soft handoff. Note that only Walsh codes in use rather than blocked are counted because the latter may be made available by reconfiguring the F-SCH preventing their use, as explained above.

9.2.7 Base Station Receive Power Management

On the reverse link, the received power from multiple users creates mutual interference. When the aggregate interference from all such users exceeds a certain level, no additional user can be served by that base station without degrading the quality of service provided to the existing users. Reverse-link power management requires estimating the maximum additional signal power that causes a tolerable increase of the interference level. Such additional power is thus a residual resource that is shared by voice and data users. A reverse-link admission control algorithm for voice traffic ensures that the residual capacity does not fall below a threshold. Low-priority data users are allowed to consume the unused capacity.

9.2.8 Residual Reverse-Link Capacity Estimation

Let P_{RX} and N_{RX} be the base station's total received power and noise power, respectively. P_{RX} comprises the background noise and the received signal power of both users within the base station of interest and those served by other base stations. Consider a base station receiver with a signal power–to–total power ratio of

$$\frac{P_{RX} - N_{RX}}{P_{RX}} = 1 - \frac{N_{RX}}{P_{RX}} \tag{9.2}$$

This ratio, representing the base station loading, is at most equal to 1. Then, a receiver with the above signal-to-total power ratio can support additional link(s) requiring a signal power to total power ratio equal to

$$C_{RX} = 1 - \frac{P_{RX} - N_{RX}}{P_{RX}} = \frac{N_{RX}}{P_{RX}} \tag{9.3}$$

The quantity C_{RX} represents the residual reverse-link capacity. The inverse of the residual capacity is equal to P_{RX}/N_{RX}, a quantity often called *raise over thermal*. The total received power may be measured by a power detector at the BTS front end. The noise floor may be simply set to a value that depends on the receiver's noise figure, which is known as a result of calibration. However, when the noise floor includes a non-CDMA interference component, or when the noise figure is affected by temperature variations and aging of hardware components, the residual capacity estimate is biased. In these cases the noise floor should be measured directly rather than extrapolated from the receiver noise figure.

An alternative method to compute residual capacity employs baseband measurements at the BTS demodulator output. After some algebra, it can be see that

$$C_{RX} = \frac{1 - \sum_i \gamma_i / (1 + \gamma_i)}{1 + I_{oc}/N_0} \tag{9.4}$$

where N_0 is the thermal noise power spectral density, I_{oc} is the interference from users in other cells, and γ_i is the signal-to-interference noise ratio (SINR) of the i-th mobile station whose active set includes the base station being considered.

In Eq. (9.4), the user's SINR depends on the radio bearer profile (e.g., R-FCH for voice, or R-FCH and F-SCH for data users). The SINR can be computed based on the reverse-pilot received chip energy–to–total noise density ratio, R-PICH E_c/N_t, and the traffic channels' transmit power offset relative to the pilot:

$$\gamma(R) = \text{F-PICH}\frac{E_c}{N_t}\left[1 + \frac{\text{F-FCHE}_c}{\text{F-PICHE}_c} + \frac{\text{F-SCHE}_c(R)}{\text{F-PICHE}_c}\right] \tag{9.5}$$

Eq. (9.5) lends itself to practical implementation because it is expressed in terms of known quantities. The R-PICH E_c/N_t is measured by the BTS demodulator (and also used for inner-loop power control). The power offsets are known a priori because their setting is controlled by the base station via Layer 3 signaling. The power offsets depend on the channel data rate and other configuration parameters.

The other cell interference term in Eq. (9.4) is proportional to the loading caused by the mobile stations in handoff with the reference base station. Namely, increasing the load in the reference base station also increases the other cell interference of the adjacent cells, which in turn is reflected as the increased outer-cell interference in the reference cell. This positive feedback mechanism is captured by modeling the outer-cell interference as proportional to the load. The residual capacity can be rewritten as

$$C_{RX} = 1 - (1 + \varphi)\sum_i \frac{\gamma_i}{1 + \gamma_i} \tag{9.6}$$

where φ represents the effective load increase due to other cell interference, which varies depending on mobile station locations and radio propagation characteristics. A typical average value is $\varphi = 0.5$. Note that as the load increases and the residual capacity approaches 0, the received power increases without bound and the system becomes unstable. For example, if $\varphi = 0.5$ and $\sum_i \gamma_i / (1 + \gamma_i) \to 2/3$, then the $P_{RX}/N_{RX} \to \infty$. In practice, the P_{RX}/N_{RX} is maintained below ~7 dB, corresponding to C_{RX} above 20%.

C_{RX} computation using Eq. (9.5) and (9.6), unlike explicit computation using Eq. (9.3), does not rely on calibration data to estimate the noise floor. However, its drawback is that it requires estimation of the outer-cell to inner-cell interference ratio.

The residual capacity is filtered to remove the effects of power control and voice activity. The filter response should not be too slow to respond to time-varying load. The BTS sends the BSC an indication of the current residual capacity, either periodically or on request or both.

9.2.9 Reverse-Link Call Admission Control

When a request to establish or reconfigure a traffic channel is received, the BSC estimates the average SINR required to sustain that channel, $\hat{\gamma}$. The estimate may be simply tabulated as a function of the requested data rate. The effective load increase due to the mutual effect of the additional traffic channel and the outer-cell interference, $\hat{\varphi}$, may be estimated based on mobile station location or set to a nominal value. The BSC grants the request if the residual capacity is sufficiently larger than the load increase arising from the additional traffic channel, that is,

$$C_{RX} - (1+\hat{\varphi})\frac{\hat{\gamma}}{1+\hat{\gamma}} \geq C_{RX,\min} \tag{9.7}$$

The threshold $C_{RX,\min}$ is used to prevent operating close to pole capacity, which may lead to system instability. The threshold may depend on the traffic QoS requirements.

9.2.10 Reverse-Link Congestion Control

Despite call admission control, the residual capacity may drop below the minimum threshold, an event that we call congestion. At the onset of congestion, the BSC can restore normal operation using the same methods discussed for forward-link congestion control.

9.3 Scheduling Algorithms

Real-time traffic does not tolerate any delay. It relies on admission control to maintain the probability of occurrence of congestion events below a tolerable level. For delay-tolerant traffic, admission control does not have to be as conservative because short-term congestion events can be mitigated with scheduling. More importantly, scheduling allows efficient utilization of resources. The benefits of multiuser scheduling for point-to-multipoint links are discussed in [1]. With certain channels and schedulers, all resources are allocated to a single mobile station for a given scheduling period. This scenario is applicable to F-PDCH. In the text that follows, we describe most common scheduling algorithms.

9.3.1 Round Robin

The round-robin scheduler allocates the same number of scheduling periods, or slots, to all the mobile stations, one at a time, according to the following policy. Given N mobile stations in the system, the i-th slot is allocated to the n-th mobile for which

$$n = \left\{(i-1)/N - \left\lfloor (i-1)/N \right\rfloor\right\} N + 1 \tag{9.8}$$

In Eq. (9.8), $\lfloor \cdot \rfloor$ denotes a real number rounded to the nearest smaller integer. If a mobile station has no data to send at the time it is scheduled for transmission, the scheduler skips over to next mobile station. If the slot duration is the same, the scheduler policy is called *equal-time* round robin. Even though the number of slots allocated to mobile stations is the same, this policy does not provide the same throughput to all the mobiles, because mobile stations are granted a data rate that depends on their channel conditions, which differ among mobile stations in different locations within the cell's coverage area. The mobile stations that experience better radio conditions have higher throughput. The scheduler functioning is illustrated in Figure 9.4 for $N = 4$ mobile stations.

The alternative to equal-time round-robin rule is the policy that serves the same amount of data to all the mobiles in a round-robin fashion, referred to as *equal-data* round robin. With equal-data scheduling, the slot duration is inversely proportional to the assigned data rate. Thus, the allocated slot duration is shorter for mobile stations with good channel conditions than for mobile stations with bad radio conditions. An example of a round-robin scheduler using equal-data policy is shown in Figure 9.5. Note that the equal-data round-robin policy provides the same throughput to all mobile stations, but the average cell throughput is lower than for the equal-time rule.

The third round-robin policy is the *equal-access* rule. According to this rule, all mobile stations access the channel in a round-robin fashion, but the length of time and amount of data can be defined on a per-user basis. This policy provides fairness in the throughput tradeoff between the equal-time and equal-data rules.

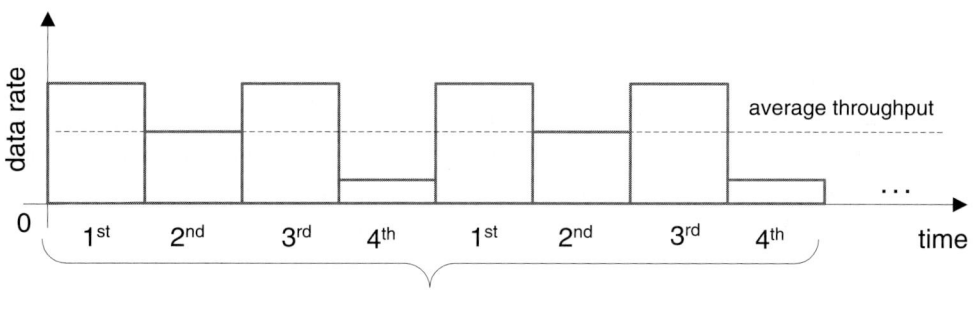

Figure 9.4 Round-robin scheduler, equal-time policy.

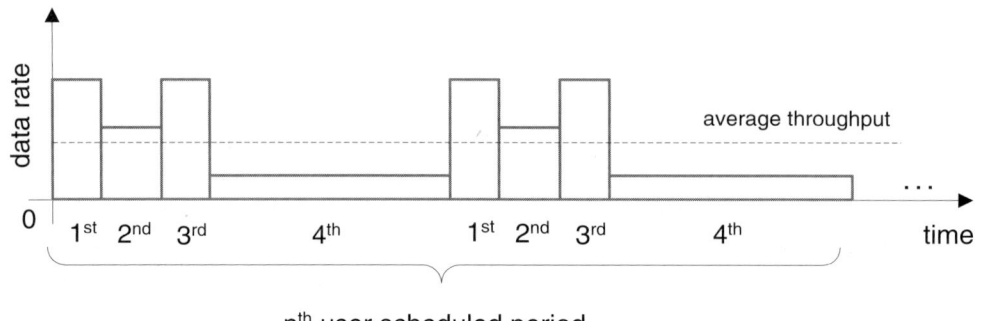

nth user scheduled period

Figure 9.5 Round-robin scheduler, equal-data policy.

9.3.2 Maximum Throughput

The maximum throughput scheduler policy attempts to maximize the cell throughput. The mobile station that is in the best radio condition among all mobiles in the system is always scheduled. A major disadvantage of this scheduler is that mobile stations that experience poor channel conditions may never be served. Moreover, a single mobile station with good channel conditions may starve all other mobile stations in the cell, provided it continuously has data to send.

The maximum throughput scheduler does not provide any degree of fairness and is, therefore, impractical. However, it is conceptually interesting because it exemplifies the multiuser diversity gain achievable with a scheduler that accounts for channel conditions when allocating the transmission slot. That leads us to search for a scheduler that achieves both a high degree of multiuser diversity gain and fairness among users.

9.3.3 Proportional Fair

The *proportional fair* algorithm exploits the variability of channel condition experienced by the mobile stations in the cell to achieve a multiuser diversity gain, but at the same time provides fairness in a proportionally fair sense. Proportional fairness was first proposed in [2] and can be summarized in the following. The data rates assigned to the mobile stations in the cell are proportionally fair if by an alternative scheduler we are able to increase the throughput of user i by $x\%$, but the summation of percentage decreases in throughput of all other mobiles is more than $x\%$.

Mathematically, we can express proportional fairness as follows. Assume that there are N mobile stations in the system. Denote with S_i the throughput allocated to user i where $i = 1, 2, ..., N$, $\mathbf{S}=[S_1, S_2, ..., S_N]$ and $\mathbf{S}$ is a feasible throughput vector obtained by the proportional fair scheduler. Then, any other throughput distribution $\mathbf{S}^*$ obtained by an alternative scheduler satisfy the following inequality:

$$\sum_{i=1}^{N} \frac{S_i^* - S_i}{S_i} \leq 0 \qquad (9.9)$$

The radio channel conditions experienced by mobile stations vary independently with time. The throughput feasibility region is varying too, and hence the proportional fair solution. In practice, there may not be enough time to compute a new throughput vector **S**. Moreover, short-term rate allocations are not perceivable by a user, but average throughput values over a period of time of, say, 1 s or more are perceivable. Therefore, what a scheduling algorithm can do in practice is gradually update the solution in the direction of the proportional fair solution with the steepest ascend. The scheduler operates as follows.

At the n-th time slot the scheduler considers the mobile stations' average throughput, $\mathbf{S}(n)$ = $[S_1(n), S_2(n), ..., S_N(n)]$ and current supportable data rates, $\mathbf{R}(n) = [R_1(n), R_2(n), ..., R_N(n)]$. Then the scheduler assigns the slot to the i-th mobile station with the largest scheduling metric, or $R_i(n)/S_i(n)$ ratio. If there are any resources left, the slot is also assigned to the mobile station with the second largest scheduling metric. Greedy filling is continued if necessary. Once the slot is assigned, the scheduler updates the average throughputs for the next time slot as

$$S_i(n+1) = \begin{cases} (1-\alpha)S_i(n) + \alpha R_i(n) & \text{if scheduled} \\ (1-\alpha)S_i(n) & \text{if not scheduled} \end{cases} \qquad (9.10)$$

The parameter $\alpha < 1$ determines the filter equivalent time constant, τ. If the scheduling period is T, then $\tau = T/\ln(1-\alpha)$. Similar to round-robin equal-time policy, the proportional fair algorithm allocates approximately the same number of slots to all the mobile stations. However, the proportional fair scheduler exploits the multiuser diversity effect and attempts to transmit to a mobile station only when its radio channel condition is near the peak.

9.4 Forward Supplemental Channel Operation

The F-SCH supports data services that require a high data rate radio bearer. A single F-SCH supports a data rate of up to 153.6 kbps when using RC3 and up to 307.2 kbps when using RC4. Each base station may transmit multiple F-SCHs simultaneously if it has enough transmit power and Walsh codes available. The F-SCH is assigned to a mobile station together with the F/R-FCH, because the latter is needed to carry signaling. User data may be sent both on the F-FCH and the F-SCH. Optionally, the F/R-DCCH can be used in lieu of the F/R-FCH.

The F-SCH is a power-controlled channel and thus most suited for real-time services, such as two-way video conferencing, streaming, or conversational class traffic. Best-effort traffic is instead more efficiently transported over the F-PDCH. However, the F-SCH may be used to transport best-effort packet data services when the F-PDCH is not available. Mechanisms to operate the F-SCH for best-effort or non-real-time and real-time services are described separately in the sections that follow.

9.4.1 F-SCH for Non-Real-Time Service

Unlike voice traffic that is forward- and reverse-link symmetric and has a nearly constant bit rate, packet data traffic is in many cases asymmetric and characterized by small periods of high offered load followed by periods with little or even no offered load (see also Chapter 2 for a description of traffic models). Due to such sharp variations, an efficient operation of the F-SCH in support of packet data services is one aiming at assigning resources to a mobile only when necessary and reallocating resources to other users at all other times. That is in sharp contrast to voice applications that require assignment of a bidirectional dedicated channel, the F/R-FCH with data rate matching the peak vocoder bit rate.

In light of the above, the Layer 3 signaling procedure enables F-SCH operation in burst mode. Burst-mode operation means that the F-SCH can be assigned, reconfigured, and de-assigned multiple times throughout the life of the packet data call without requiring release of the connection and the associated F/R-FCH (or F/R-DCCH). However, operating the F-SCH in burst mode poses several challenges, as outlined below.

- The burst duration must be carefully selected to account for assignment latency and the induced signaling load.
- The data rate selection must account for the amount of data available for transmission, and with time-varying channels, it must be continuously adapted to the channel conditions experienced by the data user. The data rate selection must account for the residual resource availability. That is, a burst admission policy is needed in addition to the call admission policy enforced at call setup.
- Power control must account for the different QoS requirements of the F-SCH and the associated F-FCH. The F-SCH does not carry signaling and relies on RLP and upper-layer protocols for error recovery and therefore can be operated at a much higher frame error rate than the F-FCH. The above implies that the F-SCH should be power-controlled separately, but not without dependencies from the F-FCH.
- When F-SCH carries latency-insensitive data, it is not subject to the same stringent outage requirements imposed on the F-FCH, which is one of the motivations for soft handoff. If the F-SCH is operated at a higher frame error rate, the diversity gain provided by soft handoff is diminished. Therefore, the handoff strategy adopted for the F-SCH differs from that of the F-FCH.
- Finally, the bursty nature of the F-SCH assignments, tailored to the bursty nature of the data traffic, calls for scheduling mechanisms in order to increase efficiency of capacity utilization.

In the sections that follow we analyze the these issues and explore mechanisms that the base station may employ for efficient operation of the F-SCH. Note that such mechanisms are not specified by the standard and are instead implementation-dependent.

9.4.1.1   Assignment Policies

F-SCH assignment is controlled by the *Extended Supplemental Channel Assignment Message* (ESCAM). The ESCAM relevant fields are summarized in Table 9.1.

Table 9.1   F-SCH Assignment Parameters*

Field	Message	Description
FOR_SCH_DURATION	ESCAM	F-SCH assignment duration.
FOR_SCH_START_TIME	ESCAM	F-SCH processing start time.
SCCL_INDEX	ESCAM	Index of the record in the F-SCH code list corresponding to this assignment.

*Not all F-SCH assignment related fields are listed.

The base station triggers the initial assignment of the F-SCH whenever the packet transmission queue length exceeds a certain threshold. For the initial assignment, the ESCAM conveys the mobile station F-SCH configuration information and directs the mobile station to start processing the assigned F-SCH. The F-SCH configuration record contained in the ESCAM conveys the data rate, Walsh code index, pilot pseudonoise (PN) offset of the corresponding CDMA channel, and power-control parameters (described in details in Section 9.4.1.3). An F-SCH may be assigned only once it is configured. Each record, indexed by the SCCL_INDEX field, is added to the SCH code list maintained by both the mobile and base stations. The indexing of configured F-SCH is useful when the base station decides to reassign a preconfigured channel by sending the ESCAM. In such cases, the base station can simply point to a configured F-SCH by means of its SCCL_INDEX, thereby reducing the size of the ESCAM to one frame (20 ms). A small message size is desirable because it minimizes probability of message erasure and signaling delay.

F-SCH assignment is either for a finite duration ranging from 1 to 256 frames, as specified in the FOR_SCH_DURATION field, or infinite, that is, until the base station explicitly de-assigns the F-SCH. The mobile station starts processing the assigned F-SCH at the time specified by the FOR_SCH_START_TIME field. To allow flexibility in controlling F-SCH assignment, an ESCAM may be sent preempting the current assignment. An exemplary assignment sequence is shown in Figure 9.6. The first ESCAM is received at time t_0, assigning the F-SCH at rate R_1 starting at time t_1 for duration T_1. However, prior to expiration of the current F-SCH assignment, the base station sends a second ESCAM for an F-SCH at rate R_2 whose assignment extends beyond that of the current one. The new assignment may have been triggered, for example, by a change in channel conditions or packet transmission queue length. The second ESCAM specifies a start time $t_2 < t_1 + T_1$. Therefore, at time t_2 the mobile station aborts processing of the previous assignment and starts processing the new one. Prior to expiration of the second assignment, the base station sends a third ESCAM changing the data rate to R_3. Although the new assignment does not extend beyond the current one, once an assignment is cancelled it cannot be resumed. That is, on expiration of the current assignment at time $t_3 + T_3$, processing of the previous assignment is *not* resumed even though its expiration time would have been $t_2 + T_2 > t_3 + T_3$.

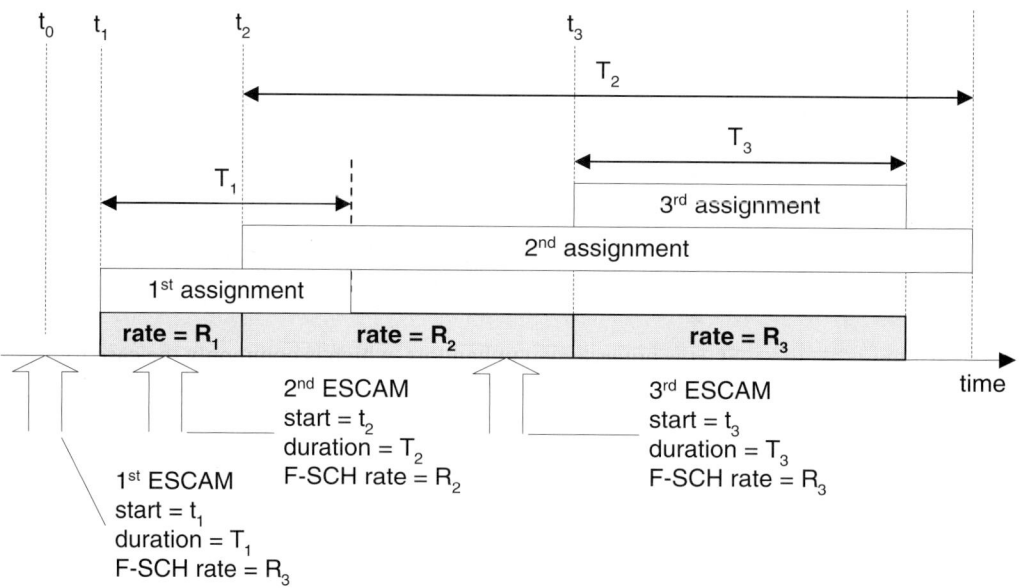

Figure 9.6   F-SCH burst assignments example.

9.4.1.2   *Data Rate Control*

When assigning the F-SCH, the base station must select the data rate based on queue length, channel conditions, and residual transmit power. The queue length can be used to determine the minimum required data rate for a given target transmission delay.[6] Assume, for example, that the queue length is L bits and the target transmission delay is D seconds. Then, the desired data rate R should be no less than L/D bps. The channel conditions dictate the minimum

---

6.   For nonreal-time application, there is no stringent delay requirement, but the base station may target certain delay.

transmit power required to achieve the desired FER for a given data rate. Conversely, the residual power available for the F-SCH dictates the maximum data rate sustainable with the given channel conditions.

The data rate selection process therefore requires the following iterations. First, the base station computes the desired data rate as $R = L/D$. Then, given the channel conditions at the time of the assignment, the base station computes the required F-SCH chip energy relative to the total base station transmit power, I_{or}, denoted as F-SCH $E_c(R)/I_{or}$. The base station grants the data rate R if its current loading is such that the required transmit power is available; otherwise it decreases the data rate and repeats the last two steps. The required F-SCH transmit power may be estimated as follows. At the time of F-SCH assignment, the F-FCH is being power-controlled and its transmit power reflects current channel condition. The F-FCH transmit power can be used to infer the required F-SCH transmit power provided we account for the different target FER and for the fact the F-SCH active set is a subset of that of the F-FCH, typically equal to one. A lower F-SCH FER target implies a lower required F-SCH E_b/N_t and proportionally lower F-SCH transmit bit energy relative to that of the F-FCH. However, the E_b/N_t difference can only be approximately estimated because it depends on mobile station speed and multipath profile. Similarly, the larger F-FCH active-set size implies a higher F-SCH transmit bit energy relative to that of the F-FCH, but the difference can only be approximately estimated because it depends, in addition to the factors above, on the relative received strength of the active-set pilots.

Assuming that the F-SCH is not assigned in handoff, which is commonly the case for best-effort traffic, the required F-SCH transmit power can be computed for a single path channel as[7]

$$\text{F-SCH}\frac{E_c}{I_{or}} = \frac{\dfrac{\text{F-PICH } E_c}{I_{or}} - \text{F-PICH}\left(\dfrac{E_c}{I_o}\right)}{\text{F-PICH}\left(\dfrac{E_c}{I_o}\right)} \cdot \frac{R_b}{R_c} \cdot \text{F-SCH}\frac{E_b}{N_t} \tag{9.11}$$

The F-PICHE_c/I_{or} represents the F-PICH transmit energy relative to the total transmit power spectral density. The F-PICH (E_c/I_o) is the received pilot chip energy to total received power-density ratio, I_o. It is measured by the mobile station and reported to the base station via Layer 3 signaling. Eq. (9.11) can practically be implemented because it is expressed in terms of quantities known to the base station.

Throughout the duration of the F-SCH assignment, channel conditions may change beyond those that can be compensated by power control, and the base station must reconsider the assigned data rate. That is the case, for example, when a mobile station that has been assigned a high-rate F-SCH moves to the fringe of the cell. Unless the base station is allowed to allocate a disproportionate amount of power to a single user, the F-SCH data rate must be

---

7.   This formula was first derived in Chapter 8 for the F-FCH.

promptly decreased. The opposite scenario is also possible. The F-SCH assigned to a mobile that moves into an area with more favorable channel conditions may be re-assigned the F-SCH at a higher rate. This process is a form of slow-rate adaptation. An effective method to trigger data rate change is to monitor the short-term average of the F-SCH transmit power. Once the transmit power falls below a lower or above an upper threshold, the data rate should be increased or decreased, respectively. Estimating the short-term average of the transmit power requires filtering, which induces a triggering delay. The triggering delay combined with signaling transmission and mobile station message-processing delays results in the overall F-SCH assignment delay. For rate control to be effective, the assignment delay must be short relative to the decorrelation time of the received-signal short-term average. In Figure 9.7, we depict a case in which the assignment delay is not sufficiently short relative to the channel decorrelation time. The data rate corresponding to the current assignment is suboptimum for about 25% of the time, which is reflected by the shaded area in Figure 9.7, leading to lower user throughput and capacity utilization.

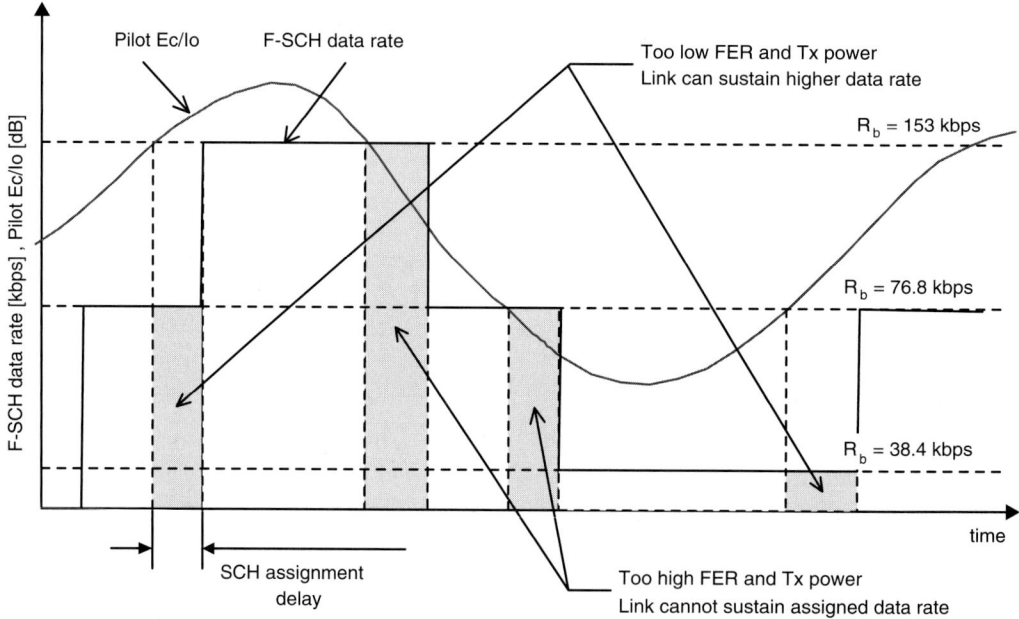

Figure 9.7   Effect of F-SCH assignment delay on throughput.

9.4.1.3   Power Control

There are seven forward-link power-control modes defined in IS-2000 [4]. They all employ a form of closed-loop power control consisting of a feedback channel, the Reverse Power Control Subchannel (R-PCSCH), transmitted by the mobile station, which is used by the base station to adjust the traffic channel transmit power. Of the seven, the five that are of practical applicability to the F-SCH are summarized in Table 9.2 and described below.

Mode 0 (FPC_MODE = 000) is the fundamental mode discussed in detail in Chapter 8. In this mode the primary traffic channel (either F-FCH or F-DCCH) is power-controlled using the inner- and outer-loop mechanisms. The mobile station maintains an outer-loop setpoint that is adjusted to achieve the target FER. At each power-control group (PCG) interval, the mobile station compares the estimated traffic channel SNR against the setpoint and generates an up or down command. The commands are mapped to power-control bits (0 for up and 1 for down command) and sent on the R-PCSCH at a rate up to 800 bps. The R-PCSCH is punctured on the Reverse Pilot Channel (R-PICH). Although primarily used for the F-FCH or F-DCCH, this mode can be used also when the mobile station is assigned the F-SCH. In this case, the F-SCH transmit power is slaved to that of the primary traffic channel with a semistatic offset. The base station can determine the offset based on the F-SCH configuration and target FER.

In mode 1 (FPC_MODE = 001) the R-PCSCH is split into a primary and a secondary R-PCSCH. The primary controls the primary traffic channel transmit power, and the secondary controls the F-SCH. Both subchannels operate at 400 bps. This mode involves two independent inner/outer power-control loops.

Mode 2 (FPC_MODE = 010) is similar to mode 1 except that more bits are allocated to the secondary R-PCSCH. In this mode the primary traffic channel is power-controlled at 200 bps, and the F-SCH at 600 bps.

Mode 5 (FPC_MODE = 101) allows independent power control of the primary traffic channel and the F-SCH, but unlike the previous modes, it does not involve the outer-loop mechanism at the mobile station. Rather than estimating the SNR every PCG, the mobile station estimates whether each received frame is good or bad. The frame quality is bad if the received frame is erased, that is, if the cycle redundancy code (CRC) does not check and it is good otherwise. Note that in the case of the F-DCCH, the mobile station cannot rely solely on the CRC because the F-DCCH may be gated off for the period of one or more frames and during that period, the mobile station must still determine whether the "frame" is a good one or not. The mobile station sets the quality indicator bit (QIB) on the primary R-PCSCH based on primary traffic channel quality, and the erasure indicator bit (EIB) on the secondary R-PCSCH based on the F-SCH quality. The QIB consists of a repetition of the eight power-control bits allocated to the primary R-PCSCH and is sent at 50 bps. The EIB consists of a repetition of the eight power-control bits allocated to the secondary R-PCSCH if the F-SCH uses 20 ms frames. If the F-SCH uses 40 or 80 ms frames, the EIB is further repeated two or four times, respectively. Thus, the EIB effective rate is 50, 25, or 12.5 bps depending on the F-SCH frame length. The transmission of the QIB or EIB starts at the second frame (20-ms frame) of the reverse traffic channel following the corresponding forward traffic channel frame in which the bit is determined. This delay is necessary because the mobile station cannot determine frame quality and instantaneously start modulating the R-PCSCH with the corresponding quality bit.

Mode 6 (FPC_MODE = 110) is a combination of modes 1 and 5. The primary R-PCSCH controls the primary traffic channel transmit power at 400 bps. The secondary R-PCSCH carries the F-SCH EIB.

Relative performance of these modes depends on channel conditions [5]. In slow-fading channels, a faster power-control mode is more effective because the inner and outer loops can more accurately track fade variations, thus achieving the target FER with lower traffic channel transmit power. The slow (50 bps) power control instead does not react to fade variations and cannot invert the channel. Moreover, the closed-loop delay is equal to several frames when the QIB or EIB is used and performance is degraded. For example, by the time the transmit power is adjusted upward in response to a bad frame indication, the channel may have already improved. The difference in performance between fast and slow power control becomes smaller in the presence of multipath that inherently combats deep fades. In fast-fading channels the slower power-control modes outperform the fast modes because neither mechanism can track fade variations. The faster modes attempt to invert the channel fades but instead cause random fluctuations in transmit power.

Furthermore, relative performance of the power control modes depend on how the F-SCH is configured and operated in handoff. In mode 0, even though the F-SCH is effectively power-controlled at the same rate as the primary traffic channel, the F-SCH's actual FER may deviate from the target because the optimum transmit power offset relative the primary traffic channel (the one that would achieve the target FER for both primary traffic channel and F-SCH) varies with varying radio channel conditions. The deviation of the F-SCH actual FER from the target increases even further in soft handoff if the F-SCH active set is smaller than the primary traffic channel active set. To mitigate this problem, the base station can request F-SCH FER measurement reports from the mobile station by setting the FOR_SCH_FER_REP flag to 1 in the ESCAM. FER measurements are sent to the base station via the *Power Measurement Report Message* (PMRM). If the FER reported in the PMRM is not within the desired limits, the base station can adjust the F-SCH power offset. Transmission of the PMRM containing the F-SCH FER report is triggered only after expiration of current F-SCH data burst assignment so that the offset adjustment is possible only on a per-burst basis.

Modes 1 and 2 are, therefore, better suited than mode 0 when the primary traffic channel and F-SCH active sets are not identical, because with the former modes the inner and outer loops for the primary traffic channel and F-SCH are operated independently, albeit at reduced rate, therefore allowing the target FER for both to be achieved. Mode 1 typically outperforms mode 2 because the loss incurred from a slower power control of the primary channel outweighs the gains due to faster F-SCH power control, especially when the latter is operated at higher target FER.

Table 9.2 Reverse Power Control Subchannel Configurations*

FPC_MODE	Primary R-PCSCH	Secondary R-PCSCH
000	800 bps	N/A
001	400 bps	400 bps
010	200 bps	600 bps
101	1 QIB per 20 ms (50 bps)	1 EIB per F-SCH frame period (50, 25, or 12.5 bps)
110	400 bps	1 EIB per F-SCH frame period (50, 25, or 12.5 bps)

* Only modes applicable to the F-SCH are shown.

9.4.1.4 Cell Switching

Although the F-SCH can be operated in soft handoff, for best-effort packet data service it is typically assigned only on the base station corresponding to the strongest pilot of the primary traffic channel active set. This scenario is called the *F-SCH reduced active set operation*. When an active-set pilot becomes stronger then the one corresponding to the base station currently transmitting the F-SCH, the F-SCH is reassigned onto the strongest one. This procedure is called *F-SCH cell switching*. The motivation for F-SCH reduced active set is that the cost incurred with a larger F-SCH active-set seize outweighs the benefits. The micro-diversity gain achieved with soft combining is smaller for F-SCH than it is for F-FCH because the former is typically operated at higher FER, and error recovery is handled by upper-layer protocols, such as RLP and Transmission Control Protocol (TCP). Cell macro-diversity, key in meeting the stringent F-FCH outage requirements, is not equally beneficial to the F-SCH when the latter does not carry real-time services.

The mobile station must monitor and report the strongest pilot to trigger cell switching. The base station enables the pilot monitoring and reporting procedure when it transmits the *Service Option Control Message*[8] (SOCM) [3]. The SOCM fields relevant to this procedure are summarized in Table 9.3. If the procedure is enabled, the mobile station sends a *Periodic Pilot Strength Measurement Message* (PPSMM) when the strongest pilot becomes weaker than another pilot by a difference specified by SP_MIN_DELTA for an interval specified by SP_INTERVAL. Once the mobile station has sent the PPSMM, it begins to monitor the pilot reported in the PPSMM as the strongest pilot.

---

8.   The service option-specific control record in the SOCM is specified in IS-707.

Table 9.3   Pilot Monitoring and Reporting Procedure

Field	Message	Description
SP_MR_CNTL	SOCM	Flag set to 1 to enable strongest pilot monitoring and reporting.
SP_MIN_DELTA	SOCM	Minimum difference, in dB/2, between the current strongest pilot and the last reported strongest pilot to set PPSMM transmission timer.
SP_INTERVAL	SOCM	Duration of the PPSMM transmission timer, in 80 ms. PPSMM transmission is triggered when timer expires.

The base station, on receipt of the PPSMM, sends the ESCAM assigning the mobile station the F-SCH on the CDMA channel corresponding to the strongest pilots, possibly at a different data rate than that of the previous assignment, depending on the channel conditions and available base station transmit power. The base station may also command the mobile station to trigger PPSMM transmission periodically and not just on change of the active-set pilot ranking.

9.4.1.5   Scheduling

When F-SCH is not placed in soft handoff, burst mode operation allows for scheduling. For bursty traffic, scheduling allows better utilization of system resources because more mobile stations can be admitted in the system and congestion events can be efficiently handled by scheduling algorithms.

A common F-SCH scheduling algorithm is round robin. One or more mobile stations are commonly scheduled for a few hundred ms at a time. Algorithms that account for channel conditions are possible but are more difficult to implement. Channel quality information reports in the form of *Pilot Strength Measurement Messages* (PSMM) are slow and in most cases cannot track short-term fading. Long-term shadowing is possible to follow because shadowing decorrelates only after a second or so. However, as discussed in Section 9.7, long scheduling intervals could negatively impact error recovery and the flow-control mechanism of TCP.

9.4.2 F-SCH for Real-Time Service

When the F-SCH is utilized to provide real-time service, such as streaming video, stringent outage requirements prohibit placing the F-SCH in reduced active set. In this case the active-set size for the F-SCH follows that of the F-FCH, because both channels have essentially the same QoS constraints.

The assignment policy procedure follows the one described in Section 9.4.1.1. When the F-SCH is allocated to real-time traffic, the duration field, FOR_SCH_DURATION, in the ESCAM is commonly set by the base station to indicate infinite allocation. The F-SCH is then explicitly deallocated once the channel is no longer needed. Finite-length F-SCH channel assignment can also be used, but it is essential that the new channel assignment arrive before the current assignment expires.

Since the active-set size of the F-SCH is the same as that of the F-FCH, FPC_MODE=000 is possible. It is still true that the F-SCH-to-F-CH transmit power offset can only be approximately estimated, but the accuracy is much better than when the F-SCH is operated in reduced active set.

9.5 Reverse Supplemental Channel Operation

The R-SCH is used in addition to the R-FCH (or R-DCCH) when the data rate provided by the latter is not adequate for the connected data service. The R-SCH only carries user data, while the R-FCH carries signaling and may carry user data. The R-SCH is operated in burst mode; that is, it can be assigned, reconfigured, and de-assigned multiple times throughout the life of the packet data call without requiring release of the connection and the associated F/R-FCH. Unlike the F-SCH, the R-SCH is never placed in reduced active set because there is no overhead (except for demodulating elements at the BTS) associated with reverse-link handoff, and reduced active set would compromise reverse-link capacity. The only distinction between R-SCH operation for real-time and best-effort service is that R-SCH scheduling is possible for latency-insensitive data. When R-SCH is scheduled, the burst duration must be carefully selected to account for assignment latency and the induced signaling load as well as impact on TCP. These and other issues such as power control and data rate control are discussed in the following sections.

9.5.1 Assignment Policies

Since the reverse-link data originates at the mobile and the channel allocation is control by the base station, the mobile station must first request R-SCH assignment by sending the *Supplemental Channel Request Message* (SCRM). SCRM transmission is commonly triggered when the packet transmission queue length exceeds a certain threshold, which may be a function of the level of QoS that the mobile station requested.

The SCRM contains the SCRM_REQ_BLOB[9] that indicates the characteristics of the requested R-SCH. The SCRM parameters relevant for the discussion are shown in Table 9.4. Two parameters are shown, the preferred data rate, PREFERRED_RATE, and the duration, DURATION. The preferred data rate may take values from 9.6 kbps up to the maximum data rate supported on R-SCH. The value of 0 indicates that mobile station requests to release the R-SCH.

Table 9.4　R-SCH Request Parameters

Field	Message	Description
PREFERRED_RATE	SCRM	Preferred R-SCH rate.
DURATION	SCRM	Requested R-SCH duration.

*Not all R-SCH request related fields are listed.

9.　The SCRM_REQ_BLOB format is specified in IS-707.

On receipt of the SCRM, the base station decides if resources can be allocated to the mobile station for burst transmission. The base station uses the methods already discussed in Section 9.2.7 to estimate residual capacity and decide on the data rate that can be assigned. Similarly as for the F-SCH, the R-SCH assignment is controlled by the ESCAM. The relevant ESCAM fields are summarized in Table 9.5. For the initial assignment, the ESCAM conveys to the mobile station R-SCH configuration information and instructs the base station to start decoding R-SCH. Besides the duration of the R-SCH assignment and the action time at which the R-SCH assignment become valid, the R-SCH assignment message also contains the number of information bits per frame. The number of information bits defines the data rate on R-SCH. The duration of the R-SCH assignment can be finite, between 1 and 256 frames, or infinite, which means that R-SCH is allocated until the base station explicitly deallocates the channel. The ESCAM also includes the R_SCH_DTX_DURATION, which indicates the maximum time duration the mobile station is allowed to pause transmission on the R-SCH. If the mobile station pauses the R-SCH transmission for a period larger than allowed discontinuous transmission (DTX) duration, it must release the R-SCH by sending the SCRM. Setting of the R_SCH_DTX_DURATION is the result of a trade-off between two conflicting objectives. As data traffic is bursty, the DTX duration must be large enough to allow the mobile station to pause R-SCH transmission without losing the current assignment. The DTX duration must be short enough to ensure quick channel release once the data transfer is complete, which allows the scheduler to reassign the R-SCH to other users. A good setting for R_SCH_DTX_DURATION is around 10 frames. Another parameter carried by the ESCAM is the USE_T_ADD_ABOT flag. If the flag is set to 1, and the strength of a neighbor-set or remaining-set pilot is found to be above a given threshold,[10] then the mobile station shall terminate any active transmission on R-SCH at the end of the current 20-ms frame. This mechanism is useful to prevent the R-SCH, especially when operated at high rate, to create excessive interference to a neighboring base station that is not yet in the active set but that is deemed sufficiently close to the mobile station (as inferred from the its relatively high pilot strength) and thus is likely to suffer from such interference.

Table 9.5   R-SCH Assignment Parameters

Field	Message	Description
REV_SCH_DURATION	ESCAM	R-SCH assignment duration.
REV_SCH_DTX_DURATION	ESCAM	Maximum allowed DTX period.
REV_SCH_START_TIME	ESCAM	R-SCH processing start time.
REV_SCH_NUM_BITS_IDX	ESCAM	Index that indicates the number of information and CRC bits per frame.
USE_T_ADD_ABORT	ESCAM	If turned on, it indicates that the mobile station must release the R-SCH once the T_ADD abort condition is met.

*Not all R-SCH assignment related fields are listed.

---

10.   The threshold is set equal to T_ADD, the same parameter used for pilot-set maintenance discussed in Chapter 7.

9.5.2 Power Control

Unlike the F-SCH, the R-SCH cannot be power-controlled separately from the primary traffic channel, such as the R-FCH. Both R-FCH and R-SCH have fixed transmit power relative to the R-PICH, which is power-controlled at 800 bps. Separate R-SCH and R-FCH inner-loop power control would reduce the effective power-control rate on both channels. Fast power control is crucial to achieve high reverse-link capacity and, therefore, a reduced power-control rate is not permitted. Moreover, unless a stringent FER constraint is imposed on the R-SCH, separate power control would have little benefit, since the R-SCH and R-FCH have always the same active set.

The R-PICH overhead can consume a significant portion of the reverse-link capacity. Unlike the F-PICH that is a common channel whose overhead is shared by all users in the cell, the R-PICH is a dedicated channel. It is therefore important to limit the overhead by optimizing the traffic channel (R-FCH, R-DCCH, and R-SCH) transmit power relative to that of the R-PICH. The ratio is typically referred to as the traffic-to-pilot (T/P) ratio.

The reverse traffic channel's transmit power is given by Eq. (9.12). The mean output power for both the traffic channel of interest and the R-PICH are expressed in dBm. The last three terms of the right-hand side of Eq. (9.12) depend on the traffic channel type, and for a given channel type, they also depend on the presence of other channels. Their sum in decibels represents the T/P ratio. We now examine each of the terms determining the T/P ratio.

$$
\begin{aligned}
\text{mean traffic channel output power} \ = \ & \text{Mean R-PICH output power} \\
& + \text{Nominal\_Attribute\_Gain} \\
& - \text{Multiple\_Channels\_Adj\_Gain} \\
& + \text{possible other corrections}
\end{aligned}
\tag{9.12}
$$

The first and most important is nominal attribute gain, which is tabulated in IS-2000. The table defines the nominal T/P ratios for the reverse traffic channels as a function of their type, data rate, frame duration, and channel coding type. In the case of the R-SCH the nominal gain also depends on the desired FER, because the outer loop is driven by the R-FCH FER only. Table 9.6 shows exemplary values for the nominal attribute gain tabulated in the physical layer specification. The pilot reference level is the R-PICH reference level used to compute the multiple channel adjustment gain.

To explain how the multiple channel adjustment gain is computed, consider the following example. Assume that only R-FCH with rate 9.6 kbps is active and the mobile station has received the ESCAM that assigns the R-SCH with data rate equal to 38.4 kbps. The pilot reference level for the R-FCH is 0 dB, and for the R-SCH at 38.4 kbps is 1.25 dB. The outer-loop setpoint is driven by the R-FCH. However, when R-SCH is set up, the setpoint needs to be adjusted to account for the R-SCH pilot reference level. In this example the R-SCH pilot reference level is 1.25 dB higher for the R-SCH than for the R-FCH; hence, the outer-loop power-control setpoint needs to be increased by 1.25 dB to account for the highest pilot reference level of the active reverse-link traffic channels. The mobile station then computes the multiple channel adjustment in order to adjust the T/P ratios of the active channels. The adjustment for the channel with highest pilot reference level is zero, while the adjustment for all other channels is the difference between the largest pilot reference level of all active channels and the pilot reference level of the channel in question. For the given example, the R-SCH multiple channel adjustment is equal to 0 dB, while for the R-FCH it is equal to 1.25 dB.

Table 9.6   Nominal Attribute Gain Table

Data rate [kbps]	Frame length [ms]	Coding	Nominal attribute gain [dB]	FER	Pilot reference level [dB]
9.6	5 ms	Convolutional	7.25	0.01	0
9.6	20	Convolutional	3.75	0.01	0
19.2	20	Turbo	5.5	0.05	0.25
38.4	20	Turbo	7.0	0.05	1.25
76.8	20	Turbo	8.5	0.05	2.375
153.6	20	Turbo	9.5	0.05	4.125
307.2	20	Turbo	11.0	0.05	6.25

*Not all related fields are listed.

Table 9.6 also shows the nominal attribute gain for 5-ms frames. Recall that the 5-ms frames are used to transport mini-messages that allow faster Layer 3 signaling. Note that even though the pilot reference level for 5-ms frames is the same as for 20-ms frames, for the same FER, the T/P ratio is a full 3.5 dB higher. Short frames require more power because of lower interleaver gain.

Corrections to the nominal attribute gain table are permitted in case the default values do not provide the desired FER. These values can be communicated to the mobile station at channel setup or during the operation. The base station can monitor the FER and upon detection of the discrepancy between the desired and actual FER, using Layer 3 signaling procedures, signal to the mobile station the correction to the T/P ratio. The correction to the nominal values can be communicated through the *Power Control Message* (PCNM), which then populates the attribute adjustment gain table. The attribute adjustment gain table is a function of the same parameters as nominal attribute gain table. Channel-based adjustments are also permitted. One example of a parameter allowing channel-based adjustment is the RLGAIN_TRAFFIC_PILOT, which allows the adjustment of each of the reverse traffic channels. These adjustments, however, are common for all frame lengths, data rates, coding types, or target FER. Other adjustments work in a similar fashion.

9.5.3 Data Rate Control

As explained in Section 9.5.1, the mobile station requests the R-SCH preferred rate and duration by populating PREFERRED_RATE and DURATION fields in SCRM. The PREFERRED_RATE can be computed based on the maximum transmit power of the mobile station and an implementation-dependent power headroom. The DURATION is computed from the requested rate and data queue information. When the mobile station acts as a modem with the data application running on an external device such as a laptop computer, the mobile station commonly requests an infinite duration assignment, as it is unable to predict how much data will be generated by the application.

As the radio channel conditions vary, so does the maximum rate supported by the mobile station. The mobile station may request a data rate change by sending a new SCRM that indicates the new PREFERRED_RATE. Once the mobile station has sent a SCRM, however, it must wait for at least one second before sending another SCRM, for example, to request an extension of the current assignment or a different data rate. An exception to this rule is the case in which, following SCRM transmission, the mobile station is assigned the R-SCH with a data rate that it cannot support given the current channel conditions. When the mobile station is reassigned the R-SCH with a different data rate, a correction to the reverse-link power-control outer-loop setpoint must be adjusted to compensate for the differences in the pilot reference level. That is necessary because the pilot reference level depends on the data rate. Figure 9.8 illustrates the timing of the setpoint adjustments relative to the corresponding data rate changes.

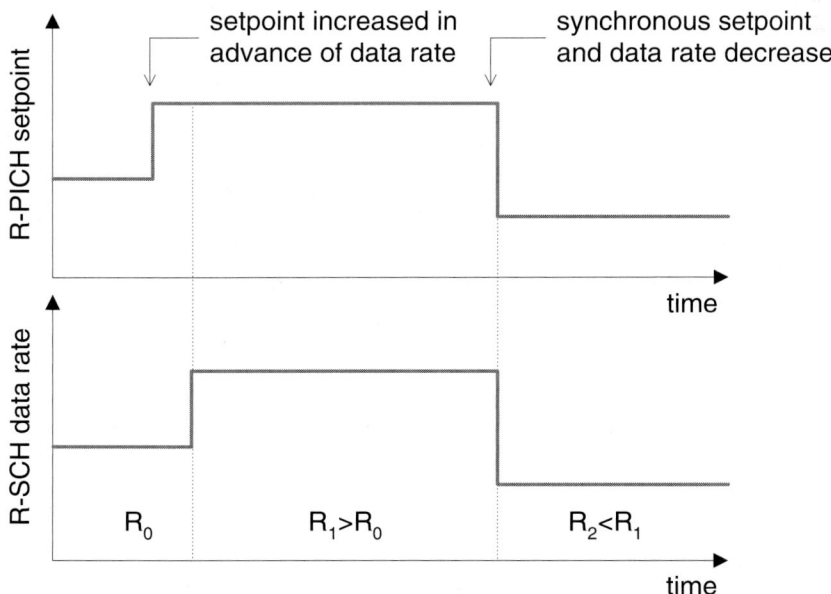

Figure 9.8 R-PICH setpoint corrections due to R-SCH data rate changes.

9.5.4 Scheduling

For the data traffic that does not impose stringent delay requirements, such as best-effort traffic, the R-SCH can be scheduled. The scheduling allows for better utilization of the residual reverse-link capacity and also provides a mechanism to recover from congestion. A suitable scheduling algorithm for the R-SCH is round robin. It is common that several mobile stations are scheduled to transmit in the same time slot because a single mobile station typically cannot consume the residual capacity due to its maximum transmit power limitation. Similarly as for F-SCH, care must be taken when selecting the scheduling duration. Too short a scheduling interval requires excessive Layer 3 signaling. Too long a scheduling interval, however, as we discuss in Section 9.7, can have a negative impact on TCP.

The scheduling algorithms for R-SCH can try to take advantage of the radio conditions experienced by the mobile station. The idea is to give higher priority to mobile stations with lower power-control setpoints and to those that are closer to the base station. The mobile stations that are close to one of the base stations in the active set in general create less outer-cell interference, thus increasing the overall system capacity. Refer to [6] for more details on the physical layer performance of several scheduling algorithms suitable for the R-SCH.

9.6 Forward Packet Data Channel Operation

F-PDCH enables efficient utilization of base station resources for delay-tolerant traffic. It is a shared channel, and short-term radio channel variations due to fading can be exploited at the expense of delay jitter. The traditional view of fading is that it is detrimental and that it needs to be compensated for. However, fading is also a source of randomness and it can be opportunistic. With smart scheduling, fading can actually significantly improve the efficiency of the air interface.

The efficiency improvement is due to multiuser diversity. In point-to-multipoint links, such as the one that exists between a single base station and the mobile stations in a given cell, the radio propagation channel varies independently. The base station can then choose to allocate its resources to a mobile station that experiences the best radio propagation environment among all mobiles in that cell and thus maximize the throughput. Selecting a mobile station among a group of mobiles is commonly called multiuser diversity. The scheduling interval must not be larger than the mean fade duration or channel coherence time in order to exploit short-term radio channel fluctuations. The magnitude of the multiuser diversity gain is quantified in Chapter 10.

The multiuser diversity gain comes at a cost and that is delay jitter. Moreover, if there is no fairness constraint imposed on the base station scheduler, the base station always schedules mobile stations that can support the highest data rate, and some mobile stations may not receive any data at all. It is, therefore, necessary that the scheduling algorithm that enables multiuser diversity effect satisfies some fairness criteria and minimizes physical layer throughput variation.

In practice, a common scheduling algorithm is proportional fair, which is explained in Section 9.3.3. The scheduling interval corresponds to a single frame, which can be 1.25 ms, 2.5 ms, or 5 ms. The data rate is determined from the channel quality information and the available base station resources. The time constant of the filter that estimates the throughput must be short enough to smooth *variable bandwidth effect*, which can negatively impact TCP throughput (see Section 9.7). However, the shorter the time constant, the smaller the multiuser diversity effect. A filter constant value that provides good trade-off is 1.25 sec.

Besides delay and throughput variation, multiuser diversity gain comes at the expense of overhead. The overhead is twofold. There is the reverse-link overhead in the form of a feedback channel that transports radio channel quality information, which helps the base station with the scheduling decisions. There is also the forward-link control channel that informs the mobile station when it has been scheduled on the F-PDCH.

The channel quality information may be inaccurate. For example, at low SNRs, the measurement itself is not accurate. In vehicular channels, due to finite delay between the actual channel measurement and decoding at the base station and high speed of the mobile station, the channel quality information could be obsolete. In these scenarios, rate adaptation alone produces high packet-loss rate and needs to be accompanied by Hybrid ARQ (H-ARQ) type II techniques. H-ARQ refers to joint decoding of the current and previous transmission, and type II refers to error correction as opposed to error detection only, which is a type I characteristic.

The entity that controls F-PDCH operation is called PDCHCF. This entity controls link adaptation, scheduling, and H-ARQ type II operation, and it is responsible for mapping f-pdch logical channels to the corresponding physical channels. PDCHCF is considered part of the MAC Layer [7] and it is always implemented at the BTS.

9.6.1 Control Signaling

F-PDCH is a shared channel. The base station grants access on F-PDCH for up to 5 ms. Such a short access requires a fast signaling protocol to alert the mobile station. For that purpose, the base station uses F-PDCCH, which operates in parallel to the F-PDCH. The F-PDCCH frame length is equal to the F-PDCH frame length. Since the number of bits in the F-PDCCH frame is constant, its data rate varies with the frame length. The mobile station buffers the signal received on F-PDCCH and F-PDCH. The F-PDCCH is decoded considering all three possible frame lengths, 1.25, 2.5, and 5 ms. The F-PDCCH frame is protected by a CRC and, neglecting the rare occurrence of undetected errors, only one of the possibilities can produce a valid CRC checksum. Figure 9.9 illustrates F-PDCCH transmission timing.

The control message, illustrated in Figure 9.10, contains 21 bits. In addition, there are 16 bits for CRC and 8 encoder tail bits. In the control message itself, the first 8 bits are reserved for the MAC_ID, which identifies a mobile station. The scope of the MAC_ID is per CDMA channel, or pilot. When the mobile station performs handoff to another pilot, the MAC_ID can change. The MAC_ID is initially communicated to the mobile station via *Extended Channel Assignment Message* (ECAM) and can be subsequently updated in case of a soft handoff event by sending the *Universal Handoff Direction Message* (UHDM).

There are six different values that the encoder packet size can take. The mobile station needs to know the encoder packet size in order to successfully decode the incoming packet. The

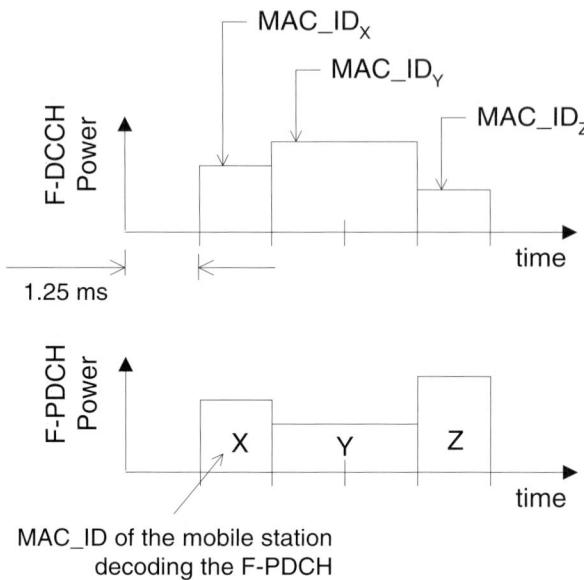

Figure 9.9   F-PDCCH message timing.

MAC_ID	EP_SIZE	ACID	SPID	AI_SN	LWCI
(8 bits)	(3 bits)	(2 bits)	(2 bits)	(1 bit)	(5 bits)

Figure 9.10 F-PDCCH message format.

encoder packet size is contained in a 3-bit EP_SIZE field. ER_SIZE=111 refers to extended message types used for F-PDCCH control messages. Control F-PDCCH messages contain only MAC_ID, EP_SIZE=111 and the 10-bit control information (eight of them are currently reserved).

The mobile station can simultaneously receive four parallel physical layer data streams, transmitted on four separate ARQ channels. To discriminate between the channels, F-PDCCH message contains a 2-bit field called an ARQ Channel Identifier (ACID). Each ACID supports independent H-ARQ type II operation.

Up to seven retransmissions or eight transmissions of the same encoder packet are permitted. To support Adaptive and Asynchronous Incremental Redundancy (AAIR), four different subpacket types are supported. As explained in Section 9.6.4.1, the 2-bit field, Subpacket Identifier (SPID), in conjunction with the selected data rate, determines which coded bits are selected for transmission. During the AAIR operation, the sequence of SPIDs is not mandated by the standard. It is only mandatory that the first transmission contains SPID = 00, which includes all systematic plus some parity bits.

Since the sequence is not mandatory and SPID = 00 can be repeated, the mobile station must be notified when the new encoder packet begins. For that purpose, ARQ Identifier Sequence Number (AI_SN) bit is added to the F-PDCCH message. The AI_SN bit is toggled whenever new encoder packet transmission begins.

Finally, the F-PDCCH message contains the Last Walsh Code Index (LWCI) that identifies the last Walsh code in the Walsh code tree. The Walsh codes that are preassigned for overhead channels, such as F-PICH, F-SYNCH, one or more F-PCHs, and few others, are excluded. Only 28 Walsh codes of length 32 are potentially available for F-PDCH. Exactly which 28 codes are available is signaled to the mobile station through WALSH_TABLE_ID. The scope of WALSH_TABLE_ID is per pilot. It is a 3-bit field and is part of both ECAM and UHDM. Walsh code space can become fragmented due to, for example, assignment and tear down of F-SCH. The F-PDCCH message, as shown in Figure 9.10, cannot address fragmented Walsh space because the LWCI indicates the last code in a contiguous Walsh code space. Underutilized Walsh space leads to a dimension limitation problem, which typically leads to high usage of high-order modulation, such as 8-PSK or 16-QAM. These modulations are less efficient than QPSK and should be avoided whenever possible.

To alleviate Walsh fragmentation problem, a special F-PDCCH broadcast message is designed to signal the fragmented Walsh code space available for F-PDCH. The MAC_ID in this message is set to 00000000, which indicates that the message is addressed to all the mobiles in the cell. The remaining 13 bits are bitmapped to 13 Walsh codes of length 16. The limited space in the F-PDCCH message prevents indexing length-32 Walsh code space. Recall that actually up to 14 Walsh codes of length 16 could be available for F-PDCH. An index called WALSH_TABLE_ID indicates which two Walsh codes of length 16 are consumed by the overhead channels. The Walsh code space is ordered, and the first 13 Walsh codes are bitmapped. If any of the descendents of the Walsh code are not free, the bitmap shows that the parent Walsh code is not available. This means that one occupied descendent of length 32 prevents the other from being used for F-PDCH even though one of them could potentially be free. The availability of the last two Walsh codes of length 32 (descendents of the length-16 Walsh code that is not bitmapped) is signaled through LWCI.

The standard allows simultaneous operation of two F-PDCHs; that is, up to two mobiles can be scheduled at the same time. This flexibility is added for more efficient support of WAP traffic and Layer 3 signaling. Each F-PDCH is accompanied by its own F-PDCCH. Let us denote the two channels as F-PDCH0 and F-PDCH1. The corresponding control channels are then labeled F-PDCCH0 and F-PDCCH1 respectively. The concurrent operation of two channels is illustrated by Figure 9.11. Assume that the mobile station MS0 is scheduled over F-PDCH0 and the mobile station MS1 over F-PDCH1. Both MS0 and MS1 decode F-PDCCH0. MS0 determines that it is scheduled over F-PDCH0. For MS0, the LWCI determines Walsh code space on F-PDCH0. As explained above, LWCI indicates the last Walsh code. The procedure for MS1, however, is different. After decoding F-PDCCH0, MS1 determines that it is not scheduled over F-PDCH0. MS1 also decodes F-PDCCH1 and determines that it is scheduled over F-

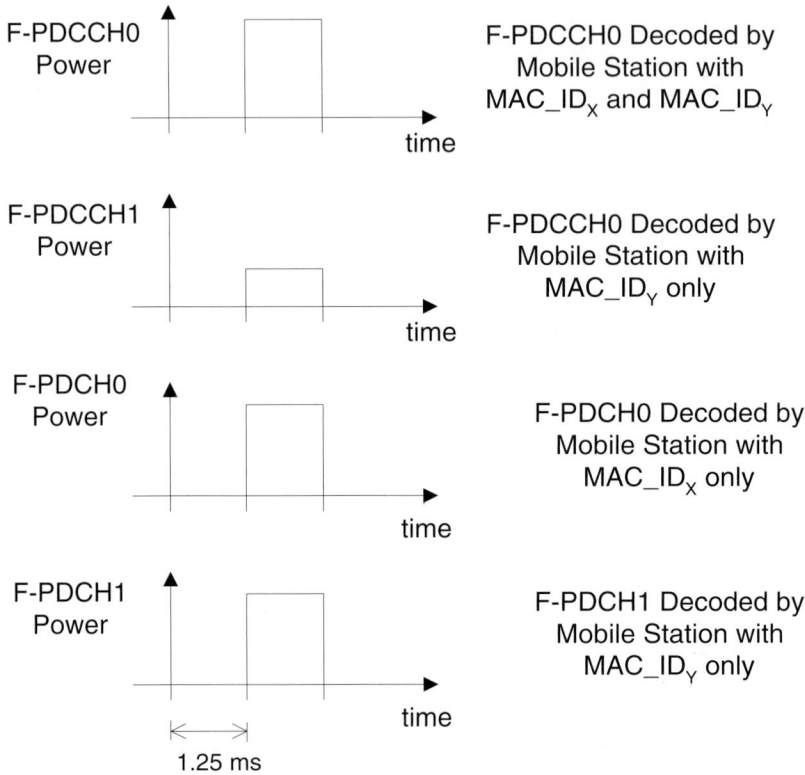

Figure 9.11 Scheduling two mobiles at the same time.

PDCH1. Then, the LWCI decoded from F-PDCCH0 determines the beginning, while LWCI decoded from F-PDCCH1 determines the end of the Walsh code space. The frame length must be equal on both channels. Blind rate detection is only performed over F-PDCCH0, not over F-PDCCH1.

9.6.2 Link Adaptation

The F-PDCH allows adaptive modulation and coding to improve spectral efficiency. The radio channel state information at the mobile station is communicated to the base station through the Reverse Channel Quality Indicator Channel (R-CQICH). Use of this feedback channel is twofold. It allows the base station to implement a scheduler that exploits channel variability to achieve multiuser diversity gain, such as the proportional fair scheduler discussed in Section 9.3.3. It also allows the selection of the optimal F-PDCH data rate given the current channel conditions. Link adaptation must be used with care in order not to negatively impact TCP performance (see Section 9.7).

9.6.2.1   Feedback of Channel Quality

Mobile stations report their channel quality over the R-CQICH. There are two modes allowed, full and differential carrier to interference (C/I) reporting. The full C/I report represents a result of the measurement of pilot chip energy to total noise plus interference ratio Pilot E_c/N_t, mapped to a 4-bit channel quality indicator. In the full mode, the current Pilot E_c/N_t is reported every PCG or 1.25 ms. The differential update is interpreted as a ±0.5 dB correction relative to the most recent accumulated C/I value. The differential scheme itself consists of one full report every 20 ms and 15 ±0.5 dB updates in between. The accumulator sums up the differential updates every PCG and it refreshes itself every 20 ms when the full report is received.

In Figure 9.12, differential C/I reporting is shown. One full report is followed by 15 differential reports. The full report can be repeated in order to improve reliability. In that case, the receiver soft-combines the repeated symbols. The number of times the full C/I report is repeated can be configured by the base station using the REV_CQICH_REP field, part of ECAM, UHDM, and *Rate Change Message* (RATCHGM). REP_CQICH_REP can take the following values: 1, 2, and 4. Typically, values larger that 1 are common for soft handoff scenarios when the serving base station, selected by the mobile station as the one with the strongest forward link, may not effectively power-control the R-PICH. Remember that forward and reverse links use disjointed frequency bands, commonly 40 MHz apart, and short-term channel fluctuations, or fading, is practically uncorrelated between the two. Hence, it is not uncommon in soft handoff that for short periods, R-PICH is effectively power controlled, through down power-control commands, by the non-serving base station. Since R-CQICH needs to be decoded by the serving base station, R-CQICH symbols are repeated and soft-combined to improve reliability. The improved detection comes at the expense of a decreased R-CQICH symbol rate. The actual mapping between C/I codes, accumulator, and measurement values is shown in Table 9.7.

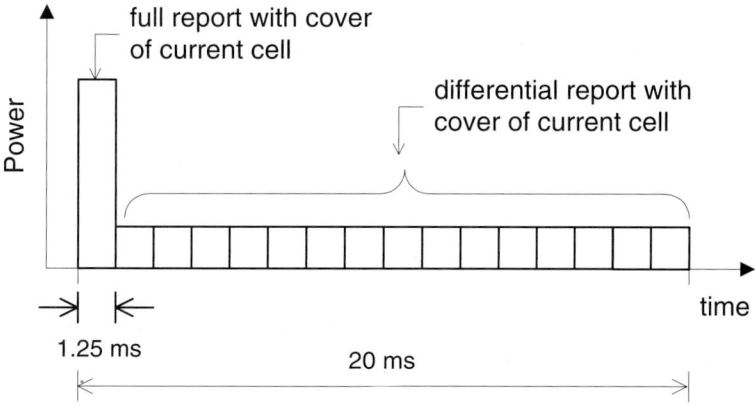

Figure 9.12 Differential C/I reports.

Table 9.7   Accumulator and Measurement Value Mappings for C/I Codes

Channel Quality Indicator value	Measured Pilot Ec/Nt [dB]	Accumulator value [dB]
0000	< -15.5	-16.25
0001	-15.5 to -14.0	-14.75
0010	-14.0 to -12.5	-13.25
0011	-12.5 to -11.0	-11.75
0100	-11.0 to -9.5	-10.25
0101	-9.5 to -8.0	-8.75
0110	-8.0 to -6.5	-7.25
0111	-6.5 to -5.0	-5.75
1000	-5.0 to -3.5	-4.25
1001	-3.5 to -2.0	-2.75
1010	-2.0 to -0.5	-1.25
1011	-0.5 to 1.0	0.25
1100	1.0 to 2.5	1.75
1101	2.5 to 4.0	3.25
1110	4.0 to 5.5	4.75
1111	> 5.5	6.25

The full C/I reporting is more accurate than the differential, but it can create significant reverse-link overhead. The R-CQICH mean output power is given by

$$
\begin{aligned}
\text{R-CQICH mean output power} \;=\; & \text{mean R-PICH output power} \\
& + \text{Nominal\_RCQICH\_Attribute\_Gain} \\
& - \text{Multiple\_Channel\_Adjustment\_Gain} \\
& + \text{possible other corrections}
\end{aligned}
\tag{9.13}
$$

The power levels in Eq. (9.13) are expressed in dBm. The nominal R-CQICH attribute gain is given in Table 9.8 as a function of the CQI_GAIN parameter. As can be seen from the table, the differential report is transmitted at 9 dB lower than the full report because of the lower effective R-CQICH bit rate.

Table 9.8 Nominal R-CQICH Attibute Gain Table

CQI_GAIN	Nominal Attribute Gain [dB]	Pilot Reference Level [dB]
LOW – Differential, 1 bit	−7	0
HIGH – Full, 4 bits	2	0

9.6.3 Cell Selection and Switching

The F-PDCH is never placed into soft or softer handoff. The mobile station indicates the serving base station by "covering" the R-CQICH symbol with one of six different Walsh functions of length-8 chips. In the context of R-CQICH operation, such Walsh functions are called Walsh covers. At the call setup, with ECAM or when the active set is updated through UHDM, the network signals the PILOT_PN to Walsh cover mapping using the REV_CQICH_COVER field. The mobile station expects to be served from the base station associated with the Walsh cover used when C/I measurement is reported.

The mobile station achieves selection diversity by selecting the base station with the strongest received pilot signal, that is, the highest Pilot E_c/N_t. However, the mobile station cannot instantly change the serving base station, because cell switching requires queue synchronization for the outstanding data. Queue synchronization between BTSs involves an A_{bis} interface and it is, thus, subject to delay, typically about 50 ms. The network communicates the soft handoff switching delay to the mobile using PDCH_SOFT_SWITCHING_DELAY. The parameter is part of ECAM and UHDM. Its purpose is to inform the mobile station how much interruption in service can be expected due to cell reselection. The values the parameter can take are between 10 ms and 2.56 sec. The same parameter is provided for softer handoff. Switching delay is typically shorter for softer handoff because it involves the cells belonging to the same BTS.

The mobile station indicates switching by using a predetermined switching pattern, shown in Figure 9.13, which illustrates a single frame for the case when the number of switching slots is equal to 2. The slot refers to PCG, or 1.25 ms. The number of switching slots is also a configurable parameter called NUM_SOFT_SWITCHING_SLOTS. It is part of ECAM and UHDM, and it can take values of 2, 4, and 7 slots. The same parameter is also provided for softer handoff.

The number of frames that include the switching pattern is configurable. The switching pattern is an indication to the serving base station to complete the transmission of the outstanding encoder packets and prepare for switching to the target base station after the transmissions are completed. The length of the switching procedure is configured by the

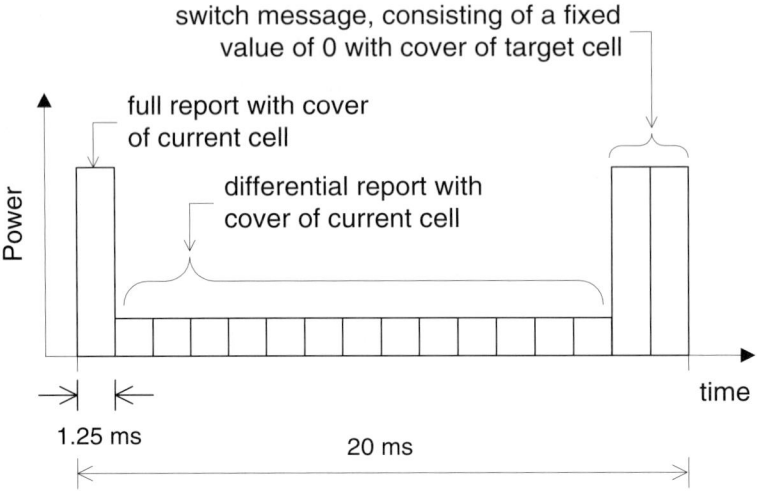

Figure 9.13 Cell-switching frame example.

NUM_SOFT_SWITCHING_FRAMES parameter, contained in ECAM or UHDM. The unit is one 20-ms frame, and allowed values are anywhere between 1 and 16. The network can configure different values for soft and softer handoff. Note that the number of switching frames, NUM_SOFT_SWITCHING_FRAMES, configures the switching procedure in the mobile, while switching delay, PDCH_SOFT_SWITCHING_DELAY, simply informs the mobile about the possible interruption in service due to switching and network delays because of issues like queue synchronization. As we mentioned earlier, typically softer switching delay is shorter than soft switching delay. Note that the base station has an option to terminate the transmission of the switching pattern before the configured switching period expires by using a special F-PDCCH control message. For control F-PDCCH messages, EP_SIZE = 111.

9.6.4 Asynchronous Adaptive Incremental Redundancy

H-ARQ type II uses AAIR. The encoder packet is initially encoded with rate 1/5 turbo code and, as explained in Chapter 4, interleaved. The coded symbols are then transmitted to the mobile station in parts, referred to as subpackets.

The first subpacket contains all of the systematic bits and some parity bits. The remaining subpacket may or may not contain systematic bits. The subpacket data rate is determined from the reported Pilot E_c/N_t so that the rate matches the radio channel. If the packet is not decoded successfully, which is indicated to the base station over R-ACKCH, additional subpackets are transmitted and soft-combined at the mobile. The subsequent subpacket usually contains parity bits other than the ones transmitted with the initial transmission, effectively reducing the aggregate code rate after each retransmission. Due to this feature, this H-ARQ type II scheme is commonly called incremental redundancy. The subpacket can be transmitted at any point in time and at any rate, not necessarily at the initial subpacket rate. This is the reason this scheme is also called asynchronous and adaptive incremental redundancy, or AAIR. In total, eight subpacket transmissions are allowed per encoder packet. AAIR must not be abused, or it can create excessive delay that may negatively impact TCP throughput (see Section 9.7).

9.6.4.1 *Subpacket Transmission*

The subpacket data rate is determined as follows. Given the number of available Walsh codes and the number of 1.25-ms slots, the modulation scheme for each of the available encoder packet sizes is determined. The encoder packet size can be chosen for the initial transmission, but for all subsequent subpacket transmissions, the encoder packet size must remain the same as for the initial transmission. Table 9.9 summarizes the code rate guidelines for selecting the modulation scheme. The modulation and coding scheme are chosen so that the resulting data rate matches the radio conditions state reported by the mobile station.

Table 9.9 Modulation and Code Rate Ranges

Modulation	Code rate
QPSK	<3/4
8-PSK	1/2 < 2/3
16-QAM	1/2 < 4/5

The code rate, R, can be computed as

$$R = \frac{N_{EP}}{48 \cdot N_W \cdot N_s \cdot M_{ORDER}} \tag{9.14}$$

where N_{EP} is the encoder packet size, N_W is the number of available Walsh codes, and M_{ORDER} represents the modulation order, or the number of coded bits per modulation symbol. M_{ORDER} is equal to 2, 3, and 4 for QPSK, 8-PSK, and 16-QAM, respectively. The smaller modulation order scheme that matches the most recent R-CQICH is selected.

The number of available Walsh codes, number of slots, modulation order, and SPID can change from subpacket to subpacket. The coded symbols for each subpacket are selected with the following rule. Denote with k the subpacket index, and let us label the coded bits starting with index 0. Denote with $S_{k,i}$ the index of the i-th coded bit of the k-th subpacket, and with P_{max} the maximum coded packet length. The maximum length is up to 7,800 coded bits and determined by the following formula: $P_{max} = \max (7{,}800, 5N_{EP})$. Then,

$$S_{k,i} = (F_k + i) \bmod (P_{max}) \tag{9.15}$$

where $F_k = (\text{SPID}_k\, L_k)\, \text{mod}(P_{max})$, and L_k is the number of coded bits per subpacket k, $L_k = 48$ $N_{W,k}\, N_{S,k}\, M_{\text{ORDER},k}$.

An example of AAIR operation is shown in Figure 9.14. If $5\, N_{EP} > 7{,}800$, the sequence of the coded bits wraps around, and systematic bits are transmitted with SPID other than SPID = 00. Due to adaptive operation, certain coded bits may be omitted, which corresponds to puncturing operation, while certain coded symbols may be repeated. When channel measurements are accurate and the mobile is moving at low speeds, retransmissions are rarely needed. However, for vehicular channels, retransmissions are essential. The retransmissions do not have to be at the same rate as the initial transmission. The rate for the subsequent subpackets is typically determined based on the most recent channel measurement, but it could also take into account the rate of previous subpackets and radio environment at those instants. The base station must first transmit the subpacket with SPID = 00. If retransmissions are needed, the base station may choose a subpacket with an arbitrary SPID.

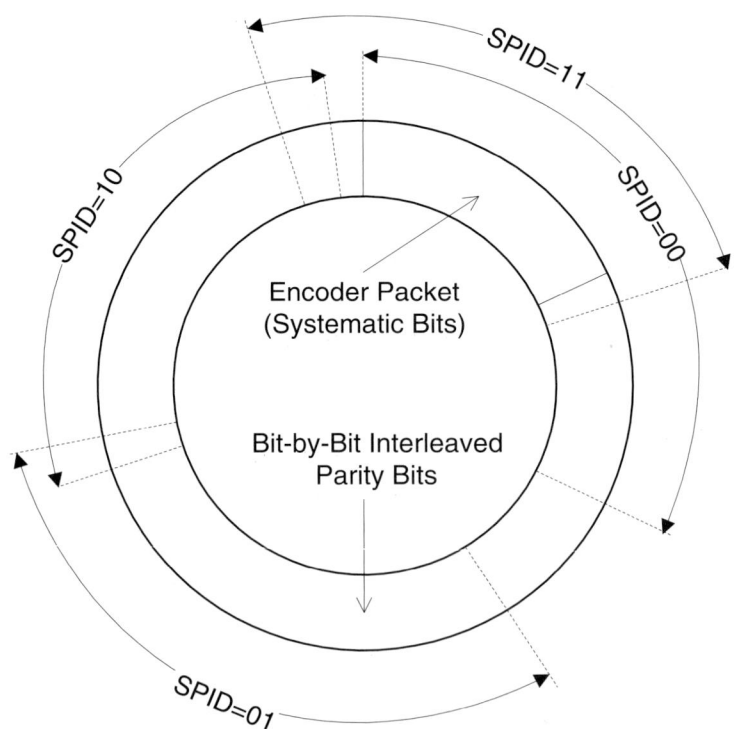

Figure 9.14 Subpacket generation.

9.6.4.2 *Reverse Acknowledgment Channel (R-ACKCH)*

The R-ACKCH supports AAIR operation. When enabled, the R-ACKCH carries one bit of information, which signals successful or unsuccessful encoder packet decoding. In case the F-PDCCH frame did not pass the CRC check, or in case the MAC_ID did not match the MAC_ID assigned to the mobile station, the R-ACKCH is not transmitted. The R-ACKCH transmission timing is synchronous with subpacket reception. The delay is identified by the mobile station through the ACK_DELAY field as part of the *Origination Message* (ORM). This field indicates mobile station capability to decode a frame in less than 1.25 ms; that is, ACK_DELAY can indicate 1.25-ms or 2.5-ms delay. A typical length of the acknowledgment channel bit is 1.25 ms. However, the bit can be repeated in order to improve reliability. Whether or not the bit is repeated is specified by the REV_ACKCH_REPS parameter, which can take values 1, 2, and 4 in 1.25-ms units. This parameter is part of ECAM. Figure 9.15 illustrates the synchronous R-ACKCH operation and acknowledgment channel repetition. The power level of R-ACKCH is given by

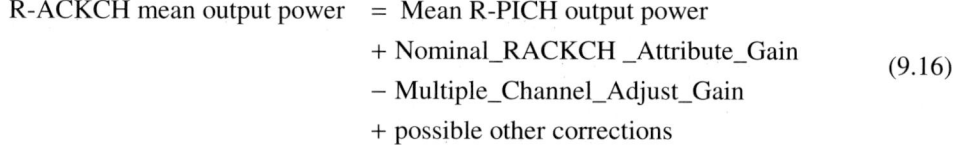

$$
\begin{aligned}
\text{R-ACKCH mean output power} \;=\;& \text{Mean R-PICH output power} \\
&+ \text{Nominal\_RACKCH \_Attribute\_Gain} \\
&- \text{Multiple\_Channel\_Adjust\_Gain} \\
&+ \text{possible other corrections}
\end{aligned}
\tag{9.16}
$$

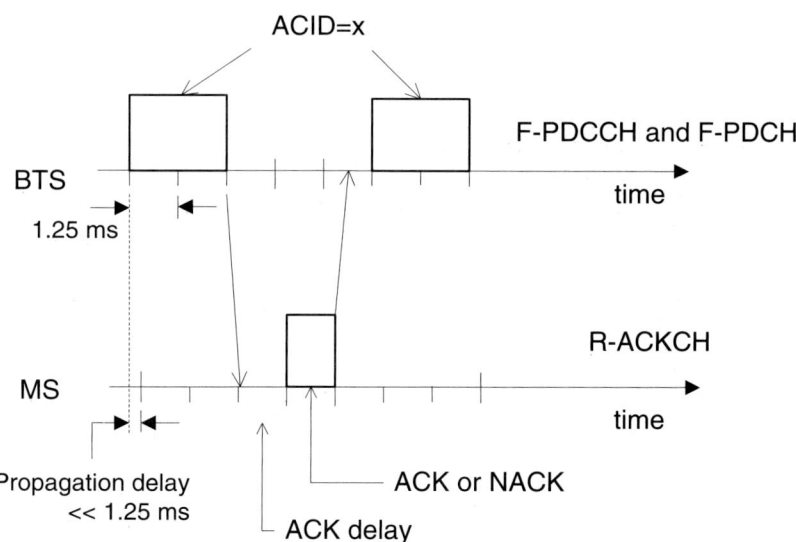

Figure 9.15 R-ACKCH timing.

where the nominal R-ACKCH attribute gain is given in Table 9.10.

Table 9.10 Nominal R-ACKCH Attribute Gain Table

Nominal R-ACKCH Attribute Gain [dB]	Pilot_Reference_Level [dB]
−3	0

9.6.5 Multiple ARQ Channels

The AAIR operation represents a form of stop-and-wait protocol. The base station transmits to a mobile in a given slot and waits to receive the positive or negative acknowledgment before it can transmit again. The time the base station needs to wait is equal to a sum of ACK_DELAY, round-trip propagation time rounded up to the nearest integer number of 1.25-ms slots or PCGs, and REV_ACKCH_REPS. At the minimum, this time is equal to 3 PCGs. The stop-and-wait nature of AAIR could therefore significantly limit the maximum throughput experienced by a mobile station.

Multiple ARQ channel mode allows a base station to serve a single mobile over multiple parallel channels, therefore increasing the channel utilization to its maximum. Multiple ARQ channels operation is illustrated in Figure 9.16. While the base station awaits positive or negative acknowledgment on one of its ARQ channels, it is allowed to serve the same mobile station on an alternative ARQ channel. The F-PDCCH message contains ARQ_ID field that identifies the ARQ channel. Four such channels are permitted.

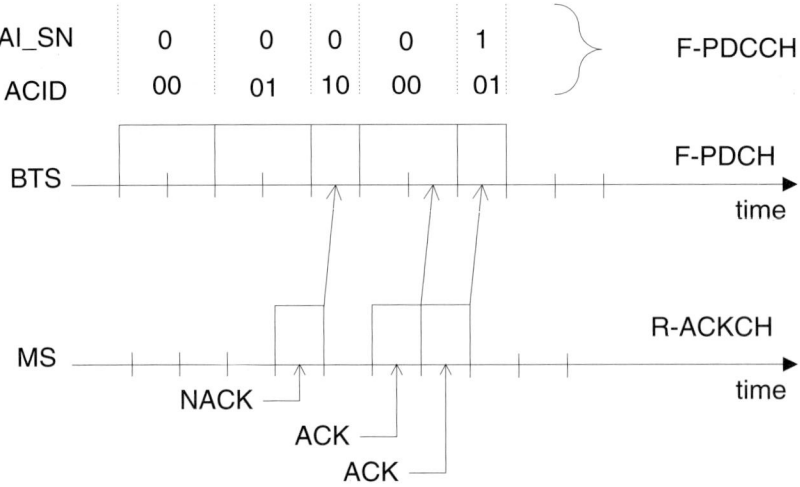

Figure 9.16 Multiple ARQ channels operation.

9.6.6 Channel Configurations

There are six different permitted channel configurations with F-PDCH. Table 9.11 lists all the possibilities. In all of the allowed channel configurations the reverse dedicated channel is present, such as R-FCH or R-DCCH. However, as the first two rows of the table indicate, in the forward direction a primary traffic channel (F-FCH or F-DCCH) is not required.

Table 9.11 Allowed Channel Configurations with F-PDCH Assigned

Forward channels	Reverse channels
F-PDCH+F-CPCCH	R-FCH
F-PDCH+F-CPCCH	R-DCCH
F-PDCH+F-FCH	R-FCH
F-PDCH+F-DCCH	R-DCCH
F-PDCH+F-FCH	R-FCH+R-DCCH
F-PDCH+F-FCH+F-DCCH	R-FCH+R-DCCH

Indeed, the F-PDCH and F-PDCCH are commonly operated without a primary traffic channel to minimize Walsh code consumption. In absence of a primary traffic channel, Layer 3 signaling is carried by the F-PDCH. In this case, the setting of the MuxPDU type 5 header is used by the mobile station to distinguish between F-PDCH signaling frames and those carrying user data (as discussed in Chapter 5, sr_id = 000' indicates signaling). Operation without F-FCH and F-DCCH also requires different channel supervision procedure, as discussed in Chapter 6. Power-control procedures are also affected. In lieu of the forward-link power-control subchannel multiplexed with the primary traffic channel, the mobile station is assigned a subchannel on the F-CPCCH. The F-CPCCH is used for closed power control of the reverse link. The F-CPCCH is more efficient than the F-FCH in terms of Walsh resource consumption because it can support simultaneously several mobile stations, each assigned to one of its subchannels. The F-CPCCH itself needs to be power controlled. Power control is only possible at the serving base station by utilizing R-CQICH information.

9.6.7 Impact on RLP

F-PDCH impacts RLP [3] operation. For F-PDCH, RLP typically forms fixed-size 378-bit blocks, including overhead, and supplies the data to the multiplex sublayer. Recall from Chapter 5 that the multiplex sublayer can utilize MuxPDU type 5 with a fixed-length payload indication. The fixed-size blocks are not mandated, but they are the most common implementation as long as there is enough data in the buffer to fill up one 378-bit RLP frame. Unlike SCH where the data rate is negotiated with Layer 3 signaling, the data rate and hence the encoder packet size on F-PDCH is variable and can change every 1.25 ms. The encoder packet size or physical layer Service Data Unit (SDU) is selected by PDCHCF, which, as we mentioned before, is implemented at the BTS. To exploit reverse-link soft handoff, RLP is commonly implemented at the BSC. This means that the information about the most recent radio conditions cannot possibly be passed to RLP. Hence, RLP needs to accommodate its frame size to fit the smallest possible physical layer SDU. The multiplex sublayer adds the header to each RLP frame, and PDCHCF then selects enough MuxPDUs type 5 to utilize a physical layer SDU. The physical layer frame sizes allow efficient packing of MuxPDUs type 5 into the physical layer frame. Figure 9.17 shows how encapsulation of RLP frames into the fixed-size MuxPDU type 5 and physical layer packet is performed.

To efficiently support asynchronous parallel ARQ channel operation, RLP needs to delay negative acknowledgments. Figure 9.18 illustrates why delay detection is needed. One RLP instance can be mapped to as many as four ARQ channels, and in error-prone links, out-of-order delivery is quite likely. Due to H-ARQ type II operation, the frame may be transmitted up to eight times before error recovery at the physical layer is abandoned. Since retransmissions are asynchronous, it is hard to estimate how long it is going to take to finish with the retransmissions once the initial subpacket is transmitted. The delay can be different on each of the ARQ channels, which requires delayed detection Without delayed detection, RLP would in many instances send negative acknowledgments, or NAKs, for the data that would anyway be correctly decoded

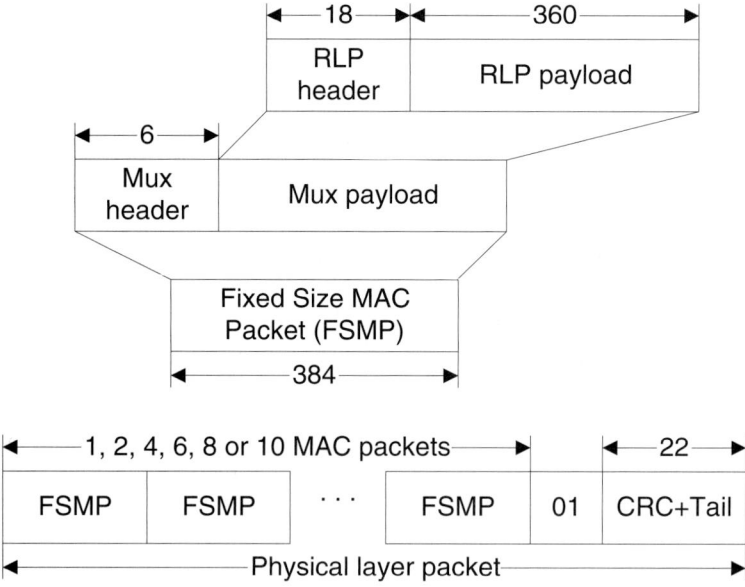

Figure 9.17 Figure 9–17:Encapsulation of fixed-size RLP type 5 frames into physical layer frame.

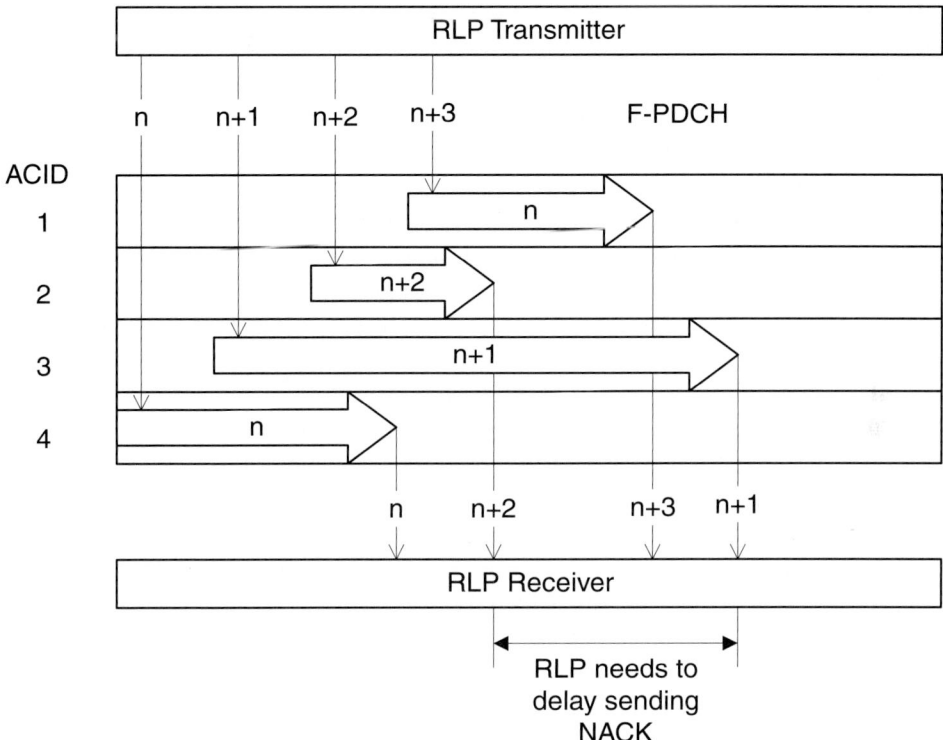

Figure 9.18 Delayed detection for RLP over F-PDCH.

shortly afterward. Therefore, when RLP detects a missing frame, RLP waits for a period of time specified by DELAY_DETECTION_WINDOW and a NAK is sent only if, after the specified time, the RLP frame is still missing. Note that RLP can send a NAK earlier if it can conclude that the retransmissions of a missing frame are abandoned. One such indication could be the reception of a toggled AI_SN bit. The physical layer packet error rate after eight transmissions is reasonably low so that an RLP instance mapped to F-PDCH normally requires only a single NAK round and only one NAK per round.

9.7 Impact on TCP Performance

TCP implements a flow-control mechanism in order to prevent buffer overflow at the receiving side and network congestion. The flow control employs a sliding window mechanism, where the number of unacknowledged bytes at the transmitter side is controlled by the receiver advertised window. The number of bytes the transmitter can send to the receiver is also bounded by the congestion window, maintained at the transmitting side. The purpose of the congestion window is to avoid congestion in the network.

The congestion-control mechanism dynamically adjusts the window size based on perceived network state. It consists of a slow start and a congestion-avoidance phase. During the slow-start phase, the window increases exponentially, while during the congestion-avoidance phase, the window increases only linearly. The congestion-avoidance phase begins when the window reaches a congestion threshold. During the congestion-avoidance phase, the congestion threshold follows the window. A loss of an acknowledgment within a dynamically adjusted timeout period is interpreted as a signal of the network congestion. The timeout period is computed from a filtered estimate of a round-trip time and the variance of the estimate. To recover from network congestion, TCP reduces the congestion threshold by half, and the congestion window is reduced to one segment. After the congestion window is reduced to one segment, the slow-start phase begins. An example of the behavior of the congestion window is illustrated in Figure 9.19.

In wireless communications, a missing acknowledgment is not necessarily a signal of network congestion. It is more common that TCP segments are lost due to errors on the air interface rather than congestion in the network. To address this issue, current versions of TCP (e.g., Reno) attempt segment recovery with fast retransmissions before reducing the congestion window to a single segment. Fast retransmissions are triggered when the TCP transmitter detects multiple duplicate ACKs, which are a strong indication that the following unacknowledged segment has

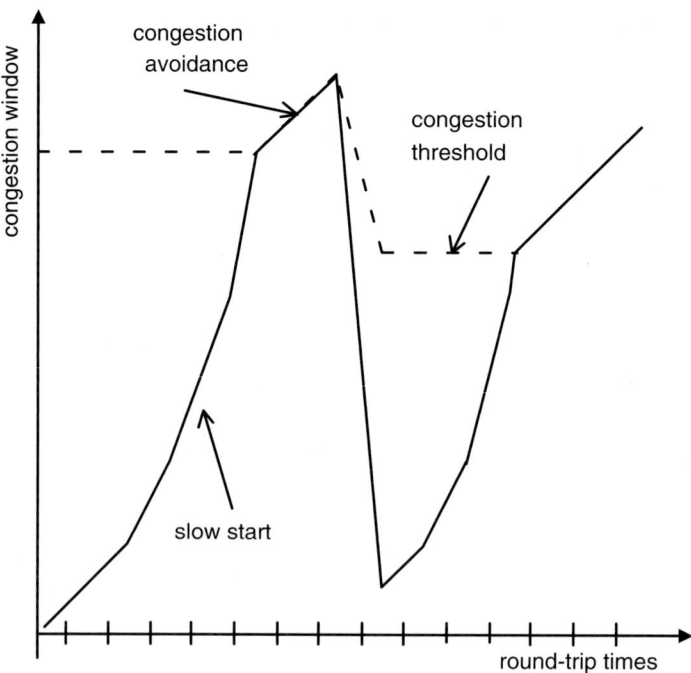

Figure 9.19 TCP congestion window.

been lost and that the network is not congested. Fast retransmission is followed by a fast recovery phase. Fast recovery performs congestion avoidance but not slow start.

Three major factors affect the TCP performance on the cdma2000 air link: variable bandwidth and asymmetric bandwidth effects and impact of HARQ and RLP.

The variable bandwidth effect could be created with scheduling. For example, F-SCH burst assignment followed by F-SCH de-assignments can create a variable bandwidth effect for TCP, which may cause segment timeout. TCP segment timeout is followed by segment retransmission and aggressive reduction of the congestion window. The reason for the segment timeout can be explained in the following. TCP updates its estimate of the round-trip time and the standard deviation whenever it receives an ACK. Each value is recorded, and the mean and standard deviation are estimated using a low-pass filter. Denote with m the mean round-trip time or mean time elapsed between transmission of TCP segment and reception of the acknowledgment, and with σ the round-trip time standard deviation. The timeout time, T_o, is commonly implemented as

$$T_o = m + 4\sigma \tag{9.17}$$

If the scheduling interval is long enough, the estimate of the round-trip time is solely based on the experience during the scheduled time. Abrupt increase in round-trip time, which happens when high-rate F-SCH is substituted by low-rate F-FCH, can cause timeout. Several segments need to be transmitted and acknowledged before new round-trip parameters are learned. Segment timeout drastically reduces the window size and forces slow-start phase of TCP, which limits the throughput. Therefore, it is very important from the TCP perspective that mobile stations are scheduled on a regular basis without large bandwidth swings. The variable bandwidth effect could be created with link adaptation on F-PDCH. However, in the case of a proportionally fair scheduler with a proper choice of the time constant τ ($\sim$ 1s) that is used to estimate the user's throughput, the variable bandwidth effect on TCP is not present.

The asymmetric bandwidth effect could be created when a high-speed channel such as F-PDCH is paired with low-speed reverse channel such as R-FCH. Consider a file download scenario from a server to the mobile station. To achieve high TCP throughput, it is not only sufficient that the rate supported over F-PDCH is high. The R-FCH could also limit the TCP throughput because the low data rate on R-FCH limits the TCP ACKs throughput, which in turn limits the rate at which the TCP transmit window opens. Therefore, even though it is not necessary that the forward and reverse links are symmetric, large asymmetry could negatively impact TCP throughput.

The physical and link layer recovery mechanisms, such as HARQ and RLP, increase the transmission delay of TCP segments. Eq. (9.18) shows the upper bound on sustainable TCP throughput, S, as a function of a maximum segment size (MSS), round-trip time (RTT) and segment loss rate, p:

$$S < \frac{MSS \cdot 0.93}{RTT \cdot \sqrt{p}} \tag{9.18}$$

Since the upper bound on a sustainable TCP throughput is inversely proportional to the round-trip time, increase in delay negatively impacts TCP throughput. HARQ and RLP reduce error rate but at the expense of the delay. Obviously, there is an optimum trade-off between the delay and packet loss rate. HARQ and RLP should, therefore, take as little time as possible and produce a tolerable segment loss rate; typically it is required that $p \leq 0.01$. A robust RLP scheme for F/R-SCH is one retransmission in the first round, two in the second, and three in the third. A popular alternative that does not unnecessarily retransmit multiple copies of the same packet is with multiple rounds (up to four) and a single NAK per round. If error recovery is performed at the physical layer with HARQ, a common RLP implementation is with only one round and a single NAK. Refer to [8] and references therein for more details on this topic.

REFERENCES

[1] Tse, David, "Optimal Power Allocation over Parallel Gaussian Channels." *In Proc. ISIT,* 1997.

[2] Kelly, Frank, "Charging and Rate Control for Elastic Traffic," *European Transactions on Telecommunications 8: 33–37,* 1997.

[3] "TIA/EIA-IS-707-B Data Service Options for Spread Spectrum Systems," *Telecommunications Industry Association,* 2004.

[4] "TIA/EIA-IS-2000.2 Physical Layer Standard for cdma2000® Spread Spectrum Systems, Release C," *Telecommunications Industry Association,* 2003.

[5] Yeh, Jessica, A. Khan, L. Aydin, and W. Hamdy, "Performance Comparison of cdma2000 Forward Power Control Modes." *In Proc. of PIMRC,* September 2003.

[6] Damnjanovic, Jelena, Avinash Jain, Tao Chen, and Sandip Sarkar, "Scheduling for cdma2000 Reverse Link." *In Proc. of IEEE VTC,* October 2002.

[7] "TIA/EIA-IS-2000.3 Medium Access Control (MAC) Standard for cdma2000® Spread Spectrum Systems, Release C," *Telecommunications Industry Association,* 2002.

[8] Chaponniere, Etienne, Sunil Kandukuri, and Walid Hamdy, "Effect of Physical Layer Bandwidth Variation on TCP Performance in cdma2000," *IEEE VTC,* April 2003.

CHAPTER 10

Radio Access Network Performance

In this chapter we describe a modeling framework for estimating key performance indicators relevant to planning and operation of cdma2000 wireless networks. The modeling framework adopts a semi-analytical approach. The obtained results are approximate because the models are based on simplifying assumptions and only limited Monte-Carlo simulations. However, the semi-analytical models give good insight into how various parameters impact the system performance. Monte-Carlo simulations can produce more accurate results. However, simulations are tedious and tend to obscure the key factors affecting the system performance.

Using the semi-analytical approach, we estimate some performance indicators for network planning and operation, such as voice and data capacity of a single embedded cell. We also discuss the methodology to compute the forward- and reverse-link budgets.

10.1 Models for Capacity Estimation in Cellular CDMA

Both forward and reverse link of a cdma2000 system are strictly interference-limited. Therefore, we need to define the various sources of interference and develop statistical models to characterize them.

10.1.1 Cell Layout Model

The cell layout we consider in our modeling framework is illustrated in Figure 10.1. The base station transceivers (BTSs) are located in the center of the hexagons forming an idealized grid. Each BTS comprises three cells. The arrows indicate the antenna bore-sight of each cell. The shaded area represents the service area where the eastward facing cell of the center BTS is the best serving cell. The actual cell service area depends on the antenna radiation pattern. Note

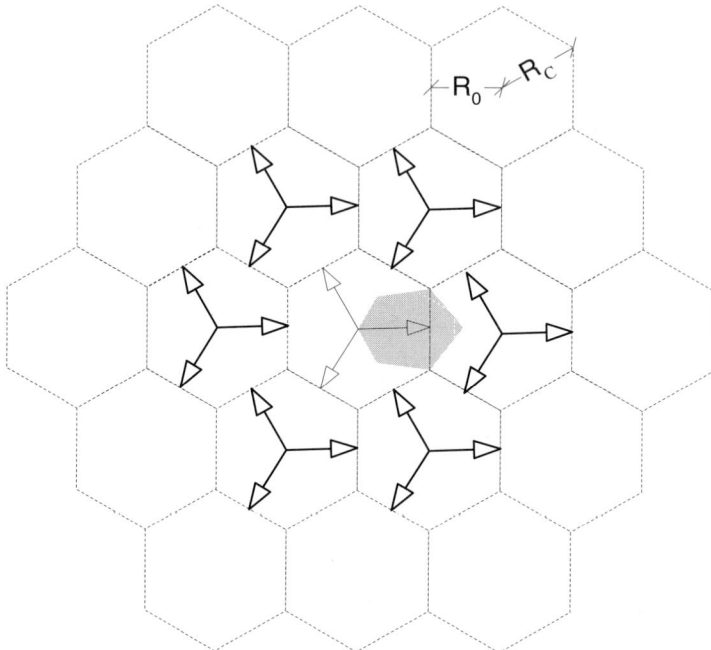

Figure 10.1 BTS layout. The shaded area represents a cell's coverage area.

that the antennas belonging to adjacent BTSs do not point directly to each other. Rather, there is always 60 degrees offset between them. The reason behind such an orientation is to maximize the signal to interference and noise ratio (SINR) at the mobile station.

For later reference, we need to define some quantities related to the cell layout. We denote with R_0 the BTS-to-edge distance, which is equal to a half of the minimum distance between two BTSs, and we denote with R_C the BTS-to-corner distance. The ratio of the two is then $R_C/R_0 = 2/\sqrt{3}$.

10.1.2 Propagation Model

Commonly used propagation model is the extended Hata model, also referred to as COST-231. According to the model, the propagation loss L_p in a logarithmic domain can be represented in terms of distance-independent component δ and distance-dependent component γ as

$$L_p = \delta + 10\gamma \log_{10} R \tag{10.1}$$

where R is the distance in Km between the mobile station and the BTS. The model is valid for $R > 35$ m.

The distance-independent component δ is a function of carrier frequency, mobile station and base station antenna height, and cell morphology (for example, urban or suburban):

$$\delta = C_1 + C_2 \log_{10} f_c - 13.82 \log_{10} h_{BTS} - a\left(h_{MS}\right) + C_M \tag{10.2}$$

where C_1 and C_2 are carrier frequency-dependent constants, f_c is carrier frequency in MHz, h_{BTS} is the antenna height in meters, h_{MS} is mobile station antenna height in meters, $a(h_{MS})$ is the correction factor for the mobile station antenna height, and C_M is the cell morphology correction. The values for C_1 and C_2 are summarized in Table 10.1. The correction factor for the mobile station antenna height is given by

$$a\left(h_{MS}\right) = 3.2\sqrt{\log_{10}\left(11.75 h_{MS}\right)} \tag{10.3}$$

and the cell morphology correction for the urban environment is $C_M = 0$ dB. For suburban environments, $C_M < 0$ dB.

Table 10.1 C_1 and C_2 as a Function of Carrier Frequency

Parameter	Model	Carrier frequency	Value
C_1	Hata	$150 \leq f_c \leq 1500$ MHz	69.55
	COST-231	$1500 \leq f_c \leq 2000$ MHz	49.3
C_2	Hata	$150 \leq f_c \leq 1500$ MHz	26.16
	COST-231	$1500 \leq f_c \leq 2000$ MHz	33.9

The second component is distance-dependent, where the propagation factor γ is a function of the base station antenna height and it is equal to

$$\gamma = 4.49 - 0.655 \log_{10} h_{BTS} \tag{10.4}$$

Assuming, for example, $h_{BTS} = 30$ m and $h_{MS} = 1.5$ m, and using the parameters given in Table 10.1, the propagation loss equation in an urban environment at 1.9 GHz reduces to

$$L_p = 133.6 + 35 \log_{10} R \tag{10.5}$$

For the purpose of capacity estimation, in addition to the propagation loss, it is also necessary to account for shadowing loss caused by large objects, such as buildings, obstructing the line of sight between mobile station and base station. The propagation loss that accounts for shadowing, $L_p^{'}$, is modeled as a lognormal distributed random variable equal to

$$L_p^{'} = \delta + 10\gamma \, \mathrm{Log}_{10} R + \zeta \qquad (10.6)$$

where ζ is the zero mean normal variable with standard deviation σ. In Eq. (10.6), we do not distinguish between forward and reverse links. They are assumed perfectly correlated because shadowing is largely independent of the carrier frequency. The random shadowing loss can be modeled as a sum of two components. The first component, ξ, is due to objects close to the mobile station obstructing the line of sight to all base stations. The second component, ξ_i, is due to objects far from the mobile station that obstruct the line of sight to a given receiving BTS, and it is independent from one BTS to another. Then, ζ can be modeled as $\zeta = a\xi + b\xi_i$, where $a^2 + b^2 = 1$, and $a^2 \le 1$. Moreover,

$$
\begin{aligned}
E\{\zeta_i\} &= E\{\xi\} = E\{\xi_i\} = 0 \\
E\{\zeta_i^2\} &= E\{\xi^2\} = E\{\xi_i^2\} = \sigma^2 \\
E\{\xi\,\xi_i\} &= 0 \quad \forall i \\
E\{\xi_i\,\xi_j\} &= 0 \quad \forall i \ne j
\end{aligned}
\qquad (10.7)
$$

The correlation coefficient of the propagation loss to the two BTSs is then given as

$$E\{\zeta_i\,\zeta_j\}/\sigma^2 = a^2 = 1 - b^2 \quad \forall i \ne j \qquad (10.8)$$

It is common to assume that the correlation coefficient is equal to one half, which corresponds to an equal standard deviation of the near field and BTS-specific shadowing (i.e., $a^2 = b^2 = \frac{1}{2}$). The shadowing losses between the collocated base stations, associated with collocated antennas but distinct coverage areas, is correlated with the correlation coefficient equal to 1.

10.1.3 Antenna Radiation Pattern Model

The antenna radiation pattern shapes the coverage and affects cell capacity because it determines the relevant interference statistics on both forward and reverse links. An antenna is characterized by its isotropic gain, that is, the gain relative to an ideal isotropic radiating element. The antenna gain in a given direction is a function of the angle relative to the antenna azimuth. The antenna azimuth is the angle corresponding to the direction of maximum gain and is taken as a reference. Other parameters of interest are the front-to-back ratio, the 3-dB beam-width, and the antenna roll-off. A useful model of the horizontal antenna radiation pattern that captures all these parameters is given by [1].

$$G(\theta) = -\min\left\{ 3 \cdot 2^{\alpha} \left| \frac{\theta}{\theta_{3\text{-dB}}} \right|^{\alpha}, L \right\} \quad \text{dB} \tag{10.9}$$

The angle $\theta \in [-\pi, \pi]$ represents the offset from the azimuth. The antenna 3-dB beam-width, $\theta_{3\text{-dB}}$, is such that the antenna gain, G, satisfies the constraint $G(\theta = \theta_{3\text{-dB}}/2) = -3$ dB. The parameter α is the antenna roll-off factor and determines how quickly the antenna gain tapers off for increasing values of the offset from the azimuth. L is the front-to-back ratio, which determines the minimum antenna gain. If we compare the antenna radiation pattern of real antennas against our model, we find that the best match in a mean square error sense is achieved when the beam-width is set equal to that specified by the antenna manufacturer, the roll-off is set to 1.7~2, and the front-to-back ratio is set to 15~20 dB.

10.1.4 Voice Activity

The use of vocoders with variable bit rate (VBR) coding is crucial for efficient spectrum utilization of CDMA systems. As explained in Chapter 3, VBR vocoders adapt the encoding rate to the speech activity. During a period of silence, the vocoder delivers small number of bits per frame interval (low rate), primarily to synthesize comfort noise. When the user talks, the vocoder delivers large number of bits per frame interval (high rate). The low data rate frames can be transmitted at lower power than the high-rate frames. The average transmitted power relative to the transmitted power at full rate is then less than one, and it is commonly called average voice activity, or *voice activity factor*. Smaller average transmitted power corresponds to less interference and, in an interference-limited system such as CDMA, increased capacity.

Selectable mode vocoder (SMV) is a VBR codec that operates in one of four selectable modes, from mode 0 to mode 3. The modes differ in the average data rate used to encode the same speech pattern, with mode 0 yielding the highest average rate and also the best voice quality. At the other extreme, mode 3 yields the lowest average rate, thus providing for the largest capacity at the expense of voice quality. However, reduction of interference is not proportional to the reduction of average voice activity due to the presence of constant-rate overhead channels on both forward and reverse links. In the sections that follow we estimate the first- and second-order statistics for voice activity for both reverse and forward links conditioned on the SMV mode. These results turn out to be handy later on.

10.1.4.1 *Reverse-Link Voice Activity*

The reverse-link voice activity, $v \le 1$, is defined as the average reverse traffic channel transmitted power relative to the transmitted power at full rate. A reverse traffic channel, such as Reverse Fundamental Channel (R-FCH), must be accompanied by the Reverse Pilot Channel (R-PICH). The R-PICH also carries the time-multiplexed power-control subchannel. The R-FCH is operated with variable data rate, and its transmitted power relative to the pilot is varied depending on the data rate. The IS-2000 physical layer standard specifies the default fundamental to pilot channel power ratios, referred to as traffic-to-pilot T/P ratio. The T/P ratio is a function of data rate. Let us denote the T/P ratio with $\rho(R)$ and show the default values in Table 10.2.

Table 10.2 Standard Values for R-FCH T/P Values

Data rate [kbps]	R-FCH transmit power [dB-pilot]
$R_1 = 9.6$ (full rate)	3.75
$R_2 = 4.8$	–0.25
$R_3 = 2.7$	–2.75
$R_4 = 1.5$	–5.875

The average voice activity factor in terms of the T/P ratios is given by

$$\bar{v} \quad E\{v\} = \sum_{i=1}^{4} \frac{1+\rho(R_i)}{1+\rho(R_1)} \Pr\left[R = R_i\right] \tag{10.10}$$

while its second-order moment can be computed as

$$\overline{v^2} \quad E\{v^2\} = \sum_{i=1}^{4} \left[\frac{1+\rho(R_i)}{1+\rho(R_1)}\right]^2 \Pr\left[R = R_i\right] \tag{10.11}$$

The vocoder rate probabilities summarized in Table 10.3 were obtained in [2] from SMV clean-speech statistics normalized by those obtained from measurements in a commercial network [2]. Using Eqs. (10.10) and (10.11), we can compute the first and the second moments of the voice activity. The results are summarized in Table 10.4.

Table 10.3 Data Rate Probabilities for SMV Modes 0, 1, 2, and 3

Data rate [kbps]	Mode 0	Mode 1	Mode 2	Mode 3
9.6 (full rate)	0.370	0.203	0.084	0.037
4.8	0.045	0.094	0.212	0.260
2.7	0.0	0.098	0.092	0.091
1.5	0.585	0.605	0.612	0.612

Table 10.4 Reverse-Link Voice Activity for SMV Modes 0, 1, 2, and 3

Voice activity	Mode 0	Mode 1	Mode 2	Mode 3
$E\{v\}$	0.614	0.528	0.476	0.456
$E\{v^2\}$	0.45	0.33	0.26	0.23
$Var\{v\}$	0.07	0.05	0.03	0.02

10.1.4.2 Forward-Link Voice Activity

The forward-link voice activity is defined as the average F-FCH transmitted power relative to the power transmitted at full rate. Similarly as on reverse link, the power reduction at lower rates is not proportional to the reduction in data rate. Unlike the reverse link, however, the forward-link subrate transmitted power adjustments are not standardized. An exemplary setting is given in Table 10.5.

Table 10.5 F-FCH Subrate Power Adjustment

Data rate [kbps]	F-FCH transmit power relative to full rate frame[dB]
$R_1 = 9.6$ (full rate)	0
$R_2 = 4.8$	−4
$R_3 = 2.7$	−7
$R_4 = 1.5$	−10

As with the reverse link, we must account for an overhead channel in the forward link too. In the forward direction, the overhead channel is the forward-link power-control subchannel. Recall that the power-control subchannel is time-multiplexed with the F-FCH. Every power-control bit punctures four of the F-FCH coded symbols. Since there are 16 power-control bits and 768 coded symbols per frame in Radio Configuration (RC) 3, one out of $(16 \times 4)/768 = 12$ symbols are punctured with power-control subchannel symbols. The remaining 11 are the traffic channel symbols. The power-control symbols are typically transmitted at a power level equal to that of the full rate F-FCH symbols. The first two moments of the voice activity in terms of the relative subrate transmit power ratios are then given as

$$\bar{v} \quad E\{v\} = \sum_{i=1}^{4} \left[\frac{1}{12} + \frac{11}{12} \rho\left(R_i\right) \right] \Pr\left[R = R_i \right] \tag{10.12}$$

$$\overline{v^2} \quad E\{v^2\} = \sum_{i=1}^{4} \left[\frac{1}{12} + \frac{11}{12} \rho\left(R_i\right) \right]^2 \Pr\left[R = R_i \right] \tag{10.13}$$

The vocoder rate probabilities, summarized in Table 10.6, were obtained in [2] from SMV clean-speech statistics normalized by those obtained from extensive measurements in a commercial network.[1] Using Eqs. (10.12) and (10.13), we can compute the moments of the voice activity. The results are summarized in Table 10.7.

Table 10.6   Forward-Link Data Rate Probabilities for SMV Modes 0, 1, 2, and 3

Data rate [kbps]	Mode 0	Mode 1	Mode 2	Mode 3
9.6 (full rate)	0.425	0.235	0.097	0.042
4.8	0.055	0.109	0.245	0.301
2.7	0.0	0.113	0.106	0.105
1.5	0.52	0.543	0.551	0.551

Table 10.7   Forward-Link Voice Activity for SMV Modes 0, 1, 2, and 3

Voice activity	Mode 0	Mode 1	Mode 2	Mode 3
$E\{v\}$	0.519	0.378	0.291	0.256
$E\{v^2\}$	0.44	0.27	0.15	0.10
$Var\{v\}$	0.17	0.12	0.06	0.03

10.1.5 Models of Reverse-Link Interference

To model reverse-link interference, consider the k-th user. Denote with $I_o^k\left(s, r, j\right)$ the total received power spectral density at the s-th softer handoff leg of the r-th antenna and the j-th recovered path of the k-th user. The total received signal density can then be written as

---

1.   You may have noticed that the probability of using full rate on the forward direction is considerably larger than on the reverse link. The skewed statistic is mainly due to users retrieving voice mail. That results in a one-way conversation with forward-link voice activity close to 1 and reverse-link activity close to 0.

$$I_0^k(s,r,j) = \sum_m E_{c,m}^{k,j}(s,r,j)\left[1 + \sum_l \rho(R_{m,l})\right] + N_0 \qquad (10.14)$$

where $E_{c,m}^{k,j}(s,r,j)$ is the m-th user reverse-pilot chip energy at the tap position of the s-th softer handoff leg, on the r-th antenna and the j-th path of the k-th user. $\rho(R_{m,l})$ is the T/P ratio of the l-th traffic channel of the m-th user, and N_0 is thermal noise power spectral density. Similarly, the total interference and noise density experienced by the k-th user, denoted as $N_t^k(s,r,j)$, can be expressed as

$$
\begin{aligned}
N_t^k(s,r,j) \ = &\sum_{i \neq j} E_{c,k}^{k,j}(s,r,i)\left[1 + \sum_l \rho(R_{k,l})\right] \\
&+ \sum_{m \neq k} E_{c,m}^{k,j}(s,r,j)\left[1 + \sum_l \rho(R_{m,l})\right] + N_0
\end{aligned}
\qquad (10.15)
$$

The first term in Eq. (10.15) represents the self-interference of user k that is present in case of a multipath channel. The second term represents interference from the other users. The BTS receiver typically combines the energy using maximum ratio combining. The effective pilot chip energy is then given by

$$\frac{E_{c,k}}{N_t^k} = \sum_s \sum_r \sum_j \frac{E_{c,k}^{k,j}(s,r,j)}{N_t^k(s,r,j)} \qquad (10.16)$$

The interference from the other users can be decomposed into two parts: interference from the users effectively power-controlled by the same BTS as the desired user, called same-cell interference; and interference from users power-controlled by the other BTSs, referred to as outer-cell interference:[2]

$$
\begin{aligned}
\sum_{m \neq k} E_{c,m}^k(s,r,j)\left[1 + \sum_l \rho(R_{m,l})\right] \ = &\sum_{m \in C, m \neq k} E_{c,m}^k(s,r,j)\left[1 + \sum_l \rho(R_{m,l})\right] \\
&+ \sum_{m \notin C} E_{c,m}^k(s,r,j)\left[1 + \sum_l \rho(R_{m,l})\right]
\end{aligned}
\qquad (10.17)
$$

2.   Note that in Chapter 9, the same-cell interference referred to interference from users that do not have the reference base station in their active sets. The distinction is made in Chapter 10 for the convenience of analyzing reverse-link performance.

Denote the same-cell interference as $I_{sc}^k(s,r,j)$ and other or outer-cell interference as $I_{oc}(s,r,j)$. We can then express the total interference and noise as

$$N_t^k(s,r,j) = \sum_{i \ne j} E_{c,k}^{k,j}(s,r,i)\left[1 + \sum_l \rho(R_{k,l})\right] + I_{sc}^k(s,r,j) + I_{oc}^k(s,r,j) + N_0 \qquad (10.18)$$

and the total received power spectral density as

$$I_0^k(s,r,j) = \sum_i E_{c,k}^{k,j}(s,r,i)\left[1 + \sum_l \rho(R_{k,l})\right] + I_{sc}^k(s,r,j) + I_{oc}^k(s,r,j) + N_0 \qquad (10.19)$$

The difference between the total received power and interference is then given as

$$I_0^k(s,r,j) - N_t^k(s,r,j) = E_{c,k}^{k,j}(s,r,j)\left[1 + \sum_l \rho(R_{k,l})\right] \qquad (10.20)$$

and it can be neglected when the number of simultaneous users and/or the number of resolvable paths is large. If we denote the received power spectral density of the k-th user as $I_s^k(s,r,j)$, we can rewrite Eq. (10.19) as

$$I_0^k(s,r,j) = I_s^k(s,r,j) + I_{sc}^k(s,r,j) + I_{oc}^k(s,r,j) + N_0 \qquad (10.21)$$

When the number of simultaneous users is sufficiently large, it is reasonable to model both the same-cell and outer-cell interference as a stationary Gaussian process and drop the dependence on r and j. The power spectral density of the desired signal can also be assumed independent of r and j, so that the sum on the right-hand side of Eq. (10.21) becomes independent of k too. Eq. (10.21) is now only a function of the softer handoff leg, s. When the load is equal among cells associated with collocated antennas (same BTS), Eq. (10.21) is also independent of s. In practice, however, that may not always be the case. Nevertheless, it is insightful to simplify the analysis, drop index s, and assume that the signal energy is fully concentrated in one of the cells. Under these assumptions, the received signal power over a bandwidth W can be written as

$$I_0 W = I_s^k W + I_{sc}^k W + I_{oc} W + N_0 W \qquad (10.22)$$

10.1.5.1 Same-Cell Interference

In this section we quantify the same-cell interference for some important special cases. Consider a reference cell and the set of K active users for which the reference cell is the best serving cell. The average same-cell interference power created by other users to the k-th user, as indicated in Eq. (10.17), can be written as

$$I_{sc}^k W = W \sum_{\substack{i=1 \\ i \neq k}}^{K} E_{c,i}^k \left[1 + \sum_l \rho\left(R_{i,l}\right) \right] \tag{10.23}$$

Note that the pilot chip energy in Eq. (10.23) is not combined energy, but energy per antenna. For the remainder of the chapter, all signal energy values are assumed to be *per antenna* unless stated otherwise. It is common that on the reverse link there are two receive antennas.

Consider the case when all active users have a single traffic channel, the same data rate, and undergo the same radio channel conditions and therefore require the same average received energy. Eq. (10.23), after dropping the mobile station index i, reduces to

$$I_{sc}^k W = (K-1) E_b R \tag{10.24}$$

where $E_{b,t}$ is total traffic channel bit energy, including reverse-pilot overhead, and it is equal to

$$E_b = E_c \left[1 + \rho(R) \right] \frac{W}{R} \tag{10.25}$$

The product of the total traffic channel energy and data rate R is commonly called the receiver sensitivity, $S = E_{b,t} R$.

For voice service, due to variable rate operation, Eq. (10.24) becomes

$$I_{sc}^k W = (K-1) E\{v\} E_b R_b \tag{10.26}$$

where R_b corresponds to the maximum data rate (e.g., 9.6 kbps for voice), v is the voice activity factor, and $E_{b,t(R\text{-}FCH)}$ is the total R-FCH bit energy for a full-rate frame, including the reverse-pilot overhead. Eq. (10.26) can also be expressed in terms of the R-FCH activity factor, $E\{v_{R\text{-}FCH}\}$, as

$$I_{sc}^k W = (N-1) \left[E\{v_{R\text{-}FCH}\} + 1/\rho\left(R_1\right) \right] E_{b,t(R\text{-}FCH)} R_b \tag{10.27}$$

where the R-FCH activity factor is defined by

$$E\{v_{\text{R-FCH}}\} = \sum_{i=1}^{4} \frac{\rho(R_i)}{\rho(R_1)} \Pr\left[R = R_i\right] \tag{10.28}$$

We can use either Eq. (10.26) with (10.10) or (10.27) with (10.28), provided the bit energy is defined consistently. In the reminder of this chapter we use the former approach mainly because of legacy of using Eq. (10.26) to define the same-cell interference.

In addition to the same-cell interference, a user's signal could experience self-interference due to multipath, as indicated by Eq. (10.18). Moreover, in practical systems, due to nonideal pulse shape and transmitter and/or receiver implementation, a user's signal also suffers from self-interference even in nondispersive channels. However, the impact of self-interference is relatively small, and it can be neglected particularly for large K, which is a common case for voice. When K is small, it is more important to account for self-interference. But even for small K, the outer-cell interference typically dominates the self-interference. Self-interference has more significance in the forward link when a user is assigned the Forward Packet Data Channel (F-PDCH) at high data rates.

10.1.5.2 Other-Cell Interference

Due to the universal frequency reuse, as we said before, interference arrives not only from users power controlled by the same BTS, but also from the users controlled by other BTSs. Although the soft handoff guarantees that the signal received from an other-cell mobile station is at a power level lower than the sensitivity, the aggregate received signal power from all users in surrounding cells can be considerable. It is of interest to estimate the average other-cell interference relative to same-cell interference, because the ratio plays an important role in the estimation of the reverse-link capacity. We denote the ratio of inner-cell to outer-cell interference with f. We estimate f by simulation using the cell layout and propagation loss models introduced in Sections 10.1.1 and 10.1.2, respectively. A realistic antenna radiation pattern model of Section 10.1.3 with $\theta_{\text{dB}} = 90$ degrees is considered.

The definition of outer-cell interference in Chapter 10 differs from the definition of other-cell interference used in Chapter 9. For the purpose of estimation of reverse-link capacity, it is convenient to model outer-cell interference as interference received by all users effectively power-controlled by other BTSs. However, some of those users' handoff active set may include the target cell, the one whose capacity we estimate.

The results are summarized in Figure 10.2. The relative interference decreases with the increasing values of the path loss exponent and increases with the increasing shadowing loss variance due to the opposite effects the two have on the effective separation of adjacent cells.

Note that the relative other-cell interference is related to analogous quantities often cited in the literature in the context of capacity. Two such quantities are the reuse factor and the reuse efficiency that can be defined in terms of the relative other-cell interference as

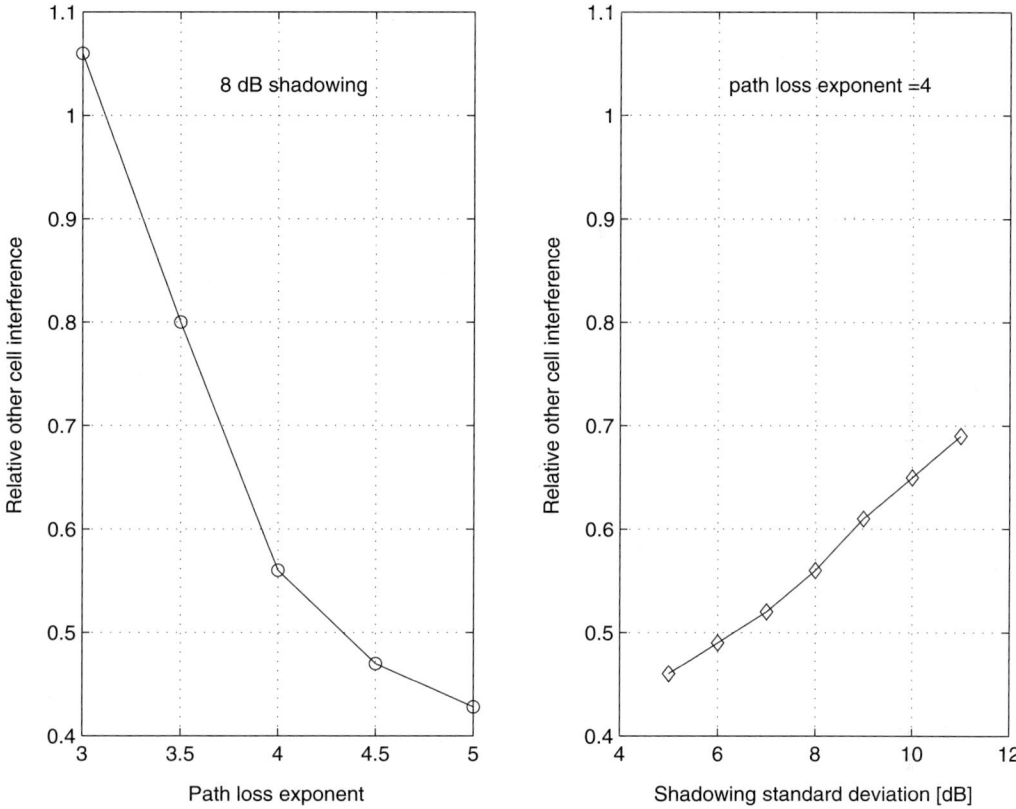

Figure 10.2 Relative other-cell interference. The correlation coefficient of shadowing to two BTSs is $a^2 = \frac{1}{2}$.

$$\text{frequency reuse factor } = 1+f = \frac{\text{total interference}}{\text{same-cell interference}} \qquad (10.29)$$

$$\text{frequency reuse efficiency} = \frac{1}{1+f} = \frac{\text{same-cell interference}}{\text{total interference}} \qquad (10.30)$$

10.1.6 Forward-Link Interference

Similarly as for the reverse link, the forward-link interference can be divided into same-cell and other-cell interference. However, when compared to the reverse-link interference, the forward-link interference is different in nature. As we discussed in Chapter 4, the forward-link channel separation is achieved with orthogonal Walsh codes. The idea behind orthogonal channel design is to eliminate interference among channels belonging to the same cell. The interference from the same cell is nonexistent in the absence of multipath and imperfections in transmitter and/or receiver implementation. A user receives interference from only several relatively strong interferers. This is in contrast to reverse link, where a user receives interference from a relatively large number of sources. We assume a common scenario where the base station and the mobile station have a single transmitter and receiver antenna, respectively.

10.1.6.1 *Same-Cell Interference*

Due to Walsh code channelization, there is no interference among forward-link channels using the same pilot pseudonoise (PN) offset. However, the assumption of a nondispersive channel is unrealistic. A dispersive channel causes multipath that, although exploited by the Rake receiver as a form of spatial diversity, causes loss of orthogonality. The signals received through multiple paths will no longer be orthogonal to one another, and they will appear as interference at the matched filter output.

Consider now a user with a single pilot in its active set and a dispersive radio channel. Denote with α_k the normalized complex fading amplitude for the k-th path, where $\sum_k \|\alpha_k\|^2 = 1$. The carrier-to-interference and noise ratio (CINR), assuming ideal transmitter and receiver implementation per Rake finger, is then given by

$$CINR_k = \frac{\|\alpha_k\|^2}{\dfrac{N_0 + I_{oc}}{I_{or}} + \sum_{\substack{n \\ n \neq k}} \|\alpha_n\|^2}$$ (10.31)

where $\hat{I}_{or}$, N_0, and I_{oc} are the carrier, thermal noise, and the power spectral density of the other-cell interference respectively. At the mobile station, the combining of Rake fingers is typically based on pilot signal strength, which sets the weights to the complex conjugate of α_k and does not account for different levels of interference. Pilot-based combining is easier to implement than maximum ratio combining, and the loss in performance relative to maximum ratio combining is not significant. After pilot-based combining, the *CINR* can be computed as

$$CINR = \frac{\left(\sum_k \|\alpha_k\|^2 \right)^2}{\sum_k \|\alpha_k\|^2 \left(\dfrac{N_0 + I_{oc}}{I_{or}} + \sum_{\substack{n \\ n \neq k}} \|\alpha_n\|^2 \right)}$$ (10.32)

The self-interference created by loss of orthogonality bounds the *CINR* when the background noise is vanishingly small.

The self-interference can also be created by nonideal transmitter and/or receiver implementation, and it consists of the interchip interference, transmitter radio noise floor, and receiver analog-to-digital (A/D) quantization. The interchip interference arises due to the non-Nyquist pulse shape specified in IS-2000 [3]. The transmitter noise floor is the effect of signal distortion caused by nonlinear elements present in the transmitter chain of the BTS. A measure of the noise floor is given by the waveform quality defined as the normalized correlated power between the ideal waveform and the actual transmitted waveform after compensation for timing and frequency errors. The A/D converter is typically implemented in the form of a b-bits uniform midtread quantizer.[3] Using such a device, the relative quantization error power is equal to $2^{-(b+1)}/12$. Such self-interference is proportional to the received carrier power and therefore represents a bound on the maximum achievable carrier-to-interference ratio (CIR) even in the absence of multipath and other-cell interference. Since it represents a maximum achievable CIR, we can characterize this type of self-interference with the parameter ε $1/CIR_{MAX}$. Then, the effective CIR becomes

$$CIR_{effective} = \frac{1}{1/CIR + \varepsilon} \qquad (10.33)$$

With a typical transmitter/receiver implementation, the maximum achievable CIR is in the range of 13 to 14.8 dB, equivalent to ε in the range 3.3% to 5%, as detailed in Table 10.8. The self-interference has minimal impact on the forward-link voice capacity. However, self-interference plays a more important role when using rate control in the context of F-PDCH transmission, because it limits the maximum sustainable data rate.

Table 10.8 Computation of Self-Interference

Index, i	Self-interference component	$\hat{I}_{or}/I^{i}_{self}$	Notes
1	Interchip interference	16.5 dB	Using the IS-2000 baseband pulse shape
2	Radio noise floor	17.5 to 20 dB	Corresponding to 98% to 99% transmit waveform quality
3	Analog to Digital Conversion (ADC) quantization noise	20, 31.9 dB	For 4-bits and 6-bits ADC, respectively
	Maximum CIR = I^{i}_{self}	13 to 14.8 dB	

3. If we picture the quantizer input-output amplitude characteristics like a staircase function, a midtread quantizer is one in which the origin lies in the middle of a tread.

10.1.6.2 Other-Cell Interference

In this section we quantify the interference from other cells and also consider impact of soft handoff. Assume a cell layout described in Section 10.1.1. Denote with $I_{or,i,k}$ the average power spectral density received from the i-th cell by the k-th user. The assumption is that the network is fully loaded and that the power spectral density of the transmitted signal, I_{or} is the same across the entire network. Denote with $\alpha_{i,k,j}$ the normalized complex fading amplitude for the j-th ray, from the i-th cell and the k-th user. Then the *CINR* per Rake finger can be written as

$$
\begin{aligned}
CINR_{i,k,j} &= \frac{I_{i,k}\left\|\alpha_{i,k,j}\right\|^2}{N_0 + \sum_{\substack{j,l \\ l\neq i}} I_{l,k}\left\|\alpha_{i,k,j}\right\|^2 + \sum_{\substack{n \\ n\neq k}} I_{i,k}\left\|\alpha_{i,n,j}\right\|^2} \\[2em]
&= \frac{\left\|\alpha_{i,k,j}\right\|^2}{\dfrac{N_0}{I_{i,k}} + \sum_{\substack{j,l \\ l\neq k}} \dfrac{I_{l,k}}{I_{i,k}}\left\|\alpha_{i,k,j}\right\|^2 + \sum_{\substack{n \\ n\neq k}}\left\|\alpha_{i,n,j}\right\|^2}
\end{aligned}
\tag{10.34}
$$

After pilot-based combining, for a mobile station in h-way soft (and/or softer) handoff, $i\in H$, $H\in\{1,2,\dots,h\}$, the *CINR* can be computed as

$$
CINR_k = \frac{\left(\sum_{j,i\in H}\left\|\alpha_{i,k,j}\right\|^2\right)^2}{\sum_{j,i\in H}\left\|\alpha_{i,k,j}\right\|^2\left(\dfrac{N_0}{I_{i,k}} + \sum_{\substack{j,l \\ l\neq i}}\dfrac{I_{l,k}}{I_{i,k}}\left\|\alpha_{i,k,j}\right\|^2 + \sum_{\substack{n \\ n\neq k}}\left\|\alpha_{i,n,j}\right\|^2\right)}
\tag{10.35}
$$

The outer-cell interference, for k-th user, $I_{oc,k}$, now refers to the interference created by the base station using the pilot PN that is not in its active set, that is, $i \notin H$:

$$
I_{oc,k} = \sum_{\substack{j,l \\ l\notin H}} I_{l,k}\left\|\alpha_{i,k,j}\right\|^2
\tag{10.36}
$$

Eq. (10.35) can be rewritten as

$$CINR_k = \frac{\left(\displaystyle\sum_{j,i\in H}\|\alpha_{i,k,j}\|^2\right)^2}{\displaystyle\sum_{j,i\in H}\|\alpha_{i,k,j}\|^2\left(\dfrac{N_0+I_{oc,k}}{\hat{I}_{i,k}}+\displaystyle\sum_{\substack{j,l\in H\\l\neq i}}\dfrac{\hat{I}_{l,k}}{\hat{I}_{i,k}}\|\alpha_{i,k,j}\|^2+\displaystyle\sum_{\substack{n\\n\neq k}}\|\alpha_{i,n,j}\|^2\right)} \qquad (10.37)$$

where $I_{oc,k}$ is given by Eq. (10.36).

10.1.6.3 Effective CINR

Obviously, the other-cell interference is larger for users located at the cell border than for those located in the proximity of the base station. However, the interference from the same cell depends solely on the multipath spread and is largely independent[4] of distance from the base station. We consider the *CINR* as a function of the normalized distance from the base station for different multipath spread profiles and show results in Figure 10.3. As an illustration, it is assumed that the mobile station is not in handoff and that the radio channel is time-invariant and has two paths. The ratio between the two paths is $|\alpha_2|^2/|\alpha_1|^2 = -3$ dB. The received power is obtained by simulation sampled at a normalized distance from the base station along the boresight of the transmit antenna. The decibel difference between the CINR and the carrier to other-cell interference ratio can then be seen as the degradation due to orthogonality loss. Such degradation is small for mobile station located near the cell boundary where the other-cell interference is dominant. The dominant source of interference for mobile stations located near the serving base station is multipath.

We now turn to the estimation of the *CIR* statistics using the cell layout, propagation loss, and antenna radiation pattern models described in Sections 10.1.1 to 10.1.3, respectively. For this purpose let us neglect the impact of thermal noise and short-term fading. The received power spectral density from the *i*-th cell, $\hat{I}_{or,i}$, can be ordered starting from the strongest. The *CIR* can be written as

$$CIR_1 = \frac{\hat{I}_{or,1}}{\displaystyle\sum_{i>1}\hat{I}_{or,i}} = \frac{r_1^{-\gamma}10^{\zeta_1/10}}{\displaystyle\sum_{i>1}r_i^{-\gamma}10^{\zeta_j/10}} \qquad (10.38)$$

The *CIR* computed in this fashion is also commonly called *cell geometry*. It is a random variable that depends on the ratio of the ordered and correlated log-normal distributed random variables. We resort to simulations to obtain its distribution because it is impractical to do so analytically. The *CIR* distribution depends on antenna beam-width, path loss exponent, and to a smaller extent shadowing loss standard deviation. In Figure 10.4 we plot the empirical dis-

---

4.   In practice, mobile stations located far away from the serving base station are less likely to be in line of sight of the transmitter and typically experience larger multipath spread.

tribution of the effective *CIR* using the antenna radiation pattern discussed in Section 10.1.3 with $\theta_{3dB} = 90°$.

10.2 Cell Capacity

Now that we have established the system model, we can proceed with the estimation of the cell voice and data capacity. The computed capacity plays an important rule in network planning because for a given subscriber density, it determines the maximum cell size.

10.2.1 Reverse-Link Voice Capacity

The task of capacity estimation is impaired by the numerous assumptions must be made to obtain a mathematically treatable problem. Nevertheless, the process of deriving capacity models built on each other by progressively relaxing, eliminating, or simplifying assumptions is insightful. We follow the approach from [4] and begin with a simple model, and then move to probabilistic models that are somewhat more advanced.

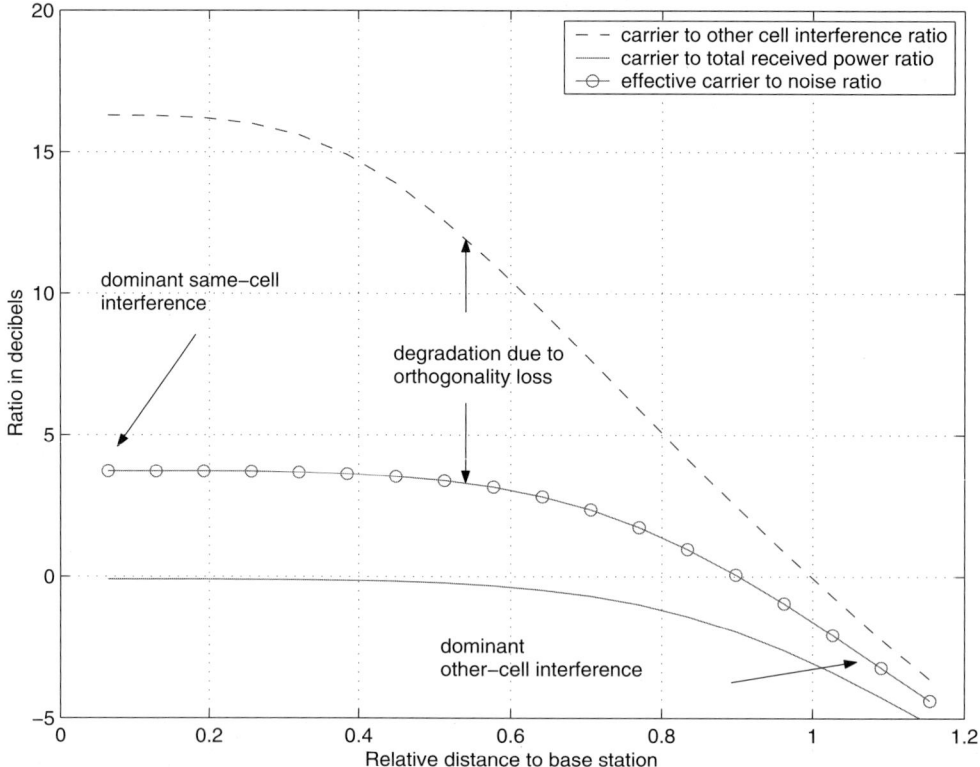

Figure 10.3 Impact of same- and other-cell interference on the effective carrier-to-noise ratio. Two-path channel with $\|a2\|^2/\|a1\|^2 = -3$ dB.

10.2.1.1 Pole Capacity

Consider an isolated cell and assume that each active user is perfectly power-controlled so that per-user received power is equal to the receiver sensitivity, S. Also assume that all active users, K, undergo identical channel conditions and therefore require identical received SNR. We do not account for the softer handoff. The received SNR per receiver antenna, E_b/N_t, that includes reverse-pilot overhead can be written as

$$\frac{E_b}{N_t} = \frac{S/R_b}{N_0 + (K-1)S/W} = \frac{W}{R_b} \cdot \frac{S/(N_0W)}{1 + (K-1)S/(N_0W)} \qquad (10.39)$$

where N_0 denotes the thermal noise, R_b is the data rate, and W is the bandwidth. We neglected self-interference due to nonideal transmitter and/or receiver implementation and possible multipath.

We can solve Eq. (10.39) for the receiver sensitivity relative to the background noise power and obtain

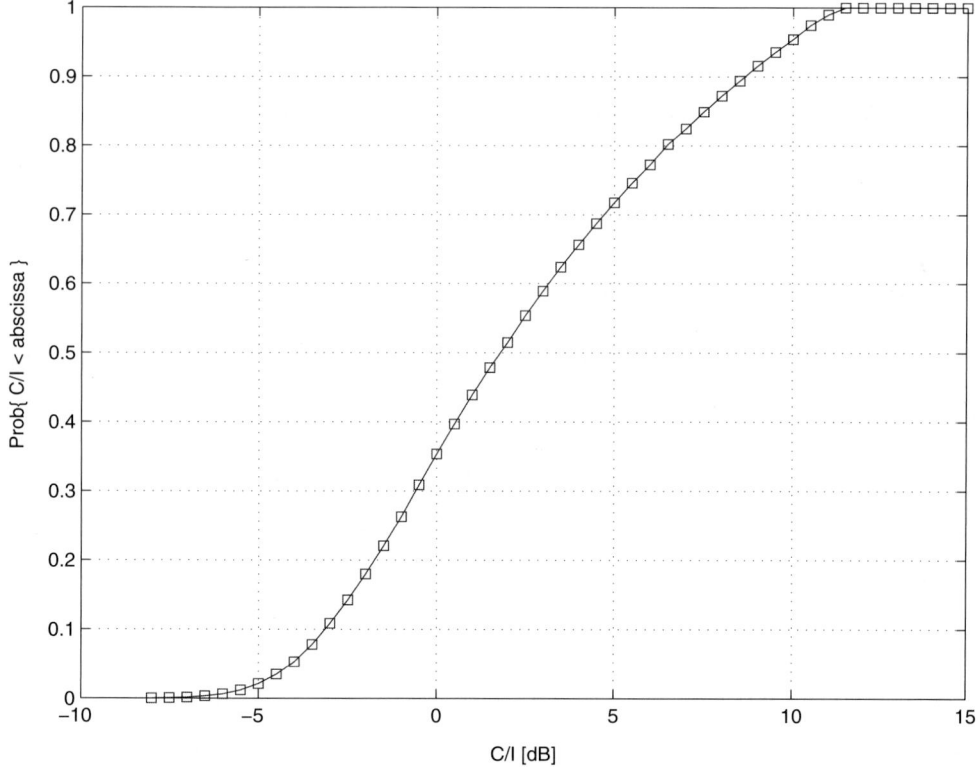

Figure 10.4 Effective *CIR* distribution for different transmit antenna radiation patterns: shadowing standard deviation = 8 dB; shadowing correlation to two BTSs = 0.5; path loss exponent = 3.5; self-interference parameter ε = 0.05.

$$\frac{S_a}{N_0 W} = \frac{\dfrac{R_b}{W} \cdot \dfrac{E_b}{N_t}}{1 - \dfrac{R_b}{W} \cdot \dfrac{E_b}{N_t}(K-1)}$$

(10.40)

Bounded receiver sensitivity implies that a mobile station is able to close the reverse link with a finite amount of transmitted power. From Eq. (10.40) we note that the receiver sensitivity is bounded if the denominator is greater than zero; that is,

$$1 > \frac{R_b}{W} \cdot \frac{E_b}{N_t}(K-1)$$

(10.41)

which in turn yields

$$K < 1 + \frac{W/R_b}{E_b/N_t} \cong \frac{W/R_b}{E_b/I_0}$$

(10.42)

where we approximated the total interference and noise density, N_t, with the total received power spectral density, I_0. The right-hand side of Eq. (10.42) is dubbed *pole capacity* because it represents the maximum, or asymptotic, achievable number of users per cell. In practice, to avoid operation near the pole and accounting for the limited dynamic range of the receiver, it is desirable to limit the total received power, $I_0 W$ relative to the receiver noise floor, $N_0 W$. Let us denote with η the ratio of receiver noise to total received power density, $\eta = N_0/I_0 < 1$. Note that the inverse of η is called rise over thermal (RoT). We can write that as

$$KE_b R_b + N_0 W = I_0 W$$

(10.43)

and after imposing the limit on η,

$$KE_b R_b = (I_0 - N_0)W < I_0 W (1 - \eta)$$

(10.44)

obtain the upper bound on the maximum number of users:

$$K < \frac{W/R_b (1 - \eta)}{E_b/I_0}$$

(10.45)

Now consider voice activity. If we neglect the random nature of voice activity and simply consider that all mobile stations' transmitted power is reduced by a factor equal to the voice activity factor, v, we obtain

$$K < \frac{W/R_b\,(1-\eta)}{v\,E_b/I_0} \tag{10.46}$$

Finally, consider an embedded cell rather than an isolated one. The base station receiver in an embedded cell experiences interference from all active users served by other cells. If the ratio of the other-cell interference to the aggregate received power from the user served by the cell is f, the contribution of all other-cell users on average is equal to $f\,K$. The interference seen by the receiver is then equivalent to $K\,(1+f)$ active users, which leads to

$$K < \frac{W/R_b\,(1-\eta)}{v\,E_b/I_0\,(1+f)} \tag{10.47}$$

Eq. (10.47) represents a simple capacity model, but it is also based on numerous unrealistic assumptions (e.g., nonprobabilistic treatment of voice activity, ideal power control).

10.2.1.2 Outage Probability with Perfect Power Control

After accounting for the other-cell interference, the equivalent number of active users is $K' = K\,(1+f)$. For the ideal power control, Eq. (10.43) can be rewritten to account for VBR vocoding as

$$\sum_{k=1}^{K'} v_k E_b R_b + N_0 W = I_0 W \tag{10.48}$$

The quaternary random variable v represents voice activity and takes on the values from Table 10.2 for a given vocoder mode. We now have a probabilistic capacity model. The outage constraint and the probability of outage become

$$\Gamma \quad \frac{E_b}{I_0} \frac{R_b}{W} \sum_{k=1}^{K'} v_k \le 1-\eta \tag{10.49}$$

$$P_{\text{outage}} = \Pr\left(\Gamma > 1-\eta\right) \approx Q\left(\frac{1-\eta-E\{\Gamma\}}{\sigma_\Gamma}\right) \tag{10.50}$$

where $Q(y) = 1/\sqrt{2\pi}\int_y^\infty e^{-x^2/2}dx$, $y>0$.

In Eq. (10.50) we have used the Gaussian approximation for the random variable Γ. The Gaussian approximation is reasonably accurate for a relatively large number of active users, K'. The average deviation, $E\{\Gamma\}$, and standard deviation, σ_Γ, can be readily computed as

$$E\{\Gamma\} = \frac{E_b}{I_0} \frac{R_b}{W} \left[K(1+f) \right] E\{v\}$$

$$\sigma_\Gamma = \frac{E_b}{I_0} \frac{R_b}{W} \sqrt{K(1+f)\operatorname{var}[v]}$$

(10.51)

10.2.1.3 Erlang Capacity with Perfect Power Control

We are now interested in computing the outage probability when the effective number of active users is a random variable. We assume for simplicity that there is no admission-control mechanism. All incoming resource requests are granted, and all users are admitted into the system. Under these assumptions, the system can be seen as one with an infinite number of servers. The system occupancy, or the effective number of active users, under Poisson arrivals with rate λ and exponential distributed call-holding time with the mean of $1/\mu$ is also a Poisson random variable, and its probability mass function, P_n, is given by

$$P_n = \frac{(\lambda/\mu)^n}{n!} e^{-\lambda/\mu} \quad n \geq 0$$

(10.52)

Both the mean and variance of the effective number of active users is given by $(1+f)\lambda/\mu$. The probability of outage can be tightly upper bounded if we include the desired user's received signal into the interference power. Then, assuming Γ to be approximately Gaussian, the outage constraint can be written as

$$\Gamma \quad \frac{E_b}{I_0} \frac{R_b}{W} \sum_{k=1}^{K'} v_k \leq 1 - \eta$$

(10.53)

K' is Poisson random variable and the sum $\sum_{k=1}^{K'} v_k$ can be seen as a compound process, with the mean given by $E[K']E[v]$ and the variance computed as $E\{K'\}\operatorname{var}\{v\} + \operatorname{var}\{K'\}E\{v\}^2$. Now it is easy to show, as it is done in [5], that the average of Γ is given by

$$E\{\Gamma\} = \frac{E_b}{I_0} \frac{R_b}{W} E\{K'\}E\{v\} = \frac{E_b}{I_0} \frac{R_b}{W} (1+f)\frac{\lambda}{\mu}\bar{v}$$

(10.54)

and the standard deviation by

$$
\begin{aligned}
\sigma_\Gamma &= \frac{E_b}{I_0} \frac{R_b}{W} \sqrt{E\{K'\}\mathrm{var}\{v\} + \mathrm{var}\{K'\}E\{v\}^2} \\
&= \frac{E_b}{I_0} \frac{R_b}{W} \sqrt{\frac{\lambda}{\mu}(1+f)\overline{v^2}}
\end{aligned}
\tag{10.55}
$$

The probability of outage is again given by Eq. (10.50).

10.2.1.4 *Erlang Capacity with Imperfect Power Control*

Let us now account for the power-control inaccuracy. Due to power-control inaccuracy, the received E_b/N_t is not constant but rather exhibits fluctuations around the desired setpoint. These fluctuations are primarily the result of the combined effect of the inner-loop power-control bit errors and the energy estimation errors. In addition to inner-loop inaccuracy, the outer-loop power control also contributes to the fluctuations of the received E_b/N_t. Outer-loop contributions are the combined effect of the intrinsic feedback-loop variance and the fluctuations in radio conditions causing the required $E_b/I_0 \approx E_b/N_t$ to change. The received E_b/I_0 must, therefore, be treated as a random variable, and the outage constraint equation becomes

$$
\Gamma \quad \frac{R_b}{W} \sum_{k=1}^{K'} v_k \frac{E_{b,k}}{I_0} \le 1 - \eta
\tag{10.56}
$$

Experimentally, it has been shown that the received SNR can be closely approximated by a lognormal distributed random variable. Therefore, consider the decibel value of the SNR, $\gamma = 10 \mathrm{Log}\, E_b/I_0$, where γ is a Gaussian random variable with mean ε [dB] and standard deviation σ [dB]. We can write E_b/I_0 in terms of its decibel value as

$$
E_b/I_0 = 10^{\gamma/10} = e^{\beta\gamma}
\tag{10.57}
$$

where β is a shorthand notation for $\log_e 10/10$. The mean second-order moment and variance of E_b/I_0 are

$$
\begin{aligned}
E\{E_b/I_0\} &= \int_{-\infty}^{\infty} e^{\beta\gamma} \cdot \frac{e^{-(\gamma-\varepsilon)^2/2\sigma^2}}{\sqrt{2\pi}\sigma} \, d\gamma \\
&= e^{(\beta\sigma)^2/2 + \beta\varepsilon}
\end{aligned}
\tag{10.58}
$$

$$E\left\{ \left(E_b / I_0 \right)^2 \right\} = \int_{-\infty}^{\infty} e^{2\beta\gamma} \cdot \frac{e^{-(\gamma-\varepsilon)^2 / 2\sigma^2}}{\sqrt{2\pi}\sigma} d\gamma$$

$$= e^{2(\beta\sigma)^2 + 2\beta\varepsilon} \tag{10.59}$$

$$\mathrm{var}\left\{ E_b / I_0 \right\} = E\left\{ \left(E_b / I_0 \right)^2 \right\} - E\left\{ E_b / I_0 \right\}^2$$

$$= e^{(\beta\sigma)^2 + 2\beta\varepsilon} \left[e^{(\beta\sigma)^2} - 1 \right] \tag{10.60}$$

The probability of outage is again given by Eq.(10.50) but with the variable Γ's average and the standard deviation equal to

$$E\{\Gamma\} = \frac{R_b}{W} E\{K'\} E\{E_b / I_0\} E\{v\}$$

$$= \frac{R_b}{W} \frac{\lambda}{\mu} (1+f) e^{(\beta\sigma)^2 / 2 + \beta\varepsilon} \overline{v} \tag{10.61}$$

$$\sigma_\Gamma = \frac{R_b}{W} \sqrt{ E\{K'\} \mathrm{var}\{v \, E_b / I_0\} + \mathrm{var}\{K'\} E\{v \, E_b / I_0\}^2 } \tag{10.62}$$

and since K' is a Poisson random variable, $E\{K'\} = \mathrm{var}\{K'\} = (1+f)\lambda / \mu$ and Eq. (10.62) reduces to

$$\sigma_\Gamma = \frac{R_b}{W} \sqrt{ \frac{\lambda}{\mu} E\left\{ \left(v \cdot E_b / I_0 \right)^2 \right\} }$$

$$= \frac{R_b}{W} \sqrt{ \frac{\lambda}{\mu} (1+f) \overline{v^2} \, e^{2(\beta\sigma)^2 + 2\beta\varepsilon} } \tag{10.63}$$

The decibel average of E_b / I_0 depends on channel conditions. A practical value corresponding to full-rate transmission in a pedestrian outdoor environment with a frame error rate (FER) of 1% is about 2 dB per receiver antenna, or 5 dB combined (for two receiver antennas). The E_b / I_0 value includes reverse-pilot overhead. The standard deviation is typically equal to 2~2.5 dB.

10.2.1.5 Outage Probability with Imperfect Power Control

With the same methodology as that used in Section 10.2.1.4, the impact of imperfect power control on the outage probability can be calculated in the case of a fixed number of users. The average of the random variable Γ is can be written as

$$
\begin{aligned}
E\{\Gamma\} &= \frac{R_b}{W} K' E\{E_b/I_0\}E\{v\} \\
&= \frac{R_b}{W} K'(1+f)e^{(\beta\sigma)^2/2+\beta\varepsilon}\overline{v}
\end{aligned}
\tag{10.64}
$$

while the standard deviation is given by

$$
\begin{aligned}
\sigma_\Gamma &= \frac{R_b}{W}\sqrt{K'(1+f)\mathrm{var}\left\{(v\cdot E_b/I_0)^2\right\}} \\
&= \frac{R_b}{W}\sqrt{K'(1+f)\left[E\left\{(v\cdot E_b/I_0)^2\right\}-E\left\{(v\cdot E_b/I_0)\right\}^2\right]} \\
&= \frac{R_b}{W}\sqrt{K'(1+f)\left[\overline{v^2}\,e^{2(\beta\sigma)^2+2\beta\varepsilon}-\overline{v}^2\,e^{(\beta\sigma)^2+2\beta\varepsilon}\right]} \\
&= \frac{R_b}{W}\sqrt{K'(1+f)\left[e^{(\beta\sigma)^2+2\beta\varepsilon}\left(\overline{v^2}e^{(\beta\sigma)^2}-\overline{v}^2\right)\right]}
\end{aligned}
\tag{10.65}
$$

The outage probability is again given by Eq. (10.50).

10.2.1.6 Summary of Reverse-Link Voice Capacity Results

From the analysis above, it is apparent that the reverse-link capacity depends on many parameters: cell layout, which affects frequency reuse efficiency; vocoder mode of operation, which affects the voice activity factor; maximum tolerable receiver desensitization, which affects setting of the outage threshold; user mobility and quality of service (QoS), which affect average and variance of the received E_b/I_0. The reuse efficiency, average E_b/I_0, and voice activity significantly affect capacity. Therefore, in Figure 10.5, Figure 10.6, and Figure 10.7 we provide numerical examples for the cell outage probability as a function of the offered load for different settings of reuse efficiency, mean E_b/I_0, and vocoder mode. Dual antenna diversity is considered.

In Table 10.9 we summarize cell capacity results using different modeling assumptions with the most often quoted parameters' settings for dual receiver antenna diversity.

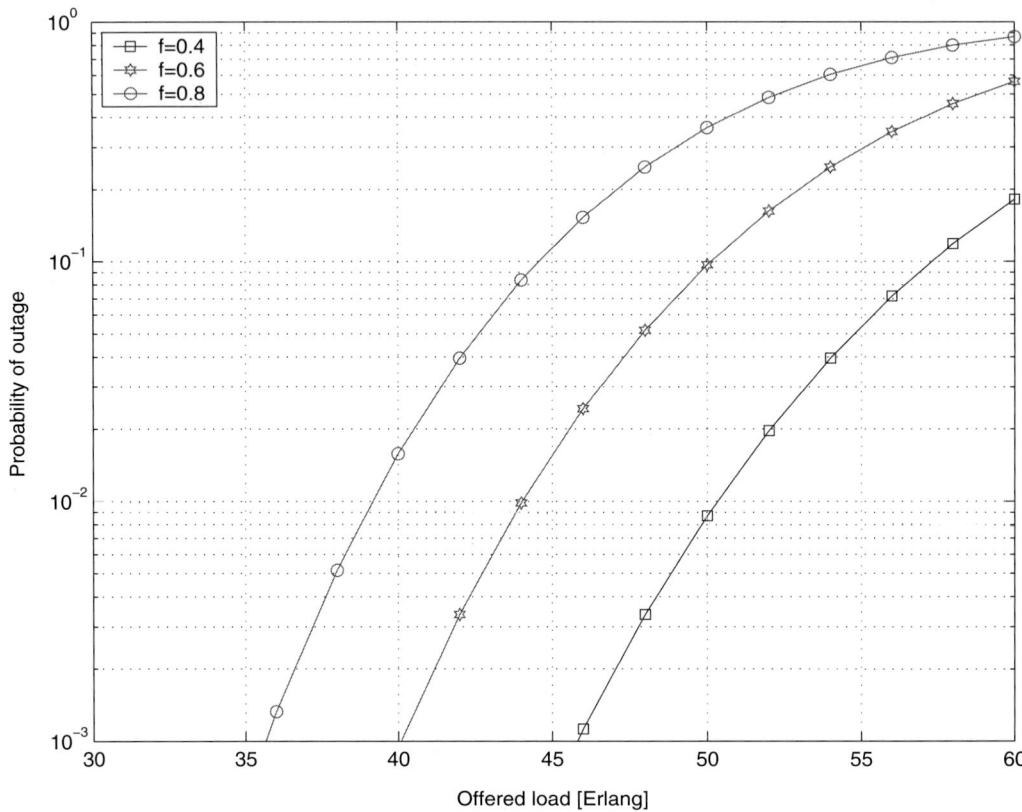

Figure 10.5 Outage probability versus Erlang load: reuse efficiency is a parameter; mean E_b/I_o = 2 dB per antenna; SMV Mode = 1, $1/\eta$ = 5 dB; standard deviation of E_b/I_o = 2 dB.

Table 10.9   Calculated Reverse-Link Voice Capacity per 1.25 MHz

	Isolated cell[*]	Embedded cell[*†]	Embedded cell[*†], imperfect power control[‡]
Asymptotic max. number of active users	104 users	65 users	58 users
Erlang capacity at 2% outage	84 Erlangs	52 Erlangs	46 Erlangs
Max. (fixed) number of mobiles at 2% outage	96 users	60 users	52 users

[*] mean E_b/I_o = 2 dB per antenna, SMV Mode=1, $1/\eta$ = 5 dB; [†] reuse efficiency, f = 0.6; [‡] standard deviation of E_b/I_o = 2 dB.

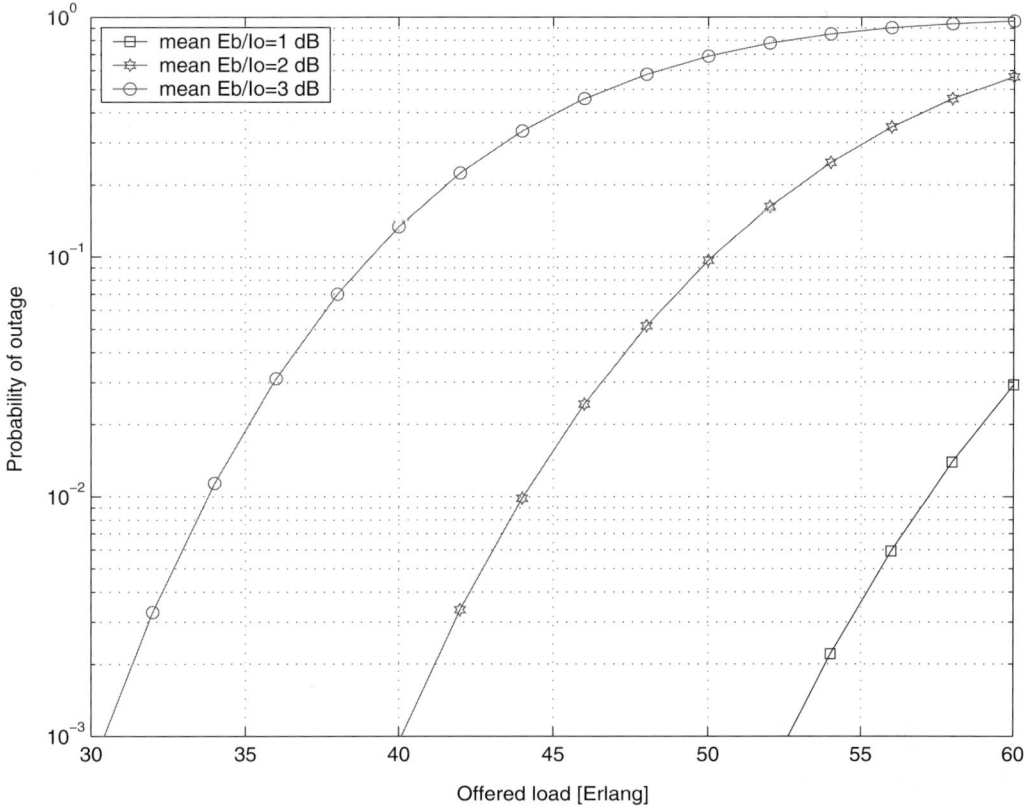

Figure 10.6 Outage probability versus Erlang load: mean required E_b/I_0 is a parameter; SMV Mode = 1, $1/\eta$ = 5 dB; standard deviation of E_b/I_o = 2 dB; frequency reuse efficiency, f = 0.6.

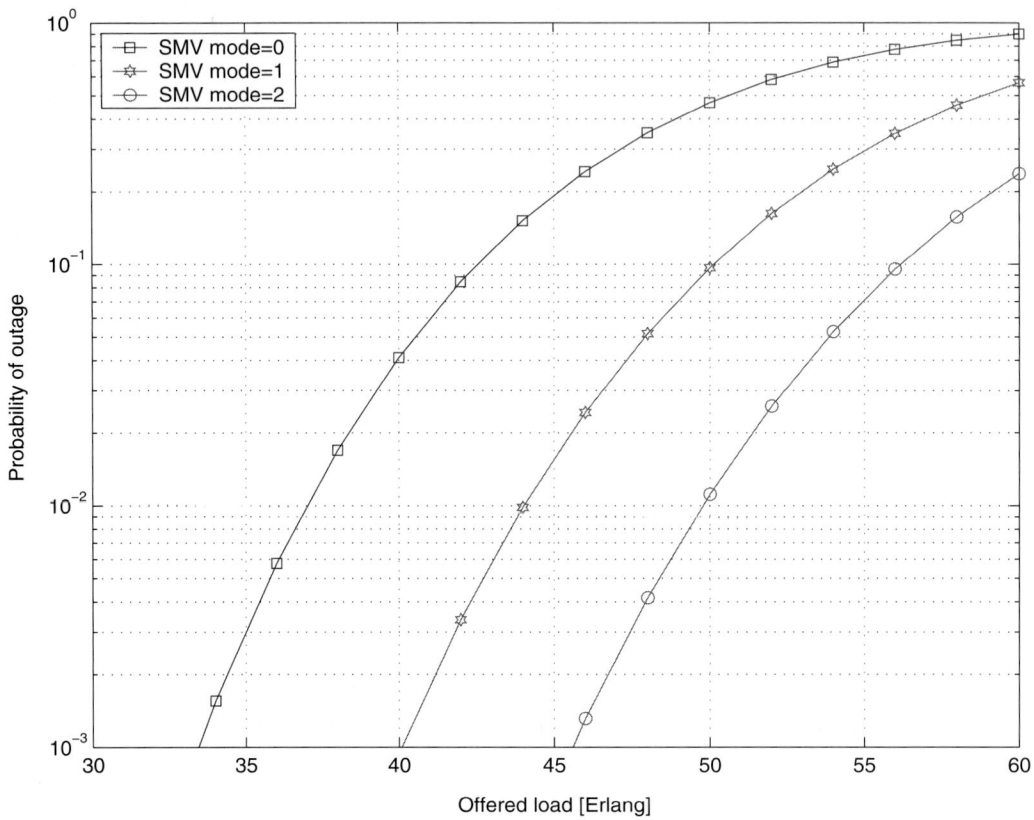

Figure 10.7 Outage probability versus Erlang load: SMV mode is a parameter; mean $E_b/I_o = 2$ dB per antenna; $1/\eta = 5$ dB; standard deviation of $E_b/I_o = 2$ dB; frequency reuse efficiency, $f = 0.6$.

10.2.2 Reverse-Link Capacity with Mixed Voice and Data Traffic

In this section, for simplicity, we assume that a reverse-link data user continuously transmits data at a constant rate. Given the nature of the reverse-link traffic, this is a reasonable approximation for an FTP upload application (see Chapter 3 for traffic models).

To account for N_d simultaneous data users transmitting the Reverse Supplement Channel (R-SCH) with data rates R_i, Eq. (10.56) can be modified as

$$\Gamma = \frac{R_b}{W} \sum_{k=1}^{K'} v_k \frac{E_{b,k,t(\text{R-FCH})}}{I_0} + \sum_{k=1}^{K_d} \left[\frac{R_k}{W} \frac{E_{b,k,t(\text{R-SCH})}}{I_0} + \frac{R_{\text{R-FCH}}}{W} \frac{E_{b,k,t(\text{R-FCH})}}{I_0} \right] \leq 1 - \eta \qquad (10.66)$$

where $E_{b,k,t(\text{R-SCH})}$ is total signal bit energy per receiver antenna over R-SCH, including the pilot overhead:

$$E_{b,k,t(\text{R-SCH})} = E_c \left[1 + \rho \left(R_{\text{R-SCH}} \right) \right] \frac{W}{R_k} \qquad (10.67)$$

It is easy to show that assuming all data users support the same rate, $R_{\text{R-SCH}}$, and undergo the same radio conditions, Eqs. (10.61) and (10.62) can be rewritten to account for data traffic as

$$
\begin{aligned}
E\{\Gamma\} &= \frac{R_{\text{R-FCH}}}{W} E\{K'\} E\{E_{b,t(\text{R-FCH})}/I_0\} E\{v\} + K_d \left[\frac{R_{\text{R-SCH}}}{W} E\{E_{b,t(\text{R-SCH})}/I_0\} + \frac{R_{\text{R-FCH}}}{W} E\{E_{b,t(\text{R-FCH})}/I_0\} \right] \\
&= \frac{R_{\text{R-FCH}}}{W} \frac{\lambda}{\mu} (1+f) e^{(\beta\sigma)^2/2 + \beta\varepsilon_{\text{R-FCH}}} \bar{v} + K_d (1+f) \left[\frac{R_{\text{R-SCH}}}{W} e^{(\beta\sigma)^2/2 + \beta\varepsilon_{\text{R-SCH}}} + \frac{R_{\text{R-FCH}}}{W} e^{(\beta\sigma)^2/2 + \beta\varepsilon_{\text{R-FCH}}} \right]
\end{aligned} \qquad (10.68)
$$

and

$$
\begin{aligned}
\sigma_\Gamma &= \sqrt{ \left(\frac{R_{\text{R-FCH}}}{W} \right)^2 E\{K'\} E\{(v \cdot E_b/I_0)^2\} + K_d \left[\left(\frac{R_{\text{R-SCH}}}{W} \right)^2 \text{var}\{E_{b,t(\text{R-SCH})}/I_0\} + \left(\frac{R_{\text{R-FCH}}}{W} \right)^2 \text{var}\{E_{b,t(\text{R-FCH})}/I_0\} \right] } \\
&= \sqrt{ \left(\frac{R_{\text{R-FCH}}}{W} \right)^2 \frac{\lambda}{\mu} (1+f) \overline{v^2} e^{2(\beta\sigma)^2 + 2\beta\varepsilon_{\text{R-FCH}}} + K_d (1+f) \left(e^{(\beta\sigma)^2} - 1 \right) e^{(\beta\sigma)^2} \left[\left(\frac{R_{\text{R-SCH}}}{W} \right)^2 e^{2\beta\varepsilon_{\text{R-SCH}}} + \left(\frac{R_{\text{R-FCH}}}{W} \right)^2 e^{2\beta\varepsilon_{\text{R-FCH}}} \right] }
\end{aligned} \qquad (10.69)
$$

We assumed that a certain percentage of reverse-link capacity is reserved for data and the remainder is to be used for voice. However, due to stringent grade of service (GoS) and QoS requirements (see Chapter 9), voice traffic inefficiently utilizes reserved capacity. In practice, temporarily unused capacity could be dynamically reassigned to data traffic, thus increasing the reverse data throughput.

Care should be taken when using Eqs. (10.68) and (10.69). The reverse-pilot overhead must be accounted for into the corresponding energy per bit, E_b. Note also that the R-FCH increases the aggregate data rate per data user. Also, for high data rates when the number of simultaneous users is small, the difference between I_0 and N_t may not be negligible. Note that from Eq. (10.20), we can bound the ratio of I_0 and N_t as

$$1 < \frac{I_0}{N_t} \leq 1 + \frac{E_b}{N_t}\frac{R}{W} \tag{10.70}$$

When the number of users is small, Γ may not be as accurately modeled as a Gaussian random variable. As a consequence, given the mean and standard deviation of Γ, the outage criteria may not be accurately computed. In Table 10.10 we show numerical examples for mixed voice and data capacity. The assumption made in Table 10.10 is that the data users have enough power to support rate $R_{\text{R-SCH}}$. In practice, there is a limit on the reverse-link transmitter power, and if the user experiences high propagation delay (located at the edge of the cell), rate $R_{\text{R-SCH}}$ may not be feasible.

Table 10.10 Reverse-Link Mixed Voice and Data Cell Capacity per 1.25 MHz

Embedded cell[*†], imperfect power control[‡]	$R_{\text{F-SCH}} = 76.8$ kbps, $K_d = 2$	$R_{\text{R-SCH}} = 76.8$ kbps, $K_d = 3$
Erlang capacity at 2% outage	22 Erlangs	10 Erlangs

*	median R-FCH $E_b/I_o = 2$ dB per antenna, SMV Mode = 1, median SCH $E_b/I_o = 0$ dB per antenna, $1/\eta = 5$ dB; † reuse efficiency, $f = 0.6$; ‡ standard deviation of $E_b/I_o = 2$ dB.

10.2.3 Forward-Link Voice Capacity

As already mentioned, the forward link differs from the reverse link. For the purpose of cell capacity estimation, the following differences must be accounted for:

- Forward channel transmission is one-to-many, that is, from one transmitter, the base station, to many receivers geographically dispersed. The implication is that the interference suffered by a mobile station depends on its distance from the base station.
- A forward channel consists of dedicated channels that are individually power-controlled and common channels that are transmitted at constant power.
- The forward-link code channels are transmitted synchronously using orthogonal signature waveforms. In the absence of channel dispersion or multipath, the interference is largely caused by the signal received from the base stations other than the serving one.
- Soft handoff on the forward link requires simultaneous transmission from multiple base stations.

Therefore, far more variables must considered in order to estimate forward-link capacity. We must typically resort to Monte-Carlo simulations to achieve reasonably accurate results. Unfortunately, simulations provide little insight into key factors affecting capacity. Here, we choose to pursue a semi-analytical approach. Even though the method relies on many simplifying modeling assumptions, it leads to an insightful and treatable model.

10.2.3.1 *Forward-Link Outage Based on Power Constraint*

The forward-link capacity can be computed based on the finite amount of available power at the transmitter. We do not take into account the Walsh code availability in our analysis. Therefore, the obtained results should always be checked against the available Walsh codes. The actual capacity is the smaller of the two values. Keep in mind that that the forward link, like the reverse link, is an interference-limited system. Increasing the maximum transmit power does not increase the forward-link throughput because the interference is also proportionally increased.

Let us denote with o a fraction of the maximum transmit power allocated to the overhead channels. The residual power is then available for allocation to traffic channels. Assuming that all users are granted identical QoS, the power allocation among mobile stations is nonuniform because transmitted power to each mobile depends on the required received SNR, path loss, and the amount of interference received by the mobile station. An obvious outage constraint is that the sum of the transmitted power required by all active mobiles must be less than the residual available power. Denote with $\phi = (\text{F-FCH } E_c)/I_{or}$ the F-FCH transmitted chip energy relative to the total transmitted power spectral density, I_{or}. Note that the pilot overhead is not included in F-FCHE_c because the pilot is shared among all users. The outage constraint can then be written as

$$\sum_{k=1}^{K} v_k \phi_k \leq 1 - o \tag{10.71}$$

where K is the number of active F-FCH in a given cell[5] and v represents voice activity. Then, on average, the maximum number of traffic channels is equal to

$$K = \frac{1-o}{E\{v\}E\{\phi\}} \tag{10.72}$$

Assume that the base station transmitted power of each cell is constant. Then the relative traffic channel transmitted energy for the k-th user, $\phi_k = (\text{F-FCH } E_{c,k})/I_{or}$, is equal across all h handoff legs that belong to the active set of user k. Given the result in Eq. (10.34), we can express the traffic channel *SINR* per Rake finger as

---

5.   Note that the number of active F-FCHs is different from the number of active users. The two are different when accounting for soft/softer handoff overhead.

$$\left(E_b/N_t\right)_{i,k,j} = \phi_k \frac{W}{R_b} CINR_{i,k,j} \qquad (10.73)$$

where R_b denotes the full-rate F-FCH, $R_b = 9{,}600$ bps. After pilot-based combining, the traffic channel SNR can be computed as

$$\left(E_b/N_t\right)_k = \phi_k \frac{W}{R_b} CINR_k \qquad (10.74)$$

where $CINR_i$ is given by Eq. (10.37).

The inner-loop power-control mechanism adjusts the ϕ_k every power control group in order to compensate for changes in $\alpha_{k,j,i}$ to meet the target E_b/N_t. From Eq. (10.74), we can express, the relative traffic channel transmitted energy for the k-th user, ϕ_k, as

$$\phi_k = \frac{R_b}{W} \left(\frac{E_b}{N_t}\right)_k \left(\frac{1}{CINR_k}\right) \qquad (10.75)$$

Of interest is the average value of ϕ_k. $CINR_k$ depends on the handoff state, which impacts ϕ_k. Thus, it is convenient to express its average value as the weighted sum of its expectations conditioned on the handoff state:

$$E\{\phi_k\} = \sum_h E\{\phi_k | h\} P_h \qquad (10.76)$$

where P_h is the probability of the call being in h-way handoff. After dropping the user's index, we can write

$$E\{\phi | h\} = \frac{R_b}{W} E\{(E_b/N_t | h)(1/CINR | h)\} \qquad (10.77)$$

The average E_b/N_t is largely independent of the inverse of $CINR$ so that we can rewrite Eq. (10.77) as

$$E\{\phi | h\} \approx \frac{R_b}{W} E\{(E_b/N_t | h)\} E\{(1/CINR | h)\} \qquad (10.78)$$

Eq. (10.78) considers perfect inner power control that can invert the radio channel. Recall from Chapter 8 that perfect inversion of a single-path Rayleigh fading requires infinite power. In practice, due to the power constraints imposed by the base station and the finite inner-loop power-control rate and step size, the excess power is limited. It depends on the handoff state and the multipath profile, while its dependence on the short-term average of the interference to carrier ratio is much weaker. We obtain the excess transmit power conditioned on the handoff state, Γ_h, by simulating the effect of fast fading and power control. Hence, in order to account for the transmit power increase, the average of the inverse of SNR is expressed as

$$E\{1/CINR \mid h\} \approx \Gamma_h / E\{CINR \mid h\} \tag{10.79}$$

Eq. (10.79) is a convenient way to account for imperfect power control. We can now rewrite Eq. (10.78) as

$$E\{\phi \mid h\} = \frac{R_b}{W} E\{(E_b/N_t \mid h)\} \frac{\Gamma_h}{E\{CINR \mid h\}} \tag{10.80}$$

We have all the elements needed to compute the maximum number of traffic channels. Eq. (10.72) can be rewritten so that the maximum number of traffic channels is given as

$$K \approx \frac{1-\eta}{E\{v\} \sum_h P_h \dfrac{R_b}{W} E\{(E_b/N_t \mid h)\} \Gamma_h / E\{CINR \mid h\}} \tag{10.81}$$

From Eq. (10.81), the maximum number of users, N, can be found after the handoff reduction factor is accounted for:

$$N = \frac{K}{E\{h\}} \tag{10.82}$$

Forward-link capacity in terms of the maximum number of users can be estimated using Eqs. (10.81) and (10.82). However, to provide an acceptable outage probability, $\Pr\left(\sum_{k \in \text{cell}} \phi_k < 1-o\right) < \delta$, it is necessary to factor an additional *backoff factor*, ρ, into Eq. (10.81):

$$K \approx \frac{1-o-\rho}{E\{v\} \sum_h P_h \dfrac{R_b}{W} E\{(E_b/N_t \mid h)\} \Gamma_h / E\{CINR \mid h\}} \tag{10.83}$$

The backoff value depends on outage probability δ, mobile station velocity, and multipath profile, and it is hard to evaluate analytically. The backoff also depends on the actual number of users in the cell. The larger the number, the smaller the backoff. The Erlang capacity on the forward link is commonly computed using the Erlang B formula. Assuming a cell can support N users, given the arrival rate λ call/sec and the average call duration of $1/\mu$ sec, the probability of blocking a user is given by

$$\Pr_{outage} = \frac{(\lambda/\mu)^N / N!}{\sum_{n=0}^{N} (\lambda/\mu)^n / n!}$$
(10.84)

10.2.3.2 Example of Forward-Link Capacity Computation

In Table 10.11, we give exemplary settings for the variables affecting the forward-link capacity. For the purpose of this analysis, we assume that $\rho = 0.1$ and overhead due to pilot, paging, and sync channel is $o = 0.25$. We assume a low-velocity mobile station with a single-path Rayleigh fading channel, which is typical for urban environments. Relatively large values for Γ are appropriate for no handoff case (one-way) because of high excess transmit power (see Chapter 8). SMV mode = 1 is assumed.

Table 10.11 Exemplary Setting of Variables Affecting Forward-Link Capacity

Handoff state	P_h	$E\{CINR_h\}$	$E\{10\mathrm{Log}E_b/N_t\}$	Γ_h
1	60%	4 dB	5.5 dB	3.0 dB
2	30%	1 dB	5.5 dB	1.0 dB
3	10%	−1 dB	5.5 dB	0.5 dB

For the given parameters, the estimated capacity considering only power outage is $N = 45$ voice users per cell. However, assuming that the setting is valid for Radio Configuration 3 (RC3), the number of Walsh codes limits the cell capacity to 40. For RC 3, 61 Walsh codes of length 64 are available for voice traffic. After accounting for a soft handoff reduction factor of 1.5, the maximum number of users is limited to 40. For the probability of outage of 2%, the Erlang capacity is equal to 31. Note that for SMV modes 0 and 1, the forward-link voice capacity is typically smaller than the reverse-link capacity.

10.2.4 Forward-Link Capacity with Mixed Voice and F-SCH Data Traffic

To compute mixed voice and data capacity over F-SCH, we modify Eq. (10.83) to account for power consumed by F-SCH. For data traffic, F-SCH is commonly scheduled and not placed in handoff in order to prevent bursty data traffic from uncontrollably using forward-link power. All data users, however, do have an F-FCH active, which is necessary for the Layer 3 signaling and Radio Link Protocol (RLP) control traffic. F-FCH is also used for data traffic regardless whether or not F-SCH is scheduled. Assuming that a fraction, υ, of the forward-link power is used for scheduled F-SCH data traffic, the maximum number voice traffic channels per cell can be estimated as

$$K \approx \frac{1 - \eta - \rho - \upsilon - K_d \sum_h P_h \frac{R_b}{W} E\left\{\left(E_b/N_t \mid h\right)\right\} \Gamma_h / E\left\{CINR \mid h\right\}}{E\left\{v\right\} \sum_h P_h \frac{R_b}{W} E\left\{\left(E_b/N_t \mid h\right)\right\} \Gamma_h / E\left\{CINR \mid h\right\}} \tag{10.85}$$

where K_d is the number of simultaneous data users. The actual number of voice users is then computed after accounting for soft handoff, using Eq. (10.82).

To estimate the forward data capacity, we assume a scheduled F-SCH and a round-robin scheduler. The average data throughput per cell can then be estimated as the mean data rate achievable with the allocated fraction of power. The average rate, $\overline{R}$, can be approximated as

$$\overline{R} \approx \frac{\upsilon W}{E\left\{E_b/N_t\right\} \Gamma / E\left\{CINR\right\}} \tag{10.86}$$

where we assume that all users undergo the same radio channel and, therefore, require the same E_b/N_t. Each user gets the throughput proportional to its average $CINR$.

In Table 10.12 we give exemplary settings for the variables affecting the forward-link data capacity. For the purpose of this analysis, we assume the same settings for voice as in Section 10.2.3.2 and that 30% of base station power is allocated to data, $\upsilon = 0.3$ and $K_d = 4$. Scheduled F-SCH is not placed in handoff. The required F-SCH E_b/N_t is smaller than the F-FCH E_b/N_t because the former is typically operated at higher FER. Moreover, the F-SCH employs turbo codes, which provide the higher coding gain. The F-SCH is normally power-controlled separately from the F-FCH. For the given parameters, the estimated capacity is $N = 17$ voice users per cell, or 11 Erlangs. The average aggregated scheduled F-SCH data throughput is 146 kbps. In addition, each of the four users gets 9.6 kbps over the F-FCH.

Table 10.12 Exemplary Setting of Variables Affecting Forward-Link Data Capacity*

Handoff state	$E\left\{CINR_h\right\}$	$E\left\{10 \mathrm{Log} E_b/N_t\right\}$	Γ_h	υ
1	0 dB	2.5	3.0	0.3

10.2.5 Forward-Link Capacity with F-PDCH Data Traffic

To estimate the cell capacity when the F-PDCH is used, it is crucial to estimate the multiuser diversity gain. In this section we estimate that gain for the proportional fair algorithm described in Chapter 9 by following the approach introduced in [6].

Recall from Chapter 9 that at time t, the k-th mobile's priority metric, $M_k(t)$, is given by

$$M_k(t) = \frac{R_k(t)}{S_k(t)} \quad k = 1, 2, ..., K \tag{10.87}$$

where there are K users in the cell, $R_k(t)$ is the rate supportable by the user at that time, and $S_k(t)$ is the throughput computed over a time window sufficiently long to take advantage of the channel fluctuations but short enough to control delay jitter. A mobile station with the largest metric is allocated all available BTS resources, as discussed in Chapter 9.[6] Let us now approximate the proportional fair priority metric of user k with

$$M_k(t) \approx \frac{CINR_k(t)}{S_k} \quad k = 1, 2, ..., K \tag{10.88}$$

where $CINR_k(t)$ is the maximum available carrier to interference and noise ratio after Rake combining for user k at time t, and S_k is its stationary throughput.[7] The approximation is accurate as long as the rate is a linear and continuous function of the $CINR$ ratio with no limit. This can be assumed for a single modulation and coding scheme. However, it may not be as accurate an assumption for adaptive coding and modulation, as used over F-PDCH. Nevertheless, this approach leads to a good first-order approximation of the proportional fair algorithm.

Let us now decompose $CINR_k(t)$ into static and time-variant components as

$$CINR_k(t) = a_k b_k(t) \tag{10.89}$$

where a_k is a static component due to distance from the serving BTS, shadowing, and interference from other BTSs and that $b_k(t)$ is the time-varying component due to Rayleigh or Rician fading. Assume that at time t, $b_k(t)$ are independent and identically distributed random variables with unit mean.

We can now claim that under the assumptions that the rate is linearly proportional to the $CINR$, the interference from other base stations is constant, and the fading statistics are indepen-

---

6.   The standard actually allows selecting up to two users. However, as shown in [6], selecting a single user maximizes air-link efficiency. Two users may be selected only if there is not enough data in the buffer of the user with the highest priority metric to fill the entire physical layer encoder packet.
7.   An infinite time window, $W \rightarrow \infty$, is assumed.

dent among users—asymptotically, all users transmit the same fraction of time. To support this claim, assume that for all users the ratio of the throughput and the average *CINR* is a constant:

$$\frac{S_k}{a_k} = c \qquad (10.90)$$

From Eqs. (10.90), (10.89), and (10.88), we can rewrite the priority metric as

$$M_k(t) = \frac{b_k(t)}{c} \qquad (10.91)$$

which means that all users transmit the same amount of time. The assumption that the throughput to average SNR is not constant among all users leads to a contraction. For example, if for two users *i* and *j* we can write

$$\frac{S_i}{a_i} > \frac{S_j}{a_j} \qquad (10.92)$$

it means, referring to Eq. (10.91), that the *j*-th user priority is larger than *i*-th and that the *j*-th user is scheduled more often than user *i*. The last statement implies that

$$\frac{S_i}{a_i} < \frac{S_j}{a_j} \qquad (10.93)$$

which is a contradiction.[8] This can be viewed as leveling of the priority metric, as illustrated in Figure 10.8 on a two-user example.

Note that even when the rate is not a linear function of the *CINR*, for the priority metric given by Eq. (10.88), Eq. (10.90) is still valid—only in this case, not all users will necessarily transmit the same fraction of time.

Given Eq. (10.91), at any point in time *t*, the priority metrics are random numbers dependent only on the fading channel characteristics. Denote with x_k the *k*-th user priority metric at time *t*. We can write

$$x_k = b_k(t) \qquad (10.94)$$

---

8.   This is proved as a special case of a more general result in [7].

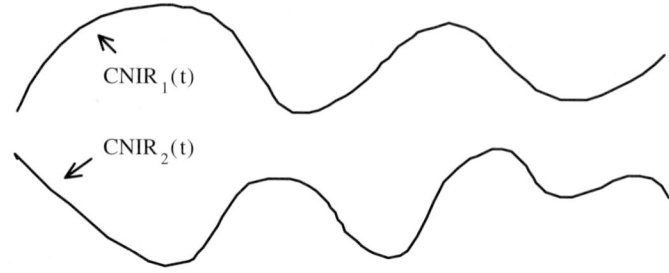

C/I of two users before leveling

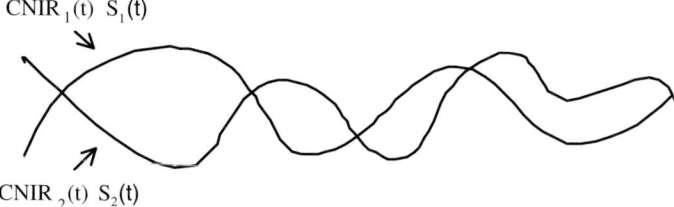

Priorities after leveling with throughputs

Figure 10.8 Leveling the priority metric in a proportional fair algorithm.

where x_k is equal to the priority metric of user k, because constant c is irrelevant. The x_k's are assumed to be independent and identically distributed random variables. Since the interference from the neighboring base stations is assumed constant, we can write $x_k = ||\alpha_k||^2$, where α_k represents the complex channel gain of the k-th user. To estimate the multiuser diversity gain, let us compute the ratio of the mean x_k and mean of x_k when selected for transmission. Consider ordered x_k as

$$x_1 \leq x_2 \leq \cdots \leq x_K \tag{10.95}$$

Then, the expected value for the largest among them, x_K, is given as

$$\mu_K = E[x_K] = K \int_0^\infty x [P(x)]^{K-1} p(x) dx \tag{10.96}$$

where $p(x)$ and $P(x)$ denote probability density and cumulative probability density functions, respectively.

For Rayleigh distributed channel gain, x is a central Chi-squared distributed random variable with two degrees of freedom, or an exponentially distributed random variable. Assuming unit mean, the distribution and the cumulative distribution are given by

$$p(x) = e^{-x}; \quad P(x) = 1 - e^{-x} \tag{10.97}$$

The mean for the largest among K is then

$$\mu_K = \sum_{k=1}^{K} \frac{1}{k} \tag{10.98}$$

Figure 10.9 shows μ_K as a function of K. It is interesting to notice that the average squared channel gain more than doubles when $K = 4$. This means that if there are only four mobile stations in the cell and they all have the same average *CINR*, but they fade independently and fading is Rayleigh, assuming that the scheduler is able to select the mobile with highest current *CINR*, the average *CINR* for the served mobile stations is more than doubled. This gain is commonly called multiuser diversity gain.

The channel variability positively impacts the multiuser diversity gain, so the gain is smaller for Ricean than for Rayleigh fading. It can be computed numerically using the same equation. For

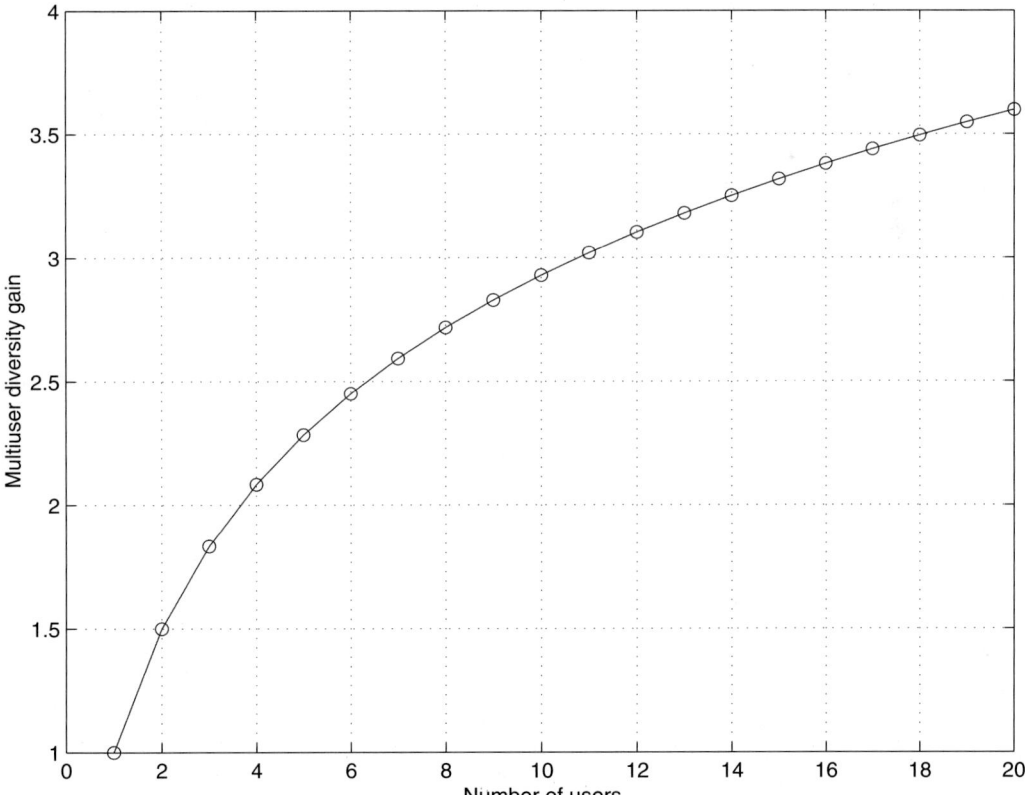

Figure 10.9 Multiuser diversity gain for Rayleigh fading.

Ricean fading, x is a noncentral Chi-squared distributed random variable with two degrees of freedom. Table 10.13 compares the average squared channel gain, for $K = 10$ with Ricean factor K as a parameter. The multiuser diversity gain does not exist in nonfading channels.

Table 10.13 Average Squared Channel Gain for Ricean Channel

Ricean Factor, K (Linear Unit)	$\mu_K = 10$
(Rayleigh)	2.93
2	2.35
5	1.98
10	1.72

The presented analysis is applicable only for low-velocity mobile stations when link adaptation is successfully operating. If the mobile station is moving at high velocity, the link adaptation is impaired by data-rate selection errors that occur when using obsolete channel quality information. In such a case, the system heavily relies on the hybrid automatic repeat request (H-ARQ) type II technique to recover from errors (see Chapter 9). Refer to [7] for the analysis of the proportional fair algorithm in the case of mobile stations experiencing a mix of channel conditions.

10.2.5.1 *Example of F-PDCH capacity*

To get a numerical example for the F-PDCH capacity, consider the scenario depicted in Table 10.14. There are 10 users in a given cell, and their mean *CINR* ratio is representative of the *CINR* distribution obtained by simulation in Section 10.1.6.3. The data rates that can be achieved for the given *CINR*s can be obtained by simulation. For the purpose of this exercise, assume the values summarized in Table 10.15. The values are obtained by interpolation from the link-level curves adopted in [1], assuming FER = 1%. We consider the case when 30% of base station transmit power is consumed by overhead channels. The table provides the F-PDCH data rates as a function of the *CINR* quantized in 2-dB steps. The data rates between the given points can be approximated by means of a linear interpolation.

Table 10.14 Average SNR for Each User

User	1	2	3	4	5	6	7	8	9	10
CINR [dB]	−4	−2.5	−1	0	1	2.5	4	5.5	7.5	10

Table 10.15 Rate as a Function of SNR Assuming 30% of Power Is Consumed by Overhead Channels

CINR [dB]	−4	−2	0	2	4	6	8	10	12
Rate [kbps]	245	385	614	865	1225	1550	1950	2450	3000

The cell capacity assuming equal-time round-robin algorithm, described in Chapter 9, can be approximated by averaging the data rates supported by the corresponding *CINR*s. For the example summarized in Table 10.14 and Table 10.15, the cell throughput using the round robin scheduling algorithm is 1034 kbps.

The cell capacity of the proportional fair algorithm can be approximately estimated using the framework described in the previous section. The scheduling algorithm effectively increases the average *CINR* of each user due to multiuser diversity gain by a certain amount dependent of the number of users and the channel type. Assuming a Rayleigh fading channel and 10 users per cell, the multiuser diversity gain is approximately equal to 2.93 or 4.66 dB. In Table 10.16, we summarize the effective *CINR* ratio when users are scheduled with the proportional fair algorithm. The throughput can now be estimated by averaging out the rates achieved by the adjusted average *CINR*. From Table 10.15 and Table 10.16, we estimate the cell throughout to be 1775 kbps.

Table 10.16 Average CINR for Each User

User	1	2	3	4	5	6	7	8	9	10
CINR [dB]	0.5	2	3.5	4.5	5.5	7	8.5	10	12	13

Note: The impact of self-interference is negligible to all users except user 10. Due to self-interference, the fading distribution is no longer Rayleigh, but in this example we approximate as if it were Rayleigh. We simply clip the CINR at 13 dB to account for self-interference.

The estimated numbers should be treated with caution. As we said in the previous section, they assume efficient link adaptation, which is only possible for low-velocity mobile stations effectively moving at pedestrian speeds. The results represent a first-order approximation. Note also that they assume full buffers. The cell throughput with the traffic models, described in Chapter 3 is lower. One of the reasons is that the user with the best radio condition may not have enough data to fill the entire physical layer frame. In this scenario the data is padded to fit into the required frame size, and the padded bits do not count toward the throughput. This is a scenario when transmitting to two users instead of one is beneficial. There are also periods when the radio resources are completely torn down and the user is in dormant state because there is no data traffic at all. These events are random, which means that the number of active users in a cell is also random. The movement of the users and the cell reselection contribute to the randomness too. And as we have seen, the number of users in the cell impacts the multiuser diversity gain and therefore the cell capacity.

10.3 Cell Coverage

In this section we discuss the framework to compute forward- and reverse-link budgets. We first explain the methodology and then present some examples.

Link budgets are an important tool for network planning that allows estimation of the maximum allowable path loss on a given link. Given the maximum allowable path loss and the channel model, we can compute the maximum cell radius. When designing for capacity-limited systems, however, the actual cell size may be limited by the cell capacity for a given subscriber density. When the link budget is computed, the propagation loss does not include shadowing, which is separately accounted for with a shadowing margin.

The maximum allowable path loss is computed as a difference between the transmitter Effective Isotropic Radiated Power (EIRP) and the receiver sensitivity, S:

$$L_p [\text{dB}] = \text{EIRP}[\text{dBm}] - S[\text{dBm}] \tag{10.99}$$

where dBm refers to decibels relative to 1 mW. The transmitter EIRP can be computed as the sum of transmitted power, P_{TX}, and transmitter antenna gain, G, reduced by the cable and other losses, L:

$$\text{EIRP}[\text{dBm}] = P_{TX}[\text{dBm}] + G[\text{dBi}] - L[\text{dB}] \tag{10.100}$$

where dBi refers to the decibel gain relative to isotropic antenna. In the logarithmic domain, the receiver sensitivity is the sum of thermal noise power spectral density, N_0; the required information bit signal to thermal noise ratio, E_b/N_0; and the data rate, R:

$$S[\text{dBm}] = N_0[\text{dBm/Hz}] + \frac{E_b}{N_0}[\text{dB}] + 10\log_{10}(R)[\text{dBHz}] \tag{10.101}$$

The sensitivity can also be expressed as a sum of total interference, N_t, and the required information bit signal to total interference, E_b/N_t, and data rate:

$$S[\text{dBm}] = N_t[\text{dBm/Hz}] + \frac{E_b}{N_t}[\text{dB}] + 10\log_{10}(R)[\text{dBHz}] \tag{10.102}$$

Yet another way to express the receiver sensitivity, particularly useful for the forward-link budget, is a logarithmic sum of thermal noise, traffic channel chip energy to thermal noise ratio I_{or} / N_0, and channel bandwidth or chip rate, R_c:

$$S[\text{dBm}] = N_0[\text{dBm/Hz}] + \frac{I_{or}}{N_0}[\text{dB}] + 10\log_{10}(R_c)[\text{dBHz}] \qquad (10.103)$$

10.3.1 Reverse-Link Budget

In this section we give an example of reverse-link budget computation. A single base station transmit antenna and a single mobile station receiver antenna are assumed. Both voice and data users are considered. The results are summarized Table 10.17.

Row (1) indicates the chip rate, which for single carrier systems is 1.2288 Mcps. The maximum transmitter power at the mobile station is limited to 200 mW, as shown in row (2). In the logarithmic domain, that power level is equal to 23 dBm. The data-bearing channel, R-FCH or R-SCH for voice and data users respectively, is operated at the data rate given in row (3). For voice services, the R-FCH data rate is equal to that of the full vocoder rate, or 9.6 kbps. For data services, the R-SCH data rate is assigned by the base station, and in this example we assume it is equal to 38.4 kbps. The total reverse transmit power is split between the data-bearing channel(s) and the applicable overhead channels. For voice users, the reverse channels consist of the R-FCH and the R-PICH. The R-FCH to total reverse channel transmit power ratio is given by

$$\frac{\text{R-FCH Tx power}}{\text{Total traffic Tx power}}[\text{dB}] = 10\log_{10}\left(\frac{\rho_{\text{R-FCH}}}{1 + \rho_{\text{R-FCH}}}\right) \qquad (10.104)$$

where $\rho_{\text{R-FCH}}$ denotes the R-FCH to R-PICH transmit power ratio. For full rate transmission, $\rho_{\text{R-FCH}} = 3.75$ dB (2.37 in linear).

For data users, we assume that the data transfer is bidirectional and the F-PDCH is assigned on the forward direction. Thus, the reverse traffic channels of data users comprises the R-SCH and R-FCH (carrying signaling and user data). The R-FCH in this case is considered to be transmitted at full rate. We consider a case when F-PDCH is assigned, which implies that the Reverse Channel Quality Indicator Channel (R-CQICH) is active. The Reverse Acknowledgement Channel (R-ACKCH) is neglected because it is a discontinuous transmission channel with a low-duty cycle. The R-SCH to total reverse channels transmit power ratio is then given by

$$\frac{\text{R-SCH Tx power}}{\text{Total traffic Tx power}}[\text{dB}] = 10\log_{10}\left(\frac{\rho_{\text{R-SCH}}}{1 + \rho_{\text{R-SCH}} + \rho_{\text{R-FCH}} + \rho_{\text{R-CQICH}}}\right) \qquad (10.105)$$

where $\rho_{\text{R-SCH}}$ and $\rho_{\text{R-CQICH}}$ denote the R-SCH to R-PICH and average R-CQICH to R-PICH transmit power ratios respectively. In this case, it is $\rho_{\text{R-SCH}} = 7$ dB (5 in linear), $\rho_{\text{R-FCH}} = 2.5$ dB (1.77 in linear), and on average $\rho_{\text{R-CQICH}} \approx \angle 7.25$ dB (0.18 in linear).

In row (4) we give the values for the fraction of total reverse traffic channel power that is consumed by the R-FCH and R-SCH according to Eq. (10.104) and (10.105). Row (5) shows the traffic channel transmit power, computed from row (2) and (4). The mobile station antenna gain is shown in row (6), and the EIRP is computed in row (7). It is common to consider a single transmit antenna at the mobile station. The transmit power increase is not significant at the mobile station side because the base station receiver commonly employs dual receive antenna diversity. However, we still account for it in row (8).

The receiver antenna gain is accounted for in row (9), and the cable and connector losses in row (10). The BTS receiver noise figure is typically 5 dB, as shown in row (11). The thermal noise density is computed in row (12), and the desired rise over thermal level is given in row (13). In row (14) we compute the total interference, while row (15) shows the required per-antenna E_b/N_t. At this point, all the necessary components are given, and we can compute the receiver sensitivity. In row (16), receiver sensitivity is computed using Eq. (10.102).

Before computing the maximum path loss, we need to account for the building or vehicle penetration losses, if any, and that is given by row (17). Also, the lognormal shadowing margin needs to be considered, which can be computed from rows (18) and (19). In this link budget, we assume the mobile station is located outdoors, and therefore the penetration loss is equal to 0. A typical value for in-building and in-vehicle penetration loss is 20 dB and 8 dB, respectively.

Recall from Section 10.1.2 that the signal is subject to lognormal shadowing with standard deviation σ. For a proper system design, it is necessary to account for shadowing with an appropriate margin.[9] In the case of hard handoff, the system is designed so that the shadowing loss, ζ, is less than a certain amount, δ, with a given probability, P_{out}. Such probability is thus called the outage probability. Its complement, $1 - P_{out}$, represents the confidence level that a mobile station located at the cell edge will be able to close the reverse link. A typical value for Pout is equal to 0.1. Hence, we can express P_{out} as

$$P_{\text{out}} = Q\left(\frac{\delta}{\sigma}\right) \qquad (10.106)$$

Now we can compute the margin, d, given s and P_{out}. In this link budget, we assume $\sigma = 8.9$ dB. For $P_{\text{out}} = 0.1$, $\delta = 11.2$ dB, as indicated in row (23). In soft handoff the required margin is substantially reduced. For a mobile station on the boundary of the two cells in soft handoff, the outage probability can be expressed as

---

9.   Short-term fading is taken into account through transmit power increase and when computing $\hat{I}_{or}/N_o$. Therefore, an additional margin is not necessary.

$$P_{\text{out}} = \Pr\left[\max\left(10^{\varsigma_1/10}, 10^{\varsigma_2/10}\right) < 10^{\delta/10}\right]$$
$$= \Pr\left[\max\left(\varsigma_1, \varsigma_2\right) < \delta\right] \tag{10.107}$$

where ζ_1 and ζ_2 are correlated normal random variables, as defined in Section 10.1.2. Using the same numerology as in Section 10.1.2, we can rewrite Eq. (10.107) as

$$P_{\text{out}} = \Pr\left[\max\left(\xi_1, \xi_2\right) < \frac{\delta - a\xi}{b}\right] \tag{10.108}$$

where ξ_1 and ξ_2 are independent normal random variables, and as we said in Section 10.1.2, it is common to assume that the correlation coefficients $a = b = 1/\sqrt{2}$. Eq. (10.108) can be evaluated numerically. In Table 10.18, we show numerical examples for the soft handoff margin, δ for $P_{\text{out}} = 0.1$, and σ as a parameter. The soft handoff gain refers to the difference between the hard and soft handoffs margins. The path loss is computed in row (20) based on Eq. (10.99). The transmit power increase is also accounted for. The maximum cell radius is the one that corresponds to the maximum tolerable path loss and is computed using Eq. (10.5). In this link budget we assume the urban propagation model and obtain the maximum cell radius in row (21).

Table 10.17 Reverse-Link Budget

Row	Item	Voice	Data	Equation
(1)	Chip rate [Mcps]	1.2288	1.2288	Input
(2)	Maximum transmitter power [dBm]	23.0	23.0	Input
(3)	Data rate [kbps]	9.6	38.4*	Input
(4)	Fraction of power used for traffic channel [dB]	−1.5	−2.0	Input
(5)	Traffic channel transmit power [dBm]	21.5	21.0	(2) + (4)
(6)	Transmitter antenna gain [dBi]	1.0	1.0	Input
(7)	EIRP per traffic channel [dBm]	22.5	22.0	(5) + (6)
(8)	Transmit power increase [dB]	0.5	0.5	Input
(9)	Receiver antenna gain [dBi]	17.0	17.0	Input
(10)	Cable and connector losses [dB]	−1.0	−1.0	Input
(11)	Receiver noise figure [dB]	5.0	5.0	Input

Table 10.17 Reverse-Link Budget (continued)

Row	Item	Voice	Data	Equation
(12)	Thermal noise density, N_0 [dBm/Hz]	−169.0	−169.0	−174+(11)
(13)	Rise over thermal, I_0 / N_0 [dB]	5.0	5.0	Input
(14)	Noise plus interference density, I_0 [dBm]	−164.0	−164.0	(12) + (13)
(15)	Required traffic channel per antenna $E_b / N_t \approx E_b / I_0$ [dB]	−0.5	−1.0	Input (channel dependent)
(16)	Receiver sensitivity [dBm]	−120.5	−115.0	(14) + (15) + 10$\log_{10}$ (3)
(17)	Penetration loss [dB]	0.0	0.0	Input
(18)	Log normal fade margin [dB]	11.2	11.2	Input
(19)	Soft handoff gain [dB]	5.0	5.0	Input
(20)	Maximum path loss [dB]	146.5	140.0	(7) − (8) − (16)
(21)	Maximum range [km]	2.34	1.52	Eq. (10.5)

* Excluding R-FCH.

Table 10.18 Reverse-Link Soft Handoff Gain

σ	Hard handoff margin [dB]	Reverse-link soft handoff gain [dB]	Reverse-link handoff margin [dB]
6	7.7	3.2	4.5
8	10.2	4.2	6.0
10	12.8	5.3	7.5
12	15.4	6.4	9.0

10.3.2 Forward-Link Budget

In this section we gave an example of forward-link budget computation. A single receiver antenna at the mobile station is assumed. We consider voice users only, data users only, and mixed voice and data users case. The results are summarized in Table 10.19. For data users, we consider only the F-PDCH. However, the presented methodology applies to the F-SCH as well.

The chip rate for a single carrier system is 1.2288 Mcps, and that is indicated in row (1). The interference from adjacent cells is represented with I_{oc}, and it is shown in row (2). Voice results assume operation is two-way soft handoff, while data over F-PDCH does not. The adjacent cell interference is naturally smaller in soft handoff. The values given in row (2) are typical assumptions for network planning purposes.

The maximum base station transmit power is assumed to be equal to 20 W, which corresponds to 43 dBm, as indicated in row (3). Not all power is available for traffic. In rows (4) through (7) we indicate what fraction of power is available for which channel. The overhead due to overhead channels is typically 25%, and it is common that a single voice call cannot consume more than 10% of the total base station power. The Walsh code availability is not taken into account in this link-budget computation. It is assumed that the use of Walsh codes is proportional to consumed power, and therefore the system is not Walsh code–limited.

In row (8) we account for the cable and connector losses, while row (9) shows transmitter antenna gain. The total and traffic channel EIRP are given in rows (10) and (11), respectively. Row (12) indicates the transmit power increase due to 800 bps inner-loop power control. The transmit power increase could be significant on the forward link because both the base station transmitter and mobile station receiver employ a single antenna.

On the mobile station side, the parameters of interest are the receiver antenna gain and the receiver noise figure, given in rows (13) and (14), respectively. The thermal noise power spectral density, based on the receiver noise figure, is computed in row (15).

In the voice-only case, as presented in row (16), the required user data rate is 9.6 kbps. For the data traffic, the requirement is for the throughput at the cell edge, as shown in row (17). We assume four data users located at the cell edge. This implies that the throughput per user is four times less than the cell throughput. For the mixed voice and data case, we assume two data users at the cell edge.

Besides the thermal noise and bandwidth, the receiver sensitivity is impacted by I_{or}/N_0, which is shown in row (19). It is most accurate to do network simulation and obtain the required I_{or}/N_0 given the desired FER and maximum transmitted power. However, the network simulations are tedious, and because the required E_b/N_t values are well understood, we use them for the link budget. I_{or}/N_0 is then estimated by Eq. (10.74). Row (18) shows the required E_b/N_t. The traffic channel I_{or}/N_0 can be estimated from Eq. (10.74). We approximate the channel gains, $\|\alpha_{i,k,j}\|^2$, with the average values, $E[\|\alpha_{i,k,j}\|^2]$, where index j denotes the path, index i the cell, and index k the user. The approximation allows us to express I_{or}/N_0 as a function of E_b/N_t. For a frequency nondispersive channel and two-way soft handoff with equal strength of the paths,

$I_{1,i} = I_{1,i} = I_{or}/2$ and $E[\|\alpha_{i,k,j}\|^2] = 1$, where $j = 1$ and, $i = 1,2$. After dropping the user index, k, Eq. (10.74) can be rewritten as

$$E_b/N_t = \phi \frac{W}{R_b} \frac{I_{or}}{N_0 + I_{oc} + I_{or}/2} \tag{10.109}$$

I_{or}/N_0 can now be expressed as

$$\frac{I_{or}}{N_0} = \frac{1}{\phi \dfrac{W/R_b}{E_b/N_t} - \dfrac{I_{oc}}{I_{or}} - \dfrac{1}{2}} \tag{10.110}$$

and row (19) can be estimated from row (18) using Eq. (10.110). Since we are actually interested in $I_{1,i}/N_0$, we subtract 3 dB from Eq. (10.110). Note that Eq. (10.110) is only valid for the non-dispersive channels in two-way soft handoff with equal path strength. However, using the same approximation, a similar equation can be derived for any channel. For example, in the case of J equal strength resolvable paths in two-way soft handoff

$$\frac{I_{or}}{N_0} = \frac{1}{\phi \dfrac{W/R_b}{E_b/N_t} - \dfrac{I_{oc}}{I_{or}} - \dfrac{1}{2} - \dfrac{J-1}{J}} \tag{10.111}$$

For data traffic, of interest is no soft handoff. In this case, for a frequency nondispersive channel, I_{or}/N_0 can be expressed as

$$\frac{I_{or}}{N_0} = \frac{1}{\phi \dfrac{W/R_b}{E_b/N_t} - \dfrac{I_{oc}}{I_{or}}} \tag{10.112}$$

The multiuser diversity gain is applicable only for F-PDCH, and it is given in row (20). The multiuser diversity gain values for a Rayleigh fading channel, estimated in Section 10.2.5, are considered.

At this point, we have all the components necessary to compute the receiver sensitivity, which is shown in row (21). Before computing the maximum path loss, similarly as on the reverse link, we need to account for the building or vehicle penetration losses, if any, given by row (22), and the lognormal shadowing margin that can be computed from rows (23) and (24). In this link budget, we assume the mobile user is outside and, therefore, penetration loss is equal to 0. We also need to consider that in soft handoff, the signal at the mobile station is received from two sources. This is taken into account in row (25).

The soft handoff margin is computed differently on the forward link. For a mobile station on the boundary of two cells in forward-link soft handoff, the outage probability can be defined as

$$P_{out} = \Pr\left(10^{\varsigma_1/10} + 10^{\varsigma_2/10} < 10^{\delta/10}\right) \tag{10.113}$$

where ς_1 and ς_2 are correlated normal random variables, as defined in Section 10.1.2. Using the same numerology as in Section 10.1.2, we can rewrite Eq. (10.107) as

$$P_{out} = \Pr\left[\left(10^{b\xi_1/10} + 10^{b\xi_2/10}\right) < 10^{(\delta - a\xi)/10}\right] \tag{10.114}$$

where ξ_1 and ξ_2 are independent normal random variables and $a = b = 1/\sqrt{2}$. Eq. (10.108) can be evaluated numerically.

In Table 10.20, we show numerical examples for the soft handoff margin, δ for $P_{out} = 0.1$, and σ as a parameter. In this link budget, we assume $\sigma = 8.9$ dB. The soft handoff gain is the difference between the hard and soft handoffs margins. The path loss is computed in row (20), based on Eq. (10.99). We also consider the transmit power increase. The maximum cell radius is computed for the urban propagation model, and it is shown in row (21).

Table 10.19 Forward-Link Budget

Row	Item	Voice	Data only	Data with voice	Equation
(1)	Chip rate [Mcps]	1.2288	1.2288	1.2288	Input
(2)	I_{oc} at cell edge	$0.25 I_{or}$	$1.25 I_{or}$	$1.25 I_{or}$	Input
(3)	Maximum power [dBm]	43.0	43.0	43.0	Input
(4)	Fraction of power used for overhead channels	0.25	0.3	0.3	Input
(5)	Fraction of power used for voice	0.75	0.0	0.35	Input
(6)	Fraction of power used for data	0.0	0.7	0.35	$1 - (4) - (5)$
(7)	Fraction of power used for traffic channel	0.1	0.7	0.35	Input
(8)	Cable and connector losses [dB]	2.0	2.0	2.0	Input
(9)	Transmitter antenna gain [dBi]	17.0	17.0	17.0	Input

Table 10.19 Forward-Link Budget (continued)

Row	Item	Voice	Data only	Data with voice	Equation
(10)	Transmitted EIRP [dBm]	58.0	58.0	58.0	$(3) + (9) - (8)$
(11)	Traffic channel EIRP [dBm]	48.0	56.5	53.5	$(10) + 10\log_{10}(7)$
(12)	Transmit power increase [dB]	1.0	N/A	N/A	Input (channel dependent)
(13)	Receiver antenna gain [dBi]	−1.0	−1.0	−1.0	Input
(14)	Receiver noise figure [dB]	10.0	10.0	10.0	Input
(15)	Thermal noise density, N_0 [dB/Hz]	−164.0	−164.0	−164	$-174 + (14)$
(16)	Voice data rate [kbps]	9.6	N/A	N/A	Input
(17)	Data throughput at cell edge [kbps]	N/A	307.2	153.6	Input
(18)	Required E_b / N_t [dB]	5.5	1.5	1.5	Input (channel dependent)
(19)	Required I_{or} / N_0 [dB]	−4.5	1.3	1.3	Eq. (10.110) for voice; Eq. (10.112) for data
(20)	Multiple user gain [dB]	N/A	3.0	1.7	Input (channel dependent)
(21)	Receiver sensitivity [dBm]	−107.6	−104.8	−103.5	$10\log_{10}(1) + (15) + (19) - (20)$
(22)	Penetration loss [dB]	0.0	0.0	0.0	Input
(23)	Log normal fade margin [dB]	11.2	11.2	11.2	Input
(24)	Soft handoff gain* [dB]	6.2	5.0	5.0	Input
(25)	Power adjustment for soft handoff [dB]	3.0	0.0	0.0	Input
(26)	Maximum path loss [dB]	151.6	154.1	152.8	$(11) - (12) + (13) - (21) - (23) + (24) + (25)$
(27)	Maximum range [km]	3.26	3.85	3.53	Eq. (10.5)

*   In the case of data only or mixed voice and data users, we account for the F-PDCH selection diversity gain arising from the cell-switching procedure. Such diversity gain is akin to that achieved with soft handoff, albeit smaller.

Table 10.20 Forward-Link Soft Handoff Gain

σ	Hard handoff margin [dB]	Forwardslink soft handoff gain [dB]	Forward-link handoff margin [dB]
6	7.7	4.9	2.8
8	10.2	5.7	4.5
10	12.8	6.4	6.2
12	15.4	7.5	7.9

The obtained maximum radius from the forward-link budget is normally larger than the radius obtained from the reverse-link budget. The actual maximum cell radius is then the minimum of the two. The radius may be further reduced after accounting for the subscriber density and cell capacity.

REFERENCES

[1] Derryberry, R. T., "1xEV-DV Evaluation Methodology." *Contribution to the 3GPP2 standard committee TSG-C number 3GPP2-C30-20030616-043R1*, June 2003.

[2] DeJaco, Andrew, "SMV Capacity Increase." *Contribution to the 3GPP2 standard committee TSG-C number 3GPP2-C11-20001016*, October 16, 2000.

[3] "TIA/EIA-IS-2000.2 Physical Layer Standard for cdma2000® Spread Spectrum Systems Release C," *Telecommunications Industry Association*, 2002.

[4] Viterbi, Andrew, *CDMA Principles of Spread Spectrum Communications*, Addison-Wesley Wireless Communications Series, Reading, Massachusetts, 1995.

[5] Viterbi, Andrew, and Audry Viterbi, "Erlang Capacity of a Power Controlled CDMA System," *IEEE Journal on Selected Areas In Communications*, 11(6), August 1993.

[6] Holtzman, Jack, "CDMA Forward Link Waterfilling Power Control." *In Proc. of IEEE VTC*, October 2000.

[7] Holtzman, Jack, "Asymptotic Analysis of Proportional Fair Algorithm." *In Proc. of IEEE PIMRC*, October 2001.

cdma2000 End-to-End Network Operations

Throughout this book we discussed the operation of the cdma2000 cellular network mainly from a radio access network (RAN) perspective and focused on the mobile station and base station interoperability procedures. We also introduced the network architecture and the services that the end user can access using his or her wireless terminal. We find it useful to conclude the book with this chapter that summarizes how these services are realized throughout the various network interfaces. The interworking between the network elements is described in a vast body of specifications. When considered in isolation, the significance of these interworking procedures, their interactions, and how they can provide an end-to-end service is sometimes obscured. Thus, by assembling all the network procedures in a single, end-to-end view of the user services, it is possible to fully appreciate the significance of each network element and how such services are realized.

We focus on the fundamental network operations and services: location state management, basic call processing, intersystem handoffs, short message service (SMS) delivery, packet data services, concurrent voice and data services, and Voice over Internet Protocol (VoIP). We use high-level flow diagrams because they are compact representations of complex network operations. However, these flow diagrams should be regarded as examples only. Failures and/or alternative scenarios are not described, since a high-level flow diagram cannot capture all aspects of the underlying protocols.

11.1 Location and State Management

Among the basic functions supported by automatic roaming are service qualification, mobile station location update, and state management. Service qualification functions are used to inform the visited system of the features and services that the roaming subscriber can and cannot access. Mobile station location and state update procedures are used to create and/or update the record corresponding to the roaming subscriber in the vaster location register (VLR) and home location register (HLR). Knowledge of current mobile station location and its state enables, for example, call delivery when roaming and specialized call treatments. Hereafter, we limit the discussion to basic location update and state management procedures.

11.1.1 Location Update

In the example, illustrated in Figure 11.1, the mobile station is in idle state when it performs an idle handoff to a base station in the service area of a new mobile switching center (MSC) and sends a registration message. The new serving MSC informs the HLR in the mobile

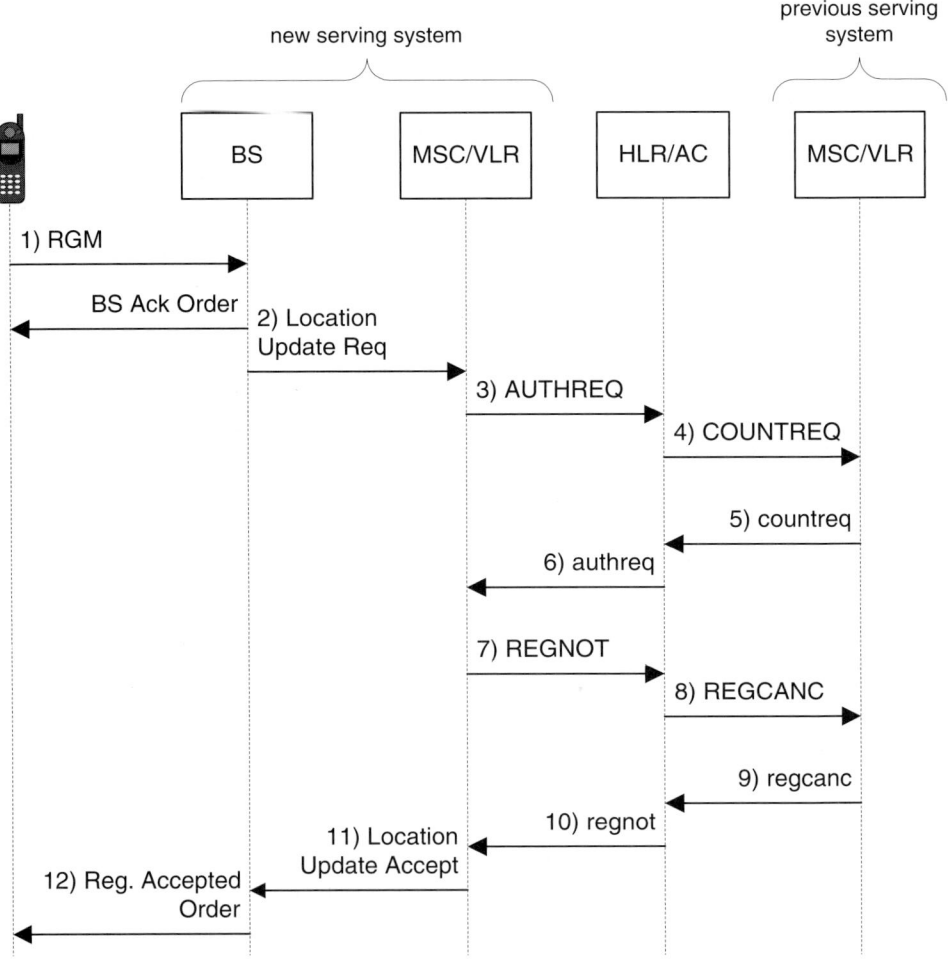

Figure 11.1 Initial registration with authentication, SSD shared with previous system.

station home systems so that it can update the mobile station location information. Then, the HLR must inform the previously serving MSC so that the mobile station can be deregistered from its associated VLR.

1. The mobile station determines from the paging channel overhead messages that it is now being served by a new system and that authentication is required on all system accesses (AUTH=1). The overhead messages also carry the random number (RAND) to be used for authentication. The mobile station executes the signature-generation algorithm, described in Chapter 6, using the currently stored SSD-A, the electronic serial number (ESN), and the RAND value to produce a registration authentication result (AUTHR). The mobile station sends the *Registration Message* (RGM) on the access channel and transitions to the Registration Access Substate of the System Access State. The RGM contains the MIN, ESN, AUTHR, the call history count (COUNT), and RANDC derived from the RAND used to compute AUTHR. On receipt of the RGM, the base station sends the *Base Station (BS) Acknowledgment Order Message* (ORDM) and the mobile station reenters the idle state.

2. The base station forwards the serving MSC the registration information in the Location Update Request Message.

3. The MSC verifies the RANDC supplied by the mobile station and sends the appropriate value of RAND in an AUTHREQ message to the HLR/Authentication Center (AC) associated with the mobile identification number (MIN). The AC validates the MIN and ESN reported by the mobile station. The AC then executes the signature-generation algorithm using the currently stored SSD-A, the ESN, and the RAND value to produce a registration AUTHR. The AC verifies that the AUTHR message received from the mobile station matches the one locally generated.

4. In this example we assume that the SSD was shared with the previous serving system so that the AC is unaware of the current value of the call history count for the mobile station. The HLR/AC retrieves the current value of COUNT from the previous serving system by sending a COUNTREQ message to the previous serving VLR/MSC.

5. The VLR of the previous serving system returns the current COUNT value to the HLR/AC in a countreq message.

6. The AC then verifies that the COUNT received from the mobile station is consistent with the value retrieved from the previous serving system. The HLR/AC sends an authreq message to the HLR. The authreq message may include SSD and directives to issue a unique challenge, to update the mobile station SSD, or to update the mobile station COUNT according to AC/HLR local administrative practices. Alternately, the authreq message may indicate that the access is denied, for example, because the serving system is excluded from the allowed serving systems based on the subscriber profile.

7. Following successful authentication of the mobile station, the new serving MSC sends a registration notification (REGNOT) message to the HLR.

8. As the mobile station was previously registered in another system, the HLR sends a registration cancellation (REGCANC) to the previous serving VLR/ MSC.

9. The previous serving MSC/VLR returns a regcanc message to the HLR.

10. The HLR records the new location of the mobile station in its local database and responds to the REGNOT with a regnot message that includes the information requested by the serving system.

11. The serving MSC sends the Location Update Accept message to the base station.

12. The base station sends the mobile station the *Registration Accepted* ORDM on the paging channel, indicating that the location update has been successful. The message may also carry the roaming indicator corresponding to the mobile station roaming condition. The indicator may be used to set the roaming indicator flag (typically, a letter R) on the mobile station display.

11.1.2 Deregistration

In the example illustrated in Figure 11.2 the user switches off the mobile station, therefore triggering transmission of the RGM (assuming that power-down registration is enabled in the serving system; see Chapter 6). The serving MSC informs the HLR, which then deletes the mobile station's location information from the corresponding record in its database. The mobile station status is changed to inactive, but the subscriber information is still retained. This procedure enables immediate application of the proper call treatment (e.g., announcement or reorder tone) to the calling party in case of a call terminated to an inactive mobile station.

1. The mobile station sends the RGM on the access channel when the user directs it to power off. The REG_TYPE filed of the RGM is set to 0011 for power-down registration. The base station sends the Layer 2 acknowledgment. The mobile station powers off.

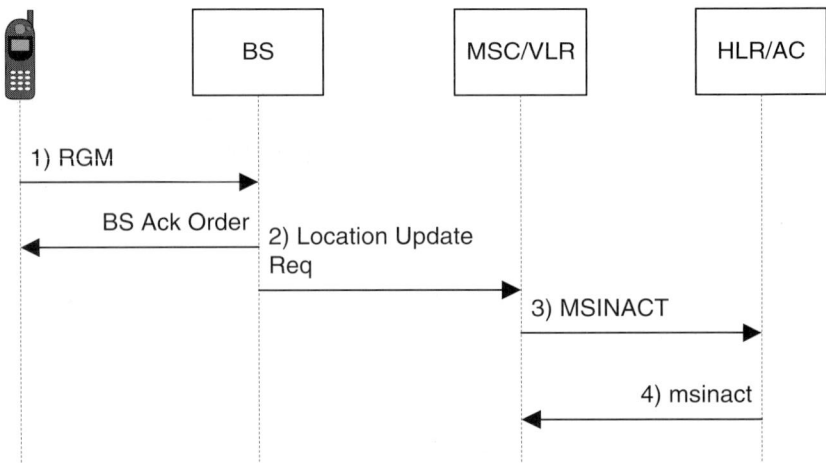

Figure 11.2   Mobile station deregistration following power-down registration.

2. The base station forwards the serving MSC the registration information in the Location Update Request Message.

3. As the mobile station has powered down, the serving system's MSC/VLR decides that the mobile station needs to be deregistered. If SSD is shared with the serving system, the serving system must also convey the updated call history count (COUNT) to the HLR/AC. The MSC/VLR sends the mobile station inactive (MSINACT) message to the HLR/AC in the mobile station's home system. The HLR accesses the record in its database corresponding to the mobile station, deletes the location information (i.e., the serving system's VLR identity), and changes the mobile station's state to inactive. The updated COUNT is stored in the AC.

4. The HLR sends the previous serving system an empty msinact message in acknowledgment to the MSINACT. The previous serving system cancels the mobile station record from its VLR.

11.2 Basic Call Processing

Call processing includes a broad category of functions that establish, maintain, and release calls to and from a mobile station. In this section we examine in detail two call-processing scenarios: call origination and call delivery to an idle roaming subscriber. While these two are basic scenarios, they exemplify end-to-end procedures because they encompass many network nodes and interfaces.

11.2.1 Voice Call Origination

In the example illustrated in Figure 11.3 a mobile station originates a voice call. It is assumed that the mobile station has already performed successful registration with the serving system, which may or may not be the home system, and that the subscriber profile has been fetched and maintained at the serving system's VLR. It is also assumed that the global challenge is enabled at the serving system and that SSD is not shared between the AC and the serving system's VLR. Then, authentication procedures need to be carried out involving the subscriber home system's AC. Although SSD may be shared in practice, we choose a scenario in which SSD is not shared because it represents the most general case.

In the example we assume that the traffic channel establishment is initiated only after explicit command from the MSC. In practice, however, the implementation of some base stations is such that traffic channel establishment is triggered immediately after receipt of the *Origination Message* (ORM) and while the core-network procedures (e.g., for authentication) are in progress. This method, also called *early traffic channel establishment*, is used when the latency of the MSC directive to establish the traffic channel is expected to be large, which has a negative effect on call setup success. That is particularly true if the access network does not support access handoffs or channel assignment in soft handoff (see Chapter 7). In such cases a long delay in traffic channel establishment is such that the quality of the cell whose paging channel is

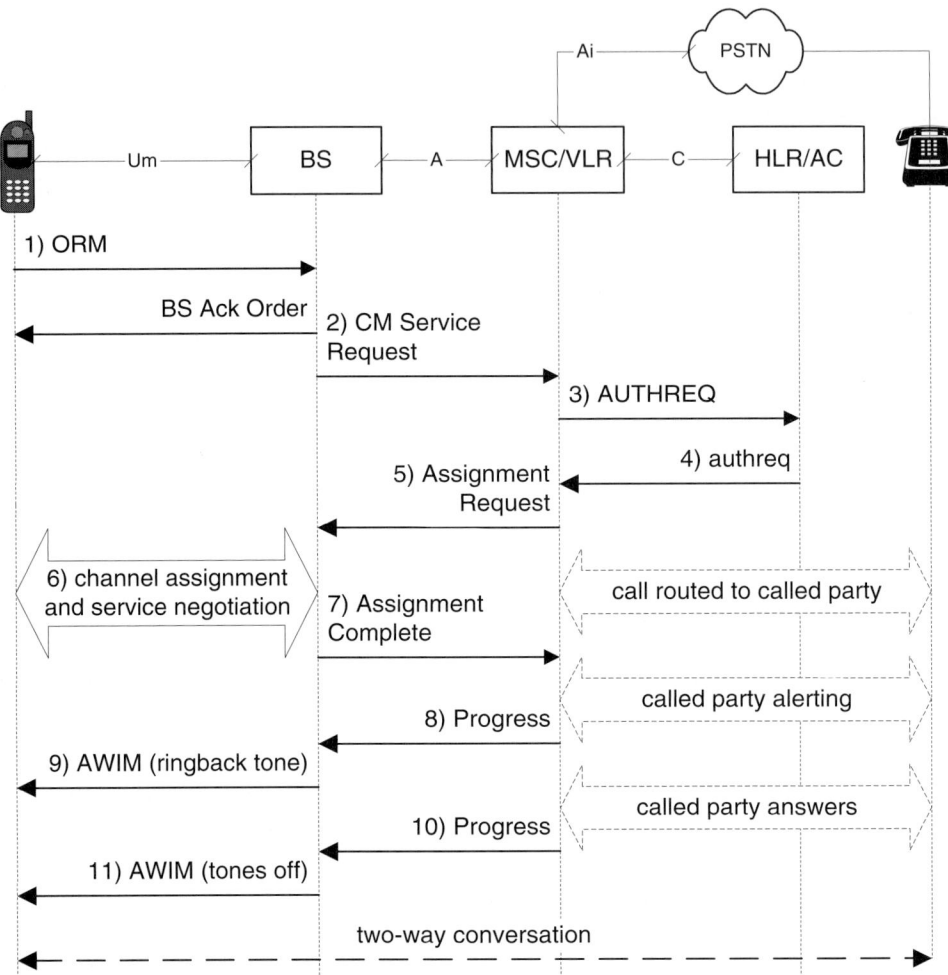

Figure 11.3   Voice call origination example.

being monitored by the mobile station while in the system-access state may deteriorate and become unusable for traffic channel establishment. Early traffic channel assignment has the drawback of consuming more capacity, because a mobile station that ends up being unauthorized for the service it requested is given access, albeit momentarily, to network resources.

The signaling flow on the air interface (U_m) in Figure 11.3 is not discussed in detail because it is presented in Chapter 6.

1. The mobile station in idle state determines from the overhead messages on the paging channel that global challenge is enabled; that is, authentication is required on all system

accesses. The random number to be used for authentication (RAND) is also obtained from the overhead messages. When the user originates a voice call, the mobile station executes the signature-generation algorithm using the dialed digits, RAND, ESN, and the currently stored SSD to produce an origination authentication signature carried by the AUTHR message. The mobile station sends the ORM providing the dialed digits, its MIN, ESN, AUTHR, the call history count (COUNT), and the RANDC from the RAND used to compute AUTHR, and enters the system access state. Upon receipt of the ORM, the base station sends the *Base Station Acknowledgment* ORDM.

2. The base station forwards the content of the ORM to the MSC. In some implementations, the base station at this point proceeds with establishing a traffic channel to expedite call setup (early traffic channel establishment). In this example it is assumed that early traffic channel establishment is not implemented.

3. If the SSD is presently shared with the serving system, the VLR performs validation of the mobile station authentication signature. In this example it is assumed that the SSD is not shared, so the serving system forwards the AUTHREQ message to the HLR/AC associated with the mobile station identification number.

4. Since in this example the SSD is not presently shared with any another system, the AC is aware of the current call history count. The AC verifies the MIN and ESN reported by the mobile station and then executes the signature-generation algorithm using the SSD-A and ESN currently associated with the mobile station along with the value of RAND and the dialed digits provided by the serving system to produce an origination AUTHR message. The AC verifies that the AUTHR received from the mobile station matches the one locally generated. The AC then verifies that the COUNT received from the mobile station is consistent with the value currently stored in its database. The HLR/AC sends an authreq message to the serving system. The authreq message may also include directives to issue a Unique Challenge, to update the mobile station SSD, or to update the mobile station COUNT according to AC local administrative practices.

5. On receipt of positive authentication results, the MSC sends the base station the Assignment Request Message to request assignment of radio resources. This message also includes the identity of the terrestrial circuit to be used between the MSC and the base station to transport user data (voice). The call is routed to the public switched telephone network (PSTN).

6. On receipt of the Assignment Request message from the MSC, the base station initiates Forward/Reverse Fundamental Channel (F/R-FCH) establishment, followed by service negotiation. The details of this U_m interface signaling procedure are described in Chapter 6.

7. After the traffic channel and circuit have both been established, the base station sends the Assignment Complete message to the MSC.

8. When the called party has been alerted, the MSC sends a Progress indication to the base station. Alternately, the MSC may decide to generate the ringback and apply it

into the audio path directly. In the latter case, this supervision procedure is transparent to the base station and mobile station and steps 8 to 11 are skipped.

9. The base station sends the *Alert with Information Message* (AWIM) directing the mobile station to locally generate a ringback tone.

10. Once the called party answers the call, the MSC sends a Progress indication to the base station.

11. The base station sends the AWIM directing the mobile station to stop playing the ringback tone. The two parties are now in two-way conversation.

11.2.2 Call Delivery

In the example illustrated in Figure 11.4 and Figure 11.5, a voice call is terminated to an idle subscriber. The subscriber is roaming; that is, the serving system is not the home system. Using IS-41 terminology, this scenario is called *call delivery* (CD). Note that CD is a network feature in the sense that, as with many other features, the user may or may not be authorized for it, and it can be enabled and disabled by the subscriber. As in the previous example, it is assumed that the SSD is not shared between the serving system and the AC, and that early traffic channel assignment is not implemented by the base station.

The call is originated by a land party and is routed to the home system's MSC. The home system does not have any record of the mobile station's current location (as the mobile station is registered with another system), and it must therefore interrogate the home system's HLR. Once given location and routing information, the home system MSC routes the call to the serving system MSC/VLR. From this point onward, the call flow is identical to that of a call terminated to an idle mobile station in its own system.

1. A call origination and the dialed mobile station address digits (i.e., directory number) are received by the originating MSC.

2. The originating MSC sends a LOCREQ message to the HLR in the mobile station's home system. The association between mobile station and HLR is made through the dialed mobile station address digits.

3. If the dialed mobile station address digits are assigned to a legitimate subscriber, the HLR sends a ROUTREQ message to the VLR in the system where the mobile station is registered.

4. In reaction to the ROUTREQ message, the serving MSC consults its internal data structures to determine if the mobile station is already engaged in a call on this MSC. In this scenario we assume that the mobile station is idle. As we assumed that successful registration and location update had taken place prior to the call delivery attempt, the serving system already has the service profile of the roaming mobile station. If it did not, then it would have to fetch it from the home system HLR. The serving MSC allocates a temporary local directory number (TLDN) and returns this information to the HLR in the routreq message.

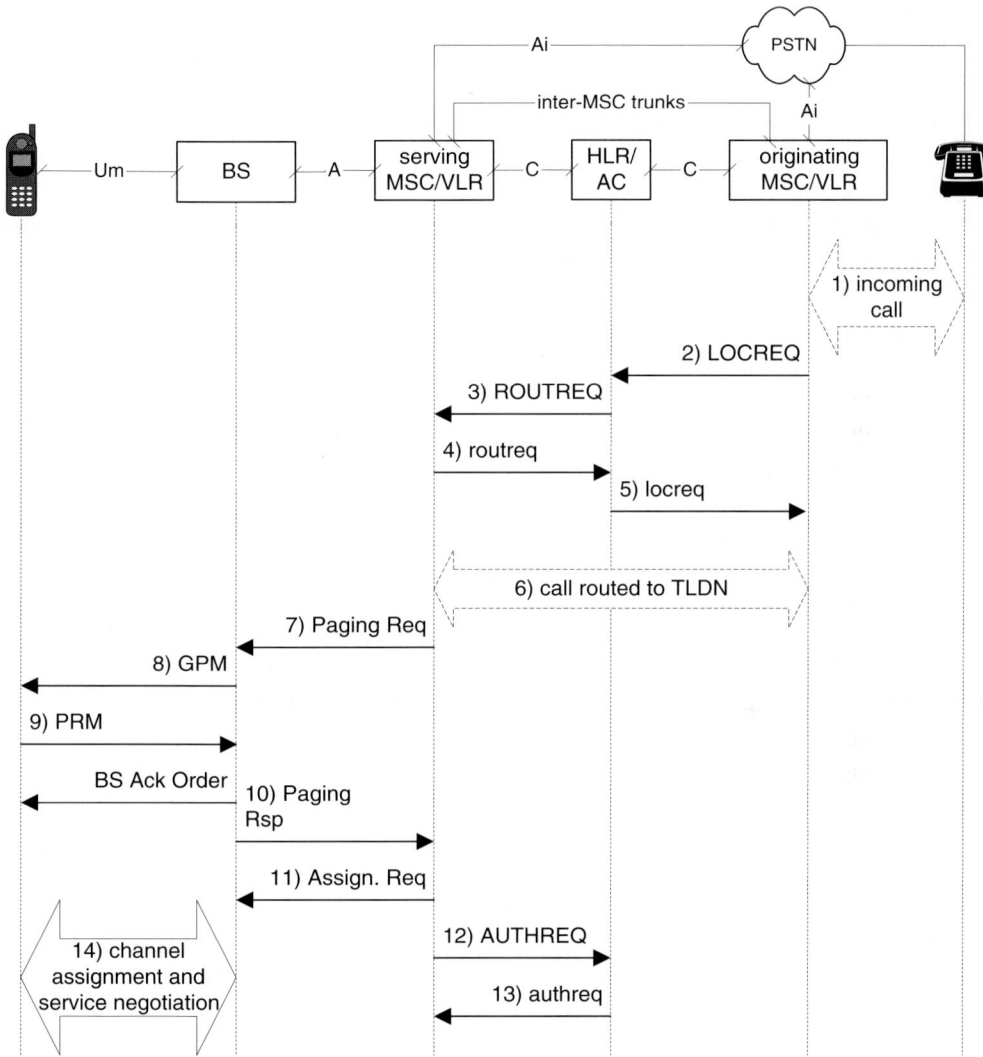

Figure 11.4   Call delivery to an idle roaming subscriber, part I.

5. When the routreq message is received by the HLR, it returns a locreq message to the originating MSC. The locreq message includes routing information along with an indication of the reason for extending the incoming call (i.e., for CD).

6. The originating MSC establishes a voice path to the serving MSC using existing interconnection protocols, such as Signaling System 7 (SS7) and the routing information specified in the locreq message.

7. The serving MSC determines that an incoming call terminates to a mobile within its serving region and sends the Paging Request message to the base station to initiate a mobile-terminated call setup.

8. The base station pages the mobile station on the paging channel of the cells in the paging area indicated by the MSC by means of the *General Page Message* (GPM).

9. The mobile station, upon receipt of the page, sends the *Page Response Message* (PRM) on the reverse-access channel and moves to the system-access state (see also Chapter 6). The mobile station determines from the overhead messages that authentication is required on all system accesses (AUTH=1). The random number to be used for authentication (RAND) is obtained on the paging channel's overhead information. The mobile station executes the authentication signature–generation algorithm using the currently stored SSD-A, ESN, and the RAND value to produce a termination AUTHR. Upon receipt of the access-channel probe containing the PRM, the base station sends the *Base Station Acknowledgment* ORDM.

10. The base station forwards the content of the PRM to the MSC. In some implementations, the base station at this point proceeds with establishing a traffic channel to hasten call setup (early traffic channel establishment). In this example it is assumed that early traffic channel establishment is not implemented.

11. The MSC sends an Assignment Request message to the base station to request assignment of radio resources. This message also includes the identity of the terrestrial circuit to be used between the MSC and the base station to transport user data (voice).

12. Steps 12 and 13 take place in parallel to, and asynchronously with, steps 14 and beyond. The serving MSC/VLR verifies RANDC supplied by the mobile station. If SSD is presently shared with the serving system, the VLR validates the mobile station. In this example it is assumed that SSD is not shared, so the VLR forwards the AUTHREQ to the HLR associated with the mobile station.

13. Since in this example the SSD is not presently shared with any other system, the AC is aware of the current call history count. The AC verifies the MIN and ESN reported by the mobile station and then executes the signature-generation algorithm using the SSD-A and ESN currently associated with the mobile station along with the value of RAND and the dialed digits provided by the serving system to produce an origination AUTHR. The AC verifies that the AUTHR received from the mobile station matches the one locally generated. The AC then verifies that the COUNT received from the mobile station is consistent with the value currently stored in its database. The HLR/AC sends an authreq message to the serving system. The authreq message may also include directives to issue a Unique Challenge, to update the mobile station SSD, or to update the mobile station COUNT according to AC local administrative practices.

14. On receipt of the Assignment Request message from the MSC, the base station initiates traffic channel (F-FCH and R-FCH) establishment, followed by service negotiation. The details of this U_m interface signaling procedure are described in Chapter 6.

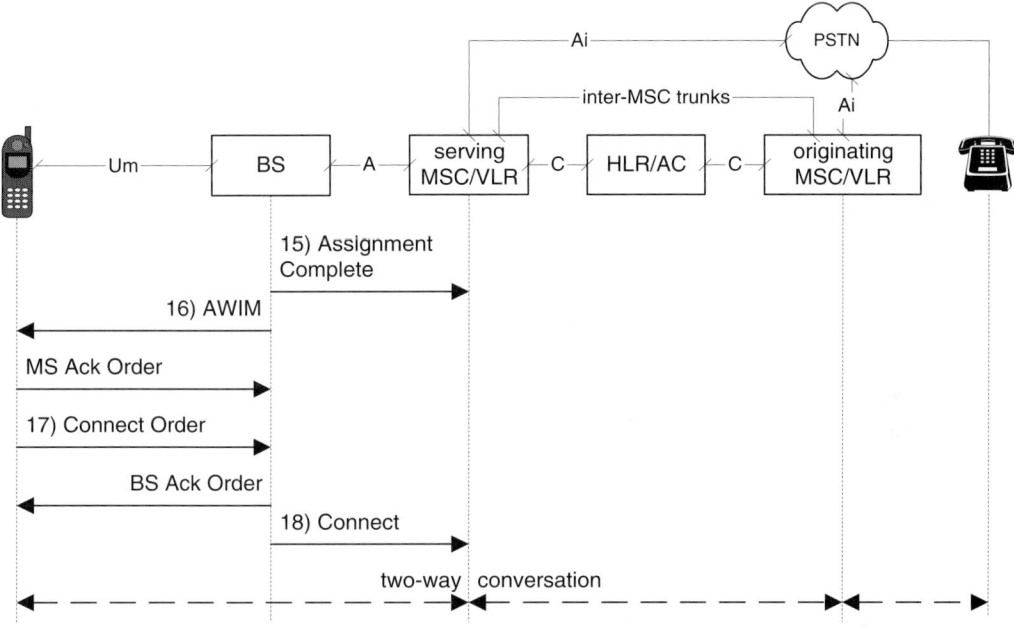

Figure 11.5 Call delivery to an idle roaming subscriber, part II.

15. After the traffic channel and circuit have both been established, the base station sends the Assignment Complete message to the MSC.

16. The base station sends the AWIM to the mobile station to alert the user of the incoming call. The mobile station acknowledges the reception of the AWIM by transmitting the Layer 2 acknowledgment order.

17. When the user answers the call, a *Connect* ORDM with acknowledgment required is transmitted to the base station. The base station acknowledges the *Connect* ORDM with the Layer 2 acknowledgment order over the forward traffic channel.

18. The base station sends a Connect message to the MSC to indicate that the call has been answered. At this point, the call is considered in conversation state.

11.3 Short Message Services

The SMS [1] allows the exchange of short messages between a mobile station and the access network, and between the access network and the message center (MC). The SMS protocol consists of the relay, transport, and teleservice layers. The relay and transport layers jointly offer a bearer service to various end-to-end teleservices. Examples of such teleservices are voice mail notification (VMN) [2], wireless paging teleservice (WPT) [3], and wireless messaging teleservice (WMT) [4]. To realize a teleservice, one or more messages specified by the SMS teleservice layer can be used. The messages are encapsulated in the SMS Point-to-Point Message or SMS Broadcast Message of the transport layer and transported end to end between the mobile station and the MC. The MC has a message store and forward functionality. Messages can be stored for postponed delivery when the destination mobile station is unavailable. The main interworking issues that arise in the end-to-end realization of the SMS service include delivering messages to roaming subscribers and routing messages from the serving system MSC to the MC in the home system of the destination mobile station. SMS message can be sent to the mobile station either through common channels or dedicated (traffic) channels, depending on the message size.

The two call flows illustrated in Figure 11.6 and Figure 11.7 exemplify end-to-end realization of the WMT, which uses point-to-point SMS as a service bearer to transfer a message between one SMS-capable mobile station and another. The WMT service comes with optional features that are either preprogrammed in the user's terminal or can be configured by the user via the keypad or other user interface. One such feature is delivery confirmation. In this example the originating user is assumed to request message delivery confirmation. Also, in the examples that follow it is assumed that the message is short enough to be carried on common channels.

11.3.1 Mobile-Originated SMS

In this example, the mobile station sends an SMS message on the control channel and receives positive receipt acknowledgment from its home MC.

1. The SMS message from the teleservice layer is delivered to the IS-2000 Layer 3 protocol for transmission. The message is encapsulated in a *Data Burst Message* (DBM) with BURST_TYPE set to 000011. In this example its length is such that it fits in the payload of the message sent on the Reverse Enhanced Access Channel (R-EACH).[1] The mobile station enters the message transmission substate of the mobile system access state (see Chapter 6) and sends the DBM on the access channel. The base station receives the DBM and sends the *Base Station Acknowledgment ORDM. The mobile station, upon receipt of the acknowledgment, reenters the idle state.*

2. The base station encapsulate the SMS application data of the DBM in the ADDS user data field of the ADDS Transfer message that is sent to the serving system MSC.

3. On receipt of the ADDS Transfer Message, the serving MSC encapsulates the SMS application data in the SMS Delivery Point-to-Point invoke message (SMDPP). The SMDPP is sent to the MC in the home system of the mobile station originating the SMS message. To route the message to the appropriate MC, the serving system MSC typically maintains a lookup table mapping IMSI to MC addresses.

---

1.   The maximum SMS message size depends on the R-EACH duration, as determined from the *Enhanced Access Parameters Message* broadcast by the base station. It also depends on the mobile station identifier type that would be used in forming the DBM.

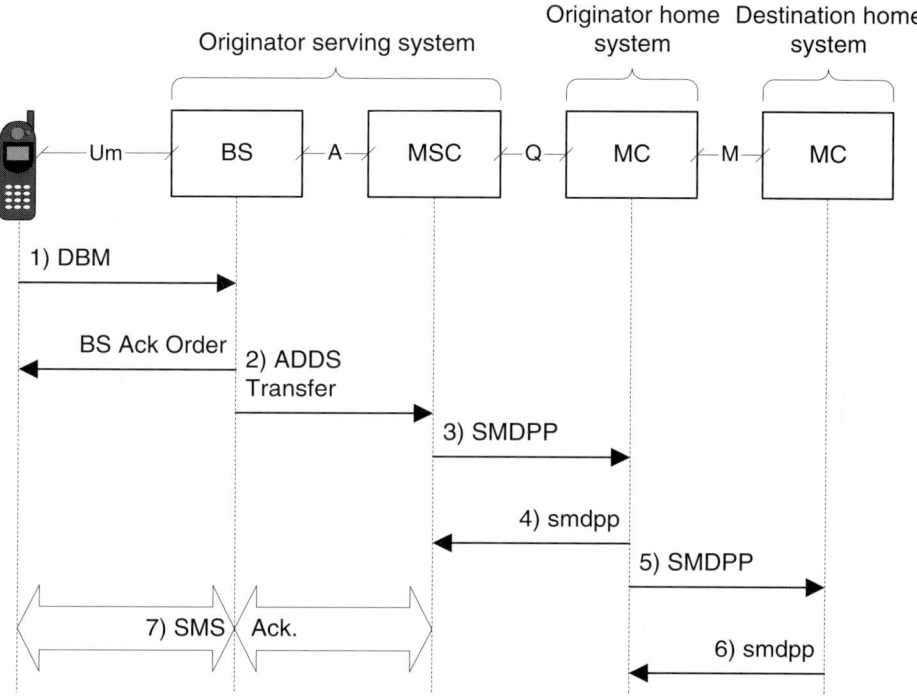

Figure 11.6   Mobile-originated SMS on the access channel.

4. The MC in the originator home system receives the SMDPP and sends the acknowledgment message (smdpp) to the serving system MSC.

5. The MC in the originator home system routes the SMDPP to the MC in the home system of the mobile station the SMS is destined to.

6. The MC in the home system of the destination mobile station receives the SMDPP, sends the smdpp acknowledgment, and initiates the procedures for SMS delivery to the destination mobile station, as explained in Section 11.3.2.

7. Step 7 takes place concurrently with steps 5 and 6. On receipt of the smdpp acknowledgment, the serving system MSC initiates the procedures to route the acknowledgment to the originating mobile station. The acknowledgment indicates that the SMS message has been successfully sent to the message center. It does not indicate that the message has been successfully delivered to the destination mobile station. The procedure for sending the SMS acknowledgment may be the same as that used for delivering an SMS message, which is explained in Section 11.3.2.

11.3.2 Mobile-Terminated SMS

In the example illustrated in Figure 11.7, the MC corresponding to the home system of the destination mobile station receives an SMS message to be delivered. The MC must first interrogate the HLR to obtain information about location, status, and address of the mobile station. In this example, the HLR is assumed to update the information by querying the serving system MSC/VLR. On receipt of the SMS message, the serving system MSC can choose one of several methods for message delivery to the mobile station. The message can be transferred on the traffic channel, in which case the mobile station is paged and a traffic channel is set up, just as in the case of a voice call. If the message is relatively short, a more efficient transfer mechanism is to send the message on the forward common channel, either Forward Paging Channel (F-PCH) or Forward Common Control Channel (F-CCCH). However, the mobile station location is only known at registration-zone level, which comprises many base stations. If the SMS message were sent on the paging channels of all base stations in a paging area corresponding to the registration area, paging channel capacity would quickly run out. Then, one possible procedure is to page the mobile station on the paging channels of all base stations in the paging area. Note that the GPM is much smaller than an SMS message. On receipt of the page response, the mobile station location is known at cell level. At that time the SMS can be sent on one and only one paging channel while the mobile station is in the system access state, and the call can be released. This procedure is considered in the following example.

1. The MC in the home system of the destination mobile station receives the SMDPP message carrying the SMS message to be delivered (see step 6 in Section 11.3.1).
2. The MC sends the HLR in the home system of the destination mobile station a query about location and status of the destination mobile station, the SMSRequest (SMSREQ) message.
3. In this example, the HLR sends an SMSREQ message to the serving system MSC/VLR, requesting an update of the mobile station location and status. Note that this step is optional, as the HLR may decide that it has sufficiently current information.
4. The serving MSC/VLR replies with the SMSRequest Return Result message (smsreq).
5. The HLR routes the smsreq message to the MC.
6. As the smsreq message indicates that the mobile station is currently available, the MC forwards the SMS message to the serving system MSC for delivery to the mobile station. The message is encapsulated in the SMDPP message.
7. On receipt of the SMDPP message, the serving system MSC initiates the procedure to identify the base station currently serving the mobile station, as sending the SMS message on the paging channels of all cells in the paging area would consume too much capacity. The MSC sends the base station controller the Paging Request message.
8. The base station controller sends a GPM on the paging channel of all base stations in the paging area.

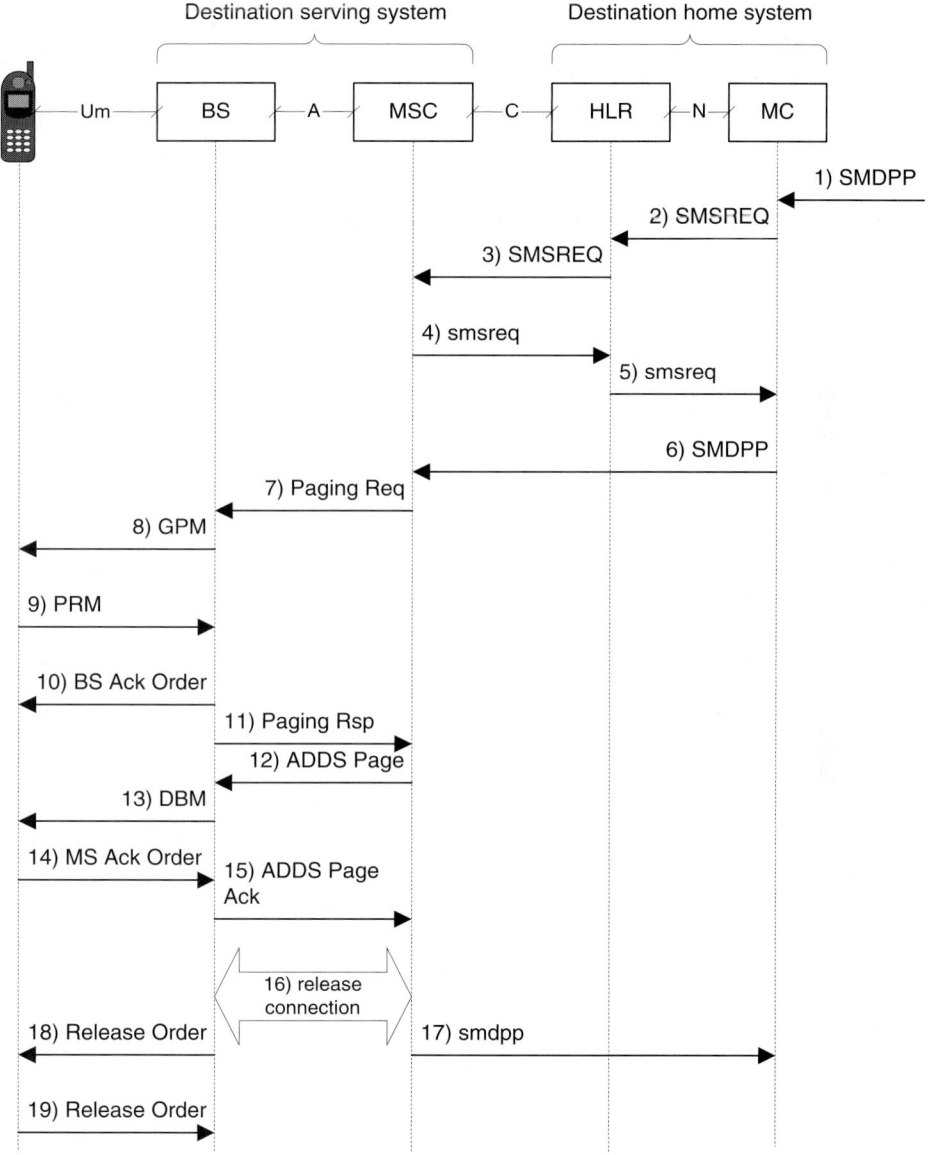

Figure 11.7   Mobile terminated SMS on the control channel.

9. The mobile station receives the page, sends the PRM, and enters the page response sub-state of the system access state.
10. The base station sends Layer 2 acknowledgment.
11. The base station forwards the page response to the MSC.

12. The serving MSC sends the ADDS Page message encapsulating the SMS message.

13. The base station sends the DBM to the mobile station on the paging channel of the cell where the page response was received.

14. The mobile station receives the DBM and sends the Layer 2 acknowledgment. The SMS message in the DBM is sent to the upper layers. The user is alerted of the incoming message.

15. The base station sends the ADDS Page Ack message to the MSC, confirming successful delivery.

16. The MSC releases the resources that had been allocated for this call.

17. The MSC sends the smdpp message to the MC, confirming successful message delivery.

18. *The base station sends the Release* ORDM to the mobile station, which is still in the system access state.

19. The mobile station sends the *Release* ORDM and transitions to the idle state.

11.4 Intersystem Handoff

Interbase-station handoff consists of a set of mobile station, base station, and network functions and procedures that enable transferring control of the radio connection from one base station to another under the domain of the same system (intra-MSC) or different systems (inter-MSC). In the following, we consider exemplary procedures for intersystem handoff forward. Handoff forward is used to transfer control of an active mobile station from the source to the target base station. Specialized forms of handoff forward are possible depending on the mobile station state at the time of handoff triggering (e.g., conversation state or waiting-for-answer state; see also Chapter 6). Discussion of intrasystem procedures is omitted, since those are simpler versions of the more general intersystem case. Intersystem handoff requires coordination of cell identification between neighboring MSC and of inter-MSC trunks carrying the user traffic.

11.4.1 Handoff Forward in the Conversation State

In the example illustrated in Figure 11.8 the mobile station is in two-way conversation state with the other party at the time handoff is triggered. The traffic channel connection is relocated from the source to the target base station under the domain of a different MSC. When the target MSC sends the source MSC a positive indication that handoff has been successfully accomplished, the source MSC initiates resource release procedures on the source base station. During and after handoff, the original connection between the PSTN and the MSC that initially served the call, the source MSC, is unaffected. Then, a new inter-MSC trunk is established to carry the user traffic from the PSTN to the target MSC via the source MSC. The source MSC then becomes the *anchor* MSC.

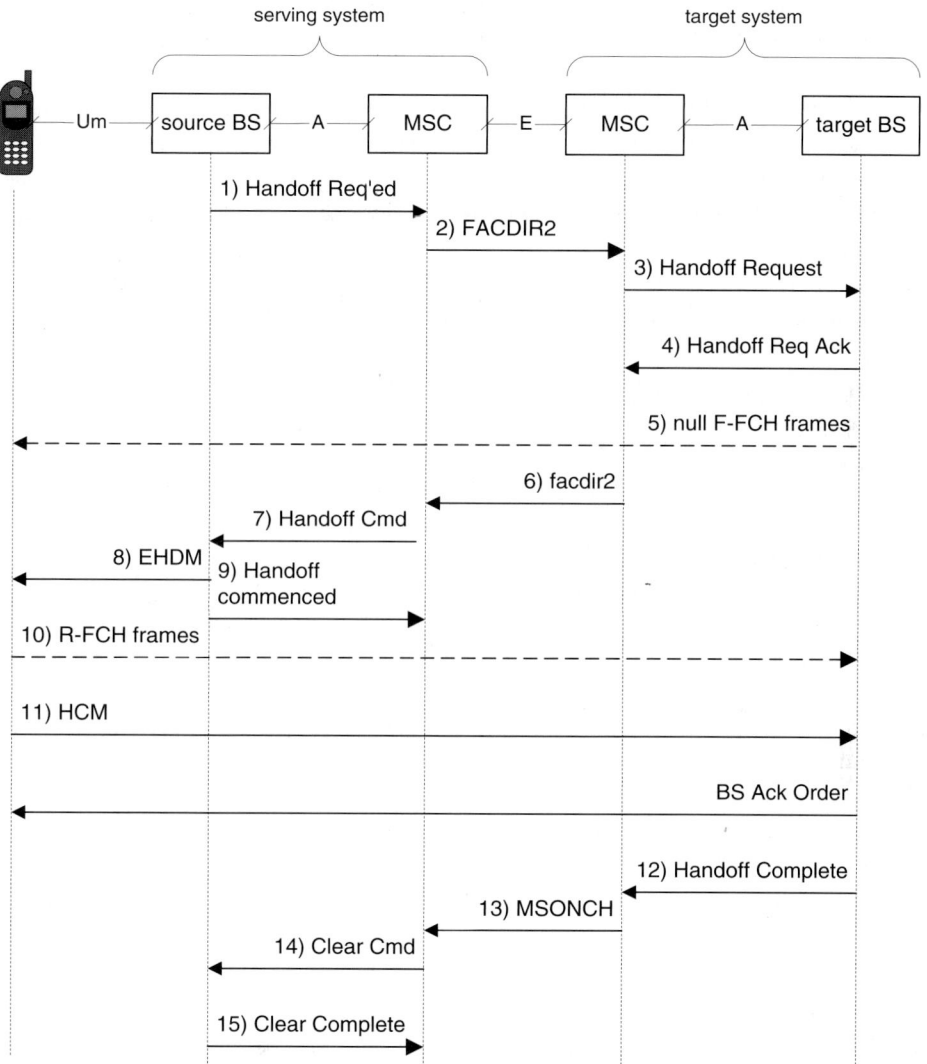

Figure 11.8   Handoff forward while in conversation state.

1. When conditions for handoff are satisfied (handoff triggering methods used for intra-base-station hard handoff also apply to interbase-station handoff; see Chapter 7), the source base station selects the target cells in the domain of the target base station and sends the serving MSC the Handoff Required message. The message carries identity and capability information of the mobile station, service configuration information, and the identity of the target cells. The cell identifier may take different formats, but the one typically used contains the MSC identity and the 2-byte cell identity. The cell identity is unique within an MSC. Also note that in IS-41 terms a cell corresponds to a geographical location.

2. Once it is established that the target cell is not within its domain, the serving MSC sends the target MSC the FACDIR2 message carrying information equivalent to that of the Handoff Required message. If, on the other hand, the handoff were interbase station-intra-MSC, the MSC would skip to step 3 below. Note that the MSC can modify or truncate the target cell list.

3. The target MSC sends the target base station the Handoff Request message.

4. The target base station allocates the proper resources on the target cells, sends the target MSC the handoff request acknowledgment, and starts a timer waiting to acquire the reverse traffic channel. The target base station may propose an alternative service option if the one in use at the source base station was not supported.

5. The target base station starts sending null frames to the mobile station.

6. The target MSC clears the acquisition timer and relays the source MSC the information contained in the handoff request acknowledgment by means of the facdir2 message.

7. The serving MSC sends the source base station the handoff command message.

8. The source base station sends the mobile station the *Extended Handoff Direction Message* (EHDM) or *Universal Handoff Direction Message* (UHDM) Although not mandated by the standard, the EHDM is typically sent in nonassured mode, possibly in quick repeats, because of the need to expedite handoff.

9. The source base station sends the serving MSC the Handoff Commenced message indicating that handoff has been initiated.

10. The mobile station tunes to the new carrier, acquires the forward traffic channel transmitted by the target base station, and starts reverse traffic channel transmission.

11. The mobile station sends the *Handoff Complete Message* (HCM). The target base station now becomes the serving base station and sends to the mobile station the Layer 2 acknowledgment via the ORDM.

12. The new serving base station sends the target MSC, now the serving MSC, the Handoff Complete message indicating successful handoff completion.

13. The new serving MSC sends the old serving MSC, now the anchor MSC, the MSONCH message to indicate completion of the handoff attempt. At this time the anchor MSC connects the land party to the mobile station by using the inter-MSC trunk.

14. The anchor MSC clears the connection with the old serving BS.

15. The old serving base station clears all resources previously allocated to the mobile station and sends the anchor MSC the Clear Complete message.

11.4.2 Answer Supervision after Handoff Forward

Handoff may be triggered just after a mobile station-originated call and before the called party has answered the call. Similarly, in case of a mobile station-terminated call, handoff may take place before the user has answered the call. As the PSTN signaling connection is terminated at the anchor MSC, we are now in a situation where the anchor MSC is responsible for call supervision toward the land party while the serving MSC is responsible for call supervision toward the mobile station. When the called party answers the call, information related to such event must then be exchanged between the two MSCs so that answer-supervisor procedures can take place. The examples that follow describe the two main scenarios for call termination and origination after handoff, respectively.

In the first example illustrated in Figure 11.9, the mobile station receives a call from a land party and the user is alerted just before handoff takes place. From the IS-2000 Layer 3 protocol standpoint, the mobile station is in the waiting-for-answer state. The handoff procedures over the core network and air interfaces are identical to those shown in Figure 11.8 with the exception that the new serving MSC must inform the anchor MSC when the user answers the call so that the anchor MSC can perform answer supervision toward the calling party.

1. The mobile station is in the waiting-for-answer state when the user answers the call. The mobile station sends the ORDM with order type set to connect.
2. The base station sends the serving MSC the Connect Message.

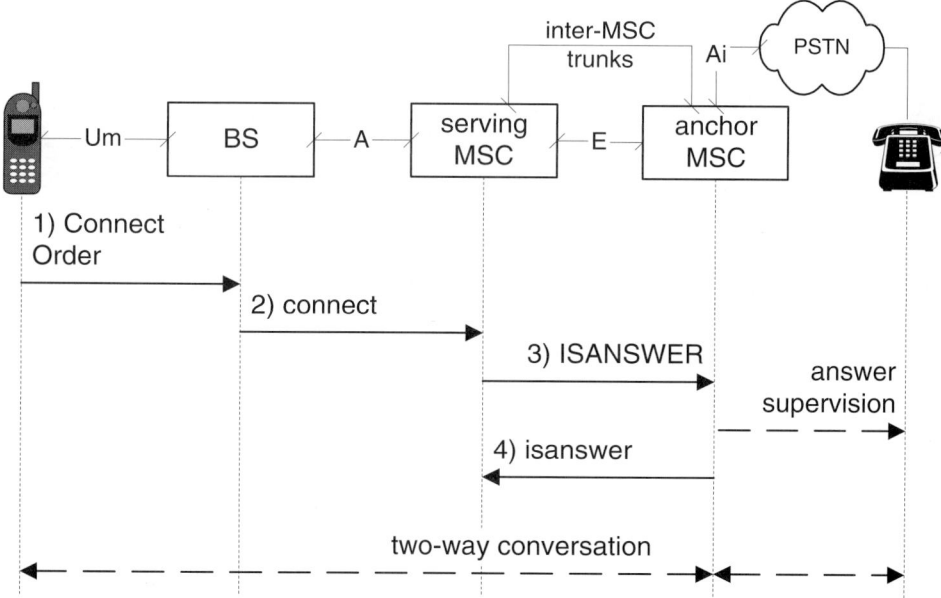

Figure 11.9 Handoff after mobile terminated call while the mobile station is alerting.

3. The serving MSC relays the Connect message to the anchor MSC via the intersystem answer (ISANSWER) message. The anchor MSC disconnects the ringback tone toward the calling party.

4. The anchor MSC sends the serving MSC the isanswer message as an acknowledgment. The mobile station and the calling party are now in two-way conversation.

In the second example, illustrated in Figure 11.10, the user originates a call just before handoff takes place. From the IS-2000 Layer 3 protocol standpoint, the mobile station is in the conversation state. The handoff procedures over the core network and air interfaces are identical to those shown in Figure 11.8, with the exception that the anchor MSC must inform the serving MSC when the called party answers the call so that the serving MSC can perform answer supervision toward the mobile station.

1. The mobile station is in the IS-2000 Layer 3 conversation state while a ringback tone is being played to the user. The called party answers the call.

2. When the called party answers the call, the anchor MSC sends the ISANSWER message to the serving MSC.

3. The serving MSC sends the base station the Progress Message.

4. The serving MSC sends the anchor MSC the isanswer message as an acknowledgment.

5. The base station sends the AWIM. Upon receipt of the AWIM, the ringback tone is interrupted. The mobile station and the calling party are now in two-way conversation.

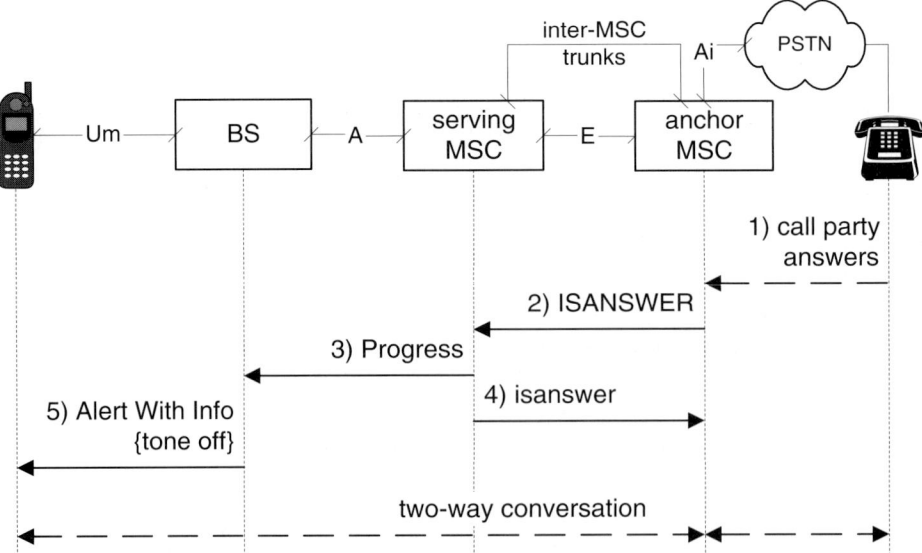

Figure 11.10 Handoff forward after call origination with the mobile station awaiting called party answer.

11.5 Packet-Switched Data Calls

The packet data service allows the mobile station to connect to the Internet. The mobile station first attaches to the network using simple or mobile IP, as explained in Chapter 2. Following a period of inactivity when no data is transmitted, the traffic channels are released to avoid unnecessary consumption of radio resources, and the mobile station transitions to the dormant state. While in the dormant state, the mobile station maintains the IP address previously obtained during the registration procedure.

In this section we show call flows that illustrate transitions in and out of the dormant state [5]. A single service, or call instance, is considered. Recall from Chapter 3 that the call instance used for best-effort data is the same as the one used for Point-to-Point Protocol (PPP) negotiation, and it is commonly referred to as the main service instance. The first call flow shows how a mobile station reactivates a dormant service. The second call flow illustrates network call reactivation from dormant state. The third call flow shows how the base station forces the mobile station into dormant state after a period of inactivity.

11.5.1 Mobile Station–Initiated Call Reactivation from Dormant State

In this example, illustrated in Figure 11.11, we show how the mobile station reactivates a packet data service. This scenario commonly happens during Web-browsing sessions, after a period of inactivity called reading time.

1. The mobile station originates a call by sending the ORM, which contains parameters such as Service Reference Identification (SR_ID) and Service Option (SO) numbers. The SR_ID corresponds to the SR_ID of the dormant service, and the service option number is SO33. The ORM also contains the Data Ready to Send (DRS) indication. DRS=1 is an indication that a mobile station has data to send. DRS=0 indicates dormant handoff. In this example we assume DRS=1.

2. The base station acknowledges receipt of the ORM message by sending the *BS Acknowledgment* ORDM.

3. The base station requests the traffic channel setup by sending the CM Service Request message to the MSC.

4. The MSC responds with the Assignment Request message that requests from the base station the setup of the radio traffic channel.

5. The base station sends the *Extended Channel Assignment Message* (ECAM) to the mobile station over the paging channel. The ECAM specifies the physical channel, the radio configuration, and many other physical layer parameters.

6. At this point the optional service negotiation phase begins. During the procedure, the service parameters are negotiated. During the service negotiation phase, the base station and the mobile station exchange the *Service Request Message* (SRQM) and the *Service Response Message* (SRM).

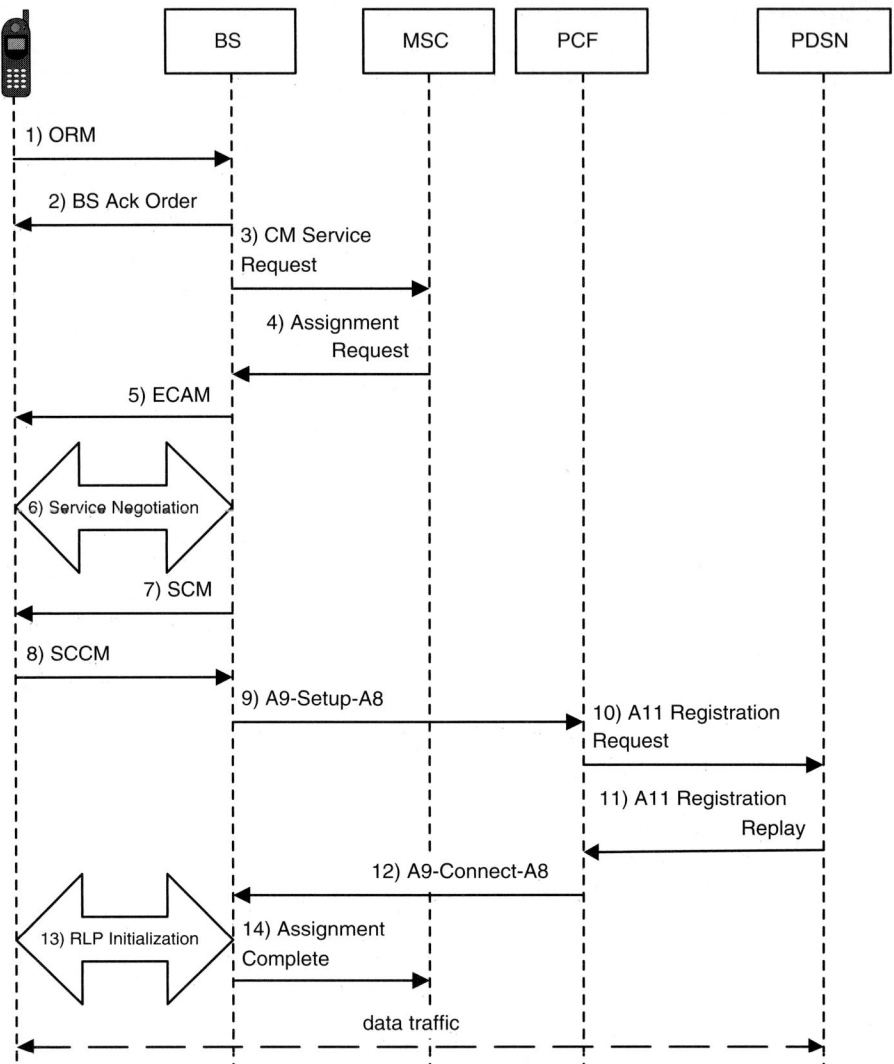

Figure 11.11   The mobile station initiated call reactivation from dormant state.

7. The base station sends the *Service Connect Message* (SCM), which contains the Service Configuration Record (SCR) that specifies the service parameters. The mobile station connects the service.

8. The mobile station acknowledges the service connection with the *Service Connect Completion Message* (SCCM).

9. The base station requests the setup of the A8 connection by sending the A9-Setup-A8 message over A9 interface.

10. In case the A10 connection does not exist, the A11-Registration Request message would set up the new A10 connection. However, the A10 connection is not torn down after the service goes dormant. Recall from Chapter 6 that information about dormant services is maintained at the Packet Control Function (PCF). When the SR_ID indicates a dormant service, the A11 Registration Requests passes only the accounting information and refreshes the A10 connections.

11. A11-Registration Response acknowledges the setup of the A10 connection. In our case the A10 connection was already there, so the message simply serves as an acknowledgment of the A11 Registration Request message. The A11 Registration Response message returns connection lifetime information that is either equal to or smaller than the proposed lifetime in the A11 Registration Request message. Before the timer initialized to the lifetime value expires, the connection needs to be refreshed.

12. The PCF acknowledges to the base station that the A8 connection is set up by sending the A9-Connect-A8 message.

13. The base station and the mobile station initialize the Radio Link Protocol (RLP).

14. The base station informs the MSC that the traffic channel is set up by sending the Assignment Complete message. The data traffic flow begins.

11.5.2 Network-Initiated Call Reactivation from Dormant State

The network-initiated call reactivation from dormant state is common for Wireless Application Protocol (WAP) push services and, as we show in Section 11.6.3, for VoIP. The call flow is illustrated in Figure 11.12.

1. The packet data session is dormant when new packet data arrive at the PCF buffer.

2. The PCF requests packet data service by sending the A9-Base Station Service Request message to the base station.

3. The base station requests service from the MSC by sending the Base Station Service Request message.

4. The MSC acknowledges the call setup request with the Base Station Service Response message.

5. The base station now responds to the PCF with the A9-BS Service Response message.

6. The MSC requests the base station to page the mobile station.

7. The base station pages the mobile station with the GPM.

8. The mobile station responds with a PRM. On receipt of the PRM, the base station acknowledges receipt of the PRM with the *Base Station Acknowledgment* ORDM.

9. The base station constructs the Page Response message, and it sends it to the MSC. This message contains complete Layer 3 information from the mobile station.

10. The MSC sends the Assignment Request to the base station, requesting the setup of radio resources and A8 connection.

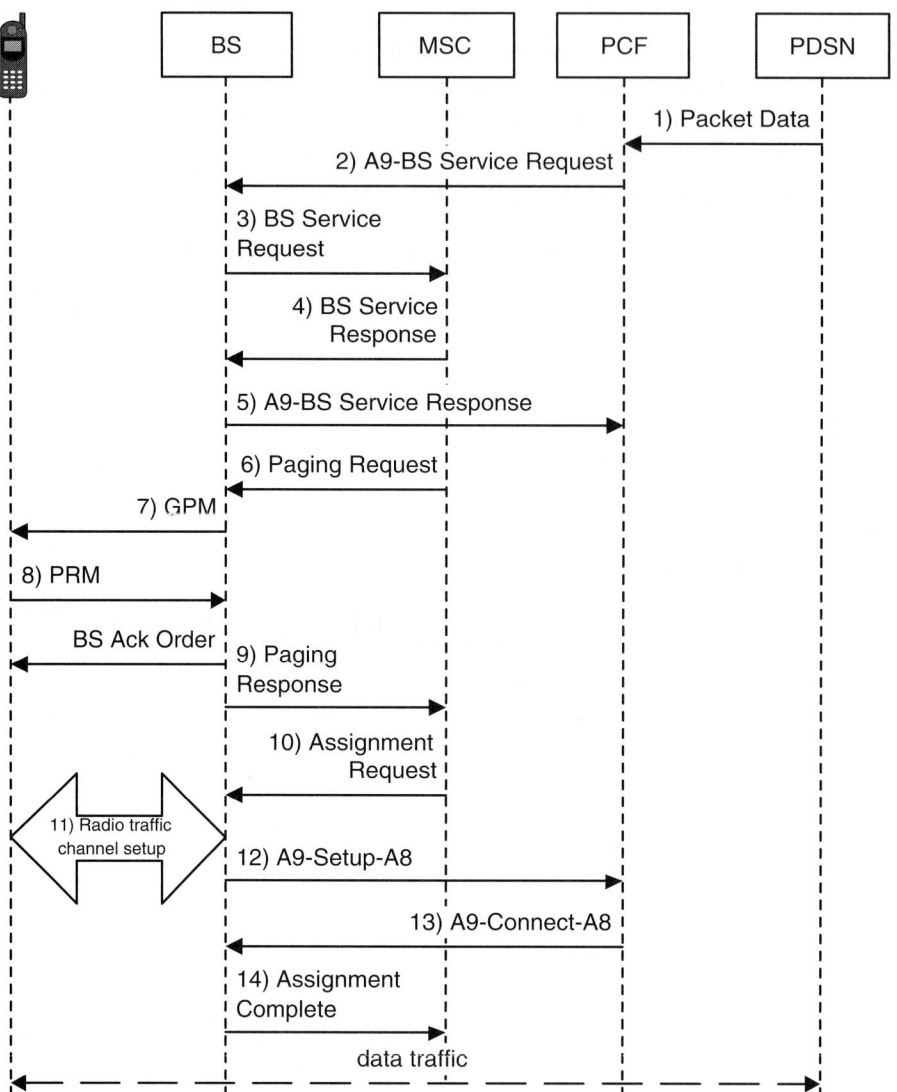

Figure 11.12 Network-initiated call reactivation from dormant state.

11. The base station and the mobile station perform the radio traffic channel setup procedure.

12. The base station requests the setup of an A8 connection with the PCF by sending the A9-Setup-A8 message to the PCF over A9 interface.

13. The PCF responds with the A9-Connect-A8 message to complete the A8 connection setup.

14. The base station informs the MSC that the radio traffic channel and A8 connection are set up by sending the Assignment Complete message to the MSC. At this time the packet data session has been activated and data can be exchanged between the mobile station and the packet core network.

11.5.3 Base Station-Initiated Transition to Dormant State

After setting up the SO33, the base station normally monitors the service for inactivity. Whenever the transmit buffer at the base station becomes empty and there is no incoming traffic from the mobile station, the base station starts the inactivity timer. The value for that timer is implementation-dependent, and it could be as little as less than a second and as much as 5 seconds. When such a timer expires, the base station forces the mobile station into dormancy and tears down the radio traffic channels. The mobile station can also request the transition into dormancy, but such a timer is much longer. The call flow in Figure 11.13 illustrates the base station-initiated transition to dormant state.

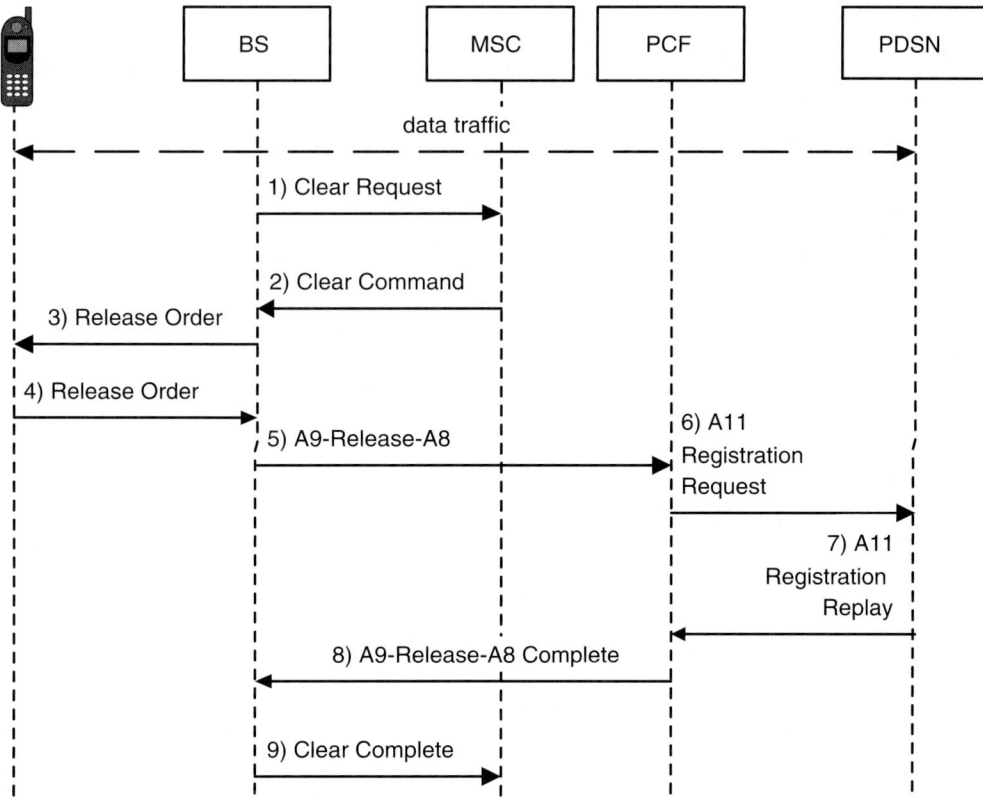

Figure 11.13 Base station initiated transition from active to dormant.

1. The mobile station is engaged in an SO33 packet data call. After some time of inactivity, the base station decides to transition the service into dormancy. Since this is only a single active service, the base station needs to tear down the associated radio traffic channel. The base station request the teardown or radio resources by sending a Clear Request message to the MSC.

2. The MSC responds with the Clear Command message that instructs the base station to tear down the radio resources.

3. The base station sends the *Release* ORDM message to the mobile station.

4. The mobile station responds with its own *Release* ORDM message that acknowledges the release of all radio resources.

5. The base station requests from the PCF the release of the associated A8 connection.

6. The PCF sends the A11 Registration Request message to the PDSN, indication zero lifetime. The purpose of the message at this point is also to pass the accounting information.

7. The PDSN acknowledges the registration with zero lifetime to the PCF in the A11 Registration Replay message.

8. The PCF maintains the service-state information. The PCF also acknowledges the teardown of the A8 connection to the base station with A9-Release-A8 Complete message. At this time the packet data service is in the dormant state.

9. The base station sends the Clear Complete message to the MSC, indicating that the radio traffic channel is torn down.

11.6 Concurrent Services

Concurrent services refer to the ability of the mobile station to be engaged in multiple calls at the same time. Of interest is a case where the user is engaged in a packet data call when it receives or originates a circuit-switched voice call. VoIP calls also require concurrent services, because the main service instance, SO33, must be active in order to initiate VoIP service with SO61.

11.6.1 Circuit-Switched Voice Call Delivery during Active Packet Data Call

In this example we assume that the packet data session is in progress and the user receives a circuit-switched voice call. We concentrate only on the serving MSC side and show the call flow in Figure 11.14. In this example we assume that while the packet data session is in progress, user data and signaling are transported on the Forward/Reverse Dedicated Channel (F/R-DCCH) and the Forward/Reverse Supplemental Channel (F/R-SCH). Alternately, the Forward Packet Data Channel (F-PDCH) may be in use in lieu of the F-DCCH and F-SCH.

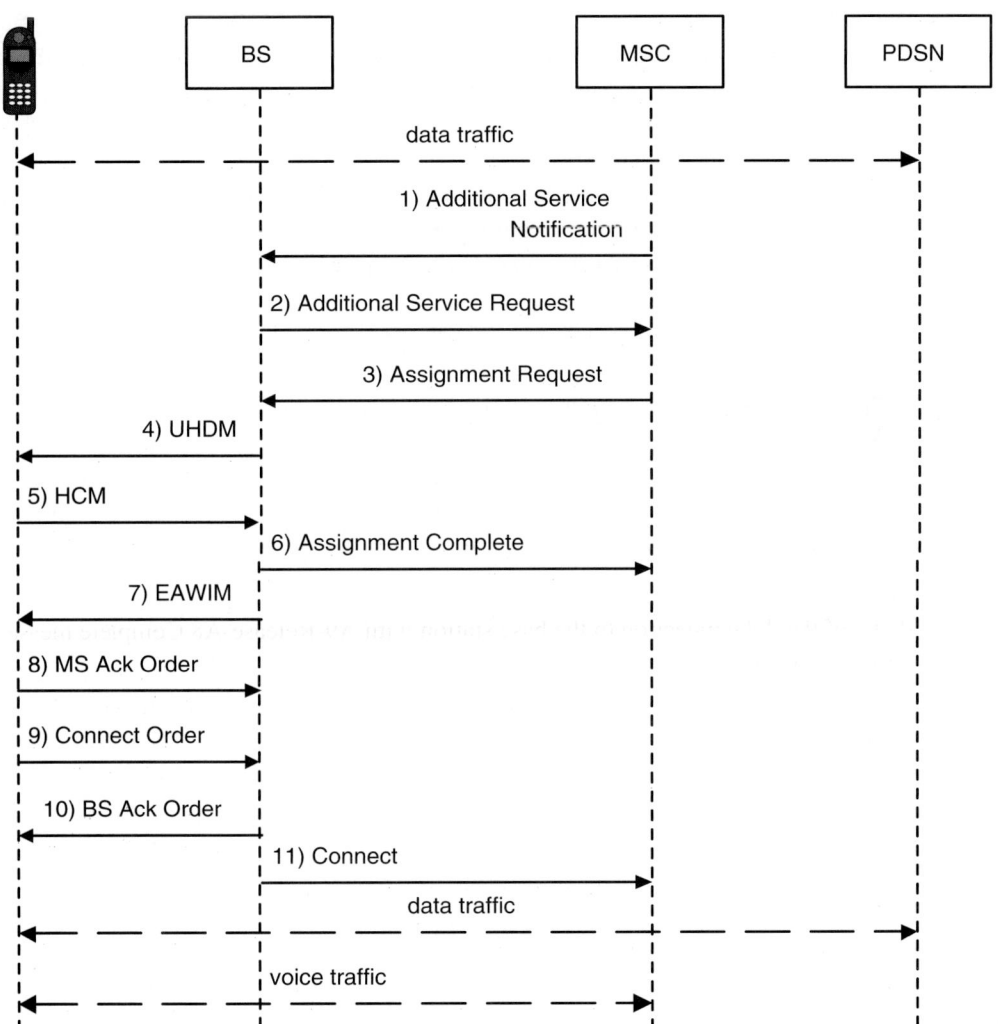

Figure 11.14 Circuit-switched voice call delivery during active packet data session.

1. The packet data session is in progress when the MSC notifies the base station of the incoming voice call with the Additional Service Notification message.
2. The base station requests the voice service to be set up by sending the Additional Service Request message to the MSC.
3. The MSC responds with the Assignment Request message, which requests the setup of the radio traffic channel suitable for voice service.

4. The base station must now reconfigure the radio bearer (see also Chapter 6) and establish traffic channels suitable for transporting voice traffic. The base station sends the UHDM, simultaneously assigning a new call and establishing the F/R-FCH to carry the voice traffic.

5. The mobile station sends the HCM in response to the UHDM.

6. On receiving the HCM, the base station notifies the MSC with the Assignment Complete message that the radio traffic channel setup has been completed.

7. The base station sends the Extended Alert with Info Message (EAWIM) to notify the mobile station of the incoming call.

8. The mobile station acknowledges the EAWIM with the Mobile Station Acknowledgment ORDM.

9. Once the user answers the call, the mobile station sends the Connect ORDM.

10. The base station acknowledges the *Connect* ORDM with the *Base Station Acknowledgment* ORDM.

11. The base station sends the Connect message to the MSC. At this time, two-way voice conversation can start while the data session proceeds undisturbed in the background.

11.6.2 Circuit-Switched Voice Call Origination during Active Packet Data Call

During an active packet data session, the user does not have to terminate the call in order to originate a circuit-switched voice call. The voice service can be set up while the packet data session is active. The call flow is illustrated in Figure 11.15. As in the previous example, we assume that while the packet data session is in progress, user data and signaling are transported on the F/R-DCCH and the F/R-SCH. Alternately, the F-PDCH may be in use in lieu of the F-DCCH and F-SCH.

1. The packet data session is in progress when the circuit-switched voice call is originated by the mobile station by sending the *Enhanced Origination Message* (EOM). The EOM contains the service option number for circuit-switched voice, that is, SO56 for Selectable Mode Vocoder (SMV).

2. The base station acknowledges receipt of the EORM with the *Base Station Acknowledgment* ORDM.

3. The base station requests the additional service setup from the MSC, using Additional Service Request message.

4. The MSC responds with the Assignment Request message that requests from the base station to set up the radio traffic channels.

5. The base station must now reconfigure the radio bearer (see also Chapter 6) and establish traffic channels suitable for transporting voice traffic. The base station sends the UHDM, simultaneously assigning a new call and establishing the F/R-FCH to carry the voice traffic.

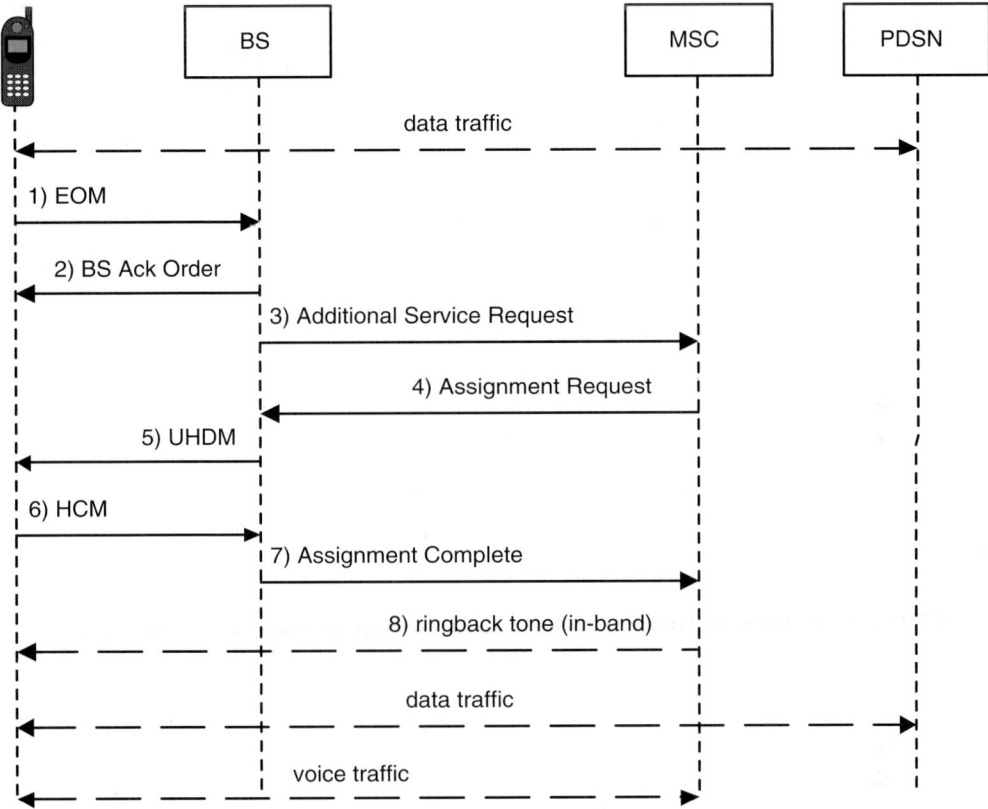

Figure 11.15 Circuit-switched voice call origination during active packet data session.

6. The mobile station sends the HCM in response to the UHDM.

7. On receiving the HCM, the base station notifies the MSC with the Assignment Complete message that the radio traffic channel setup has been completed.

8. The ringback tone is now played at the mobile station. In this example it is assumed that the ringback tone is sent in-band. Once the called party answers the call, the ringback tone is removed from the audio path. At this time, two-way voice conversation can start while the data session proceeds undisturbed in the background.

11.6.3 VoIP Call Flows

The VoIP call setup requires that the packet data service instance is active. It is assumed that Session Initialization Protocol (SIP) [6], with its extensions, is used to set up the VoIP session. The signaling flow on the calling party side is shown in Figure 11.16. The signaling flow at the called party side is shown in Figure 11.17. Note that the signaling flows are for illustrative purposes and that certain portions of the call setup procedure can be performed in parallel.

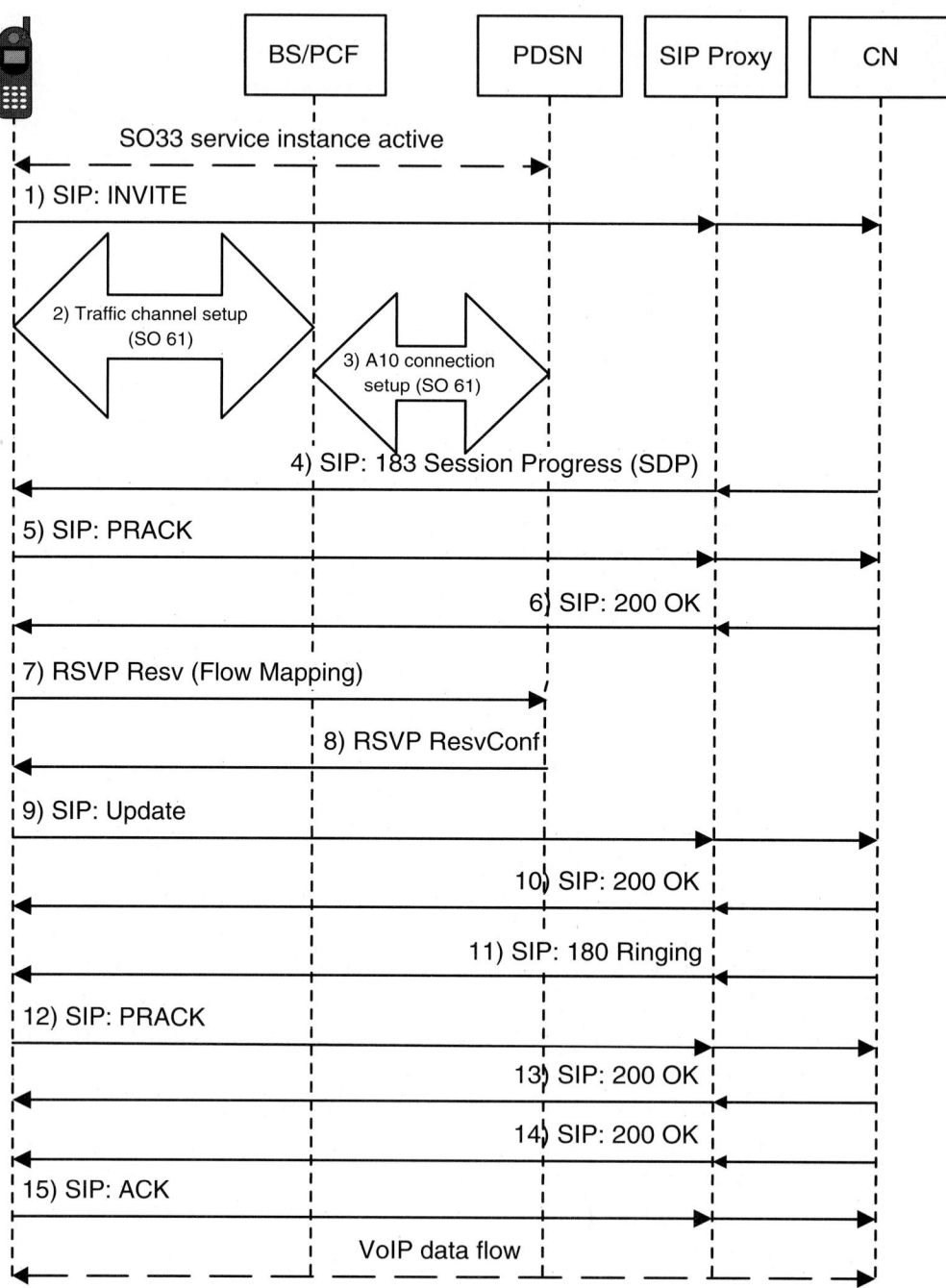

Figure 11.16 VoIP call flow—calling party side.

1. Before the VoIP voice call can be originated, the SO33 packet data call must be active. The mobile station sends the SIP INVITE message to the SIP proxy. The message is then passed to the corresponding node (CN), potentiality though a network of SIP proxies. The content of the SIP messages is encoded with Session Description Protocol (SDP) [8]. The SIP messaging uses the default service instance or the service instance set up in advance for this purpose. The SIP proxy may respond with the SIP 100 Trying message to prevent the mobile station from resending the INVITE message after the retransmission timer expires. For simplicity, the 100 Trying message is not shown.

2. The SO61 and the radio traffic channels are set up for the VoIP call. Note that VoIP requires F/R-FCH, while SIP messaging could also use F-PDCH.

3. After the traffic channels are set up, the base station needs to set up A8 and A10 connections for the VoIP service instance. Note that for simplicity we do not show a link between the base station and the PCF. We consider the base station and PCF as a single entity. Therefore, A8 connection setup is also not shown in the call flow.

4. The corresponding node sends the SIP 183 Session Progress message, indicating to the mobile station that VoIP session setup is in progress.

5. The mobile station responds with a SIP Provisional Acknowledgment (PRACK). The PRACK indicates to the corresponding node that the SIP 183 Session Progress has been received.

6. The corresponding node responds with 200 OK, acknowledging the PRACK.

7. The mobile station uses the Resource Reservation Protocol (RSVP) [9] to transport the flow-mapping parameters to the packet data serving node (PDSN). The RSVP Resv message carries the flow-mapping information. The flow-mapping parameters set up a filter that routes the incoming VoIP traffic to the appropriate service instance. The flow mapping is based on the IP addresses and port numbers.

8. The PDSN responds with the RSVP ResvConf message that acknowledges the setup of the flow treatment filters.

9. After the flow mappings are successfully established and the bearer path is completely set up, the mobile station sends the SIP Update message to the corresponding node.

10. The corresponding node sends the 200 OK message, acknowledging receipt of the SIP Update message.

11. The ringing tone is played at the mobile station.

12. The mobile station acknowledges the ringback tone to the corresponding node by sending a PRACK message.

13. The corresponding node responds with a 200 OK message, which acknowledges the PRACK message.

14. The corresponding node acknowledges the INVITE message with another 200 OK message.

15. The mobile station acknowledges the successful VoIP session setup with the ACK message. The VoIP conversation begins.

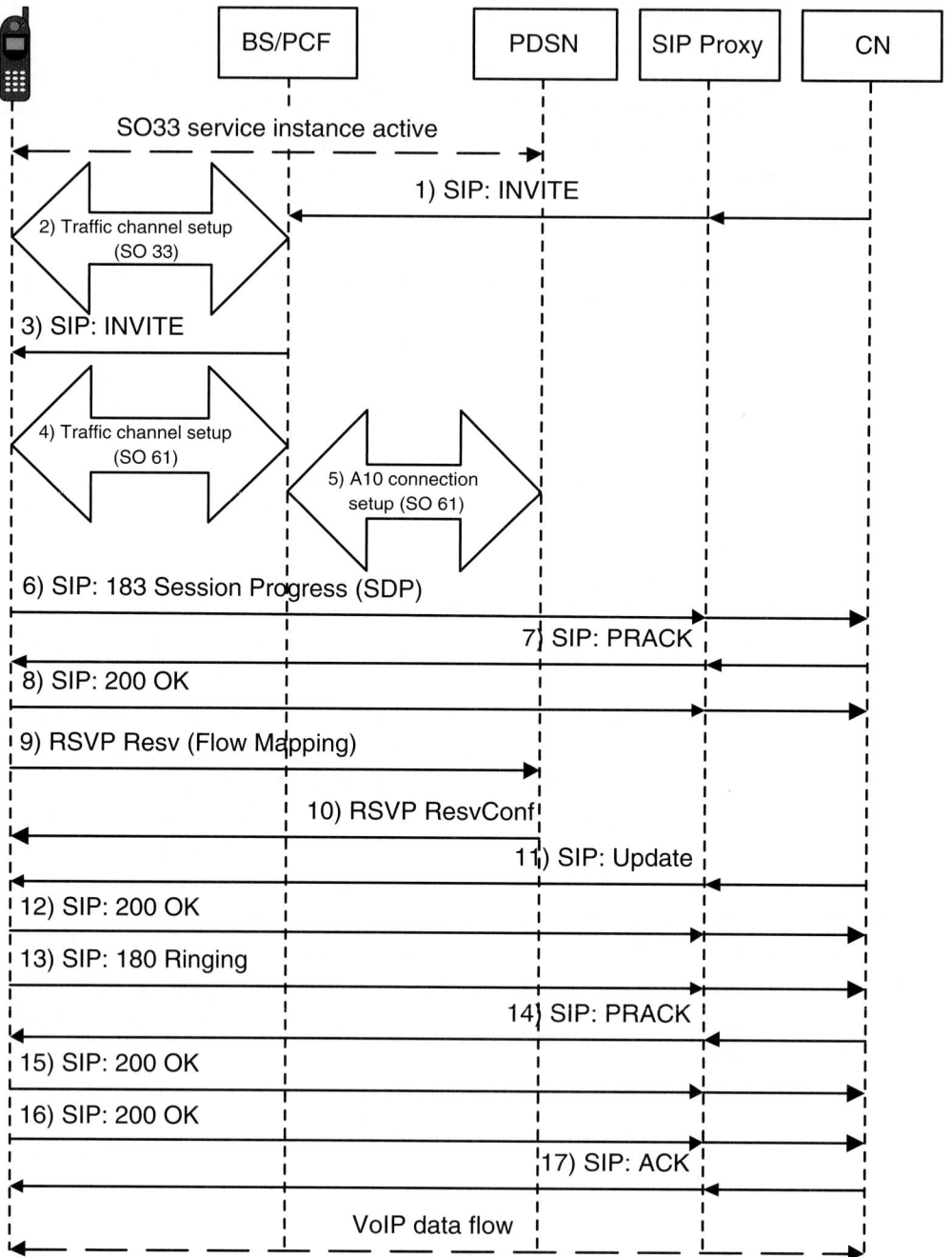

Figure 11.17 VoIP call flow, called party side.

1. Assume that the mobile station has a dormant SO33 call. The PCF receives the SIP INVITE message. The PDSN filters SIP traffic to the default service instance or an auxiliary service instance if the filters for such an instance have been set up in advance.

2. The SIP INVITE message triggers the activation of the SO33 call. For simplicity, in Figure 11.17, "traffic channel set up" refers to the network-initiated call-activation procedure shown in Figure 11.12. Note also that to simplify the call flow, we do not show a link between the base station and the PCF. We consider the base station and PCF as a single entity.

3. After the complete bearer path for the SO33 has been established, the SIP INVITE message is passed to the mobile station.

4. The mobile station and the base station set up SO61. Note that VoIP requires F/R-FCH, while SIP messaging could also use F-PDCH.

5. A8 and A10 connections are set up. The A10 connection set is shown only because the BS-to-PCF link is omitted for simplicity.

6. After the bearer path for the SO61 has been established, the mobile station sends the 183 Session Progress message indicating to the corresponding node that it has received the INVITE message.

7. The corresponding node responds with the PRACK message.

8. The mobile station acknowledges the PRACK message with a 200 OK message.

9. The mobile station sets up the flow filter for SO61 traffic at the PDSN with the RSVP Resv.

10. The PDSN confirms that the filter has been set up with RSVP ResvConf message.

11. The mobile station receives the SIP Update message, which indicates to the mobile station that the resources at the corresponding node have been set up.

12. The mobile station responds with a 200 OK message acknowledging the SIP Update message and indicating that the bearer path at its end has been established.

13. The mobile station plays the ringback tone to the corresponding node.

14. The corresponding node responds with the PRACK message.

15. The mobile station acknowledges the PRACK message with a 200 OK message.

16. The mobile station acknowledges the INVITE message with a 200 OK message.

17. The corresponding node acknowledges the last 200 OK message with the ACK message. The VoIP conversation begins.

REFERENCES

[1] "TIA/EIA-637-B, Short Message Services for Wideband Spread Spectrum Systems," *Telecommunications Industry Association*, 2002.

[2] "TIA/EIA-664-603-A, Wireless Paging Teleservice," *Telecommunications Industry Association*, 2000.

[3] "TIA/EIA-664-602-A, Wireless Messaging Teleservice," *Telecommunications Industry Association*, 2000.

[4] "TIA/EIA-664-513-A, Message Waiting Notification," *Telecommunications Industry Association*, 2000.

[5] "TIA/EIA-IS-2001, Inter-operability Specification (IOS) for cdma2000 Access Network Interfaces," *Telecommunications Industry Association*, 2002.

[6] "TIA/EIA-IS-835-C cdma2000 Wireless Network Standard," *Telecommunications Industry Association*, 2003.

[7] Rosenberg, J., H. Schulzrinne, G. Camarillo, A. Johnston, J. Peterson, R. Sparks, M. Handley, and E. Schooler, "SIP: Session Initialization Protocol," *IETF RFC 3261*, June 2002.

[8] Handley, M., and V. Jacobson, "SDP: Session Description Protocol," *IETF RFC 2327*, April 1998.

[9] Braden, R., L. Zhang, S. Berson, S. Herzog, and S. Jamin, "Resource Reservation Protocol (RSVP) Version 1 Functional Specification," *IETF RFC 2205,* September 1997.

INDEX

A

Access channel, 115, 167, 169, 195, 340

Access probe, 168, 171, 197, 341

Adaptive asyncronous incremental redundancy (AAIR), 392, 400

Adaptive coding, 176, 394

Addressing sublayer, 218

AGC, 298, 327, 336, 344, 349

Antenna,
    cross-polarized, 34
    directional, 31
    diversity, 32, 433
    radiation patten, 412

Admission control, 365, 372

Asyncronous transfer mode (ATM), 52

Authentication, 278
    key, 282
    shared secret data, 279
    signature, 279
    sublayer, 213

Authentication and key agreement (AKA), 284

Authentication, authorization and accounting (AAA), 62, 64, 65, 66, 72, 73, 74, 76, 79

Authentication center (AC), 55, 59, 61, 467, 470

Automatic gain control. (*See* AGC)

Automatic repeat request (ARQ),

hybrid, 27, 120, 177, 391, 398

ARQ sublayer, 214

B

Band class, 148, 342

Base station controller (BSC), 48, 50, 51, 52, 53 54, 61, 71

Base station tranceiver (BTS), 48, 49, 51, 61

Baseband filter, 139, 148

Basic access mode, 202

Blank-and-burst, 100, 183

C

Call,
    delivery, 468, 486
    origination, 244, 465, 466, 468, 488

informIT

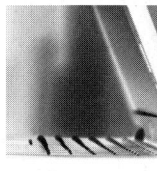

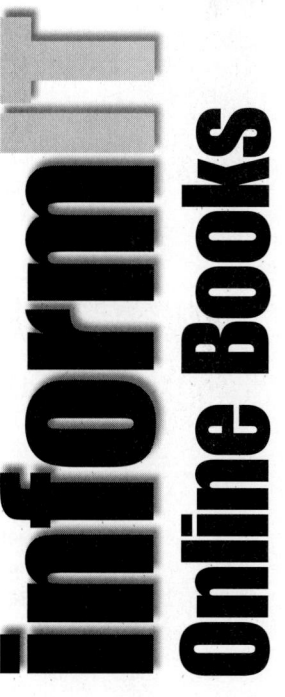

http://www.phptr.com/

Prentice Hall PTR   InformIT   InformIT Online Books   Financial Times Prentice Hall   ft.com   PTG Interactive   Reuters

TOMORROW'S SOLUTIONS FOR TODAY'S PROFESSIONALS

Prentice Hall **Professional Technical Reference**

Browse | Book Series | What's New | User Groups | Alliances | Special Sales | Contact Us

Search | Help | Home

Quick Search

PTR Favorites

Find a Bookstore

Book Series

Special Interests

Newsletters

Press Room

International

Best Sellers

Solutions Beyond the Book

Shopping Bag

Keep Up to Date with

PH PTR Online

We strive to stay on the cutting edge of what's happening in professional computer science and engineering. Here's a bit of what you'll find when you stop by **www.phptr.com**:

What's new at PHPTR? We don't just publish books for the professional community, we're a part of it. Check out our convention schedule, keep up with your favorite authors, and get the latest reviews and press releases on topics of interest to you.

Special interest areas offering our latest books, book series, features of the month, related links, and other useful information to help you get the job done.

User Groups Prentice Hall Professional Technical Reference's User Group Program helps volunteer, not-for-profit user groups provide their members with training and information about cutting-edge technology.

Companion Websites Our Companion Websites provide valuable solutions beyond the book. Here you can download the source code, get updates and corrections, chat with other users and the author about the book, or discover links to other websites on this topic.

Need to find a bookstore? Chances are, there's a bookseller near you that carries a broad selection of PTR titles. Locate a Magnet bookstore near you at www.phptr.com.

Subscribe today! Join PHPTR's monthly email newsletter! Want to be kept up-to-date on your area of interest? Choose a targeted category on our website, and we'll keep you informed of the latest PHPTR products, author events, reviews and conferences in your interest area.

Visit our mailroom to subscribe today! **http://www.phptr.com/mail_lists**